Springer Tracts in Mechanical Engineering

About this Series

Springer Tracts in Mechanical Engineering (STME) publishes the latest developments in Mechanical Engineering - quickly, informally and with high quality. The intent is to cover all the main branches of mechanical engineering, both theoretical and applied, including:

- Engineering Design
- Machinery and Machine Elements
- Mechanical structures and stress analysis
- Automotive Engineering
- Engine Technology
- Aerospace Technology and Astronautics
- Nanotechnology and Microengineering
- Control, Robotics, Mechatronics
- MEMS
- Theoretical and Applied Mechanics
- Dynamical Systems, Control
- Fluids Mechanics
- Engineering Thermodynamics, Heat and Mass Transfer
- Manufacturing
- Precision Engineering, Instrumentation, Measurement
- Materials Engineering
- Tribology and Surface Technology

Within the scopes of the series are monographs, professional books or graduate textbooks, edited volumes as well as outstanding PhD theses and books purposely devoted to support education in mechanical engineering at graduate and post-graduate levels.

More information about this series at http://www.springer.com/series/11693

Yogesh M. Joshi • Sameer Khandekar
Editors

Nanoscale and Microscale Phenomena

Fundamentals and Applications

Springer

Editors
Yogesh M. Joshi
Department of Chemical Engineering
Indian Institute of Technology Kanpur
Kanpur, UP, India

Sameer Khandekar
Department of Mechanical Engineering
Indian Institute of Technology Kanpur
Kanpur, UP, India

ISSN 2195-9862 ISSN 2195-9870 (electronic)
Springer Tracts in Mechanical Engineering
ISBN 978-81-322-2288-0 ISBN 978-81-322-2289-7 (eBook)
DOI 10.1007/978-81-322-2289-7

Library of Congress Control Number: 2015941151

Springer New Delhi Heidelberg New York Dordrecht London

Printed on acid-free paper

Springer (India) Pvt. Ltd. is part of Springer Science+Business Media (www.springer.com)

Foreword

It is my pleasure to present this volume titled *Nanoscale and Microscale Phenomena: Fundamentals and Applications*, edited by Prof. Yogesh M. Joshi, Professor, Department of Chemical Engineering, and Prof. Sameer Khandekar, Professor, Department of Mechanical Engineering, at the Indian Institute of Technology Kanpur, to the readers. Looking at the contents of this book, I am sure it will attract readership from various segments, especially chemical and mechanical engineers and scientists working in micro- and nanoscale physics, materials development, surface engineering, thermal-fluidic transport systems, and allied interdisciplinary branches of science and technology. Over 25 academicians, including young scholars as well as experienced mentors and researchers, from three premier institutions of the country, have contributed state-of-the-art articles to this volume. I congratulate all the authors and the editorial team for their excellent work.

This book is an outcome of the work carried out as part of a bunch of projects funded under the Intensification of Research in High Priority Areas (IRHPA) program of the Department of Science and Technology (DST), Government of India, during 2007–2012, in which three institutes, viz. IIT Bombay, IIT Kanpur, IISc Bangalore, participated. The administrative coordination of this group of projects was undertaken by Shri Rajeev Tayal and Shri J. B. V. Reddy. I would particularly like to complement Shri Rajeev Tayal for spearheading this activity and for bringing it to excellent fruition.

As some of the readers may be aware, IRHPA is a complementary scheme of DST, implemented through the Science and Engineering Research Board (SERB) (erstwhile Science and Engineering Research Council; SERC), with the aim to set up Centers of Excellence around core research groups comprising eminent scientists and major National Research Facilities to nucleate research activities in high-priority areas. In the last several decades, since its inception during the sixth 5-Year Plan, this scheme has contributed significantly to augment R&D capabilities at academic institutions and national laboratories, in frontier and emerging areas of science and technology. It is heartening to note that the team of authors has compiled the knowledge generated during the execution of their respective projects and have presented it for future use by the scientific community in the form of this book.

I once again congratulate all the persons directly and indirectly associated with the publication of this book. Such productive initiatives give boost to the mission of DST, which is constantly and consciously attempting to increase the efficiencies of internal processes and delivery of ongoing programs, on the one hand, and launch new initiatives with a potential impact on the science and technology landscape of the country, on the other. I believe that by working together, we can realize the vision of an S&T-enabled knowledge-based economy in the twenty-first century.

Department of Science and Technology
Government of India, New Delhi, India
2 May 2015

Ashutosh Sharma

Preface

Understanding physical phenomena at micro- and nanoscale has been a major thrust area of research and development in the past two decades. Novel insights of material behavior and transport of heat, mass, and momentum at these reduced scales has transformed a gamut of technologies and has affected different walks of our lives in an unprecedented manner. Emerging trends in micro- and nanotechnologies have the potential for pathbreaking solutions in diverse fields of engineering sciences, many of which are necessarily interdisciplinary in nature.

In this background, it gives us great pleasure to present this edited book to the readers, titled *Nanoscale and Microscale Phenomena: Fundamentals and Applications*, the genesis of which dates back to early 2007. In that year, the Department of Science and Technology, Government of India, provided funding to a total of five research groups belonging respectively to three premier institutes, viz., the Indian Institute of Science Bangalore, Indian Institute of Technology Bombay, and Indian Institute of Technology Kanpur, under their scheme *Intensification of Research in High-Priority Areas* (IRHPA). Each research group comprised of around three to five scientists from the mentioned institutes along with their graduate students and research assistants. The research projects were broadly in the areas of microscale and nanoscale materials synthesis, devices, and associated thermal–fluid transport phenomena, having specific goals and target deliverables. The projects were executed during 2007–2012, and as it turns out, all the projects in these high-priority areas were highly successful, leading to a number of publications in reputed journals and conferences and also intellectual property generation through patents. In late 2012, it was decided that the outcome of these research projects be published in the form of a book. The present edited volume is a result of that initiative.

This book is divided into four sections. In the first section, various studies on the synthesis of nanostructures are presented. Sundar and Tirumkudulu propose a single-step process to obtain monodispersed nanometric liposomes using immobilized colloidal particles for drug delivery application. On the other hand, Patil and Jadhav study the single-pass extrusion process to obtain monodisperse liposomes of the desired sizes. Gupta and coworkers report a method to synthesize nanorods from wormlike micelles, while Santhanam describes issues involved in large-scale synthesis of nanoparticles through scale-up from lab-scale processes. In the second section, various studies pertaining to instabilities in fluid–fluid and fluid–solid interfaces are discussed. Sundar and Tirumkudulu study the hole formation process caused by dewetting in spin coating of lipid bilayers and analyze the phenomenon against established theories. Patra and coworkers study various kinds of wetting transitions for different fluid–surface combinations through molecular simulations. Gambhire and Thaokar analyze instabilities at planar fluid interfaces under electric fields, while Shankar and Gaurav study the effect of a soft deformable boundary on interfacial instabilities. In the third section, applications of nanostructures are presented, wherein Mondal and Ghatak describe how limbless locomotion can be achieved by generating geometric asymmetry in elastomeric objects. They further demonstrate how geometric asymmetry can facilitate directed adhesion in soft materials. Shaukat and coworkers study the tensile flow behavior of metastable soft materials having micro- and nanocsopic structure. Sharma and Sharma discuss the development of carbon xerogels having different microstructures, with sizes varying from nano to microscales. Moreover, they discuss strategies for the development of micropatterned carbon surfaces to fabricate hierarchical carbon structures by various approaches. The last section of the book is devoted to subtleties of chemical reactions and associated thermal–fluid transport over microscopic length scales, wherein Peela and Kunzru describe the development of a microchannel reactor to produce hydrogen from ethanol, while Khandekar and Moharana study convective heat transport in microchannel flows and bring out the various nuances of estimating the transfer coefficients.

There are many stakeholders in the successful completion of this book. First and foremost is Mr. Rajeev Tayal, Adviser, Department of Science and Technology. The five research projects that form the background of the present book are an outcome of Mr. Tayal's painstaking efforts. From the inception of the idea of this edited book, he has supported the endeavor in every possible way. It gives us great pleasure to acknowledge his continuous encouragement and coordination during the execution of the projects as well as in the preparation of this book. We also acknowledge Mr. J.B.V. Reddy, Senior Scientist, Department of Science and Technology, for providing every possible help, wherever required. We thank all the contributing authors and their respective graduate students and research scholars; without their enthusiastic support, this book would not have been a possibility. Every chapter of this book has gone through a careful peer review process. We greatly acknowledge the efforts of anonymous reviewers that have contributed to enhance the quality of this work. We also acknowledge help of Ms. Khushboo Suman while correcting the proof.

Finally, we thank the Centre for Development of Technical Education (CDTE) of the Indian Institute of Technology Kanpur for partly supporting the production of this book.

Kanpur, UP, India   Yogesh M. Joshi
24 October 2014 (कार्तिक शुद्ध प्रतिपदा, बलिप्रतिपदा, विक्रम संवत २०७१) Sameer Khandekar

Message

It is heartening to witness that research and development efforts carried out under the aegis of the prestigious IRHPA funding program of the Department of Science & Technology (DST), Government of India during 2007–2012, have been successfully compiled by the team of researchers from across India into this edited volume, titled, '*Nanoscale and Microscale Phenomena: Fundamentals and Applications*'.

It was also my pleasure that I was associated during the early conceptualizing stage of these research and development projects and was given the task of reviewing and mentoring some of the ideas contained in the proposals. Series of well-planned pre-sanction activities and subsequent follow up by DST functionaries and equally enthusiastic response and initiatives of the research groups involved has translated into a well documented volume, containing state-of-the-art results.

I congratulate the editorial team of Prof. Yogesh M. Joshi and Prof. Sameer Khandekar from IIT Kanpur, who have undertaken an excellent job. Needless to say, the research teams have put in tremendous efforts in pursing their passion for furthering the science and technology frontiers of the nation.

I wish the entire team well-deserved success in all their present and future endeavors.

National Chemical Laboratory, Pune, India
December 01, 2014

B.D. Kulkarni

Contents

About the Editors

Prof. Yogesh M. Joshi is Professor and P. K. Kelkar Research Fellow in the Department of Chemical Engineering, Indian Institute of Technology Kanpur. He obtained his B.E. in Polymer Engineering from Pune University in 1996 and his Ph.D. in Chemical Engineering from the National Chemical Laboratory, Pune, and the Indian Institute of Technology Bombay in 2001. Subsequently, he worked at the Benjamin Levich Institute, New York, as a postdoctoral fellow. Dr. Joshi joined IIT Kanpur in July 2004. Prof. Joshi is a recipient of numerous awards, including the NASI Scopus Young Scientist Award (2012), the DAE-SRC Outstanding Investigator Award (2012), the Amar Dye-Chem Award (2009), the INAE Young Engineer Award (2008), and the INSA Medal for Young Scientists (2006). He is a life member of the National Academy of Sciences, India (2012) and an Associate of the Indian National Academy of Engineering (2013–2019) and was an Associate of Indian Academy of Sciences (2006–2009).

Prof. Joshi has served as a visiting faculty in the Université du Maine, France (2008), and as an INSA visiting scientist in the University of Edinburgh, UK (2010). He has over 60 research publications in international journals and has given over 50 lectures/invited talks in India and abroad. His current research interests are thixotropy, rheological behavior of glasses and gels, phase behavior of clay suspensions, and electrorheology.

Department of Chemical Engineering, Indian Institute of Technology Kanpur, Kanpur, UP, India

Prof. Sameer Khandekar is affiliated to the Department of Mechanical Engineering, Indian Institute of Technology Kanpur, India, since September 2004. Prof. Khandekar completed his master's degree course from IIT Kanpur, India, in 2000, with specialization in fluid–thermal engineering, and subsequently earned his doctoral degree from the University of Stuttgart, Germany (2000–2004). Earlier, after his undergraduate studies in Mechanical Engineering from Government Engineering College, Jabalpur (MP), in 1993, he worked as a marine power plant engineer on board sea-going merchant vessels for 4 years (1994–1998). Prof. Khandekar is a recipient of the P. K. Kelkar Research Fellowship (IIT Kanpur; 2008–2011), the DAAD Fellowship (2011), the Prof. K. N. Seetharamu Award (Indian Society of Heat and Mass Transfer, 2010), the George Grover Medal (International Heat Pipe Committee, 2007), and the Young Scientist Award (Department of Atomic Energy, India, 2005). He is a member of the International Heat Pipe Committee since 2008 and editorial board member of two international journals on the science and technology of heat pipes. He has served as an academic senate member of one centrally funded institute and one state-funded autonomous engineering institute.

Prof. Khandekar has also served as an international invited faculty member at four universities in Germany, France, Brazil, and Thailand. He has over 50 research publications in international journals and over 60 publications in international conferences including 15 keynote lectures/invited talks. His current research interests are in experimental microscale phase-change passive thermofluidic systems, boiling and condensation, pulsating heat pipes and energy systems.

Department of Mechanical Engineering, Indian Institute of Technology Kanpur, Kanpur, UP, India

Part I
Nano-particle Synthesis and Applications

Novel Method for Synthesizing Monodisperse Dispersion of Nanometer Liposomes

S.K. Sundar and Mahesh S. Tirumkudulu

Abstract Liposomes are used for drug delivery applications due to their ability to carry water-soluble and insoluble drug molecules. While there are various methods for liposome preparation, the methodology usually involves post-processing steps such as filtration and extrusion to obtain monodisperse liposomes of size less than 100 nm. Here, the focus is on a novel single-step process to obtain monodisperse liposomes in the nanometer range using colloidal particles that are immobilized inside a cylindrical column and forms a packed bed. The steps consist of introducing lipids dissolved in an organic solvent into the bed followed by solvent evaporation by passing a stream of nitrogen gas which forms a coating of dried lipid over the colloidal particles. Hydration of the lipid bilayer with an aqueous buffer causes the unbinding of lipid bilayers resulting in the formation of monodisperse liposomes. Our results indicate that the liposome size depends upon the surface roughness of the packing and not on the pore size. The process is robust and could be easily adapted for point-of-care therapeutics that involves liposomes for drug delivery.

Keywords Liposomes • Targeted drug delivery • Hydration • Encapsulation • Packed bed

1 Introduction

Liposomes are association colloids made of bilayers formed by the self-assembly of double-tailed surfactants called phospholipids. These molecules possess a hydrophilic head and a hydrophobic tail (Fig. 1). When dissolved in water and when the concentration is well above their critical micellar concentration, these molecules spontaneously rearrange into bilayers such that the head is exposed to the aqueous portion and the tails are concealed between the head groups to avoid unfavorable interactions with the surrounding aqueous environment. The bilayers then close upon themselves to form liposomes. This spontaneous vesiculation above

S.K. Sundar • M.S. Tirumkudulu (✉)
Indian Institute of Technology Bombay, Mumbai 400076, India
e-mail: sundar.bio@gmail.com; mahesh@che.iitb.ac.in

Y.M. Joshi, S. Khandekar (eds.), *Nanoscale and Microscale Phenomena*,
Springer Tracts in Mechanical Engineering, DOI 10.1007/978-81-322-2289-7_1

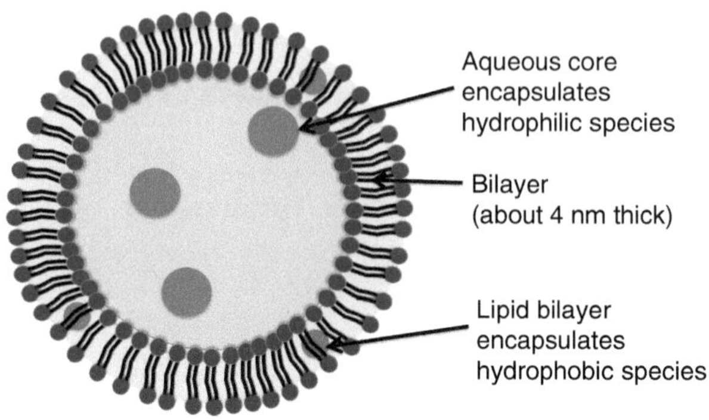

Fig. 1 The above schematic shows a unilamellar liposome which can encapsulate hydrophobic species in its bilayer and hydrophilic species inside its aqueous core

the critical concentration of lipid monomer and also at a temperature greater than the transition temperature of phospholipid occurs when the size of the bilayer exceeds a critical size so that the energetically unfavorable edges exposed to the solvent are eliminated by the bilayer closing on itself to form a vesicle at the cost of relatively lower bending energy [11].

This ability of the liposome to partition space makes it an excellent choice for drug delivery applications. The encapsulation provides protection to the drug apart from minimum drug loss and preventing damage to healthy cells [18]. It is well known that liposomes administered intravenously are cleared by macrophages of the liver and the spleen (reticuloendothelial system) if their size exceeds about 200 nm. Hence, it is required that the liposome size be preferably below 100 nm so that they can avoid detection [6]. The same can also be achieved by attaching ligands or surface-modifying polymers such as polyethylene glycol to the liposome surface. These "stealth" liposomes circulate in the bloodstream for a long duration as they avoid detection by the reticuloendothelial system and thereby increase the efficacy of the delivery system.

Liposomes of the desired size range for drug delivery are generally achieved by one or more of the following methods: by sonication, by extrusion of large liposomes through membranes with small pores, by varying the ionic concentration of the hydration medium, or by inducing changes in solubility conditions [2, 4, 25]. Most of the liposome preparation methods that require commercialization are carried out in small channels whose characteristic length scales are in millimeters or centimeters. These techniques produce a heterogeneous population of liposomes and is eliminated by subjecting them to either filtration or extrusion, thereby achieving the targeted size range of 100 nm or lower. Lasic [14] discusses the various techniques for liposome preparation along with their advantages and disadvantages. More recently, microfluidic methods for liposome synthesis have been developed that allow fine control on the liposome sizes [27]. Confinement inside small channels in an environment that provides controlled mixing enables synthesis of liposomes in the submicron range. Thus, adapting the conventional macroscale processes such as electroformation [15], hydration [16], extrusion [5], and double emulsion technique [26] to microfluidic devices prevents the loss of reagents while achieving a tight control on the liposome size. Recent microfluidic approaches by Jahn et al. [12] and modification of the same using membranes by Akamatsu et al. [1] have enabled the synthesis of monodisperse liposomes in the submicron range. A review on microfluidics by van Swaay and deMello [27] focuses on the importance of simplicity along with robustness as one of the main criteria for a process to be widely accepted in the industry. Microfluidic device fabrication and operation is difficult and often complex since designs include multiple fluid inputs, different fluid phases, and complicated flow control valves. It is because of this disadvantage that hydration remains one of the most preferred techniques followed by extrusion to produce submicron liposomes at an industrial scale.

Our work on liposome formation focuses on the thin-film hydration process to produce submicron liposomes without the need for post-processing such as extrusion and/or filtration. The hydration process typically involves the formation of dried bilayer films over a substrate surface, and subsequent hydration of the same with an aqueous solution initiates unbinding of bilayer stacks leading to the formation of liposomes [21, 28]. Many studies have attempted to produce sub-100 nm liposomes using the hydration technique. Olson et al. [20] prepared multilamellar and polydisperse liposomes from a dry deposit of a mixture of lipids composing of phosphatidylserine, phosphatidylcholine, and cholesterol in a round-bottomed flask. These multilamellar liposomes were extruded several times through filters of pore sizes ranging from 1 to 0.6 μm to reduce the size from 1,320 to 260 nm with decrease in lamellarity. The production of liposomes in the size range from 55 to 240 nm by Hope et al. [8] involved the hydration of dry lipid films followed by extrusion under moderate pressures of $\leq$500 lb/in^2 through polycarbonate filters of pore size ranging from 30 to 400 nm. Synthesis studies by Nayar et al. [19] on long-chain saturated phosphatidylcholines showed that multiple extrusion through 100 nm pore filters produces liposomes in the 60 nm size range. The above studies indicate that the liposome size depends on the filter pore size and extrusion pressure, and the process is applicable to both saturated and unsaturated lipids when it is performed above the transition temperature of the lipid. Recently, Howse et al. [9] have shown that a monodisperse suspension of giant polymer liposomes can be produced from templated surfaces made from poly(ethylene oxide)-co-poly(butylene oxide). The bilayers conform to the surface topology of the templated surface and close to form liposomes upon hydration. The final liposome size scales with the area of the templated surface. Thus the liposome size can be tailored by varying the size of the polymer island. While the aforementioned process provides a tight control on the liposome size, the scale-up of the process could be a challenge.

In this chapter, we describe a one-step method based on the hydration technique to synthesize submicron liposomes with a low polydispersity index (PDI $\leq$ 0.2). Lipid molecules, dissolved in an organic solvent, are dried in a packed bed of colloidal particles with rough surface. The solvent is then evaporated in a stream of nitrogen forming a bilayer film over the particle surface and then hydrated with an aqueous buffer to obtain a monodisperse population of liposomes. Our experiments indicate that the size distribution is independent of the superficial velocity of the aqueous medium inside the packed bed suggesting that extrusion is not the cause for the formation of liposomes. This was further confirmed by drying a preformed dispersion of large and polydisperse liposomes inside the packed bed using a stream of nitrogen and then hydrating the bed with a buffer. The resulting dispersion contained a monodisperse suspension of sub-100 nm liposomes. The size distribution was found to be independent of the colloidal particle size and the packing volume fraction in the bed, suggesting that the liposome size may be set by the particle roughness and the complex three-dimensional structure of the packing. The robustness of the single-step process along with the absence of a post-processing step produces liposomes with a tight control over its size distribution, thus making it a suitable technique to point-of-care therapeutics involving liposomal drug delivery systems.

2 Materials and Methods

2.1 Materials

1,2-Dimyristoyl-sn-glycero-3-phosphocholine (DMPC), 1,2-dipalmitoyl-sn-glycero-3-phosphocholine (DPPC), and L-α-phosphatidylcholine were obtained in solution form from Avanti Polar Lipids®, Alabama. Sodium chloride, potassium chloride, disodium hydrogen orthophosphate, and potassium dihydrogen orthophosphate were obtained from Sigma Aldrich®. Solvents (chloroform and methanol) used to dissolve lipid were of high grade and obtained from Merck®Chemicals. The α-alumina particles (AKP-15, AKP-30, and AKP-50) were obtained from Sumitomo Chemicals, Japan. Circular glass capillaries of 2 mm ID and 4 mm OD of length 100 mm were used for the packed bed experiments and were obtained from a local supplier.

2.2 Preparation of Packed Bed

The size distribution of alumina particles of each grade was monodisperse (PDI $<$ 0.2) as ascertained using dynamic light scattering (DLS). However, the particles were not spherical as visualized using a scanning electron microscope (SEM). The average sizes of the particles of the three grades, namely, AKP-50, AKP-30, and AKP-15, were 250, 350, and 600 nm, respectively (see Fig. 2). Dilute aqueous dispersion (35 % v/v) of high-purity α-alumina was prepared by dispersing them in deionized water. The pH of the dispersion was adjusted using HNO_3 and KOH. The final packing volume fraction (ϕ_f) was determined by drying the dispersions in glass capillaries. The alumina dispersion with different initial lengths (l_0) was allowed to sediment and dry in a vertical glass capillary where the bottom end was closed with a 200 nm filter paper while the upper end was kept open for drying. Once

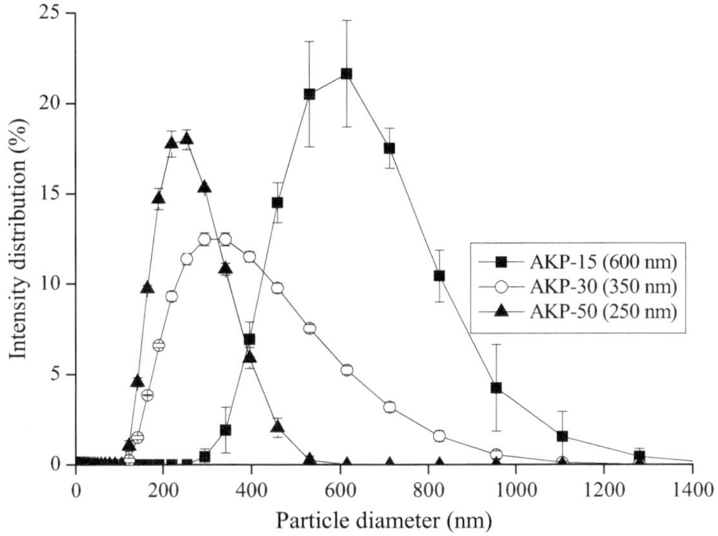

Fig. 2 Size (diameter) distribution by intensity for the three different grades of α-alumina particles (Reprinted/adapted with permission from Sundar and Tirumkudulu [24]. Copyright (2014) American Chemical Society)

the suspension dried, the final length (l_f) of the dried dispersion was measured from which the final packing volume fraction (ϕ_f) was obtained from volume balance, $l_f\phi_f = l_0\phi_0$. The particle packing was varied by choosing initial dispersions with a pH of 2, 5, and 9. Singh et al. [23] have shown that the surface charge of alumina particles is high and positive at low pH and decreases with increasing pH reaching the isoelectric point at pH 9. At low zeta potential, the particles are unstable and flocculate and the packing fraction changes from 0.7 at pH 2 to about 0.3 at pH 9. The dried capillaries were sintered at 1,100 °C for 30–60 min so that the particles become immobilized and give mechanical strength to the packing.

2.3 *Liposome Preparation and Size Distribution*

Capillaries with packed bed of alumina particles (3 cm in length) were used for liposome preparation. The experiments were carried out with DMPC lipid (gel-liquid crystal transition temperature, $T_c = 23\,°\mathrm{C}$) at a temperature, T_{exp}, of 30 °C. The experimental setup is shown in Fig. 3. The process consists of introducing a small plug (about 100 μl) of lipid solution into the bed which is then dried with nitrogen gas for about 15 min. The dried lipid inside the packed bed is contacted with a phosphate buffer that was dispensed using a syringe pump (NE-1000, New Era Pump Systems). The pressure drop inside the packed bed varied from 1 to 5 bar.

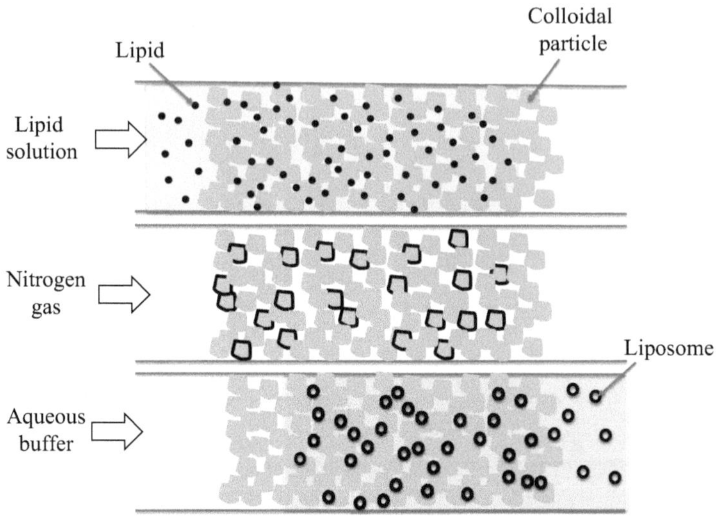

Fig. 3 Schematic representation of experimental setup. Lipid dissolved in an organic solvent is introduced in the packed bed. The solvent is evaporated in a stream of nitrogen gas to form a thin coverage of lipid bilayers over the particle surface. It is then followed by hydration using an aqueous buffer to form liposomes (Reprinted/adapted with permission from Sundar and Tirumkudulu [24]. Copyright (2014) American Chemical Society)

The liposome suspension was collected from the other end of the capillary for analysis. The size distribution was determined using dynamic light scattering while the sample was imaged via electron microscopy. The experiments were performed for buffer flow rates, Q, of 20, 60, 150, and 200 μl/min; lipid concentration, C_o, of 10, 35, 70, and 210 mM; and packing volume fraction of 0.35, 0.42, and 0.56 for packed beds of AKP-30 (350 nm) particles. For the remaining beds packed with AKP-15 and AKP-50, experiments were conducted at $Q = 20$ μl/min, $C_o = 70$ mM and $\phi_f = 0.35$.

2.4 *Transmission Electron Microscopy (TEM) and Scanning Electron Microscopy (SEM)*

The liposome size distribution was visualized using both TEM and SEM in the cryogenic mode in addition to their analysis using dynamic light scattering. Cryo-TEM analysis involves placing a small drop of the liposome sample in the grid. The maximum specimen thickness is about 1 μm and hence the excess solution is removed using a filter paper, a process known as blotting. The grid is immediately plunged into liquid ethane for vitrification after which it is stored under liquid nitrogen until analysis. For cryo-SEM analysis, a small volume of the liposome sample was placed on the sample holder and then frozen using liquid nitrogen. The frozen sample was subjected to sublimation for 5 min before imaging. In order to determine the pore size and extent of sintering, the sintered alumina packing was observed under SEM, after performing platinum coating over the particle surface, for 30 s.

3 Results and Discussion

Packed bed experiments yielded a monodisperse population of liposomes with a narrow size distribution in the sub-100 nm range for varying buffer flow rate, lipid concentration, particle volume fraction, and particle size in the packed bed and the results are shown in Fig. 4. Each experiment was repeated three times to check the reproducibility. The results show that the size distribution as measured via the

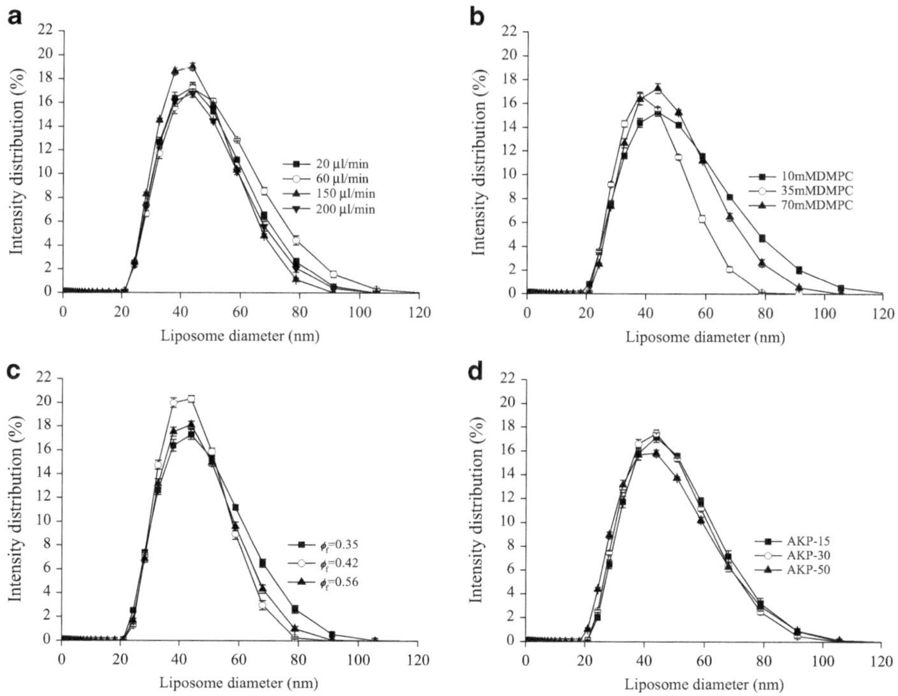

Fig. 4 Plots showing intensity distribution of liposome size (diameter) from DMPC lipid with variation in (**a**) flow rate of aqueous buffer for $C_o = 70$ mM in a packed bed of AKP-30 particles at volume fraction, $\phi_f = 0.35$, (**b**) lipid concentration for $Q = 20\,\mu$l/min in a packed bed of AKP-30 particles at $\phi_f = 0.35$, (**c**) volume fraction of particles in a packed bed of AKP-30 particles for $Q = 20\,\mu$l/min and $C_o = 70$ mM, and (**d**) particle size for a packing fraction of $\phi_f = 0.35$, $Q = 20\,\mu$l/min, and $C_o = 70$ mM (Reprinted/adapted with permission from Sundar and Tirumkudulu [24]. Copyright (2014) American Chemical Society)

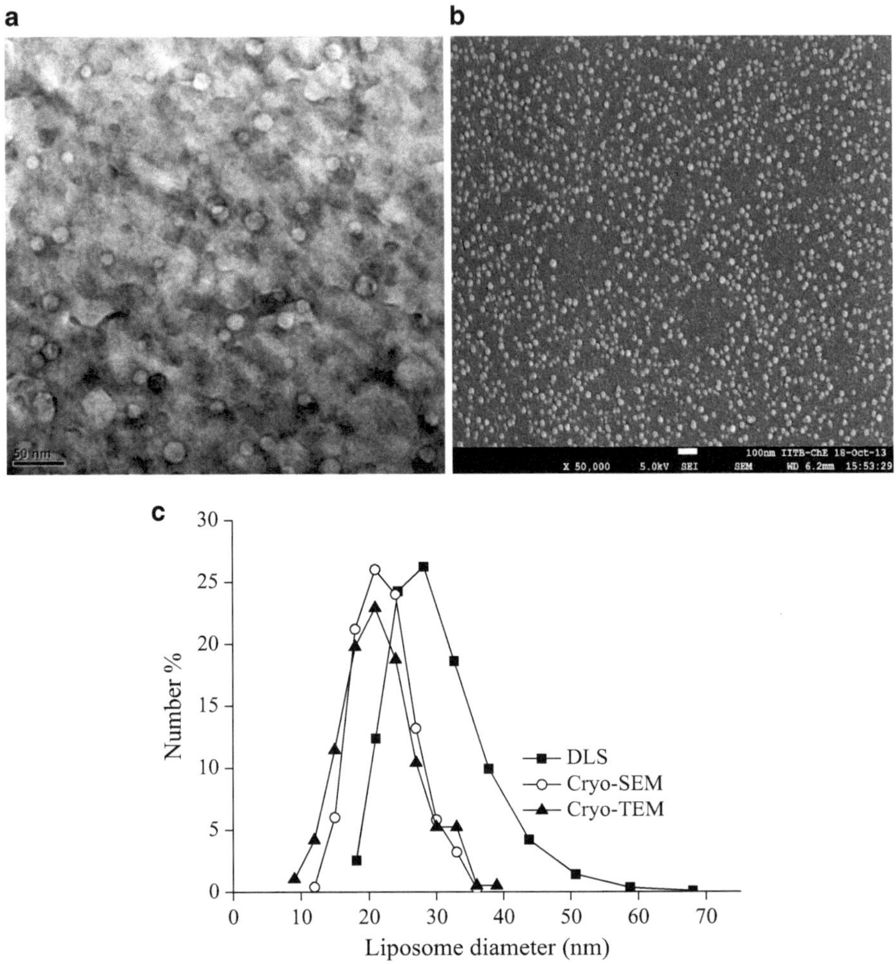

Fig. 5 (**a**) Cryo-TEM image of liposomes obtained from 70 mM DMPC lipid in bed packed with AKP-30 particles, $\phi_f = 0.35$, $Q = 20\,\mu$l/min. (**b**) Cryo-SEM image of liposomes obtained from 70 mM DMPC lipid in bed packed with AKP-30 particles, $\phi_f = 0.35$ at a flow rate of 20 μl/min and temperature 30 °C. (**c**) Comparison plot showing liposome size (diameter) distribution (by number) obtained from cryo-TEM, cryo-SEM, and DLS (Reprinted/adapted with permission from Sundar and Tirumkudulu [24]. Copyright (2014) American Chemical Society)

intensity peaks at about 45 nm with a PDI of less than 0.3. This clearly indicates that the distribution is homogeneous and monodisperse irrespective of the changes made in the process parameters.

In order to confirm the size range measured using DLS, the sample was imaged using both cryo-TEM and cryo-SEM. Figure 5a shows the image of liposomes

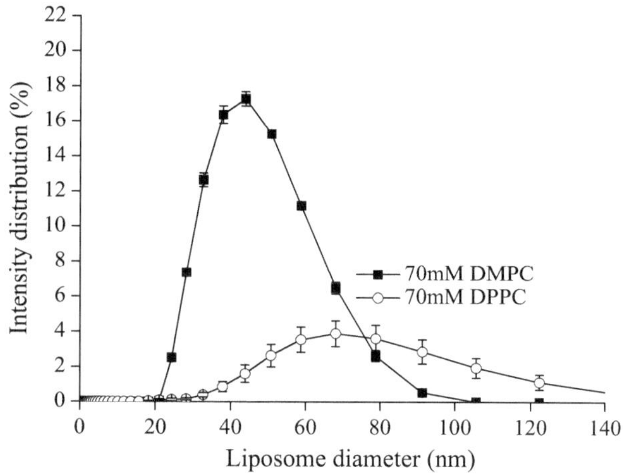

Fig. 6 Size (diameter) distribution (by intensity) of liposomes from DMPC (T_c = 23 °C, T_{exp} = 30 °C) and DPPC (T_c = 41 °C, T_{exp} = 48 °C) from 70 mM lipid concentration in bed packed with AKP-30 particles, ϕ_f = 0.35 at a flow rate of 20 μl/min (Reprinted/adapted with permission from Sundar and Tirumkudulu [24]. Copyright (2014) American Chemical Society)

formed for 70 mM DMPC at flow rate, Q = 20 μl/min in a bed packed with AKP-30 particles at ϕ_f = 0.35. The TEM images show that all the liposomes are unilamellar. Figure 5b includes a cryo-SEM image of the sample obtained under identical experimental conditions. In Fig. 5c, we compare the distribution obtained from DLS with that obtained from image analysis of both the cryo-TEM and cryo-SEM images. All three distributions indicate that majority of the liposomes fall in the range of 20–50 nm.

Experiments were also performed with DPPC, whose transition temperature (T_c = 41 °C) and acyl chain length were different from that of DMPC, in a packed bed of AKP-30 particles. Here, the bed was placed in an oven and the temperature was maintained at T_{exp} = 48 °C (Fig. 6). Interestingly, we obtained liposomes in the same size range as that observed for DMPC under identical conditions of lipid concentration and buffer flow rate. The results indicate that the technique is not specific to a particular lipid but can be extended to other lipids. However, when the experimental temperature was increased 20 °C above T_c, the liposomes were polydisperse with an average size above 200 nm.

Mayer et al. [17] have shown that extrusion of giant multilamellar liposomes through porous membranes could result in liposomes comparable to the pore size. In order to test the hypothesis that the extrusion process may cause size reduction, the packed particles in the bed were imaged using SEM. Figure 7 shows that the particles are packed uniformly and the pore size increases with increase in particle size. In addition to imaging using SEM, pore size was evaluated using porosimetry and the average pore sizes for packings of AKP-15, AKP-30, and AKP-50 were found to be 235, 115, and 93 nm. The extent of sintering, both for 30 or 60 min, does not show any variation in liposome size distribution. These results suggest that despite the pore size variation across the three grades, the final size distribution of the liposomes is very similar.

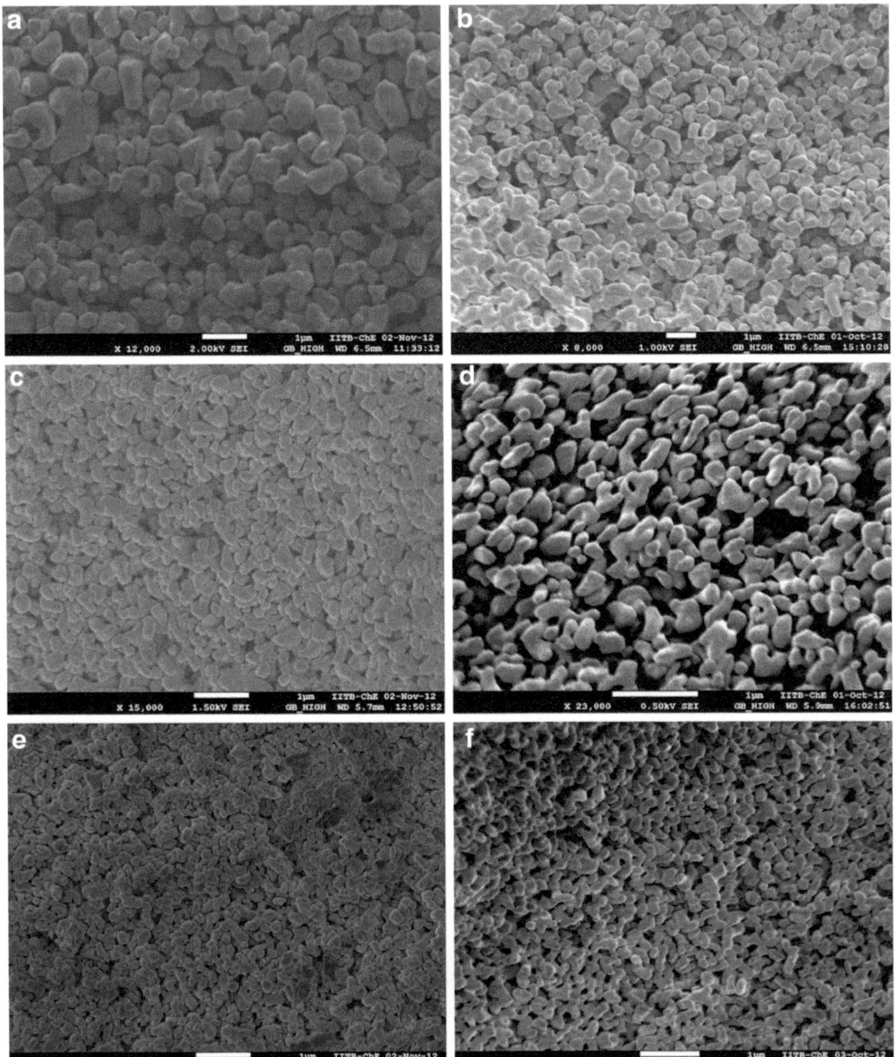

Fig. 7 SEM images of (**a**) AKP-15 (30 min sintering), (**b**) AKP-15 (60 min sintering), (**c**) AKP-30 (30 min sintering), (**d**) AKP-30 (60 min sintering), (**e**) AKP-50 (30 min sintering), (**f**) AKP-50 (60 min sintering) (Reprinted/adapted with permission from Sundar and Tirumkudulu [24]. Copyright (2014) American Chemical Society)

3.1 Extrusion Is Not the Reason for Unilamellar Liposomes

While the above results indicate that the size distribution of liposomes does not depend on the particle size used for packing, it also seems that size was not influenced by the flow of aqueous buffer through the pores inside the packed bed. This hypothesis was tested by performing two independent sets of experiments.

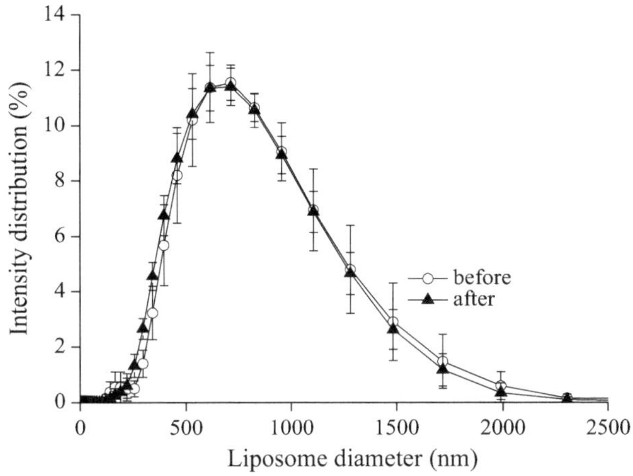

Fig. 8 Size (diameter) distribution (by intensity) of liposomes from DMPC before and after extrusion through a bed packed with AKP-30 particles, $\phi_f = 0.35$ at a flow rate of 20 μl/min and temperature 30 °C (Reprinted/adapted with permission from Sundar and Tirumkudulu [24]. Copyright (2014) American Chemical Society)

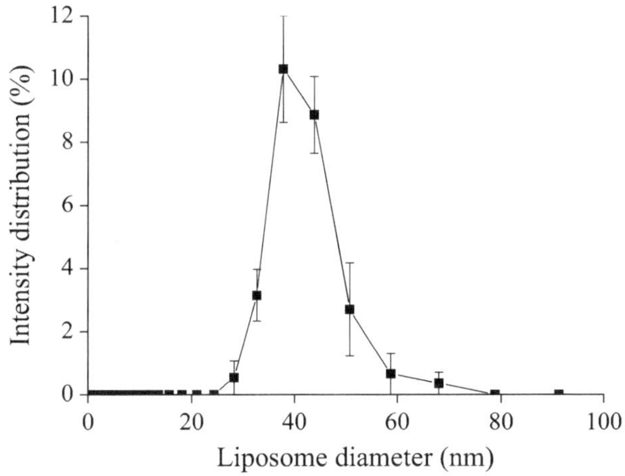

Fig. 9 Size (diameter) distribution (by intensity) of liposomes from DMPC after being rehydrated through a bed packed with AKP-30 particles, $\phi_f = 0.35$, $Q = 20$ μl/min and temperature 30 °C (Reprinted/adapted with permission from Sundar and Tirumkudulu [24]. Copyright (2014) American Chemical Society)

First, a preformed liposome suspension having an average size of around 600 nm with PDI ≤ 0.2 was allowed to pass through a packed bed of AKP-30 particles at a flow rate of 20 μl. The size distributions of liposomes before and after passage through the packed bed were very similar and in fact overlap each other as shown in Fig. 8. This further confirms that extrusion through the packed bed does not change the size distribution of the liposomes. In the second set of experiments, 100 μl of 70 mM DMPC lipid in chloroform was dried in a vial and kept under vacuum for 2 h to allow for solvent evaporation and bilayer formation. This was then followed by vigorous hydration with the aqueous buffer. The resulting milky white dispersion containing heterogeneous and polydisperse liposomes was introduced into the packed bed and dried in a stream of nitrogen gas. Upon rehydration with the aqueous buffer, a monodisperse liposome dispersion was obtained with a size distribution peaking at 45 nm (Fig. 9). Both these experiments confirm that extrusion is not the reason for the observed size distribution of liposomes.

The above results show that the liposome size is set during the hydration step itself and the flow of liposomes along with the buffer through the bed has no influence on the liposome size distribution. As the liposome size was independent of the particle size and the pore size (Fig. 4), we speculate that the size selection mechanism is a combination of the surface roughness of the packing and the bending modulus of the bilayer. Thermodynamic calculations by equating the edge energy of circular bilayer and bending energy of the bilayer to form liposome predict the smallest possible liposome radius R_l, assuming that the two structures have the same surface area [3],

$$R_l = \frac{2K_b + K_G}{\lambda},$$

where K_b is the bilayer bending modulus, K_G is the Gaussian modulus, and λ is the edge energy along the bilayer perimeter due to the unfavorable exposure of hydrocarbon tails. Molecular dynamics simulation of DMPC bilayers gives $\lambda \sim 10 - 30$ pN [13], K_b at 30 °C has been measured to be 1.15×10^{-19} J [7], and the ratio of Gaussian modulus to the bending modulus is typically taken as $K_b/K_G \sim -1$ [10]. Substituting these values in the above equation gives the minimum liposome diameter for DMPC to be 23 nm which is close to the smallest sizes observed in the measured distributions (Fig. 4).

While the thermodynamic calculations predict the smallest liposome size, the reason for the occurrence of peak at around 45 nm (by intensity) is not clear. In this respect, the investigation by Roiter et al. [22] discusses the interaction of DMPC lipid membranes with nanosurfaces formed from silica nanoparticles of different diameter. It was shown that the bilayers completely coat smooth silica particles having diameters greater than 22 nm, but for nanosurfaces with dimensions between 1.2 and 22 nm, the bilayer membrane forms holes as the formation of continuous membrane is energetically unfavorable. Hence, it is possible that the rough surface of the highly asymmetric alumina particles used in this study along with the porous nature of the packing may result in patchy coatings of membranes on the particles that subsequently unbind and close upon hydration to give sub-100 nm liposomes. However, more experiments involving alumina particles of varying size and surface roughness would be required to confirm this hypothesis. Finally, the liposome formation technique explained here can easily be extended to a continuous process to produce liposomes loaded with drug. Pumping alternate plugs of lipid solution and aqueous buffer (containing a drug) through a packed bed could enable us to generate a continuous stream of drug-loaded liposomes.

4 Conclusion

A novel technique for generating homogeneous liposome dispersion with low polydispersity in a single step suitable for scale-up has been demonstrated here. The final liposome size distribution is not affected by the process parameters such as flow rate of aqueous medium, particle size, and particle volume fraction. Experiments also indicate that liposome size reduction due to extrusion does not occur inside the packed bed. It appears that the surface roughness of the particle may play a critical role in determining the final liposome size.

Acknowledgements We acknowledge the IRHPA program of the Department of Science and Technology, India, for financial support and SAIF and Chemical Engineering Department of IIT Bombay for TEM and SEM.

References

1. Akamatsu K, Shimizu Y, Shimizu R, Nakao S-i (2013) Facile method for preparing liposomes by permeation of lipid-alcohol solutions through shirasu porous glass membranes. Ind Eng Chem Res 52(30):10329–10332
2. Bangham A, Standish M, Watkins J (1965) Diffusion of univalent ions across the lamellae of swollen phospholipids. J Mol Biol 13:238–252
3. Boal D (2001) Mechanics of the cell. Cambridge University Press, Cambridge
4. Deamer D, Bangham A (1976) Large volume liposomes by an ether vaporization method. Biochim Biophys Acta 443(3):629–634
5. Dittrich PS, Heule M, Renaud P, Manz A (2006) On-chip extrusion of lipid vesicles and tubes through microsized apertures. Lab Chip 6(4):488–493
6. Drummond DC, Meyer O, Hong K, Kirpotin DB, Papahadjopoulos D (1999) Optimizing liposomes for delivery of chemotherapeutic agents to solid tumors. Pharmacol Rev 51(4):691–743
7. Duwe HP, Kaes J, Sackmann E (1990) Bending elastic moduli of lipid bilayers: modulation by solutes. J Phys Fr 51:945–962
8. Hope M, Bally M, Webb G, Cullis P (1985) Production of large unilamellar vesicles by a rapid extrusion procedure: characterization of size distribution, trapped volume and ability to maintain a membrane potential. Biochim Biophys Acta 812:55–65
9. Howse JR, Jones RAL, Battaglia G, Ducker RE, Leggett GJ, Ryan AJ (2009) Templated formation of giant polymer vesicles with controlled size distributions. Nat Mater 8:507–511
10. Hu M, de Jong DH, Marrink SJ, Deserno M (2013) Gaussian curvature elasticity determined from global shape transformations and local stress distributions: a comparative study using the martini model. Faraday Discuss 161:365–382
11. Israelachvili JN (2011) Intermolecular and surface forces. Academic, Burlington
12. Jahn A, Vreeland W, Devoe D, Locascio L, Gaitan M (2007) Microfluidic directed formation of liposomes of controlled size. Langmuir 23:6289–6293
13. Jiang FY, Bouret Y, Kindt JT (2004) Molecular dynamics simulations of the lipid bilayer edge. Biophys J 87:182–192
14. Lasic DD (1993) Liposomes: from physics to applications. Elsevier, Amsterdam/New York
15. Le Berre M, Yamada A, Reck L, Chen Y, Baigl D (2008) Electroformation of giant phospholipid vesicles on a silicon substrate: advantages of controllable surface properties. Langmuir 24(6):2643–2649
16. Lin Y, Huang K, Chiang J, Yang C, Lai T (2006) Manipulating self-assembled phospholipid microtubes using microfluidic technology. Sens Actuators B 117(2):464–471
17. Mayer L, Hope M, Cullis P (1986) Vesicles of variable sizes produced by a rapid extrusion procedure. Biochim Biophys Acta Biomembr 858:161–168
18. Mouritsen O (2005) Life – as a matter of fat. Springer, Berlin/Heidelberg
19. Nayar R, Hope MJ, Cullis PR (1989) Generation of large unilamellar vesicles from long-chain saturated phosphatidylcholines by extrusion technique. Biochim Biophys Acta 986:200–206

20. Olson F, Hunt C, Szoka F, Vail W, Papahadjopoulous D (1979) Preparation of liposomes of defined size distribution by extrusion through polycarbonate membranes. Biochim Biophys Acta 557:9–23
21. Pozo-Navas B, Raghunathan V, Katsaras J, Rappolt M, Lohner K, Pabst G (2003) Discontinuous unbinding of lipid bilayers. Phys Rev Lett 91:028101–1–4
22. Roiter Y, Ornatska M, Rammohan AR, Balakrishnan J, Heine DR, Minko S (2009) Interaction of lipid membrane with nanostructured surfaces. Langmuir 25(11):6287–6299
23. Singh KB, Bhosale LR, Tirumkudulu MS (2009) Cracking in drying colloidal films of flocculated dispersions. Langmuir 25(8):4284–4287
24. Sundar SK, Tirumkudulu M (2014) Synthesis of sub-100-nm liposomes via hydration in a packed bed of colloidal particles. Ind Eng Chem Res 53(1):198–205
25. Szoka F, Papahadjopoulos D (1978) Procedure for preparation of liposomes with large internal aqueous space and high capture by reverse-phase evaporation. Proc Natl Acad Sci U S A 75(9):4194–4198
26. Teh S, Khnouf R, Fan H, Lee AP (2011) Stable, biocompatible lipid vesicle generation by solvent extraction-based droplet microfluidics. Biomicrofluidics 5(4):044113
27. van Swaay D, deMello A (2013) Microfluidic methods for forming liposomes. Lab Chip 13:752–767
28. Yamada N, Hishida M, Seto H, Tsumoto K, Yoshimura T (2007) Unbinding of lipid bilayers induced by osmotic pressure in relation to unilamellar vesicle formation. Europhys Lett 80:480002–p1–480002–p6

Preparation of Liposomes for Drug Delivery Applications by Extrusion of Giant Unilamellar Vesicles

Yogita P. Patil and Sameer Jadhav

Abstract The potential of liposomes as carriers in drug delivery had been predicted several decades ago and has already seen some success. One of the limitations in liposome preparation for drug delivery applications has been the need to subject suspensions of multilamellar vesicles of several membrane extrusions at high pressures so that liposomes of desired size may be obtained. In this work we have demonstrated that a single-pass extrusion of electroformed giant unilamellar vesicles at moderate pressures yields a monodisperse population of liposomes of the desired size. We also show that low-pass filtration of giant vesicle suspensions to remove smaller vesicles prior to extrusion facilitates better prediction of extruded liposome size. Our results provide new insights into the possible mechanisms involved in size reduction of giant vesicles by extrusion.

Keywords Liposomes • Multilamellar vesicles • Giant unilamellar vesicles • Electroformation

1 Introduction

Liposomes, also known as vesicles, are spherical capsules composed of one or more phospholipid bilayers enclosing an aqueous core. Liposomes may be unilamellar vesicles with a single phospholipid bilayer or multilamellar vesicles (MLVs) with several concentric bilayer shells. Based on size, unilamellar vesicles are classified as small unilamellar vesicles (SUV, 0.025–0.05 μm), large unilamellar vesicles (LUVs, 0.05–0.5 μm), and giant unilamellar vesicles (GUVs, greater than 1 μm) [27, 28]. Liposomes are usually considered as models for studying cell biomembranes in the laboratory and are used as nanocarriers for active ingredients in the cosmetic and pharmaceutical drug formulation industries. To increase the efficacy of active pharmaceutical ingredients, liposomes are increasingly used as carriers for targeted drug delivery and controlled release of drugs [2, 19]. Since drug-loaded

Y.P. Patil • S. Jadhav, PhD (✉)
Department of Chemical Engineering, Indian Institute of Technology Bombay,
Mumbai 400076, India
e-mail: srjadhav@iitb.ac.in

Y.M. Joshi, S. Khandekar (eds.), *Nanoscale and Microscale Phenomena*,
Springer Tracts in Mechanical Engineering, DOI 10.1007/978-81-322-2289-7_2

MLVs are known to exhibit unpredictable pharmacokinetics and undergo rapid clearance by the reticuloendothelial system [13], monodisperse small unilamellar vesicle populations having mean radius below 100 nm are preferred for drug delivery applications [29]. Small unilamellar vesicles (radius $<$ 100 nm) possess higher curvature which reduces opsonin recruitment and decreases reticuloendothelial system (RES) uptake [11, 15]. Encapsulation efficiency of a hydrophilic drug can be increased using unilamellar liposomes as they encapsulate a larger aqueous volume compared to MLVs.

Post-processing of MLVs by sonication, French press, detergent dialysis, or multiple extrusions is required to prepare monodisperse SUVs [1, 3, 5, 8, 12, 21]. The common limitations of these methods include a poor control of vesicle size and the necessity of additional steps such as removal of detergents to get SUVs suitable for drug delivery applications.

MLV suspensions are subjected to multiple extrusion cycles through defined cylindrical pores of track-etched polycarbonate membranes to obtain monodisperse SUVs. The mechanism of size reduction of polydisperse MLV populations into SUVs is not well understood. For instance, Clerc and Thompson suggested that MLVs subjected to extrusion through nanopores deform into cylindrical tubes which rupture and form SUVs in a manner similar to droplet formation by breaking of cylindrical fluid threads [6]. Under equilibrium conditions, the surface tension of a liposome is small. Bruinsma suggested that a vesicle can rupture when the bilayer tension exceeds the lysis tension of the vesicle while it is flowing through (or exiting from) the membrane pore [4]. On the other hand, based on their experimental work, Frisken and coworkers proposed that the rupture of the liposome occurs at the entrance of the membrane pore [14]. It has been observed that the vesicle size decreased with both pressure and flow rate [14]. Moreover, it was shown that the radius of extruded vesicles is dependent on lysis tension of the lipid bilayer and the size of pores through which the vesicles are extruded [23]. The effect of pore size has been studied, and it was observed that extrusion of MLVs through larger pores generally produced vesicles smaller than the pore size, while conversely extrusion through smaller pores yielded vesicles larger than the pore size [10, 18]. It is also well known that multiple extrusions, typically ten or more, are required to obtain monodisperse unilamellar vesicles. However, even after subjecting MLV suspensions to multiple extrusion cycles, it was observed that 30 % of extruded vesicles remained oligolamellar; the multiple extrusions led to significant loss of encapsulated drug as well as lipid [16, 24]. It was also reported that extrusion of MLVs below gel transition temperature required prohibitively large pressures [20].

In comparison to MLVs, giant unilamellar vesicles (GUVs) have been shown to encapsulate large volumes of hydrophilic materials [17]. It was shown that 80 % of electroformed GUVs are unilamellar and free from bilayer defects, while gentle hydration produces significantly smaller percentage of defect-free GUVs [25]. To separate GUVs from the small vesicles formed during gentle hydration of phospholipid films, techniques such as low-pressure membrane filtration [26] and large pore dialysis [30] have been successfully employed. In this chapter, we compare extrusion of MLV and GUV suspensions with respect to several parameters such as number of extrusion cycles, transmembrane pressure, and drug encapsulation capacity.

2 Methods

GUVs and MLVs of 1,2-dipalmitoyl-*sn*-glycero-3-phosphocholine (DPPC) were prepared by electroformation [9] and thin-film hydration method [7, 18], respectively. For electroformation of GUVs, a DPPC solution in chloroform (50 μl of 2 mg/ml) was spin coated at 1,000 rpm for 1 min onto ITO slides, and vacuum was applied overnight to remove the residual chloroform. Lipid-layered faces of two ITO slides were separated by a silicone rubber spacer forming a chamber. Next, water was injected with a syringe through the silicone rubber spacer into the chamber, and an AC voltage (10 Hz, 3 Vpp, sinusoidal) was applied for 2 h across the ITO slides using a function generator at 55 °C (Model 33220A, Agilent Technologies, CA, USA). Finally, a square wave was applied for 1 min to detach the vesicles from the ITO surface.

For film hydration, DPPC solution in chloroform (0.3 ml of 2 mg/ml) was evaporated in a rotary evaporator (Roteva Equitron, Medica Instruments, Mumbai, India) at a speed of 100 rpm for 10 min and a temperature of 55 °C. The residual chloroform from the lipid film was removed by overnight application of vacuum. Next, the dried lipid film was hydrated using 3 ml water in the rotary evaporator at 100 rpm for 60 min maintained at 55 °C. Later, the resulting MLV suspension was subjected to five freeze–thaw cycles and immediately used for extrusion studies.

An extruder was fabricated in-house that consisted of a cylindrical stainless steel chamber with a holder for the polycarbonate membrane. To carry out extrusions at defined temperatures and pressures, the extrusion chamber was water jacketed and connected to a pressure-regulated nitrogen gas cylinder. In some experiments GUV suspensions were subjected to low-pressure filtration to remove SUVs (Fig. 1a). A recently published low-pressure filtration setup [26] was significantly modified and assembled in-house to separate GUVs from smaller vesicles (Fig. 1a). The electroformed vesicle suspension was introduced into the sample holder, and all air bubbles were carefully removed by manipulating the three-way stopcock. The electroformed vesicle suspension filtered through the membrane with a pressure as low as 0.29 KPa. The GUV suspension retained in the vesicle holder was directly used for extrusion studies.

GUV suspensions were subjected to single-pass extrusion through track-etched polycarbonate membranes with nominal pore radii of 25, 50, and 100 nm at 55 °C and pressures ranging from 0.007 to 2.75 MPa (Fig. 1b). In contrast, MLV suspensions were subjected to double-pass extrusion through membranes with nominal pore radius 50 nm at 55 °C and pressures ranging from 0.035 to 2.75 MPa. The mean radius and polydispersity of the extruded liposomes were estimated by dynamic light scattering at 25 °C. Encapsulation efficiency of a hydrophilic

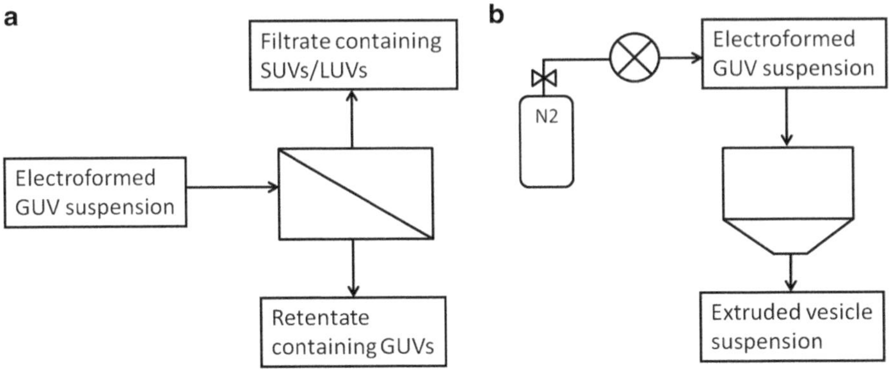

Fig. 1 Flow diagram of liposome preparation. (**a**) GUVs separated from smaller vesicles by low-pressure filtration of vesicle suspension through membrane having pore radius of 500 nm. (**b**) Liposome formation by extrusion of GUVs or MLVs at moderate pressures through track-etched membranes

fluorophore by GUVs and MLVs as well as by extruded liposomes obtained via extrusion of the GUV and MLV suspensions was measured using fluorescence spectrophotometer. Subsequent experiments were designed to compare the size distribution data of GUVs and MLVs after extrusion through a 50 nm pore radius membrane at different extrusion pressures.

3 Results and Discussion

For GUV suspensions subjected to single-pass extrusion, the vesicle radius decreased with increasing extrusion pressure. No further reduction in size was observed above 0.7 MPa extrusion pressure (Fig. 2a). In contrast, double-pass extrusion of MLV suspensions required considerably higher extrusion pressure to produce liposomes of mean radii similar to those generated from single-pass extrusion of GUV suspension (Fig. 2b). For both, GUV and MLV suspensions, the extruded liposome populations became progressively monodisperse with increase in the extrusion pressure difference.

The fraction of the fluorophore 5(6)-carboxyfluorescein (CF) present in vesicles was estimated to compare the drug encapsulation efficiency of GUVs to that of MLVs. Only ~15 % of CF was encapsulated by MLVs, whereas GUVs was found to encapsulate fourfold higher percentage of CF (Fig. 3). To evaluate whether membrane extrusion resulted in CF loss from the liposomes, CF-enclosed MLV suspensions were subjected to eleven cycles of extrusion at 20 psi (0.14 MPa), while GUV suspensions were subjected to a single-pass extrusion through a 100 nm pore radius membrane at the same extrusion pressure. The liposomes obtained from extrusion of MLVs were found to contain 12 % of the CF (Fig. 3), whereas the liposomes produced from extrusion of GUVs were found to enclose 45 %

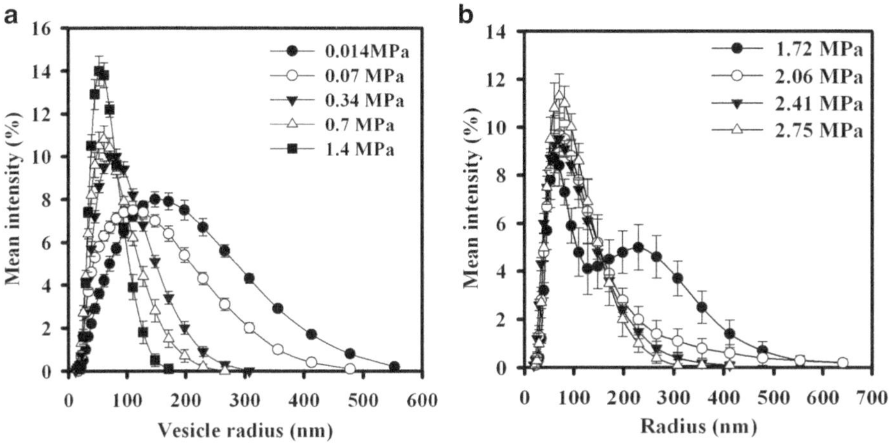

Fig. 2 Dynamic light scattering data for liposomes obtained by extrusion of GUV (**a**) and MLV (**b**) suspensions at different extrusion pressures. *Data represented as mean* $\pm$ *SEM for* $n \geq 3$ *experiments*

Fig. 3 Encapsulation efficiency of liposomes in MLV and GUV suspensions before and after extrusion. Data represented as mean $\pm$ SEM for $n \geq 3$ experiments. *NS* not significant with respect to MLVs before extrusions, $*p < 0.05$ with respect to GUVs before extrusion

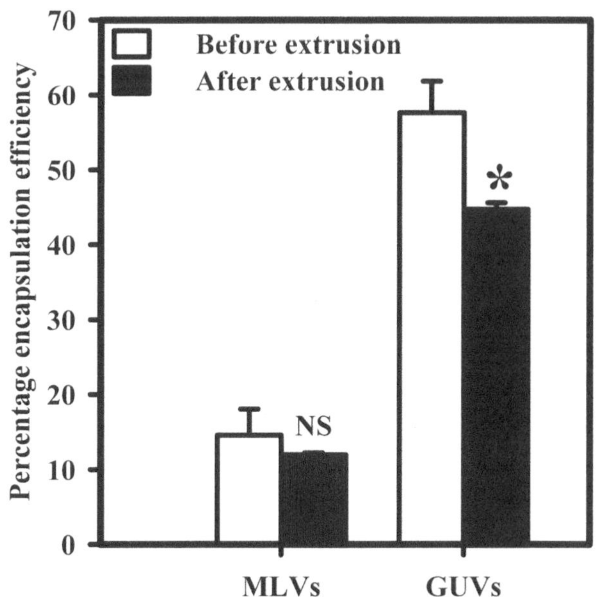

of CF (Fig. 3), indicating that liposomes obtained from single-pass extrusion of GUVs show almost a fourfold larger drug entrapment efficiency compared to those prepared by the extrusion of MLV suspensions.

FEG-TEM analysis of extruded vesicles revealed predominantly spherical vesicles. Extrusion of MLVs produced multilamellar vesicles having mean radius in the range 100–150 nm (Fig. 4a, b). In contrast, extrusion of GUVs produced unilamellar

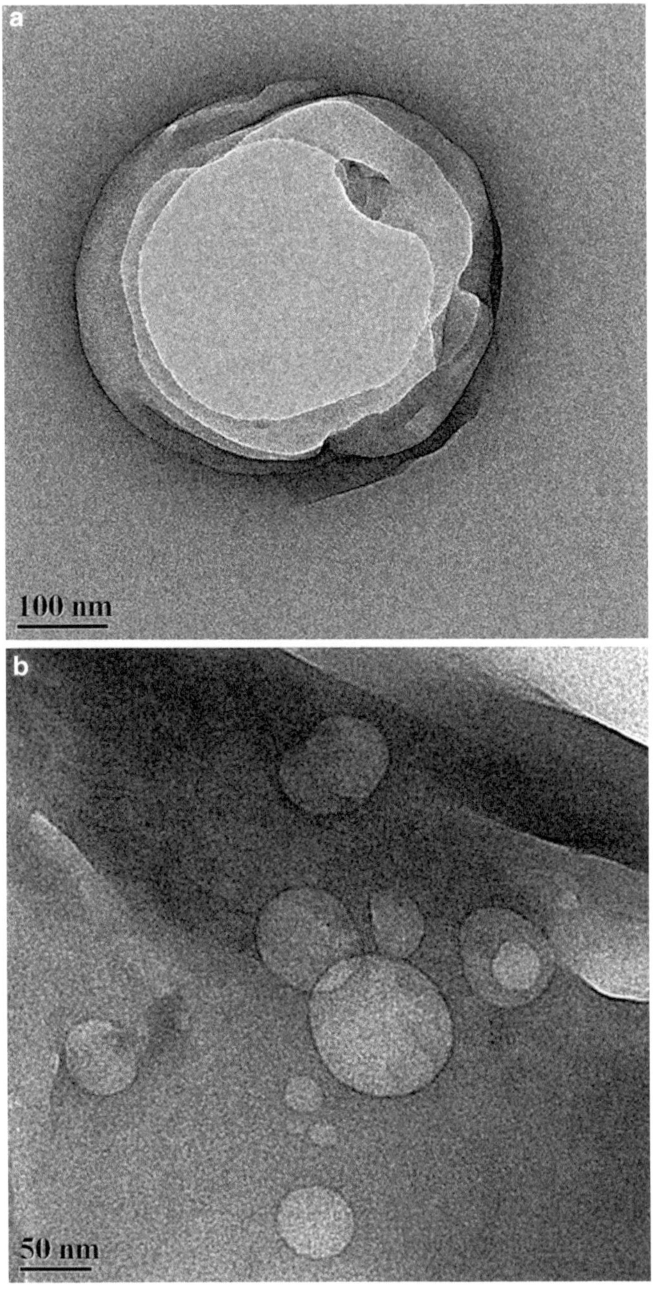

Fig. 4 FEG-TEM micrographs of extruded MLVs and GUVs. Extruded liposomes obtained by extrusion of MLV (**a**) and GUV (**b**) suspensions at 0.689 MPa through 50 nm pore radius polycarbonate membranes

vesicles having mean radius in the range 50–150 nm (Fig. 4c). The measurement of the extruded vesicle sizes, formed after extrusion of GUVs and MLVs at pressure of 0.689 MPa through 50 nm radius pore membrane, using FEG-TEM and DLS was found to be congruent.

To study the effect of membrane pore radius on the size of liposomes obtained from single-pass extrusion of GUV suspensions, we estimated the ratio of the mean radius of the extruded liposomes (R_v) to nominal pore radius of the membranes (R_p). Our data show that increase in extrusion pressure difference resulted in an asymptotic decrease in the mean radius of the extruded liposomes to a critical value beyond which there is no change in size with further increase in extrusion pressure. Interestingly, the minimum value of mean liposome radius obtained by single-pass GUV extrusion was found to be smaller than the pore radius when membranes with nominal pore radii larger than 50 nm were used [22]. The abovementioned observation is similar to that reported for liposomes obtained by subjecting MLVs to multiple extrusion cycles [14].

Our subsequent experiments were directed toward addressing this issue related to size of extruded vesicles with respect to that of the membrane pore. We hypothesized that the mean radii of suspended liposomes after extrusion were smaller than the membrane pore radii due to contributions from a subpopulation of small vesicles already present in the electroformed vesicle suspension prior to extrusion. To test our hypothesis, electroformed vesicle suspensions were subjected to gentle filtration at a pressure of 0.29 kPa through a track-etched polycarbonate membrane having nominal pore radius of 500 nm (Fig. 1a). Liss Rhod-PE-labeled electroformed GUVs were observed before and after filtration using a confocal laser scanning microscope and revealed that the electroformed vesicle suspension prior to filtration is highly polydisperse (Fig. 5a). After low-pressure filtration, the retentate comprised of a GUV suspension (Fig. 5b), while the filtrate contained mostly SUVs and LUVs.

A single-pass extrusion of these pre-filtered GUV suspensions was carried out through track-etched polycarbonate membranes having pore radii of 25, 50, 100, and 200 nm at transmembrane pressures ranging from 0.007 to 2.76 MPa. As can be seen in Fig. 6 (a–d), the mean radius of the extruded liposomes decreased with increase in the applied pressure, until further increase in applied pressure did not significantly affect the extruded liposome radius (Fig. 6).

The minimum attainable value of the mean radius (R_{min}) of extruded liposomes was compared with mean radius of liposomes obtained by direct extrusion of electroformed vesicle suspensions without the filtration step. Absence of pre-filtration step prior to extrusion through pores having radii 100 and 200 nm resulted in R_{min} less than the nominal pore radius of the membrane (Fig. 7). In contrast, the mean radius of liposomes obtained by extrusion of pre-filtered GUV suspensions was always larger than the nominal pore radius of the membrane (Fig. 7).

It was shown that the measured mean pore radius of the track-etched polycarbonate membranes differed significantly from the nominal pore radius provided by the membrane manufacturer [10]. Therefore, we estimated the mean pore radius

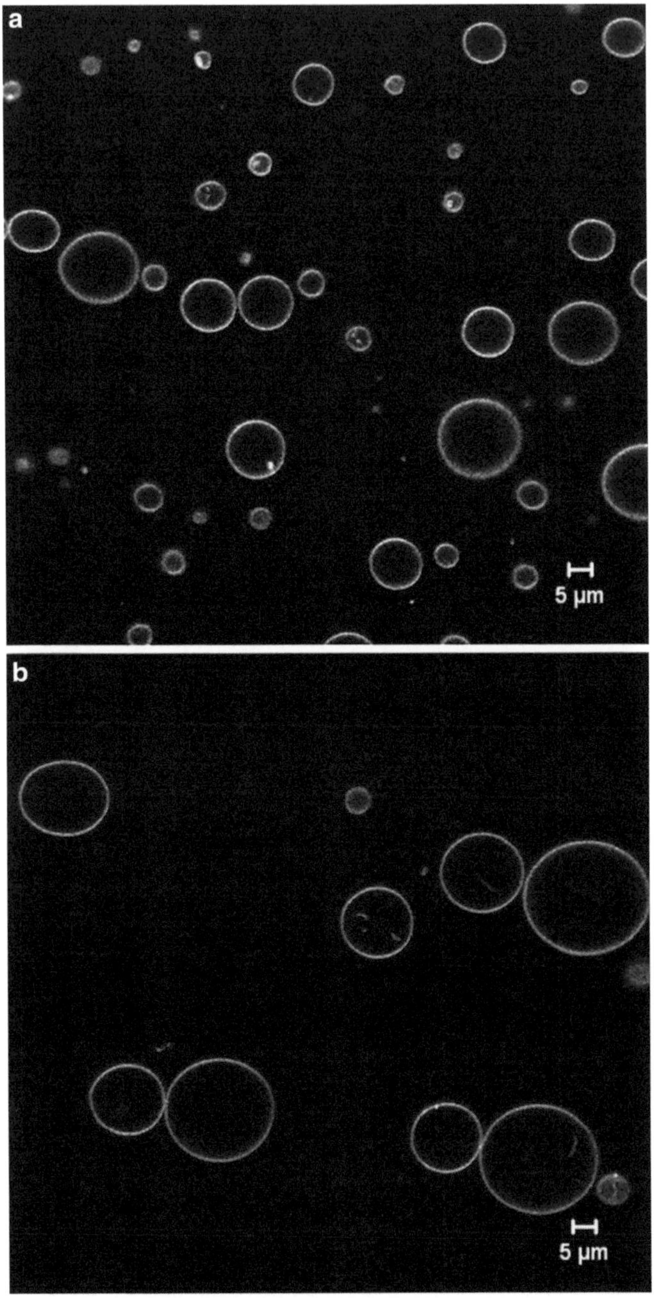

Fig. 5 Confocal laser scanning micrographs of vesicle suspensions. CLSM images of GUV suspension prior to filtration (**a**), GUVs in retentate (**b**)

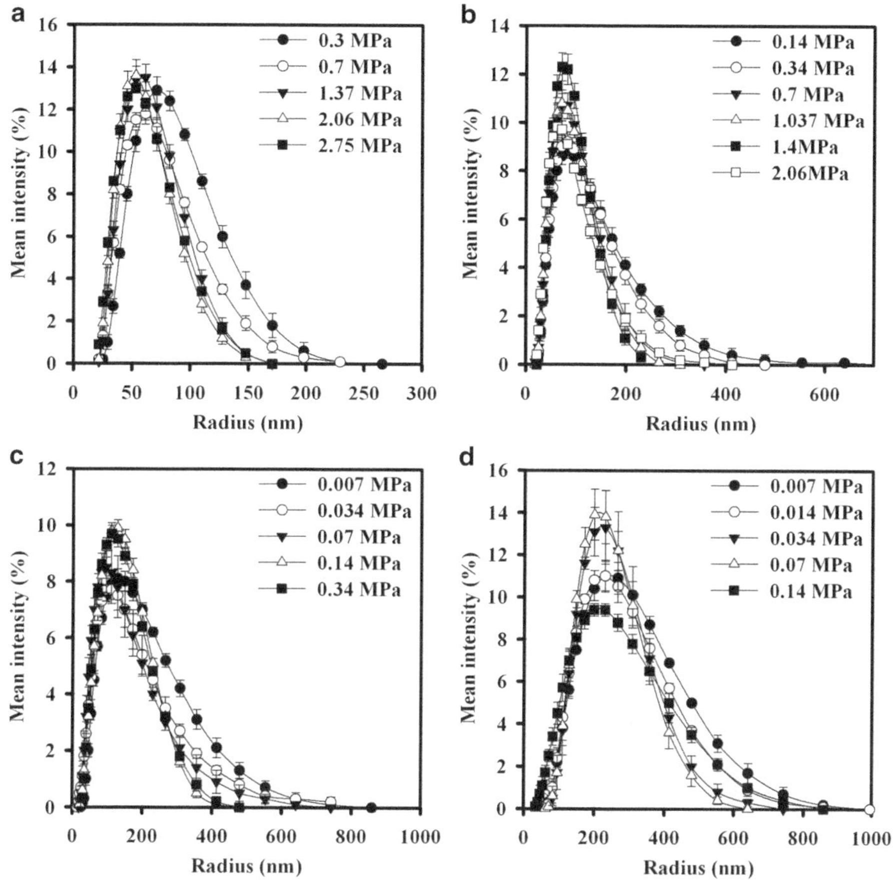

Fig. 6 Effect of membrane pore size on liposome size. Liposomes obtained by extrusion of GUV at different pressures through polycarbonate membranes of pore radii (**a**) 25 nm, (**b**) 50 nm, (**c**) 100 nm, and (**d**) 200 nm

for membranes having nominal pore radii of 25, 50, 100, and 200 nm, by scanning electron microscopy, and found the mean pore radii to be 24, 38, 72, and 143 nm, respectively.

The minimum attainable value of mean radius (R_{min}) of extruded liposomes was found to be linearly related to the mean pore radius of the polycarbonate membrane. Moreover, the mean liposome radius (R_v), normalized with this linear function ($mR_p + R_o$, with $m = 1.4$, $R_o = 25.3$ nm), was plotted as a function of the average velocity of the vesicle suspension in the membrane pore (U). As can be seen from Fig. 8, data obtained for the entire range of extrusion pressures and membrane pore sizes collapsed onto a single curve, indicating that $R_v/(mR_p + R_o)$ is a function of U. In our previous work, we had proposed a phenomenological model relating the extruded liposome radius to the average velocity in the membrane pore [22]. The

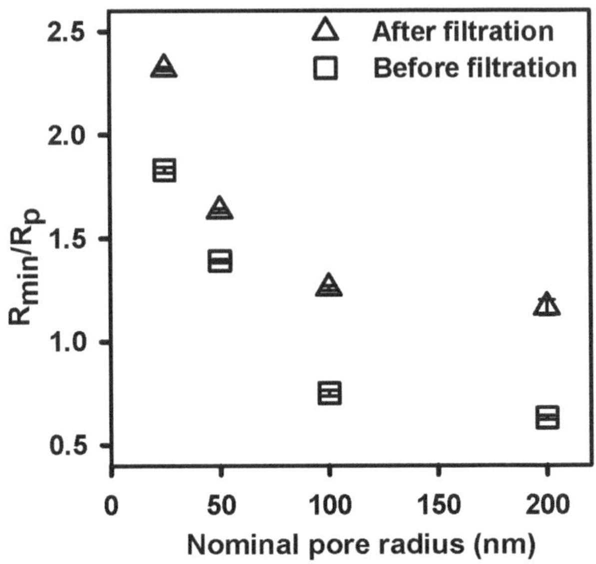

Fig. 7 Effect of nominal pore radius on mean liposome radius. Data represented as mean $\pm$ SEM for $n \geq 3$ independent experiments

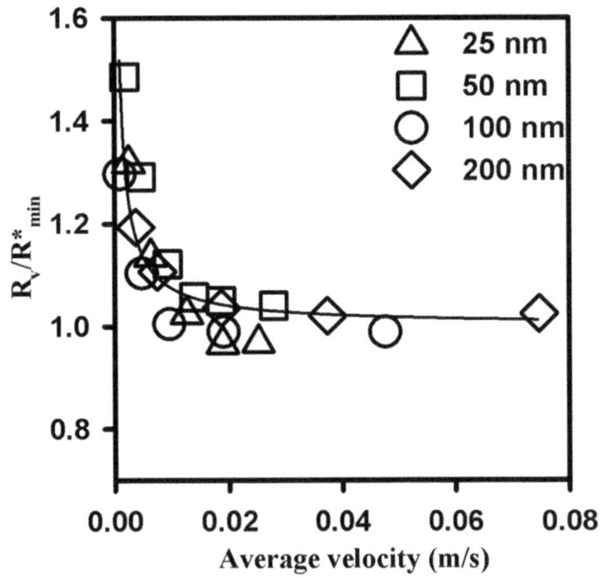

Fig. 8 Functional dependence of R_v/R^*_{min} on average velocity, U. The proposed model was fitted to the R_v/R_{min} Vs average velocity data. The parameters k_1 and k_2 were found to be 1.7×10^{-5} and 6,000 N/m^2, respectively

model has been updated to include the functional dependence of R_{min} on R_p and was fitted to $R_v/(mR_p + R_o)$ versus U data using the Levenberg–Marquardt nonlinear least squares algorithm. The model fitted the data well (Fig. 8), and the parameters k_1 and k_2 were found to be 1.7×10^{-5} and 6,000 N/m^2, respectively.

We also examined the cellular uptake of liposomes obtained by extrusion of electroformed GUVs. Liposomes containing rhodamine-PE in the bilayer and

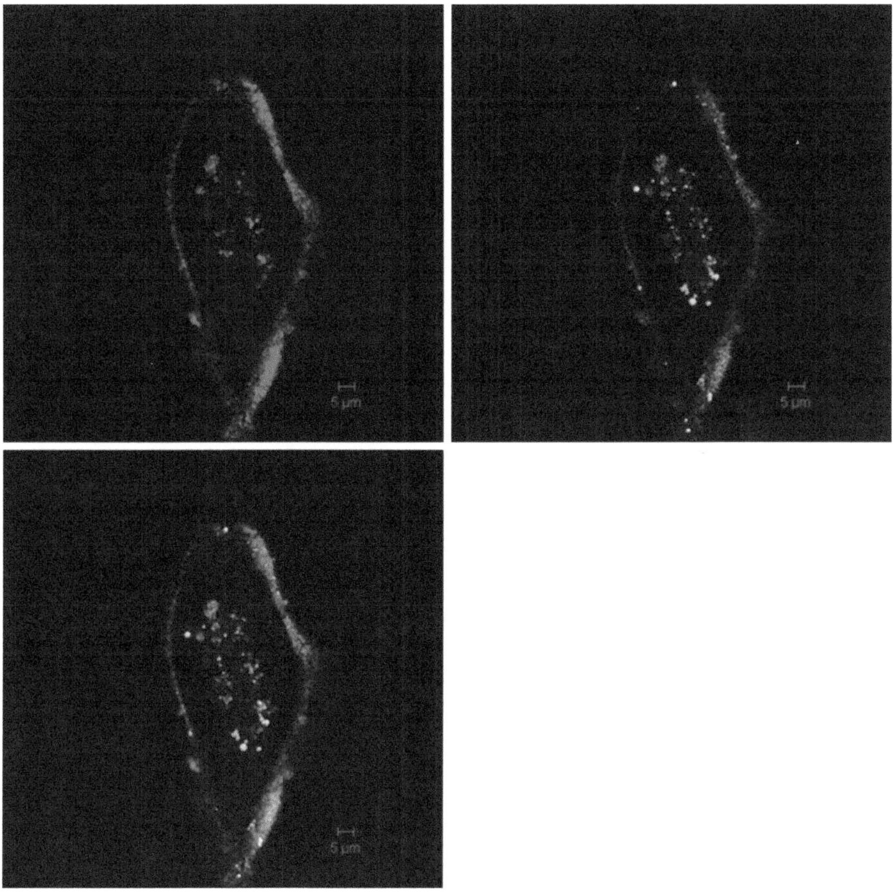

Fig. 9 Confocal laser scanning microscopy of MDA-MB-231 human breast adenocarcinoma cancer cells incubated with liposomes in DMEM at 37 °C after 30 min

carboxyfluorescein encapsulated in their aqueous core were incubated with MDA-MB-231 human breast adenocarcinoma cancer cell line for 30 min at 37 °C. Following incubation with liposomes, cells were washed and observed under confocal laser scanning microscope. Confocal imaging of 0.5 % Rh-PE-labeled liposomes encapsulating carboxyfluorescein showed that carboxyfluorescein and some Rh-PE were found inside intracellular compartments, while the remaining Rh-PE was localized on the cell membrane surface (green fluorescence) as shown in Fig. 9.

Taken together, we present a novel method where electroformed giant vesicles may be extruded through track-etched membranes to produce small unilamellar liposomes having a narrow size distribution and possessing high drug encapsulating potential. The single-pass extrusion of GUVs at substantially lower transmembrane pressures is a significant improvement over the multi-pass extrusion of MLVs requiring multiple extrusion cycles. We also demonstrate that a simple water column-based filtration protocol can be employed as an efficient method for separation of SUVs or LUVs from GUVs. This technique can also be used for concentration of dilute GUVs suspension. Moreover, our proposed model predicts the size distribution of extruded vesicles with respect to membrane pore radii and sheds light on the mechanisms of vesicle size reduction during extrusion process. The methods for liposome preparation and analysis provide insights for a rational design of drug delivery carriers.

References

1. Abramson MB, Katzman R, Gregor HP (1964) Aqueous dispersions of phosphatidylserine ionic properties. J Biol Chem 239:70–76
2. Allen TM, Cullis PR (2013) Liposomal drug delivery systems: from concept to clinical applications. Adv Drug Deliv Rev 65(1):36–48
3. Barenholzt Y, Amselem S, Lichtenberg D (1979) A new method for preparation of phospholipid vesicles (liposomes) – French press. FEBS Lett 99(1):210–214
4. Bruinsma R (1996) Rheology and shape transitions of vesicles under capillary flow. Physica A 234(1–2):249–270
5. Brunner J, Skrabal P, Hauser H (1976) Single bilayer vesicles prepared without sonication: physico-chemical properties. Biochim Biophys Acta 455(2):322–331
6. Clerc SG, Thompson TE (1994) A possible mechanism for vesicle formation by extrusion. Biophys J 67:475–477
7. Elzainy AAW, Gu X, Simons FER, Simons KJ (2005) Hydroxyzine- and cetirizine-loaded liposomes: effect of duration of thin film hydration, freeze-thawing, and changing buffer pH on encapsulation and stability. Drug Dev Ind Pharm 31(3):281–291
8. Enoch HG, Strittmatter P (1979) Formation and properties of 1000-Å-diameter, single-bilayer phospholipid vesicles. Proc Natl Acad Sci U S A 76(1):145–149
9. Estes DJ, Mayer M (2005) Electroformation of giant liposomes from spin-coated films of lipids. Colloids Surf B Biointerfaces 42(2):115–123
10. Frisken BJ, Asman C, Patty PJ (2000) Studies of vesicle extrusion. Langmuir 16(3):928–933
11. Harashima H, Sakata K, Funato K, Kiwada H (1994) Enhanced hepatic uptake of liposomes through complement activation depending on the size of liposomes. Pharm Res 11(3):402–406
12. Hope MJ, Bally MB, Webb G, Cullis PR (1985) Production of large unilamellar vesicles by a rapid extrusion procedure. Characterization of size distribution, trapped volume and ability to maintain a membrane potential. Biochim Biophys Acta 812(1):55–65
13. Hunt CA, Rustum YM, Mayhew E, Papahadjopoulos D (1979) Retention of cytosine arabinoside in mouse lung following intravenous administration in liposomes of different size. Drug Metab Dispos 7(3):124–128
14. Hunter DG, Frisken BJ (1998) Effect of extrusion pressure and lipid properties on the size and polydispersity of lipid vesicles. Biophys J 74(6):2996–3002
15. Ishida T, Harashima H, Kiwada H (2002) Liposome clearance. Biosci Rep 22(2):197–224
16. Jousma H, Talsma H, Spies F, Joosten JGH, Junginger HE, Crommelin DJA (1987) Characterization of liposomes. The influence of extrusion of multilamellar vesicles through polycarbonate membranes on particle size, particle size distribution and number of bilayers. Int J Pharm 35:263–274
17. Kim S, Martin GM (1981) Preparation of cell-size unilamellar liposomes with high captured volume and defined size distribution. Biochim Biophys Acta 646(1):1–9

18. Mayer LD, Hope MJ, Cullis PR (1986) Vesicles of variable sizes produced by a rapid extrusion procedure. Biochim Biophys Acta 858(1):161–168
19. Mittal NK, Bhattacharjee H, Mandal B, Balabathula P, Thoma LA, Wood GC (2014) Targeted liposomal drug delivery systems for the treatment of B cell malignancies. J Drug Target 22(5):372–386
20. Nayar R, Hope MJ, Cullis PR (1989) Generation of large unilamellar vesicles from long-chain saturated phosphatidylcholines by extrusion technique. Biochim Biophys Acta 986(2):200–206
21. Patil YP, Jadhav S (2014) Novel methods for liposome preparation. Chem Phys Lipids 177:8–18
22. Patil YP, Kumbhalkar MD, Jadhav S (2012) Extrusion of electroformed giant unilamellar vesicles through track-etched membranes. Chem Phys Lipids 165(4):475–481
23. Patty PJ, Frisken BJ (2003) The pressure-dependence of the size of extruded vesicles. Biophys J 85(2):996–1004
24. Perkins WR, Minchey SR, Ahl PL, Janoff AS (1993) The determination of liposome captured volume. Chem Phys Lipids 64(1–3):197–217
25. Rodriguez N, Pincet F, Cribier S (2005) Giant vesicles formed by gentle hydration and electroformation: a comparison by fluorescence microscopy. Colloids Surf B Biointerfaces 42(2):125–130
26. Tamba Y, Terashima H, Yamazaki M (2011) A membrane filtering method for the purification of giant unilamellar vesicles. Chem Phys Lipids 164(5):351–358
27. Vemuri S, Rhodes CT (1995) Preparation and characterization of liposomes as therapeutic delivery systems: a review. Pharm Acta Helv 70(2):95–111
28. Vuillemard JC (1991) Recent advances in the large-scale production of lipid vesicles for use in food products: microfluidization. J Microencapsul 8(4):547–562
29. Woodle MC (1995) Sterically stabilized liposome therapeutics. Adv Drug Deliv Rev 16(2–3):249–265
30. Zhu TF, Szostak JW (2009) Preparation of large monodisperse vesicles. PLoS One 4(4):e5009

Wormlike Micelles as Templates for Rod-Shaped Nanoparticles: Experiments and Simulations

Vinod Kumar Gupta, Suvajeet Duttagupta, Advait Chhatre, Anurag Mehra, and Rochish Thaokar

Abstract We report here a novel and generic method for the synthesis of rod-shaped nanoparticles of AgBr, AgCl, and Fe_2O_3 using wormlike micelles (WLMs) in aqueous solution. It is observed that the presence of a wormlike micellar phase is critical to the formation of such anisotropic nanoparticles. Spherical nanoparticles are otherwise obtained when wormlike micelles are absent. Nanoparticle precursors first form spherical primary particles at short times, which then coagulate and consolidate on a surfactant backbone to form nanorods. Interestingly, when preformed spherical nanoparticles are added to a wormlike micellar system, nanorods similar to the *in situ* method are observed. This technique has been explored for the synthesis of anisotropic iron oxide particles as well. Further a mechanism for the formation of these nanorods is proposed and is simulated using a framework of slithering snake dynamics for WLM. In this study, the wormlike micelles are represented by flexible polymers of fixed contour length. A rule-based intermicellar particle exchange protocol is formulated and simulated on a periodic lattice. Simulations reveal that the particles start accumulating slowly on few of the micellar backbones; toward the end, the fraction of micelles carrying no particles increases drastically which is a typical behavior observed in coagulation processes. The particulate masses accumulated on the WLMs are then converted to their respective lengths and diameters.

Keywords Surfactant system • Wormlike micelle • CTAB • CTAC • Sodium salicylate • Shear • Maxwellian rheology • Reptation • Entanglement • Anisotropic nanoparticles • Nanospheres • Nanorods • Template-assisted synthesis • AgBr • AgCl • Fe_2O_3 • Lattice Monte Carlo simulation • Self-

---

Vinod Kumar Gupta has contributed to the experimental part (AgBr nanorods), Suvajeet Duttagupta has contributed to the experimental part (Fe_2O_3 nanorods) and Advait Chhatre has contributed to the Simulation part of the chapter.

V.K. Gupta • S. Duttagupta • A. Chhatre • A. Mehra (✉) • R. Thaokar (✉)
Department of Chemical Engineering, Indian Institute of Technology Bombay, Mumbai 400076, India
e-mail: mehra@iitb.ac.in; rochish@iitb.ac.in

Y.M. Joshi, S. Khandekar (eds.), *Nanoscale and Microscale Phenomena*,
Springer Tracts in Mechanical Engineering, DOI 10.1007/978-81-322-2289-7_3

avoiding random walk • Kuhn length • Slithering snake dynamics • Intermicellar particle exchange

Acronyms

CTAB	Cetyltrimethylammonium Bromide
CTAC	Cetyltrimethylammonium Chloride
WLM	Wormlike Micelle
NaSal	Sodium Salicylate
UV	Ultraviolet
TEM	Transmission Electron Microscope
SEM	Scanning Electron Microscope
EDS	Energy-Dispersive X-ray Spectroscopy
XRD	X-ray Diffraction
G'	Storage Modulus
G''	Loss Modulus
CTASB	Cetyltrimethylammonium Silver Bromide Complex

1 Introduction

Surfactant molecules in solution, when present in sufficient concentration and in the right chemical environment, can self-assemble into aggregates of various shapes and sizes. The shape and size of such aggregates depend upon the molecular structure of the surfactant, nature of solvent and additives, and their molar concentration. The morphologies of these aggregates can be of different shapes and sizes, such as spherical micelles, wormlike micelles, and lamellar phases, depending upon the surfactant and counterion concentrations [1]. Such surfactant-stabilized micellar systems, like microemulsions, hexagonal phases, etc., have been successfully used as nanoreactors and templates for the synthesis of nanoparticles [2–6]. Nanoparticle synthesis in surfactant systems allows for better control on particle shape and size distribution compared to other methods. The size and shape of nanoparticles also depend upon the micellar concentration and the nature as well as the concentration of ionic or co-surfactant additives [7]. Nanoparticles of varying shapes have been synthesized in nonspherical surfactant templates. The formation of different mesoporous nanostructures, such as 2D hexagonal phase, cubic phase, and lamellar phase, by using different micellar phases as surfactant template has been reported [8–13]. Also, rodlike and reverse rodlike micellar systems have been used to produce nanorods [14–22].

Wormlike micelles are large one-dimensional self-assemblies of surfactant molecules, which are locally cylindrical with hemispherical caps at the ends. Above a critical concentration c^*, wormlike micelles entangle into a transient network, similar to a solution of flexible polymers, which display remarkable viscoelastic properties. Unlike polymers, wormlike micelles exhibit a transient nature arising due to continuous breakage and recombination. Due to their transient nature, they exhibit novel static and dynamic properties on time scales both long and short compared to their finite lifetime. Theoretical treatments of wormlike micelles include grand canonical Monte Carlo simulations on 2D and 3D lattices in which the breaking and reformation of micelles is implemented via individual monomer states [23] or by polymer growth Hamiltonian [24, 25] or by defining probabilities of bond breaking and forming and slithering snake dynamics with or without inclusions of chain stiffness in the models [23, 26–33].

In the present work, we explore the nanorod formation in the hitherto unexplored wormlike micellar phase and explore the possibility of a room temperature reaction for one-step synthesis of anisotropic nanoparticles. We examine CTAB–NaSal–$AgNO_3$ (for AgBr nanorods), CTAC–NaSal–$AgNO_3$ (for AgCl nanorods), and CTAC–NaSal–$FeCl_3$-NH_4OH (for Fe_2O_3 nanorods). Further, a mechanism for the formation of nanorods in WLM phase has been formulated. It is then simulated using the framework of slithering snake dynamics for wormlike micelles. Wormlike micelles have been represented by flexible polymers of fixed contour length. A rule-based intermicellar particle exchange protocol has been formulated and simulated on a periodic lattice.

2 Materials and Method

2.1 *Chemicals*

Cationic surfactants hexadecyltrimethylammonium bromide (CTAB, 99 %) and hexadecyltrimethylammonium chloride (CTAC, 98 %) and nanoparticle precursor silver nitrate ($AgNO_3$, 99.9 %) were purchased from Sigma Aldrich Chemicals, GmbH. Sodium salicylate (NaSal, 99 %) and ammonium hydroxide (NH_4OH, 25 %) were purchased from S.D. Fine Chem-Limited. Ferric chloride ($FeCl_3.6H_2O$, 99 %) was purchased from Merck Germany. In the preparation of all the samples, Milli-Q water of conductivity 18.2 M Ω cm^{-1} was used and the temperature was maintained at 25 $\pm$ 1 °C using water bath.

2.2 *Preparation of Surfactant System*

The surfactant system, for the synthesis of AgBr nanoparticles, was prepared by adding CTAB to Milli-Q water until the surfactant dissolved completely, followed by the direct addition of NaSal under constant stirring. The concentration of NaSal to be used in experimental studies was determined from a series of rheological measurements where CTAB concentration was kept constant (50 mM) and the ratio (S) of [NaSal]:[CTAB] was varied between 0 and 2; for all the experiments (of AgBr nanorods), the ratio S was set equal to 2. It was observed that on addition of NaSal salt to the aqueous CTAB solution, the surfactant system turned into a transparent soft gelatinous mass which was then used to synthesize anisotropic AgBr nanoparticles.

For the synthesis of AgCl and maghemite nanoparticles, CTAB was replaced with CTAC and the ratio of NaSal to CTAC was fixed at 1.6. The wormlike micellar phase formed by addition of CTAC and NaSal was lost on addition of FeCl3 and reformed immediately after addition of NH_4OH which was confirmed by rheology. To avoid this behavior, $FeCl_3$ and NH_4OH were mixed first, followed by addition of NaSal. The resultant mixtures in every case were kept under stirring for 24 h after which the samples were collected for characterization.

2.3 Characterization

The UV absorption of the AgBr nanoparticles was measured using SHIMADZU UV–VIS spectrophotometer. The wavelength range was maintained between 200 and 400 nm. The particle-free CTAB–NaSal wormlike micellar solution was used as a reference for UV measurements. For TEM investigations, AgBr or AgCl nanoparticles in the micellar solution were diluted several times, using Milli-Q water. A drop of this solution was placed on carbon-coated copper TEM grids and vacuum dried at room temperature for 12 h. Images of the nanoparticles were taken at different magnification from different locations on the grid using a PHILIPS CM200 transmission electron microscope. For EDS of AgBr and AgCl nanoparticles, a similar sample preparation method as that of TEM was used and was done using JSM-7600F field emission gun-scanning electron microscope. The gelatinous mass, containing AgBr nanoparticles, was washed with water and then used for XRD analysis. The XRD measurement of the AgBr nanoparticles was carried out on a Rigaku D-max 2000/JADE 6.0 copper rotating anode X-ray diffractometer using Cu-kα_1 radiation. Reference to the standard diffraction spectrum of AgBr (JCPDF №. 79-0149) indicates that the peaks correspond to that of AgBr. The rheology experiments were conducted using Anton Paar MCR 301 rheometer equipped with temperature control to measure the elastic and viscous responses to oscillatory shear. IR spectra for AgBr samples were obtained using a Nicolet MAGNA 550 FT-IR spectrometer. The samples were diluted with water and recording made in the transmission mode.

3 Results and Discussions: Experiments

3.1 AgBr Nanoparticles in Wormlike Micellar System

The wormlike micellar phase is best characterized by its rheology and exhibits a single relaxation time that indicates Maxwellian behavior [34]. Rheological

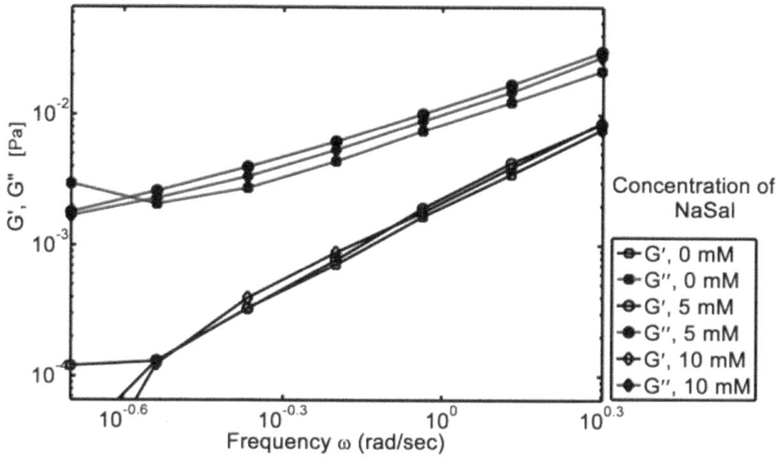

Fig. 1 Elastic modulus (G') and viscous modulus (G'') variation with frequency (ω) of CTAB–NaSal wormlike micelles with [NaSal] $< c^*_{\text{NaSal}}$, [CTAB] $= 50$ mM, and NaSal $= 0, 5, 10$ mM

properties of the CTAB solution in the presence of NaSal were first investigated as a function of S, the molar ratio of NaSal to CTAB. The CTAB concentration was fixed at 50 mM and the molar concentration of NaSal was varied from 0 to 100 mM to determine the onset of wormlike behavior. The surfactant system of CTAB–NaSal is known to form long elongated micelles when the concentration of NaSal is above a critical value, C^*_{NaSal} (50 mM for 50 mM CTAB in the present study). Below the critical concentration of NaSal, the rheological behavior is non-Maxwellian, which suggests that the solution predominantly contains individual chains in non-entangled form (Fig. 1). At a concentration of about 50 mM of NaSal, the system shows wormlike behavior with a single cross over time, in agreement with Maxwellian rheology Fig. 2. Beyond this point, the degree of entanglement of the micelles increases and the solution exhibits viscoelastic properties similar to those observed in solutions of flexible polymers in the entanglement regime [34].

In the Maxwellian rheology regime, the variation of storage modulus G' and loss modulus G'' as a function of frequency ω is given as

$$G'(\omega) = \frac{G_p \omega^2 t_r^2}{1 + \omega^2 t_r^2} \tag{1}$$

$$G''(\omega) = \frac{G_p \omega t_r}{1 + \omega^2 t_r^2} \tag{2}$$

where G_p denotes plateau modulus and t_r denotes relaxation time. Figure 2 shows the correct scaling for G' and G'' at low and high frequencies, respectively, i.e., $G' \propto \omega^2$ and $G'' \propto \omega$ at low frequencies and $G' \propto$ constant and $G'' \propto 1/\omega$ at high frequencies for NaSal concentration >50 mM (Fig. 2).

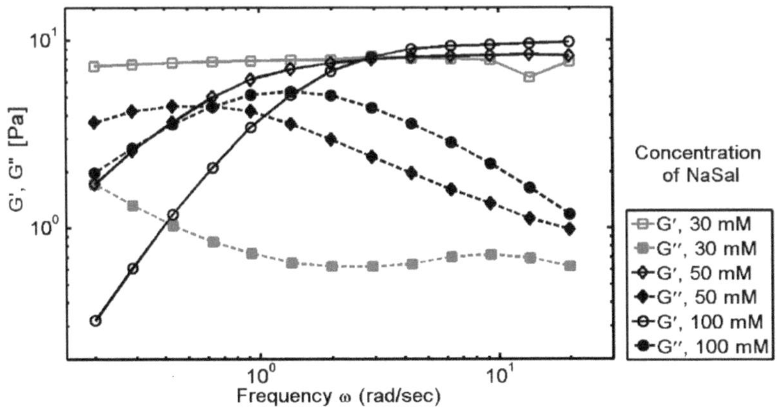

Fig. 2 Storage modulus (G') and loss modulus (G'') variation with frequency (ω) of CTAB–NaSal wormlike micelles, [CTAB] = 50 mM, and different concentration of NaSal. Transition to wormlike micelles is observed at 50 mM of [NaSal]

Figure 2 shows the variation of G', G'' with frequency for varying concentrations of NaSal. These results also are in agreement with those by Shikata et al. [35] who observed a transition from Rouse-like behavior of the surfactant system to an entangled one at a molar ratio $S = 0.6$ (Critical $S = 1.0$ in the present study). The CTAB concentration range in their case, though, was 1–10 mM which is lower than the 50 mM concentration used in the present study. At concentrations higher than C^*_{NaSal}, the G' and G'' values show reptation dynamics and rheology of living polymers [34].

NaSal salt primarily acts to bridge the CTAB micelles and forms long wormlike chains. It is known that the head group repulsion in the CTAB system can be reduced by screening the electrostatic interactions using salts like potassium bromide, thereby altering the packing factor of the surfactant molecule [6]. Small molecular weight inorganic salts are known to be in dynamic equilibrium with CTAB, leading to poor growth of the wormlike micellar system. However, organic salts like NaSal dissociate to give salicylate ions which get absorbed on the surface of micelle, thus acting as a co-surfactant and assist in the formation of wormlike micelles [6]. The Br^- and the Na^+ ions are dislodged from the surfactant backbone. As the salt-to-surfactant molar ratio (S) is increased, electrostatic interactions are more effectively screened. This favors the formation of longer micelles that behave as flexible polymer chains. Further growth of these micelles results in a network of wormlike micelles and is suggested to be the reason for the transition of the system to the Maxwellian rheology.

The synthesis of AgBr nanoparticles was carried out at different concentrations of added $AgNO_3$. Visible precipitation and settling of particles were observed when the concentration of $AgNO_3$ was increased beyond 20 mM. In all the results reported here, the concentration of $AgNO_3$ was therefore kept below this value.

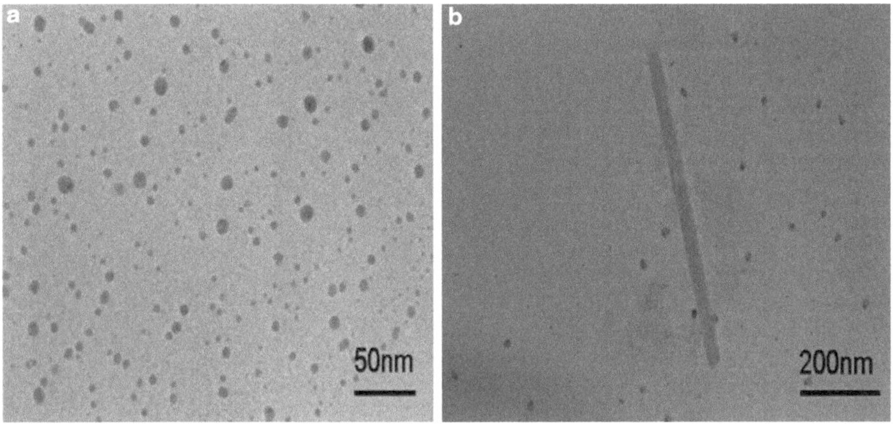

Fig. 3 TEM micrographs showing AgBr nanospheres (1 h) evolving into nanorods after 24 h. [CTAB] = 50 mM, [NaSal] = 100 mM, [$AgNO_3$] = 10 mM (**a**) After 1 h. (**b**) After 24 h

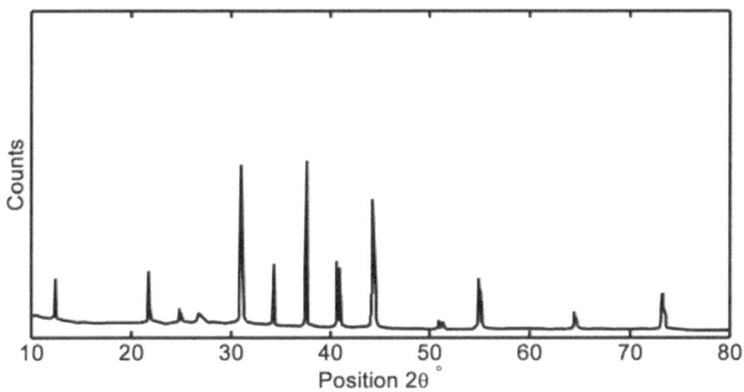

Fig. 4 XRD of synthesized AgBr nanorods

The TEM images revealed the formation of anisotropic AgBr nanoparticles in the wormlike micellar network, surrounded by spherical particles. These images were taken 24 h after the reactants were mixed (Fig. 3b). To understand the nanorod formation process, images were taken 1 h after the mixing of reactants. These showed formation only of nanospheres having the size of around 7.7 ± 3.9 nm (Fig. 3a). All nanorods formed were characterized using XRD, UV, and EDS.

The typical powder XRD pattern of the synthesized nanorods is shown in Fig. 4. The data shows diffraction peaks at $2\theta = 21.6°, 30.9°, 44.3°, 55.0°, 64.5°, 73.2°$, and 81.7° which can be indexed to (111), (200), (220), (222), (400), (420), and (422) planes of pure AgBr (JCPDF №. 79-0149) confirming the presence of AgBr particles [36]. The crystallite size was calculated from XRD data using the Scherrer formula and found to be 85.75 nm. However, along with AgBr, the analysis also

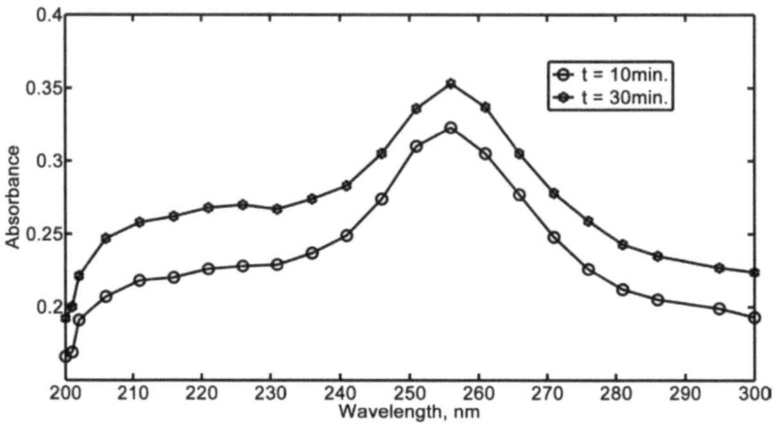

Fig. 5 UV absorption spectra of AgBr nanoparticles varying with time. $[AgNO_3] = 10\,mM$, $[CTAB] = 50\,mM$, and $[NaSal] = 100\,mM$

revealed the presence of strong lines at $2\theta = 34.3°, 37.6°, 40.8°$. These can be attributed to silver metallic particles (JCPDF №. 04-0783). Unlike Liu et al. [37], who proposed that complexes such as CTAB–AgBr are likely to form at higher surfactant concentration, we did not observe any signature of these complexes in the XRD pattern obtained in the present study (Fig. 4), thereby ruling out the probability of CTASB complexes.

Figure 5 shows the variation of UV absorption at two different values of time, i.e., 10 and 30 min. The peak wavelength was observed at 263 nm [3] for both times, and the value of the absorbance increased by approximately 15 % over time. This indicates that the reaction between Ag^+ ions and Br^- counterions continues even after 30 min, leading to an increase in the concentration of AgBr nanoparticles and a consequent increase in the absorbance values.

The energy-dispersive X-ray spectroscopy (EDS) pattern of the synthesized AgBr nanoparticles is shown in Fig. 6. The spectrum shows the presence of bromine and silver elements. The peaks located between 2 and 4 keV are related to the silver characteristic lines while peaks located between 1 and 2 keV are related to the bromide characteristic lines. The peak located at 0.5 keV represents the oxygen characteristic line. The silica-related peak at 1.75 keV appears in Fig. 6 because of the use of a silica wafer as a substrate for EDS analysis.

To confirm the role of CTAB–NaSal wormlike templates on the shape of the anisotropic nanoparticles, the synthesis of AgBr was carried out in the absence of NaSal. As mentioned earlier, CTAB does not form wormlike micelles in the absence of NaSal and it requires a critical concentration of NaSal to induce the entanglement of the wormlike chains. The nanoparticles synthesized in this regime were found to be spherical (Fig. 7a) and no indication of anisotropy was observed even after 24 h. Also, a complete absence of anisotropic nanoparticles is observed even when the concentration of NaSal is lower than the critical concentration (50 mM)

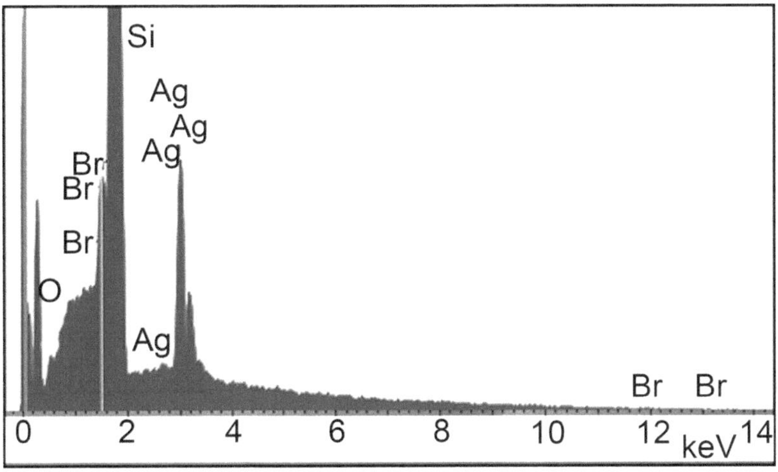

Fig. 6 The EDS pattern of synthesized AgBr nanoparticles

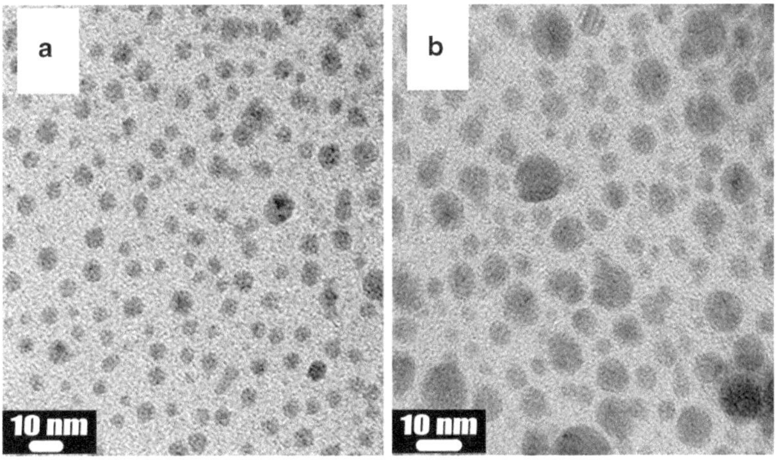

Fig. 7 TEM images showing spherical AgBr nanoparticles after 24 h, in systems with subcritical NaSal. [CTAB] = 50 mM, $[AgNO_3]$ = 10 mM, (**a**) [NaSal] = 0 mM, (**b**) [NaSal] = 10 mM

(Fig. 7b). The average sizes of the spherical nanoparticles obtained in the absence of NaSal and in 10 mM NaSal were found to be 4.88 ± 1.2 nm and 7.14 ± 1.63 nm, respectively.

To substantiate the importance of stirring on the morphology of particles, a batch of experiments were performed without stirring. The nanoparticles synthesized by this way were found to be spherical even after 24 h. The average particle size of these nanospheres was 4.32 ± 1.52 nm (Fig. 8).

Inspection of the TEM images, taken 24 h after the reagents are mixed, shows long rodlike structures having diameters in the range of 30–120 nm and lengths in

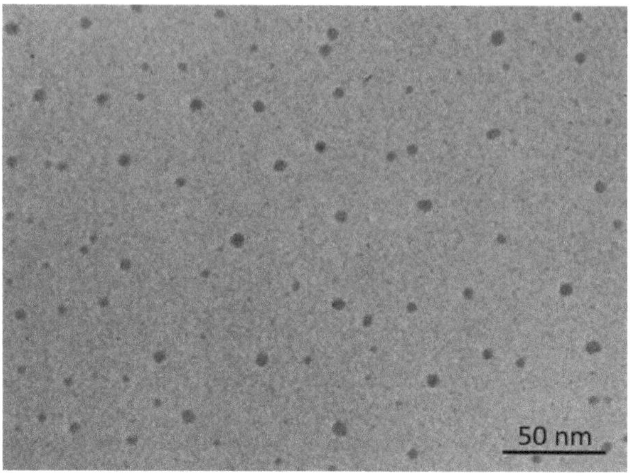

Fig. 8 TEM images showing spherical AgBr nanoparticles after 24 h, in unstirred system. [CTAB] = 50 mM, [NaSal] = 100 mM, [$AgNO_3$] = 10 mM

the range 200–2,000 nm. The size of these nanoparticles strongly depends upon the concentration of $AgNO_3$.

Figure 9a shows that at lower concentrations of $AgNO_3$ (3 mM), the anisotropic structure is an aggregation of small spherical and polyhedral AgBr nanoparticles. At such concentrations of $AgNO_3$, fewer AgBr particles are expected to form. These decorate the surfactant backbone and result in the formation of loosely aggregated structures. As the concentration of added $AgNO_3$ increases, the particles exhibit increased consolidation (Fig. 9b). The particles evolve into more regular, smaller rodlike structures as observed in Fig. 9c, d. An increase in the number of nanorods was seen although the aspect ratio (AR) decreased to 5 (in the case of 15 mM) and 2 (in the case of 20 mM) from around 8 (in case of 10 mM).

The typical lengths of nanorods were found to be 878, 186, and 74 nm for 10, 15, and 20 mM of $AgNO_3$ concentrations, respectively. The diameter also decreases with an increase in the concentration of $AgNO_3$, as shown in Table 1.

This is possibly because an increase in the concentration of $AgNO_3$ leads to a larger number of smaller diameter spherical particles which could be due to faster reaction kinetics at higher concentrations. If the total number of particles forming the nanorod is roughly the same in the three cases, it will result in smaller diameter and length of the nanorods observed at higher $AgNO_3$ concentrations. This also explains why a large number of nanorods are observed at higher $AgNO_3$ concentrations.

To shed more light on the mechanism and to confirm the effect of the template on anisotropic particle formation, synthesis of AgBr particles was first carried out in the absence of NaSal. TEM images after 1 h revealed the formation of spherical particles (12.4 ± 3.2 nm). NaSal was then added and the system was stirred for 24 h. It was found that nanorods with a typical average size of 224 nm and aspect ratio of 8 were formed (Fig. 10).

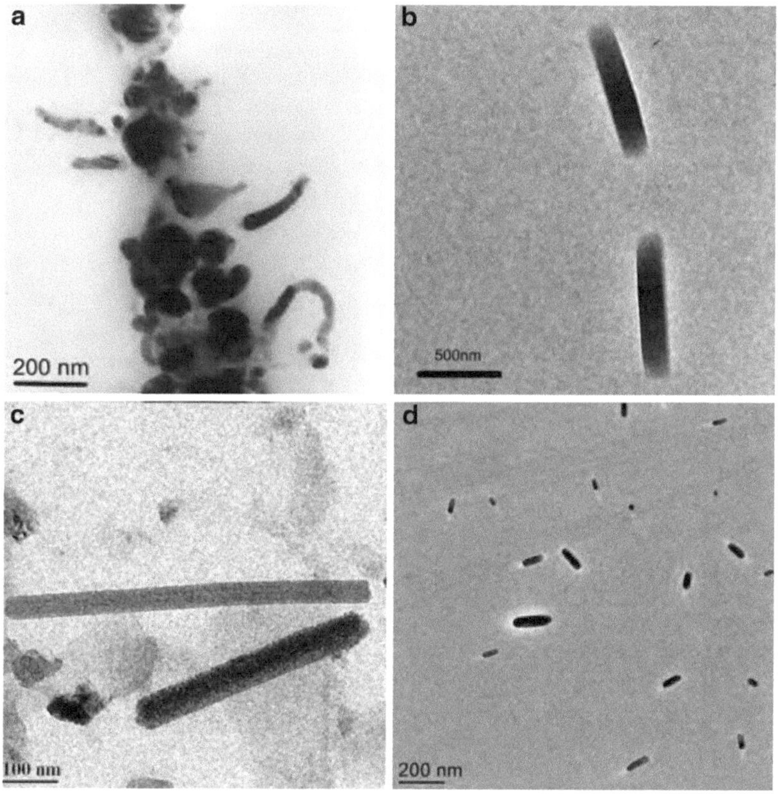

Fig. 9 TEM images of AgBr nanoparticles at 25 °C after 24 h at CTAB = 50 mM, NaSal = 100 mM, and (**a**) $AgNO_3$ = 3 mM, (**b**) $AgNO_3$ = 10 mM, (**c**) $AgNO_3$ = 15 mM, (**d**) $AgNO_3$ = 20 mM

Table 1 Effect of concentration of $AgNO_3$ on the size of AgBr nanoparticles

$[AgNO_3]$ (mM)	Length (nm)	Diameter (nm)
10	878 ± 158	99 ± 41
15	186 ± 141	36 ± 19
20	74 ± 19	26 ± 6

TEM images revealed that there is formation of nanospheres after 1 h of aging which forms a chain-like structure after 6 h. Finally, these particles consolidate to form nanorod after the aging of 24 h (Fig. 11a).

A time study revealed that nanospheres are formed after 1 h. At intermediate times, chain-like structures were observed, some of which showed signatures of consolidation, while at 24 h smoothened nanorods were observed (Fig. 11a). Further, as the stirring time is increased from 24 to 60 h, an increase in the length and

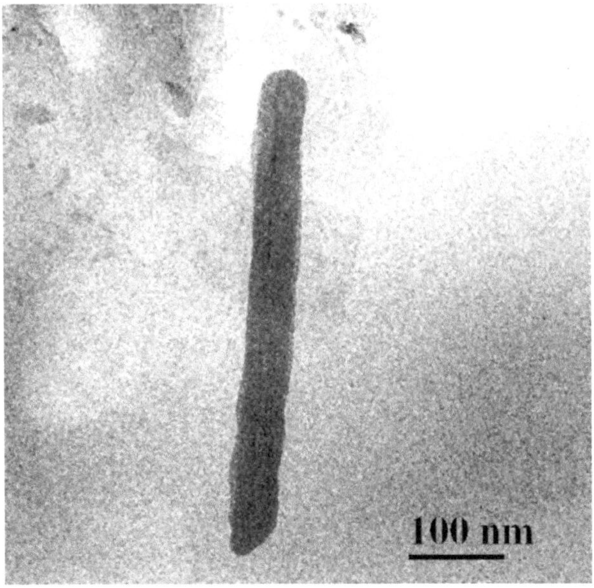

Fig. 10 TEM image showing AgBr nanorods after 24 h. [CTAB] = 50 mM, [$AgNO_3$] = 10 mM, [NaSal] = 100 mM (after 1 h)

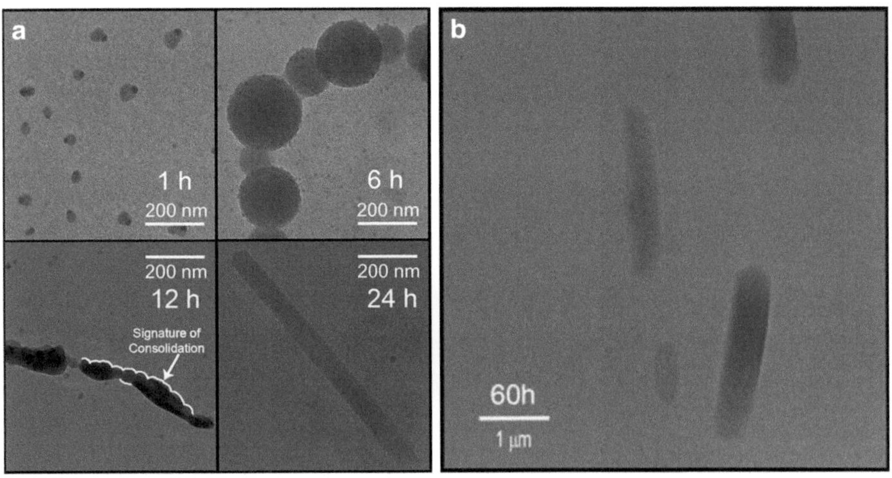

Fig. 11 Time study of AgBr nanorod synthesis, [CTAB] = 50mM, [NaSal] = 100 mM, and [$AgNO_3$] = 10 mM. (**a**) AgBr nanoparticles at different time interval of aging. (**b**) AgBr nanoparticles after the aging of 60 h

diameter of AgBr nanoparticles is observed. The dissimilar rates of increase in the diameter and the length result in a net decrease in the AR. With time, the typical length of the nanorods increases from 878 nm to 1.6 µm and a decrease in aspect ratio from 8 to 5 (Fig. 11b).

The above experimental investigations clearly indicate that

1. Wormlike micelles are critical to nanorod formation. If the surfactant system is not in wormlike micellar phase, only spherical particles will form. Thus, there is no apparent tendency of the particles to inherently form anisotropic structures (Fig. 7).
2. The nanorod formation is always preceded by spherical particles (either prepared in situ or externally added (Fig. 10)).

The above observations suggest the following possible mechanism for the formation of nanorods.

The wormlike micellar phase is characterized by a high degree of entanglement among the wormlike chains resulting in an increase in the viscosity by over four orders of magnitude (as compared with a liquid phase). This causes a substantial reduction of the particle diffusivity, and correspondingly, the frequency of collision between the (primary) spherical particles also reduces. In contrast, the high density of chains implies a larger probability of collision between the chains and particles as compared to those between the slowly diffusing particles. The van der Waals interaction between the particles and the chains may therefore lead to the adsorption of the particles on the backbone of the chains, leading to the decoration of the chains with spherical nanoparticles.

Further, in the wormlike micellar phase, the entangled chains slither past each other by a phenomenon known as reptation [34]. The particle-laden and highly entangled chains undergo reptation, carrying along with them the adsorbed particles. The contact between particles adsorbed on two different chains is thus dictated by the diffusion coefficient of the chains themselves. The consolidation of particles into a nanorod can now occur possibly by exchange of particles sitting on two different chains when the chains slither past each other and the particles on a single chain undergoing diffusion along the chain and consolidating into larger anisotropic particles.

The above mechanism is in contrast to systems like the seed-mediated growth of anisotropic nanoparticles, where anisotropic nanocrystallites formed at early stage of synthesis grow into nanorods [38].

To study the possible effect of nanoparticles on the rheology of the system, rheological characterization of the particle-laden surfactant system was carried out. The CTAB and NaSal concentrations were fixed at 50 and 100 mM, respectively. Rheological measurements were carried out after 1 and 24 h, respectively, after the addition of $AgNO_3$. These values of time were chosen to see the effect of spherical (1 h) and anisotropic nanoparticles (24 h), respectively, on the rheological behavior. The rheology of the particle-laden system indicated Maxwellian rheology at all times, confirming that the system is in the wormlike micellar phase even after the formation of spherical (short time) and anisotropic (long time) nanoparticles (Fig. 12). A marginal increase in the value of storage and loss moduli was observed for both the cases as compared to particle-free systems. This is likely because of the very low volume fraction of nanoparticles (0.02 %) which does not have any significant effect on the rheological behavior.

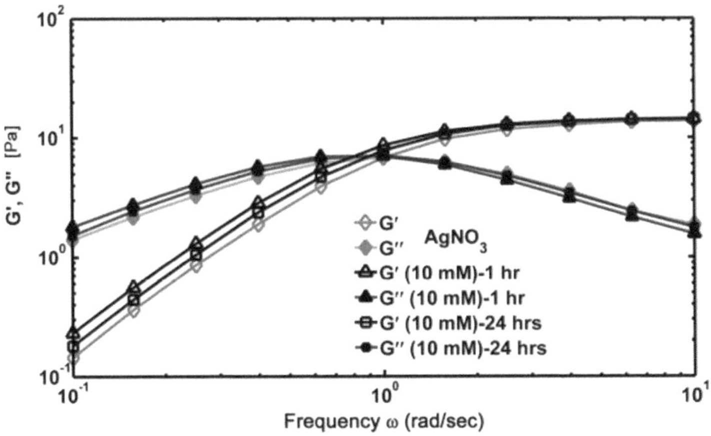

Fig. 12 Variation of storage and loss moduli G', G'' with frequency (ω) for different aging of 10 mM $AgNO_3$ at [CTAB] = 50 mM, [NaSal] = 100 mM

3.2 *AgCl Nanoparticles in Wormlike Micellar System*

For the generalization of this technique toward the synthesis of other nanoparticles, AgCl nanorods were synthesized using another wormlike micellar system, namely, CTAC–NaSal. A similar method was followed, as described for preparation of AgBr nanoparticles.

The rheological properties of the CTAC solution in the presence of NaSal were first investigated as a function of S, the molar ratio of NaSal to CTAC. The CTAC ($S = 0.3 - 2$) concentration was fixed at 100 mM and the molar concentration of NaSal was varied from 50 to 200 mM to determine the onset of wormlike behavior. In this case, C^*_{NaSal} is observed to be around 130 mM. CTAC concentration was maintained at 100 mM for various sets of experiments and NaSal concentration was varied such that the NaSal-to-CTAC ratio (S) was between 0.5 and 2 (Fig. 13). The addition of 15 mM $AgNO_3$ solution to the CTAC–NaSal transparent system results in the formation of AgCl nanoparticles. The TEM images revealed the formation of anisotropic AgCl nanoparticles in the wormlike micellar network. These images were taken 24 h after the reactants were mixed. The TEM images (Fig. 14) show long rodlike structures having diameters ranging from 30 to 60 nm and lengths ranging from 200 nm to 1.2 μm. Addition of higher concentrations of $AgNO_3$ (20 and 25 mM) solutions led to the precipitation of AgCl nanorods.

The typical powder XRD pattern of the synthesized AgCl nanoparticles is shown in Fig. 15. The data shows diffraction peaks at $2\theta = 27.6°$, 32.1°, 46.2°, 54.8°, 57.4°, 67.4°, 74.4°, 76.7°, and 81.7° which can be indexed to (111), (200), (220), (311), (222), (400), (331), (420), and (422) planes of pure AgCl (JCPDF №. 31-1238) confirming the presence of AgCl particles [39]. The crystallite size was calculated from XRD data using the Scherrer formula and found to be 52.9 nm.

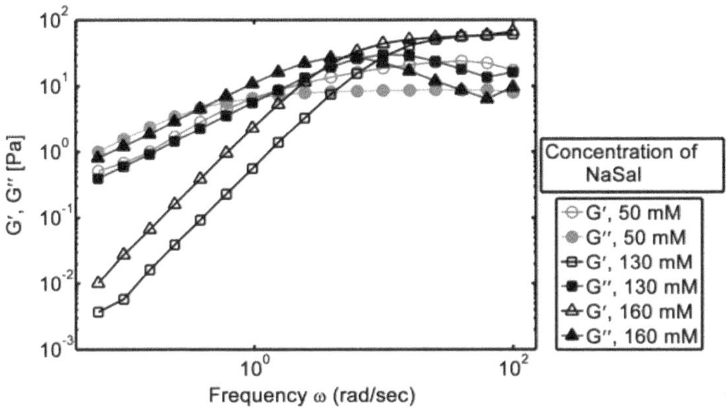

Fig. 13 Storage modulus (G') and loss modulus (G'') variation with frequency (ω) of CTAC–NaSal wormlike micelles, [CTAC] = 100 mM, and different concentration of NaSal

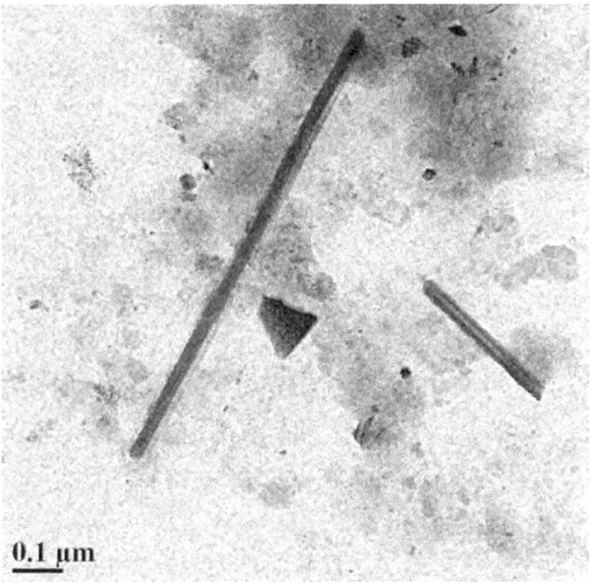

Fig. 14 TEM image showing AgCl nanoparticles after 24 h. [CTAC] = 100 mM, [NaSal] = 160 mM, [$AgNO_3$] = 15 mM

XRD also revealed the presence of strong lines at $2\theta = 37.4°$ and 41.0° which can be attributed to silver metallic particles (JCPDF №. 04-0783).

The EDS pattern of the synthesized AgCl nanoparticles is shown in Fig. 16. The data shows the presence of chlorine and silver elements. These peaks located between 2 and 4 keV represent chloride and silver characteristic lines. The peaks located on the left part of the spectrum at 0.2 and 0.5 keV represent carbon and oxygen, respectively. The peak located at 0.5 keV is connected with the oxygen characteristic line. The peaks between 7.8 and 9 keV are associated with the copper

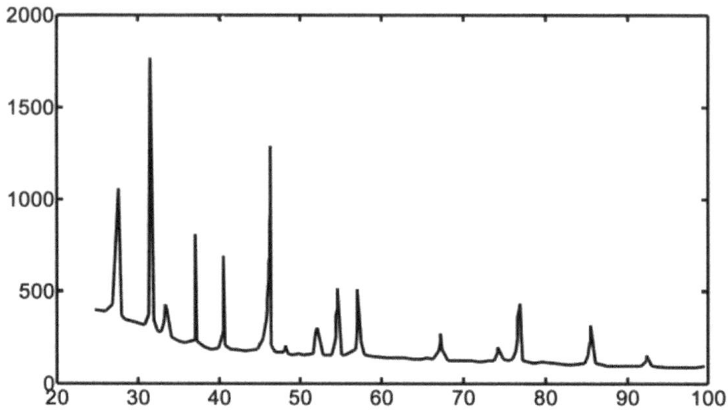

Fig. 15 XRD patterns of synthesized AgCl nanoparticles

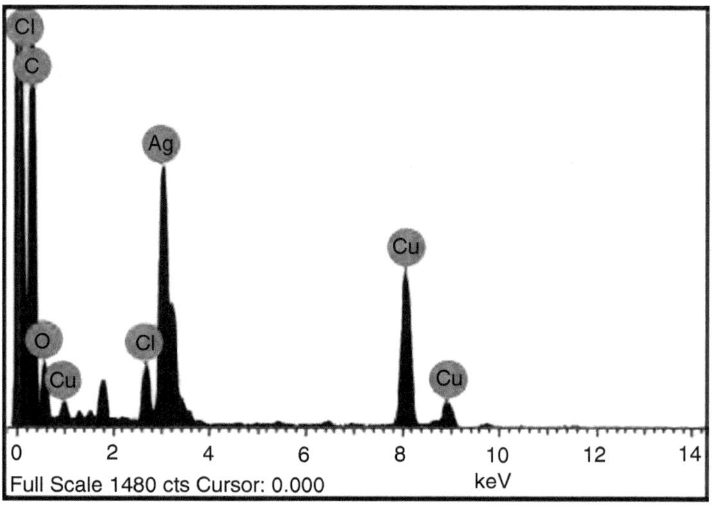

Fig. 16 The EDS pattern of synthesized AgCl nanoparticles

characteristic lines. Copper and carbon peaks are due to the carbon-coated copper grid on which the sample was analyzed.

3.3 Iron Oxide Nanorods

In the original protocol, the wormlike micelle (WLM) is prepared by adding CTAC/water/NaSal followed by the addition of a precursor. In this system, addition of iron chloride salt after formation of WLM reduces the viscosity of the system, and afterwards, the addition of NH_4OH to the same forms the WLM system again.

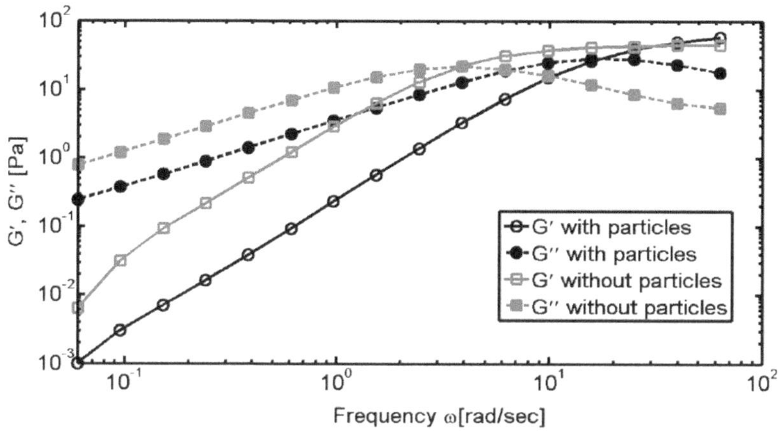

Fig. 17 Comparison of rheological property

We have termed this phenomenon as on–off rheology. Rheometric study (Fig. 17) of the samples have shown that the WLM regains similar rheological property once NH_4OH is added to the system. The change in rheology may be explained as follows:

1. WLM is formed by CTA^+ and Sal^- bridging.
2. Incorporation of Fe^{+3} results in the formation of iron salicylate complex, and this results in the breaking of WLM.
3. Added NH_4OH reacts with the Fe^{+3} and the iron salicylate complex breaks down.
4. Salicylate ion is free to form WLM along with CTA^+.

For the experimental results, we followed the protocol of adding NaSal after reacting the two precursors. This system was stirred for 24 h using a lab stirrer. Further sets of experiment were conducted using a Linkam shear cell. We notice under controlled shear (Fig. 18) the particles have a higher aspect ratio as well as there is a marked reduction in their length and diameter. The yield of particle has significantly increased.

The concentration of a precursor plays an important role in the system as well. At a lower concentration (<10 mM), we obtain spherical particles, while at a higher concentration (>20 mM), we observe the particles precipitating out. With a moderate increase in the concentration, Fig. 19, we observe there is an insignificant effect on the length of the particles as confirmed from an unpaired t-test. The average lengths obtained for 10 and 15 mM concentrations were 58.7 and 59.2 nm, respectively. However, there was an increase in polydispersity from 10 to 15 mM; the standard deviation in length of 10 mM sample was $\sigma_{L,10\,\mathrm{mM}} = 19.5$ nm while that for the 20 mM sample was $\sigma_{L,15\,\mathrm{mM}} = 58.9$ nm.

High-resolution imaging allows us to measure the d-spacing of 0.29 nm which is in agreement with the d_{220} plane. Comparing the result obtained from XRD

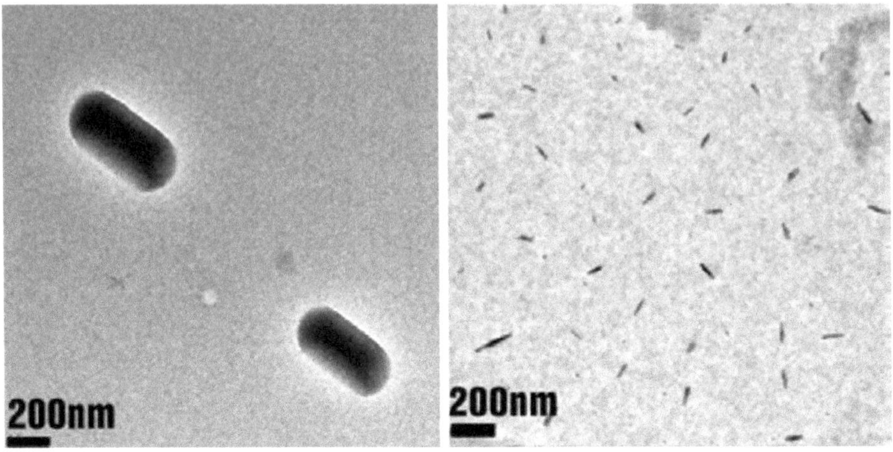

Fig. 18 Iron oxide nanorods formed in wormlike micellar system after 24 h. (L) Uncontrolled shear and (R) controlled shear

Fig. 19 Effect of $FeCl_3$ concentration on dimensions of iron oxide nanorods

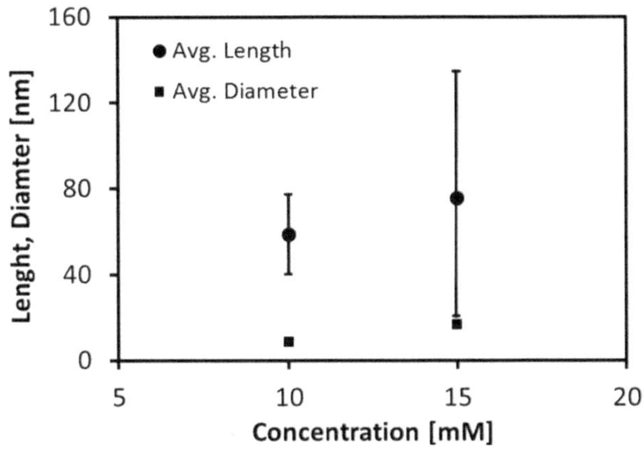

studies which indicates d_{220} has the maximum intensity, we can clearly see there is a preferred direction of growth that is assisted by the WLM system (Fig. 20).

4 Proposed Mechanism

4.1 Formation of AgBr Nanorods from Primary Particles

The wormlike micellar phase is characterized by a high degree of entanglement among the wormlike chains resulting in an increase in the viscosity by over four orders of magnitude (as compared with a liquid phase). This causes a substantial reduction of the particle diffusivity and correspondingly in the frequency of collision

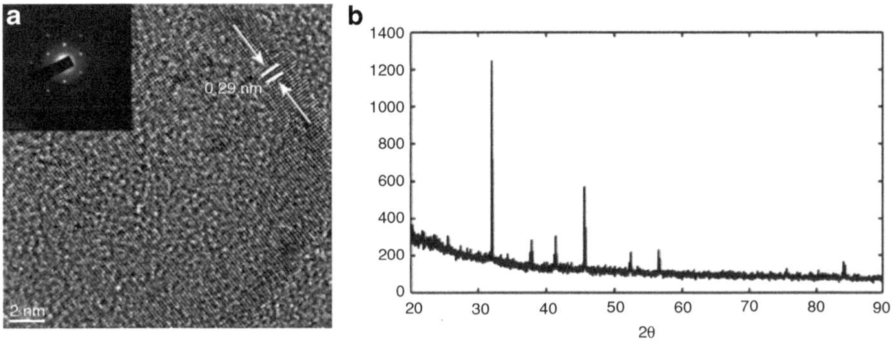

Fig. 20 Characterization of maghemite nanorods. (**a**) High-resolution TEM micrograph. (**b**) XRD analysis of maghemite nanorod

between the primary spherical particles (which are formed in very early stages of the synthesis.). Also, the high density of chains implies a larger probability of collision between the chains and particles as compared to those between the slowly diffusing particles. The van der Waals interaction between the particles and the chains may therefore lead to the adsorption of the particles on the backbone of the chains, leading to the decoration of the chains with spherical nanoparticles.

These entangled chains slither past each other. The particle-laden and highly entangled chains undergo reptation, carrying along with them the adsorbed particles. The contact between particles adsorbed on two different chains is thus dictated by the reptation of the chains. The consolidation of particles into a nanorod can now occur by exchange of particles sitting on two different chains when the chains slither past each other and the particles on a single chain undergoing diffusion along the chain and consolidating into larger anisotropic particles. The above mechanism is in contrast to systems like the seed-mediated growth of anisotropic nanoparticles, where anisotropic nanocrystallites formed at early stage of synthesis grow into anisotropic nanoparticles [38]. Figure 21 shows a cartoon of the mechanism discussed below.

5 Theoretical Analysis: Multiscale Lattice Monte Carlo Simulation

5.1 Formulation and Implementation of Monte Carlo Model

Based on the mechanism discussed in the previous section, a Monte Carlo model of the system has been formulated. The major events being modeled are slithering of the wormlike micelles and intermicellar particle exchanges. The formation of AgBr primary particles and their adsorption on the WLM backbone are assumed to

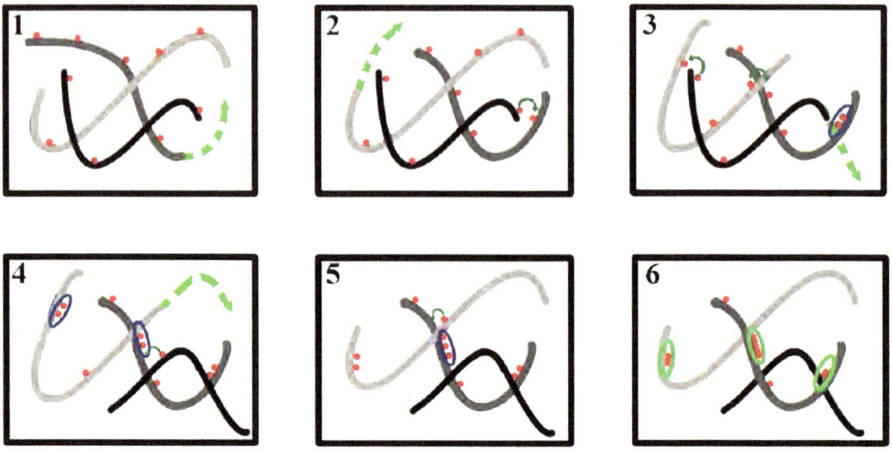

Fig. 21 Schematic of mechanism of formation of AgBr nanorods in wormlike micellar system

be instantaneous, and particle-laden WLMs are considered as initial conditions for the simulations. On the other hand, the slithering of the micelles, the intermicellar particle exchanges, and the consolidation of pearl bead chains into nanorods are the rate-controlling steps. Other assumptions made in this model are as follows:

1. The wormlike micelles are assumed to follow the excluded volume interactions. That is, at any given lattice point, only one monomer of WLM may be present at any time during the simulation. This also implies that the formation of rings is not permitted for WLMs.
2. Ensembles of wormlike micelles used in the simulation are assumed to have monodisperse contour length.
3. Wormlike micelles in reality keep on breaking and reforming continuously and there exists an equilibrium distribution of contour lengths. In the simulations, however, the breakage and reformation of the micelles are assumed to be absent.
4. All primary particles are assumed to have uniform diameter. That is, the distribution of diameter in the primary particles is neglected in the simulation.
5. Coalescence of the particles is allowed if the particles are nearest neighbors of each other. In the view of very high energy of interaction, this assumption allows for the growth of diameter along with the length of nanoparticles.
6. It is assumed that, on coalescence, individual nanospheres immediately form larger nanospheres, and volume conservations determine the radius of the newly formed particles.
7. In reality the particles may undergo surface diffusion on the surface of the micelles. In the present simulation, surface diffusion of the particles on the micellar backbone is assumed to be absent. That is, the slip between particle and micelles is assumed to be zero.

5.1.1 Initialization of Lattice

Generation of the Wormlike Micelles

The wormlike micelles are generated by carrying out self-avoiding random walk (SAW) on a multiscale 2D periodic square lattice. The unit length scale of the lattice for WLMs is chosen to represent the Kuhn length of WLM. In a typical SAW, a seed location is generated at random and checked for the validity. A location is valid if it has not been visited earlier and has got space for at least one Kuhn neighbor. Then a SAW is started by generating a random number which is used to select the direction of the next step. A SAW gets terminated if the number of Kuhn segments in a SAW has reached the number of Kuhn segments fixed for a chosen WLM (15 in the case of CTAB–NaSal WLM) or the last end has hit a site which does not have space for at least one Kuhn segment. The entire sequence of SAW is rejected if it has a contour length smaller than the number of Kuhn segments fixed for a chosen WLM. This ensures generation of population of WLM having monodispersed length and number of Kuhn segments. In the present simulation, a periodic lattice of $1{,}350 \times 1{,}350$ lattice points was used to generate 2,000 WLMs.

5.1.2 Choice of a Micelle to Slither

For choosing the micelle to undergo slithering motion for all the micelles, a number of free ends are calculated; a free end is the one which can undergo at least one SAW step from its current location. The minimum value of free ends for a micelle could be zero, while the maximum could be two. For choosing a micelle to slither, a random number (r) is generated and index i which satisfies Eq. 3 is selected for the slithering. If a chosen micelle is having 2 free ends, the end to undergo slithering is chosen by the two-choice random selection method.

$$\sum_{j=1}^{i}(n\text{freeends}(j)) < r <= \sum_{j=1}^{i+1}(n\text{freeends}(j)) \qquad (3)$$

5.1.3 Implementation of Slithering and Implementation of Shear

In slithering motion, Kuhn monomers move collectively along the chain. Effectively, this amounts to removing a Kuhn monomer on one end and connecting it to the other end of the chain and leaving the middle monomers unchanged. Therefore, density fluctuations and constraint release only occur at the chain ends. The shear necessary for the formation of particles is implemented to be a random shear or biased shear. In random shear, the direction in which micelle slithers or moves is decided randomly with equal probabilities in all the possible directions. In the biased shear, one of the directions is given arbitrary preference over the other directions.

This is in accordance with the rheological observations that WLMs tend to align in a shear field. This is considered important as AgBr particles do not grow to a cylindrical shape even in the presence of the WLMs in the absence of stirring or shear.

5.1.4 Intermicellar Particle Transfers

After a micelle has undergone slithering, the neighborhood of that particular micelle is scanned. For every site-neighbor pairs, the following set of rules are applied to find if any intermicellar particle transfers can be executed:

1. All neighborhood sites that lie on the micelle itself are filtered off.
2. All neighborhood sites that hold neither a micelle nor a particle are filtered off.
3. If none of the given pair hold a particle, the pair is filtered off.
4. If both the sites hold a particle, then a particle transfer is allowed from a less populous Kuhn segment to a more populous Kuhn segment. If both the Kuhn monomers hold equal number of particles, then particle exchange is not executed. Note that only local populations of the micelles on which both the particles are attached are used in this comparison. The local populations are the particle populations of the Kuhn monomers on which the selected particles are attached.
5. If the site is empty and the neighborhood holds a particle, then the particle is accepted only if the present Kuhn segment is having a higher population of particles; otherwise, the pair is filtered and vice versa.
6. Of all eligible site-neighbor pairs, one pair is chosen at random for the execution of intermicellar particle transfer.
7. Only individual nanospheres hop from one point to the other in a particular MC move. The particle chains are converted to the nanorods only at the end of simulation.

In a typical MC step, intermicellar particle exchange may occur only if the micelle chosen for slithering carries at least one particle. If the chosen micelle is empty, it only undergoes the reptation and this is counted as infructuous but necessary event from the view of the straightening of particle chains. MC steps are repeated until the rate of intermicellar particle exchange falls below a critical value of $1e - 5\,\mathrm{MC}^{-1}$, where MC is the number of Monte Carlo steps.

5.2 *Results and Discussion*

By using the above-described frame of the MC model, the simulation was carried for the following parameters:

1. Lattice size $= 1{,}350 \times 1{,}350$ (one lattice site corresponds to 8 nm which is the size of primary particle)

2. Number of WLMs on the lattice = 2,000
3. No. of Kuhn segments on one micelle = 15 (constant)

With this configuration, studies on the effect of initial particle concentration, Kuhn length, and shear bias were carried out.

5.2.1 Effect of Concentration

The simulations reveal that the particles start accumulating on few of the micellar backbones slowly, and toward the end, the count of particles left on the lattice falls to less than 30 % of the initial particle count on the lattice which is a typical behavior observed in any coagulation process. The particulate mass accumulated on the WLMs is then converted to length and diameter by setting length of particle = number of particles forming a straight and continuous chain of particles x the diameter of the primary particle (8 nm in our case). The diameter is then back calculated by converting the volume of particle into diameter. Figure 22a, b shows the effect of particle concentration on the length and diameter of the rods obtained by converting the chain mass in length and diameter using the protocol described. Important observations deduced from the results are

- **Effect of [AgBr] concentration on length and diameter of nanorods** – Concentrations simulated in this study were 5, 10, 15, 20, 25, 30, 40, 70, and 100 mM. The concentrations were implemented in lattice simulations in the form of particles adsorbed on WLM. A mean micellar occupancy of the particles was

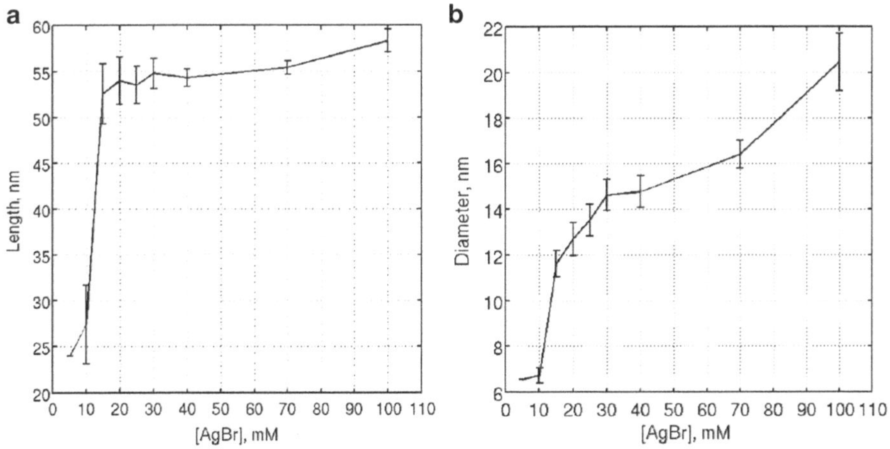

Fig. 22 Effect of [AgBr] concentration on length and diameter of nanorods. Concentrations simulated were 5, 10, 15, 20, 25, 30, 40, 70, and 100. Four distinct regimes are observed: (i) very low concentration (5 mM), no or very less number of rods are formed; (ii) below 20 mM, both diameter and length show a sharp increase; (iii) 20–40 mM, length and diameters formed in these simulations are statistically indifferent; and (iv) beyond 40 mM, both length and diameter start growing suggestive of precipitation (**a**) L (**b**) D

estimated by assuming length of WLM as 1,100 nm and CTAB head group area as 0.45 nm^2. Four distinct regimes in the length are observed.

- Very low concentration (5 mM) – No or very less number of rods are formed. This may be explained on the basis that because of scarcity of the particles on the lattice, there are very few or no interactions between the particles to induce any aggregation. This matches qualitatively to the experimental observations of synthesis of AgBr and iron oxide nanorods synthesized at precursor concentration of 5 mM.
- 20 mM – Both diameter and length show a sharp increase. In this regime, a number of interactions increase by virtue of an increase in concentration of particles and longer chains start appearing on the lattice in higher number.
- 20–40 mM – Length and diameters formed in these simulations are statistically indifferent. In this regime, the growth of the particles is capped by the length of the Kuhn segment of the WLM. This result matches qualitatively with the synthesis of iron oxide nanorods formed under controlled shear in WLM-assisted synthesis of iron oxide nanorods synthesized in 10–20 mM.
- Beyond 40 mM – Both length and diameter start growing even longer than the Kuhn length of WLM suggestive of inducing instability. This phenomenon corresponds to the precipitation of large aggregates at high concentration observed in experiments of formation of AgBr nanorods.

5.2.2 Effect of Shear Bias

The effect of shear is simulated by defining a shear bias in simulation. The shear bias is defined in terms of global probability of micelles slithering in a particular direction. For example, the shear bias of 0.25 or random shear means that all micelles move with equal probability in all directions, while a shear bias of 0.7 means that 70 % of the micelles that were moved on the lattice were moved in one preferred direction than the other directions:

- It is observed that by imposing a shear bias on micellar motion, the length and diameter of rods increases (Figs. 23 and 24). Further increase in shear bias (beyond 0.7) does not show significant effect on length or diameter of the particles. This observation is qualitatively matching with the experimental results of the controlled shear studies on iron oxide nanoparticles where variation in shear does not show significant effect on dimensions of particles.

6 Conclusion

The suitability of wormlike templates for the synthesis of anisotropic nanoparticles is presented. CTAB–NaSal and CTAC–NaSal wormlike micelles were used to form AgBr, AgCl, and iron oxide nanorods, respectively. There are no prior reports on the use of wormlike micelles as templates for nanorod formation in the literature.

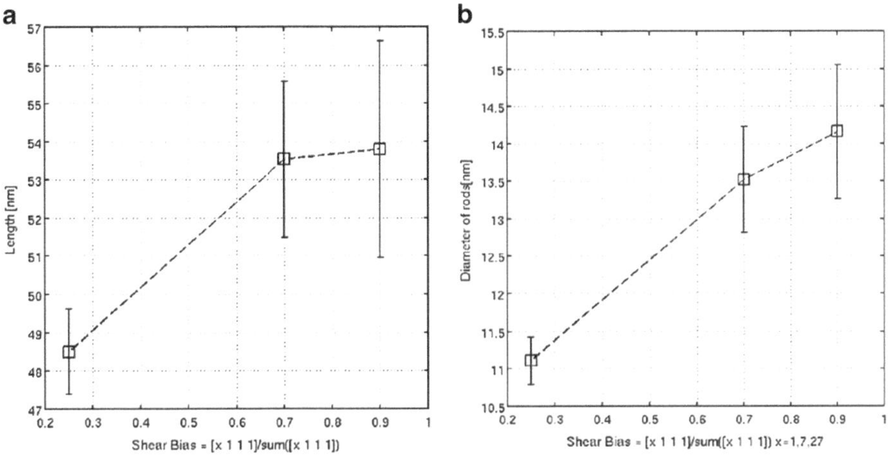

Fig. 23 Effect of shear bias on length and diameter of rods. Kuhn lengths simulated were 3, 5, and 9 (**a**) L (**b**) D

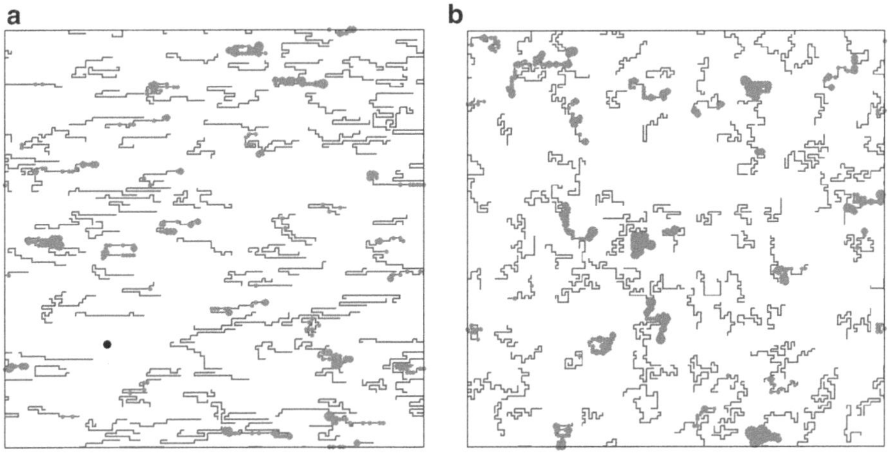

Fig. 24 Effect of shear bias on particle conformation. (**a**) Biased shear (**b**) Random shear

The transient study of the formation process indicates that the nanorods are formed by template-guided, anisotropic coagulation and consolidation of spherical nanoparticles. This was demonstrated by the lack of formation of anisotropic particles when no or subcritical NaSal concentration is used. Moreover, the addition of separately prepared spherical AgBr nanoparticles to a solution of wormlike CTAB–NaSal micelles resulted in the formation of anisotropic nanorods, clearly bringing out the role of wormlike micelles providing the backbone to facilitate anisotropic nanoparticle formation. The generality of the proposed protocol was demonstrated by applying it to the CTAC–NaSal system to obtain AgCl and iron oxide nanorods. Thus, the method can potentially be used to form nanorods of different materials by a suitable selection of the precursor materials. Further the role of the controlled shear was also highlighted. The controlled shear improves the uniformity and the number yield of the nanorods formed.

The simulations adequately demonstrate the basic mechanism of formation of nanorods in wormlike micellar phase. Several aspects of experimental trends are captured correctly in the relatively simple coarse-grained model of wormlike micelles.

Acknowledgements This study was supported by a research grant titled Engineering Aspects of Ultrafine Particle Technology Project code 07DS014/ Grant no-IR/S3/EU-03/2006, made available under the IRPHA scheme of the Department of Science and Technology, Government of India, New Delhi (India). The author also acknowledges various analytical facilities provided by the Sophisticated Analytical Instrument Facility (SAIF), FEGSEM facility and vibrating sample magnetometer (SVSM) facility provided by the Industrial Research and Consultancy Center (IRCC), and X-ray diffraction facility at the Department of Metallurgical Engineering and Material Science (MEMS), IIT Bombay, Mumbai.

References

1. Israelachvilli JN (1992) Intermolecular and surface forces, chapter 17, 2nd edn. Academic, London
2. Husein MM, Rodil E, Vera JH (2006) A novel approach for the preparation of AgBr nanoparticles from their bulk solid precursor using CTAB microemulsions. Langmuir ACS J Surf Colloids 22(5):2264–2272
3. Husein M, Rodil E, Vera JH (2004) Formation of silver bromide precipitate of nanoparticles in a single microemulsion utilizing the surfactant counterion. J Colloid Interface Sci 273(2): 426–434
4. Xu S, Li Y (2003) Different morphology at different reactant molar ratios: synthesis of silver halide low-dimensional nanomaterials in microemulsions. J Mater Chem 13(1):163–165
5. Nikolenko DYu, Brichkin SB, Razumov VF (2008) Synthesis of mixed silver halide nanocrystals in reversed micelles. High Energy Chem 42(4):305–310
6. Trewyn BG, Nieweg JA, Zhao Y, Lin VS-Y (2008) Biocompatible mesoporous silica nanoparticles with different morphologies for animal cell membrane penetration. Chem Eng J 137(1):23–29
7. Choi DG, Kim WJ, Yang SM (2000) Shear-induced microstructure and rheology of cetylpyridinium chloride/sodium salicylate micellar solutions. Korea-Aust Rheol J 12(3): 143–149
8. Kim TW, Chung PW, Lin VSY (2010) Facile synthesis of monodisperse spherical MCM-48 mesoporous silica nanoparticles with controlled particle size. Chem Mater 22:5093–5104
9. Slowing II, Vivero-Escoto JL, Wu C, Lin VS (2008) Mesoporous silica nanoparticles as controlled release drug delivery and gene transfection carriers. Adv Drug Deliv Rev 60: 1278–1288
10. Wang L, Yamauchi Y (2011) Synthesis of mesoporous pt nanoparticles with uniform particle size from aqueous surfactant solutions toward highly active electrocatalysts. Chem Eur J 17:8810–8815
11. Qiao Z, Zhang L, Guo M, Liu Y, Huo Q (2009) Synthesis of mesoporous silica nanoparticles via controlled hydrolysis and condensation of silicon alkoxide. Chem Mater 21:3823–3829
12. He Q, Shi J (2011) Mesoporous silica nanoparticle based nano drug delivery systems: synthesis, controlled drug release and delivery, pharmacokinetics and biocompatibility. J Mater Chem 21:5845–5855

13. Suzuki K, Ikari K, Imai H (2004) Synthesis of silica nanoparticles having a well-ordered mesostructure using a double surfactant system. J Am Chem Soc 126:462–463
14. Chang S-S, Lee C-L, Chris Wang CR (1997) Gold nanorods: electrochemical synthesis and optical properties. J Phys Chem B 101(34):6661–6664
15. Jana NR, Gearheart L, Murphy CJ (2001) Wet chemical synthesis of high aspect ratio cylindrical gold nanorods. J Phys Chem B 105(19):4065–4067
16. Mirgorod YA, Efimova NA (2008) The synthesis of superparamagnetic Pt/Ni nanohybrids in direct micelles of cationic surfactants. Russ J Phys Chem A 82:385–389
17. Ban I, Drofenic M, Makovec D (2006) The synthesis of iron–nickel alloy nanoparticles using a reverse micelle technique. J Magn Magn Mater 307(2):250–256
18. Xu S, Zhou H, Xu J, Li Y (2002) Synthesis of size-tunable silver iodide nanowires in reverse micelles. Langmuir 18(26):10503–10504
19. Zhang X, Cui Z (2009) Synthesis of Cu nanowires via solventhermal reduction in reverse microemulsion system. J Phys Conf Ser 152:012022
20. Yang X, Chen S, Zhao S, Li D, Ma H (2003) Synthesis of copper nanorods using electrochemical methods. J Serbian Chem Soc 68(11):843–847
21. Liu Y, Wang W, Zhan Y, Zheng C, Wang G (2002) A simple route to hydroxyapatite nanofibers. Mater Lett 56(4):496–501
22. Xi L, Lam YM, Xu YP, Li L-J (2008) Synthesis and characterization of one-dimensional CdSe by a novel reverse micelle assisted hydrothermal method. J Colloid Interface Sci 320(2): 491–500
23. Rouault Y, Milchev A (1997) A Monte Carlo lattice study of living polymers in a confined geometry. Macromol Theory Simul 6(6):1177–1190
24. Vogt M, Hernandez R (2001) A two-dimensional polymer growth model. J Chem Phys 115(3):1575
25. Vogt M, Hernandez R (2002) A three-dimensional polymer growth model. J Chem Phys 116(23):10485
26. Carl W, Rouault Y (1998) On the length distribution of semi-flexible linear micelles. Macromol Theory Simul 7(5):497–500
27. Rouault Y (1998) The effect of stiffness in wormlike micelles. Eur Phys J B Condens Matter Complex Syst 81:75–81
28. Rouault Y (1998) Change of the scaling behavior of the end-to-end square distance in a two-dimensional polydisperse system. Eur Phys J B 4(1):61–64
29. Rouault Y (1998) Living polymers in random media: a 2D Monte-Carlo investigation on a square lattice. Eur Phys J B Condens Matter Complex Syst 2:483–487
30. Rouault Y (1996) A Monte Carlo study of living polymers in 2D: effect of small chains on static properties. J Phys II 6(9):1301–1311
31. Rouault Y (1999) Equilibrium polymerization: towards a numerical description of the dynamics of wormlike micelles. Macromol Theory Simul 8(6):551–560
32. Rouault Y, Milchev A (1995) Monte Carlo study of living polymers with the bond-fluctuation method. Phys Rev E 51(6):5905–5910
33. Rouault Y (1998) Chain radius dependence on concentration in a 2D living polymer system. Eur Phys J B 4(3):321–324
34. Cates ME, Candau SJ (1990) Statics and dynamics of worm-like surfactant micelles. J Phys Condens Matter 2(33):6869–6892
35. Shikata T, Hirata H, Kotaka T (1987) Micelle formation of detergent molecules in aqueous media: viscoelastic properties of aqueous cetyltrimethylammonium bromide solutions. Langmuir 3(6):1081–1086
36. Bai J, Li Y, Zhang C, Liang X, Yang Q (2008) Preparing AgBr nanoparticles in poly(vinyl pyrrolidone) (PVP) nanofibers. Colloids Surf A Physicochem Eng Asp 329(3):165–168
37. Liu X-H, Luo X-H, Lu S-X, Zhang J-C, Cao W-L (2007) A novel cetyltrimethyl ammonium silver bromide complex and silver bromide nanoparticles obtained by the surfactant counterion. J Colloid Interf Sci 307(1):94–100

38. Murphy CJ, Sau TK, Gole AM, Orendorff CJ, Gao J, Gou L, Hunyadi SE, Li T (2005) Anisotropic metal nanoparticles: synthesis, assembly, and optical applications. J Phys Chem B 109(29):13857–13870
39. Bai J, Li Y, Yang S, Du J, Wang S, Zhang C, Yang Q, Xuesi C (2007) Synthesis of AgCl-PAN composite nanofibres using an electrospinning method. Nanotechnology 18:305601

Scalable Synthesis of Noble Metal Nanoparticles

Venugopal Santhanam

Abstract Noble metal nanoparticles possess unique size-dependent electronic and optical characteristics and are one of the foremost 'building blocks' for nanostructured device fabrication. As such, there is considerable interest in developing continuous-flow processes for large-scale synthesis of noble metal nanoparticles. In this chapter, we describe the results of our work aimed at understanding key process variables that determine particle size distribution in two popular protocols used for lab-scale synthesis of gold and silver colloids. Our understanding of the importance of aggregation and role of the pH of precursor solutions in determining the kinetics and stability of colloidal sols enabled us to propose suitable modifications in process conditions that enabled scalable synthesis of gold and silver nanoparticles. These insights also led to the development of a novel route for low-cost fabrication of silver nanostructures on paper using an inkjet printer.

Keywords Metal nanoparticles • Continuous synthesis • Effect of pH • Mode of addition • Reactive inkjet printing

1 Introduction

The expectation of an impending nanotechnology revolution, which promises faster, cheaper and portable solutions for energy, healthcare and environmental issues, has captured the imagination of laymen. The nanoscale regime (ca. 1–100 nm) corresponds to an intermediate size range, between a molecular state and the bulk state; it encompasses several characteristic length scales over which collective properties emerge. For example, answers to questions on transition from atomic to bulk behaviour, such as 'what is the size at which the electron energy levels in a collection of atoms lose their discrete nature and become delocalised over the entire structure?', have been found to lie in this size range. Such information is of considerable interest to scientists for understanding behaviour of condensed state

V. Santhanam (✉)
Department of Chemical Engineering, Indian Institute of Science, Bangalore 560012, India
e-mail: venu@chemeng.iisc.ernet.in

Y.M. Joshi, S. Khandekar (eds.), *Nanoscale and Microscale Phenomena*,
Springer Tracts in Mechanical Engineering, DOI 10.1007/978-81-322-2289-7_4

matter. The tantalising possibility of harnessing such knowledge for controlling material properties in a 'digital/discrete' manner, by varying system size, is one of the leading reasons for the buzz surrounding the term 'nanotechnology'. Nanomaterials, objects with one or more dimensions in the nanoscale regime, are expected to play a critical role as 'building blocks' in the field of nanotechnology. Metal or semiconductor or carbon-based nanostructures with different morphologies, e.g. nanoparticles, nanowires/nanotubes, nanoscale films, have been widely studied. Noble metal nanoparticles are the foremost contenders amongst nanomaterials that already have a significant market presence [1], a presence that is more likely to expand as several technologies such as drug delivery [2], nanoelectronics [3], etc. mature in the near future.

The use of noble metal nanoparticles in itself is not unique to the recent past, historical examples of their uses abound. For example, noble metal nanoparticles were used to fabricate stained glass windows in Europe, and colloids of noble metal nanoparticles were part of traditional medicines/biocides in India [4]. More recently, the use of gold nanoparticles as contrast agents for electron microscopy and point-of-care biomedical test kits are examples of nanoparticle-based commercial applications [5]. What sets apart the current research and development studies from previous efforts is the ability to characterise and manipulate the nanomaterials at the atomic scale and the resulting improved understanding of the science behind their distinctive properties.

From a process engineering perspective, the importance of research on scalable processes for producing metal nanoparticles with controlled sizes is evident. To pursue such questions, a laboratory for nanoparticle engineering was established at IISc, Bangalore, under the IRHPA scheme of DST, and this chapter summarises our research on scalable synthesis of metal nanoparticles, especially gold and silver nanoparticles. In the following, an introduction to some novel properties of noble metal nanoparticles is presented, followed by a perspective on the questions posed as we began our research on scaleup of colloidal nanoparticle synthesis. Then, the results of our investigations on the synthesis of gold and silver nanoparticles using either tannic acid or sodium citrate as both reducing and stabilising agents are presented. Finally, a summary of the key research findings is presented.

2 Properties of Noble Metal Nanoparticles

A striking attribute of colloidal dispersions of noble metal nanoparticles is their colour. Noble metal nanoparticles exhibit strong plasmon resonance absorbance bands in the visible spectrum that are size and shape dependent, resulting in the ability to tune colour without changing the material used. The surface plasmon resonance bands result from the interaction of incident electromagnetic field ('light') with the metal nanoparticles leading to coherent oscillations of the conduction band electrons. The coherently oscillating electrons behave as an entity denoted as plasmons.

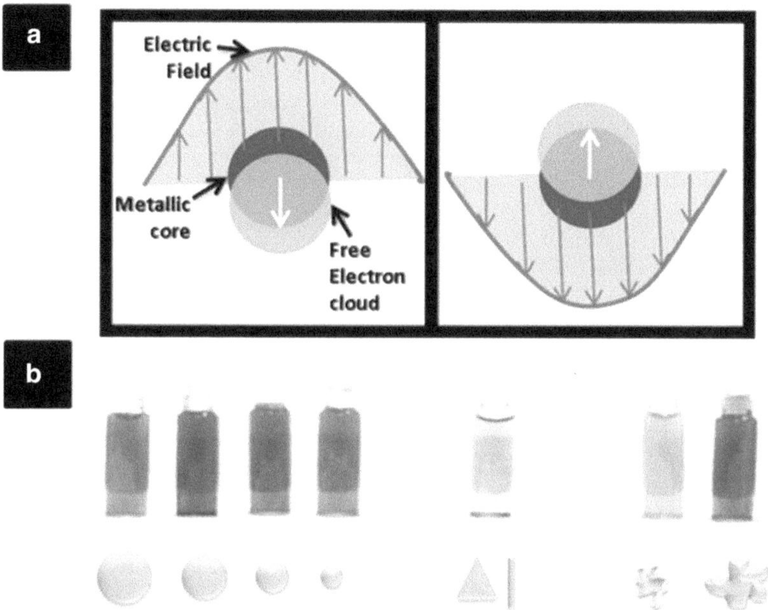

Fig. 1 (**a**) Schematic diagram representing the interaction of light with metal nanoparticles (figure not to scale). The interaction of electromagnetic field with the metal nanoparticles results in coherent oscillations of the conduction band electrons. The *circles* are a guide to illustrate the collective displacement of the mobile electrons (plasmons) within the nanoparticles from the ionic cores. The two panels illustrate the movement of electrons corresponding to two different phases of the incident radiation's electric field. (**b**) Distinct colours exhibited by aqueous sols of gold nanoparticles synthesised during the course of our research. The cartoons, underneath the sols, depict the morphology of nanostructures present in the respective sols (Color figure online)

Figure 1a shows a schematic representation of the interaction of electromagnetic light waves with metal nanoparticles that are much smaller than the wavelength of the incident light. The conduction band electrons within the nanoparticle interact with the time-varying electric field of the incident light and lead to the formation of an oscillating dipole (in the simplest case). A large enhancement in the light absorption results when the wavelength of the incident light matches the size-dependent resonant frequency of the dipole (plasmon resonance). As the restoring force is proportional to the surface area in such systems and also confined by the nanoparticle, these are commonly referred to as localised surface plasmon resonances (LSPR). Figure 1b showcases the range of colours exhibited by aqueous sols of gold nanoparticles having different sizes and shapes that were synthesised during the course of this work.

The electronic properties of small metal particles have also been investigated extensively. The simplest system studied consists of placing a single metal nanoparticle between two electrodes with tunnelling barriers separating the electrodes from the metal nanoparticle [6]. In this configuration (Fig. 2), the electrical conduction

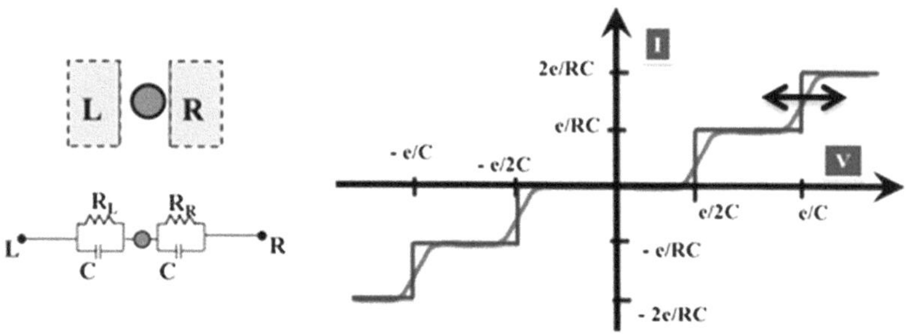

Fig. 2 Schematic representation and equivalent circuit of nanostructure consisting of a metal nanoparticle placed between two electrodes, and Ideal I-V curves (*double headed arrow* indicates the smearing of coulomb staircase at temperatures >0 K) for this nanostructure illustrating the phenomena of Coulomb blockade and Coulomb staircase

is entirely suppressed at low bias voltages. This phenomenon is called 'Coulomb blockade'. Coulomb blockade occurs when the electrostatic energy increase due to the addition of a single electron onto a capacitatively coupled metal island is much larger than the thermal energy of the electrons:

$$e^2/2C >> k_B T \tag{1}$$

where 'e' is the charge on an electron, 'C' is the effective capacitance of the metal island, 'k_B' is the Boltzmann constant and 'T' is the absolute temperature of the metal island. In the case of metal nanoparticles, the capacitance, 'C', is directly proportional to the radius of the metal nanoparticle. At room temperature, metal particles with radius less than 2–3 nm satisfy Eq. 1. Furthermore, for Coulomb blockade to be observed, the tunnelling resistance (R) to and from the metal island must also be much greater than the resistance quantum (R_Q):

$$R >> R_Q \tag{2}$$

where $R_Q = h/2e^2$ (~12.9 kΩ) with 'h' being the Planck constant.

Bulk metals typically exhibit a linear I-V response, which is characterised by the well-known Ohm's law. In contrast, when both Eqs. 1 and 2 are satisfied, the I-V curve for an asymmetric tunnelling junction [$R_R >> R_L$ or $R_L >> R_R$] shows characteristic steps in current, denoted as 'Coulomb staircase' (Fig. 2). Due to asymmetry in electron transport rates between the two junctions, electrons tend to accumulate on the isolated metal nanoparticle; the electrostatic energy associated with such charging of the 'island' has to be compensated for by increasing the external bias. This leads to a nonlinear I-V response. Such sequential tunnelling phenomena can give rise to the ability to control the charge on the island at single electron level and form the basis of the field of single electronics.

3 Synthesis of Noble Metal Nanoparticles: Prevalent Status and Identified Research Needs

Historically, the synthesis of metal nanoparticles predates modern science. The Lycurgus chalice of Rome, belonging to the fifth century, contains gold and silver nanoparticles; Mayan blue pigment from the eleventh century contains iron and chromium nanoparticles [7]; Swarna Bhasma, an Ayurvedic medicine dating back to 2000 BC, contains gold nanoparticles [4]. The report by Faraday, in 1857, on the preparation of a gold colloidal solution by reacting chloroauric acid with white phosphorus has come to be regarded as the beginning of scientific studies into the synthesis of noble metal nanoparticles [8]. Faraday postulated that the red colour of both aqueous gold solutions as well as the ruby colour of stained glasses found in medieval churches was due to the presence of very fine gold particles. At the beginning of the twentieth century, the development of the ultramicroscope led to the finding that such colloidal solutions contained 'millimicron' size particles [9]. From 1950s onwards, metal nanoparticles have been utilised as contrast-enhancing agents for the imaging of cells [10] using electron microscopes, which spurred extensive investigations into the different recipes available for the synthesis of nanoparticles.

Presently, synthesis procedures used for generating metal nanoparticles can be classified into two broad classes – (1) extruding/grinding/sculpting bulk materials (top-down approach) and (2) initiating nucleation and growth of desired atomic/reactive precursors (bottom-up approach). The top-down approaches are easily amenable to continuous-flow processing but suffer from the following disadvantages: they are energy intensive, they produce particles with a very broad size distribution, they are limited in their capability to produce particles with diameters in the range of 1–20 nm, and they can also introduce contamination due to the tools involved. Bottom-up approaches involving liquid-phase recipes are more widely prevalent in research laboratories and are routinely used for batch synthesis. To control the size of nanoparticles, parameters such as molar ratio of reducing agent to metal salt (MR) [10–12], molar ratio of capping agent to metal salt [13], and temperature [12] have been varied. Liquid-phase synthesis has typically been portrayed as simple 'recipe' for carrying out redox reactions at the lab scale, but the complexity of underlying nucleation and growth processes poses formidable challenges for scaling such recipe into continuous-flow processes.

Metal nanoparticles, especially gold and silver nanoparticles in the size range of 2–20 nm, are well suited for applications in nanoelectronics. A focus of our project was the development of scalable processes for synthesis of monodisperse metal nanoparticles for such applications. Desired characteristics were that the process should be easily scalable, and occur at room temperature, to simplify reactor design considerations. Recently, there is an increasing emphasis on using greener routes. 'Green chemistry' is defined as the design of chemical products and processes that reduce or eliminate the use and generation of hazardous substances [14]. In this context, synthesis of nanoparticles in the aqueous phase using environmentally benign reagent is more desirable for large-scale synthesis. Sodium citrate (used in 'citrate method') and a combination of sodium citrate and tannic acid (used in 'Slot and Geuze method') are the two most commonly used 'green' reagents, as these reagents can act as both reducing and stabilising agents for the synthesis of metal nanoparticles. The reported size range of particles synthesised at elevated temperatures, using these reagents, varies from 10 to 100 nm. For the rapid, room temperature synthesis of metal nanoparticles, hazardous reducing agents like sodium borohydride and hydrazine [15] were used. A common refrain for the use of such hazardous reagents was their 'strength', usually inferred from the difference in standard redox potential values between the reducing agent and bulk metal.

Interestingly, we observed that the reduction of chloroauric acid by citrate is faster when some tannic acid was added to the citrate, although both citrate and tannic acid have comparable redox potentials. This observation challenged the prevalent notions of metal nanoparticle synthesis by redox reactions in solutions. To understand the reasons behind these observations, we carried out systematic studies to elucidate the factors controlling the kinetics of nanoparticle formation using the two most widely used reagents for the synthesis of gold and silver nanoparticles, namely, citrate and tannic acid. The results of our investigations are presented and discussed in the following sections.

4 Room Temperature 'Green' Synthesis of Noble Metal Nanoparticles

4.1 Tannic Acid as Reducing Agent

Tannic acid, also known as tannin, is a polyphenolic compound derived from plants like gall nuts, tree barks, etc. Ostwald, in 1912, utilised tannin to reduce chloroauric acid at 100 °C and stabilise gold nanoparticles [16]. Ostwald reported that tannin can reduce chloroauric acid at neutral pH to form stable gold nanoparticles at 100 °C, even if tap water is used to prepare the aqueous solutions. Turkevich [12] replicated Ostwald's protocol and reported the particle size to be 12.0 ± 3.6 nm based on TEM images. Mulphrodt [17] used tannic acid along with citrate to reduce the size of nanoparticles synthesised to values lower than that obtained by citrate method. Slot and Geuze [10] further modified the protocol and controlled the size of nanoparticles from 3.5 to 15 nm by varying the concentration of tannic acid in the synthesis, while the concentration of citrate remained constant. Tannin has also been used to synthesise silver nanoparticles at 60 °C [18]. Aqueous solution of tannic acid is found to be weak reducing agent, at room temperature, which can only grow seeds into nanoparticles [19]. Hence, higher temperatures were used for the nucleation and growth of silver and gold nanoparticles using tannic acid. Thus, although tannic acid has been known to reduce and stabilise metal nanoparticles, neither were the reported procedures performed at room temperature nor were monodisperse nanoparticles obtained in the size range of 1–10 nm.

4.2 Factors Affecting Kinetics of Silver Nanoparticle Formation [20]

The pKa of tannic acid solution falls between seven and eight, depending on the extent of dissociation of tannic acid [21], which is known to partially hydrolyse under mild acidic/basic conditions into glucose and gallic acid units [22]. Gallic acid at alkaline pH reduces silver nitrate into silver nanoparticles rapidly at room temperature [23], but the particles form aggregates in solution as gallic acid is a poor stabilising agent. Glucose is a weak reducing agent at room temperature, but it is an excellent stabilising agent at alkaline pH [24]. These findings suggested to us that tannic acid could be an ideal reducing and stabilising agent under alkaline conditions at room temperature.

A few experiments were performed to understand the role of pH of tannic acid in the synthesis of metal nanoparticles. Five millilitre of 2.95 mM solution of $AgNO_3$ was added to 20 mL of tannic acid solution, maintained at the desired pH. Figure 3 shows the hydrodynamic diameter of silver nanoparticles synthesised by varying the initial pH of tannic acid used in the reaction. At acidic pH values, nanoparticle formation was not observed even after waiting for several hours. From a pH value of 6 onwards, nanoparticle formation was observed to occur in a matter of minutes, with both size and time for appearance of colour (due to SPR band of silver colloids) decreasing rapidly till a pH of 8. Further increase in the pH did not cause a significant decrease in the hydrodynamic size.

To understand the kinetics, the stoichiometric ratio required for the completion of the reaction between tannic acid and silver nitrate was investigated. Figure 4 shows UV-Vis spectra of 0.25 mL aliquots sampled during the stepwise addition

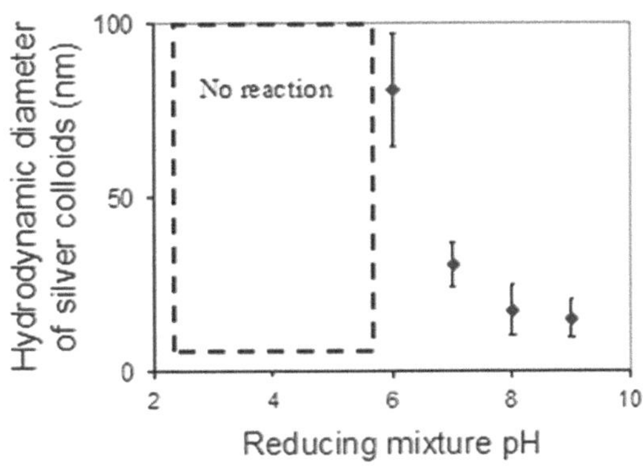

Fig. 3 Effect of pH at room temperature. Variation of hydrodynamic diameter of silver nanoparticles as a function of initial pH of tannic acid solution. There is no noticeable reaction for pH values <6. [$AgNO_3$] = 0.59 mM, MR = 0.05

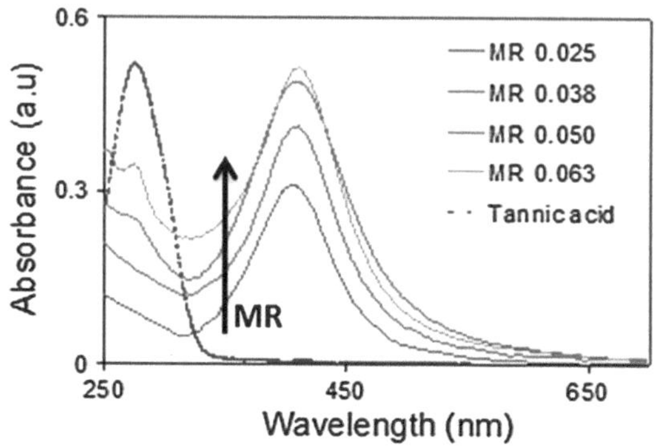

Fig. 4 UV visible spectra of silver nanoparticle solutions as a function of the molar ratio of tannic acid to silver nitrate (MR). The spectrum of a concentrated (2× original) tannic acid solution is also shown for comparison (Reproduced with permission from [24])

(corresponding MR values provided in the legend) of tannic acid solution, maintained at a pH of 8, to 5 mL of 2.95 mM silver nitrate solution. It is seen that the surface plasmon peak of silver nanoparticles at 420 nm increases initially and then saturates at an MR value of 0.05, indicating that silver nitrate is completely reduced. Also, for MR values $\geq$0.05, a small shoulder at 270 nm is seen and is attributed to the presence of excess tannic acid. The spectrum of a concentrated solution of pure tannic acid at pH 8 is also shown for comparison. A stoichiometric ratio of tannic acid to silver nitrate of 0.05 indicates that one mole of tannic acid can reduce 20 moles of silver nitrate.

Figure 5a shows the variation in hydrodynamic size of silver nanoparticles with the molar ratio of tannic acid to silver nitrate; the MR was varied by increasing the tannic acid concentration at a constant value of the silver nitrate solution. An increase in the MR value increases the hydrodynamic diameter of silver nanoparticles. To analyse the role of excess tannic acid, the nanoparticle synthesis was monitored by studying the evolution of absorbance at SPR peak wavelength with millisecond resolution in stop-flow module (Fig. 5b). From Mie theory, the absorbance is directly proportional to the volume of nanoparticles (for size <20 nm). The induction time, which is inversely proportional to the nucleation rate, increases with increasing MR. Also, the initial slope of the absorbance profile is higher for smaller MR indicating that the growth rate is higher at smaller MR. Figure 5c–e shows representative Field Emission Scanning Electron Microscope (FESEM) and TEM images for these samples. The trends of lower nucleation and growth rates, corresponding to increasing particle size, with increasing MR are counterintuitive. Intuitively, one expects that increasing reducing agent concentration should increase redox reaction rates and, therefore, result in higher growth and nucleation rates. Also, the observed trends cannot be interpreted in terms of the role of tannic acid as a stabiliser, as increasing stabiliser concentration is also expected to result in smaller sizes. This implies that the role of tannic acid is not limited to that of reducing/stabilising agent.

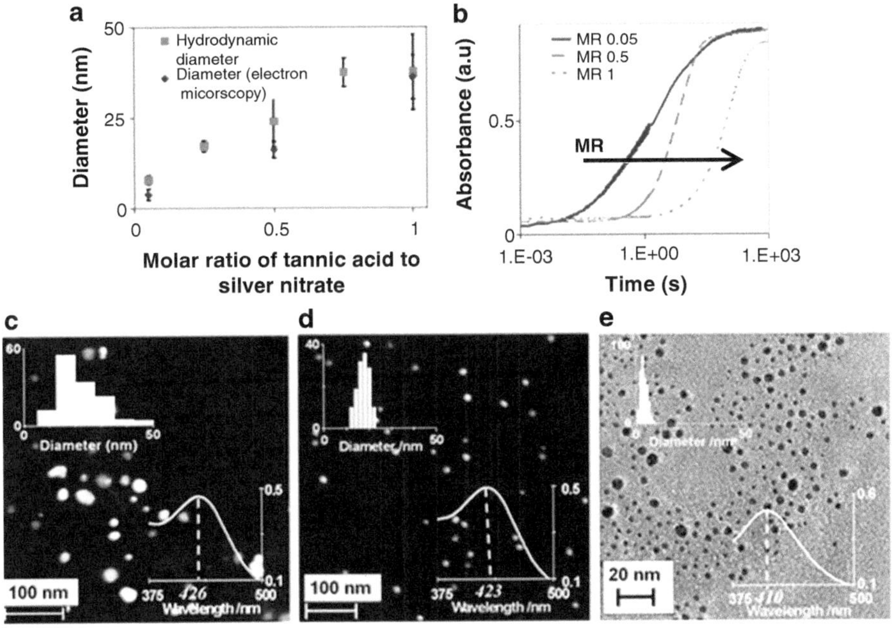

Fig. 5 (**a**) Change in the hydrodynamic diameter of silver nanoparticles with the molar ratio of tannic acid to silver nitrate. (**b**) Time evolution of absorbance at SPR peak wavelength for different initial molar ratios of tannic acid to silver nitrate (MR). The absorbance rises faster for lower MR values. Representative electron microscopy images of silver nanoparticles synthesized at various MR values. (**c**) MR = 1 (FESEM), (**d**) MR = 0.5 (FESEM), (**e**) MR = 0.05 (TEM). The inset graphs in (**c**–**e**) depict the corresponding particle size distribution and UV–Vis spectrum, with ordinates representing the actual number of particles counted and absorbance (a.u.) respectively (Reproduced with permission from [24])

4.3 *Tannic Acid as Organiser*

A representative structure of tannic acid, corresponding to its average formula weight, is shown in Fig. 6. It consists of a central core of glucose that is linked by ester bonds to polygalloyl ester chains. Tannic acid has 25 phenolic hydroxyl (–OH) groups in its structure, but only ten pairs of 'ortho'-dihydroxyphenyl groups are capable of taking part in redox reactions to form quinones and donate electrons, because of the chelating action of adjacent hydroxyl groups and constraints on carbon valency. Hence, each tannic acid molecule is capable of donating 20 electrons; this value matches well with the number deduced experimentally. At alkaline pH, the deprotonation of phenolic hydroxyl groups will enhance their chelation with cations and can thus account for their enhanced reactivity. But the ability to nucleate nanoparticles cannot be explained by this fact alone, as their redox potentials are still lower than that required to completely reduce isolated metal ions in solution. This suggests a third role for tannic acid as an 'organiser' of ions/atoms

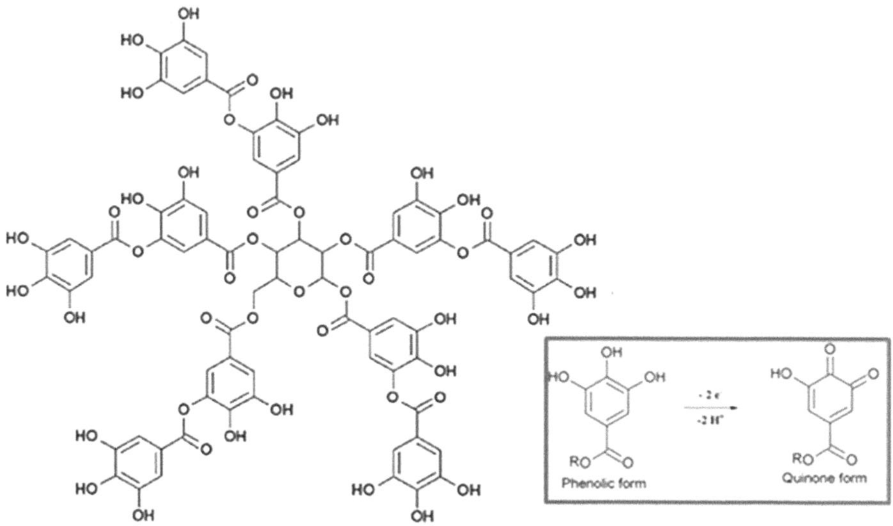

Fig. 6 Representative structure of tannic acid ($C_{76}H_{52}O_{46}$). The *inset* on the right shows the phenol to quinone conversion upon oxidation of tannic acid

for facilitating nucleation, given that a single tannic acid molecule can chelate with multiple ions and reduce them. Such a mechanism is consistent with the fact that the redox potential value for metal clusters varies rapidly with increasing cluster size, from values corresponding to a strong reducing agent for a single atom (i.e. thermodynamically unfavourable conditions for reduction by tannic acid) to that of an oxidant (similar to the bulk, and reducible by tannic acid) for clusters having a few tens of atoms [25], thereby enabling 'weak' reducing agents to nucleate metal nanoparticles. The mechanism by which reduction and atomic reorganisation occur within such chelation complexes is a matter for further investigation.

The role of 'organiser' is also consistent with the observed trends of nucleation and growth rates of silver nanoparticles as a function of MR. Tannic acid has five units of gallic acid and each gallic acid unit can combine with four silver ions. Each tannic acid molecule can be thought of as a five-armed chelator. Thus, at an MR of 0.05, tannic acid complexes will be saturated with 20 silver atoms enabling rapid nucleation (faster induction time) that results in smaller particle size. At MR of 1, each tannic acid is on average ligated to only one silver atom and so the nucleation rate will be decided by the interaction of such 'unsaturated' compounds in solution leading to slower nucleation rate (larger induction times) that results in larger particle sizes. The reduction in the initial growth rate with an increasing MR can also be accounted for by considering that growth occurs due to collision of nuclei/particles with chelated silver atoms. On an average, rate of collision between chelated silver atoms and nuclei/particle will be similar in all cases due to opposing effects of increasing concentration and decreasing ligation; however, the probability of incorporation per collision will be higher for compounds ligated with a higher

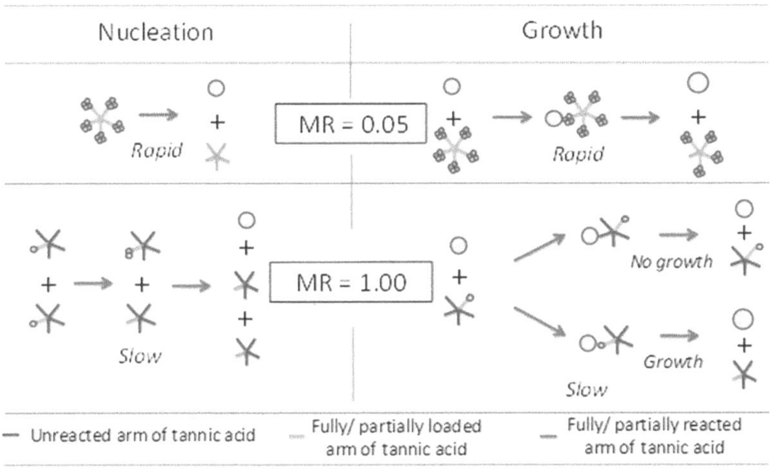

Fig. 7 Schematic representation of organiser-based nucleation and growth processes at MR values of 0.05 and 1. Tannic acid is represented as a five-armed molecule with each arm capable of reducing and chelating with four silver atoms. Both nucleation and growth processes are faster at MR value of 0.05 as compared to MR value of 1 resulting in the rapid formation of smaller nanoparticles at the lower MR value. Note: the number of loaded atoms depicted is representative of the mean value, given that all the silver ions are reduced and chelated (Color figure online) (Reproduced with permission from [20])

number of silver atoms. Thus, at smaller MR values, the incorporation efficiency of atoms onto nuclei/particles (i.e. growth) will be higher per collision resulting in higher growth rates. Figure 7 illustrates these concepts at MR values of 0.05 and 1.

4.4 *Interplay Between Reactivity and Stabilisation in the Synthesis of Gold Nanoparticles [26]*

Chloroauric acid, the most common gold precursor used, possesses pH-dependent reactivity. At acidic pH, gold ion complexes with chloride ligands are predominant, while at basic pH, gold ions complexed with hydroxyl ligands are predominant. The chloride ligand can be easily displaced as compared to the hydroxyl ligand, resulting in higher reactivity of chloroauric acid at acidic pH values. On the other hand, nucleation and growth of nanoparticles by tannic acid occur rapidly at alkaline pH. Faced with these conflicting requirements and to identify optimal conditions for rapid, room temperature synthesis of gold nanoparticles, a series of experiments were carried out by varying the initial pH of the two reagents (prior to mixing). Three millilitre of 4.4 mM tannic acid solution at desired pH was added to 22 mL of 0.288 mM chloroauric acid solution at desired pH. The resulting colloidal samples were characterised using TEM to determine particle size, using a stop-flow reactor (SFR) to determine induction time, and ζ-potential measurements. These results are summarised in Table 1.

Table 1 Experimental conditions and characterisation results for synthesis of gold nanoparticles

Experiment	Initial pH of chloroauric acid solution	Initial pH of tannic acid solution	pH of reaction mixture	Particle diameter ($\mu \pm \sigma$), nm	Induction time, *s*	ζ-potential, mV
A	3.2	3.1	3.2	14.1 ± 4.8	3	-20 ± 2
B	3.2	7.1	6.4	7.1 ± 1.6	1.5	-49 ± 2
C	7.0	6.3	6.4	10.7 ± 3.8	66.8	-69 ± 2
D	9.1	3.1	6.4	14.2 ± 5.4	163.5	-60 ± 2
E	2.1	7.1	5.0	10.0 ± 2.9	0.9	-18 ± 7
F	2.1	9.0	7.1	5.8 ± 1.0	0.5	-35 ± 7

Comparison of the results of these experiments in terms of the kinetics (induction time), stability (ζ-potential) and particle size shows the following: (1) the initial pH of chloroauric acid solution mainly controls the kinetics (A-D; B-C), whereas the pH of tannic acid has only a slight influence on kinetics (A-B; E-F). The fact that initial pH of the reagents, prior to mixing, has a significant impact implies that the kinetics of forming less reactive hydroxy-chloroauric species, after a step change in pH from acidic to neutral conditions, is slower than that of the redox reaction; (2) the pH of the reaction mixture determines the stability, and the final particle size is more sensitive to variations in stability (B-E; A-D) rather than kinetics (B-C-D;C-E). This suggests that aggregation is an important pathway in the formation of the nanoparticles and that it is minimised at reaction mixture pH values of 6.4 and above. All the colloidal sols were stable upon storage for more than a year, indicating that aggregation is active only during the early growth stages.

Overall, these results show that gold nanoparticle size distribution is determined by a fine balance between the reactivity of precursors and coalescence in the initial period, which can be manipulated by controlling the initial pH of reactants and the reaction mixture pH, respectively. Figure 8 illustrates the various pathways involved. A key outcome is that independent control over reactivity and stabilisation can be achieved by manipulating the pH of the precursor solutions. The detailed investigation of the role of molar ratio and the pH of tannic acid showed that it is plausible to synthesise nanoparticles with various morphologies at room temperature. Surprisingly, the polydispersity of nanoparticles synthesised under various conditions remained high ($>$15 %).

Given that alkaline pH is more favourable for preventing aggregation, an experiment ('G') was carried out to minimise coalescence during mixing by altering the order of addition. This was achieved by simply reversing the order of addition (i.e. chloroauric acid into tannic acid) while maintaining the initial pH of reagents, overall amount of reagents, the volume of the added reagent and the volume of the total reaction mixture at values used in experiment 'B'. For comparison, a 1 mL sample was prepared by rapidly mixing ($\sim$ within 3 ms) the two reagents in stop-flow module (experiment H), while maintaining the concentration, initial pH and volumetric ratio similar to experiment 'B'. The corresponding size distributions

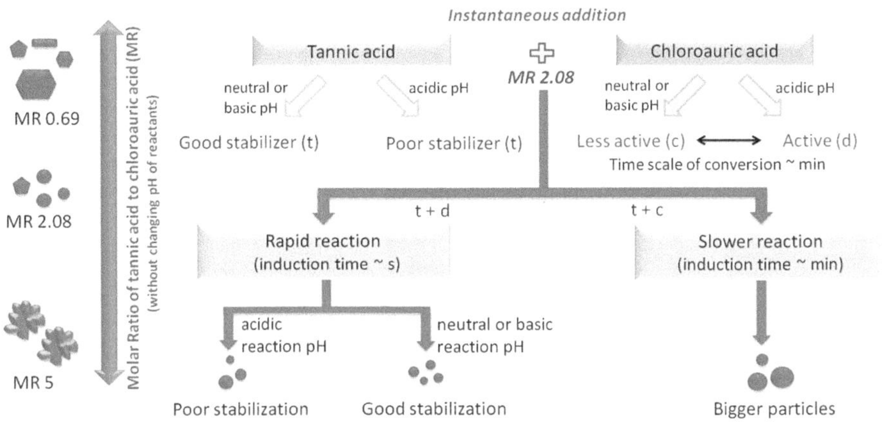

Fig. 8 Schematic summary of the effect of MR and the roles of reagent and the reaction mixture pH in determining gold nanoparticle size and shape (Color figure online) (Reprinted from [26] under CCA terms)

were determined to be 5.9 ± 1.6 nm (G) and 6.2 ± 1.9 nm (H), respectively. The effect of mode as well as speed of mixing on size distribution is minimal indicating that stabilisation of nanoparticles by adsorption of tannin requires a finite time during which particles may coalesce.

Attempts to minimise coalescence by using an excess of tannic acid did not yield the desired result. So, the next best strategy was to add diluted chloroauric acid slowly into tannic acid to provide adequate time for adsorption of tannin. Experiment 'I' was conducted by adding chloroauric acid at a rate of ~1 mL/min into tannic acid, while maintaining overall concentration and total volume of the reaction mixture constant as in experiment 'G'. Representative TEM images and corresponding particle size distributions of aliquots collected during different stages of addition are shown in Fig. 9. These results clearly show that the slow addition of chloroauric acid into tannic acid is the best strategy to synthesise monodisperse gold nanoparticles. This process can be easily extended to form size-controlled nanoparticles with larger mean size, if care is taken to ensure that the reaction mixture pH never falls below 6.4.

The constant value of standard deviation estimates (0.6–0.8 nm) indicates that the growth mode is polynuclear surface reaction controlled (i.e. the growth rate is independent of particle size). Mass balance requirement for surface reaction-controlled growth dictates that increase in mean nanoparticle diameter must be proportional to 1/3 power of the volume of chloroauric acid added. Figure 10 shows a plot of the predicted and the observed mean diameters as a function of the 1/3 power of the volume of chloroauric acid added. The observed values are seen to agree exceptionally well with the predicted values, which is a clear indication that the slow (dropwise) addition protocol is akin to one-shot process for seed formation followed by their controlled growth into nanoparticles.

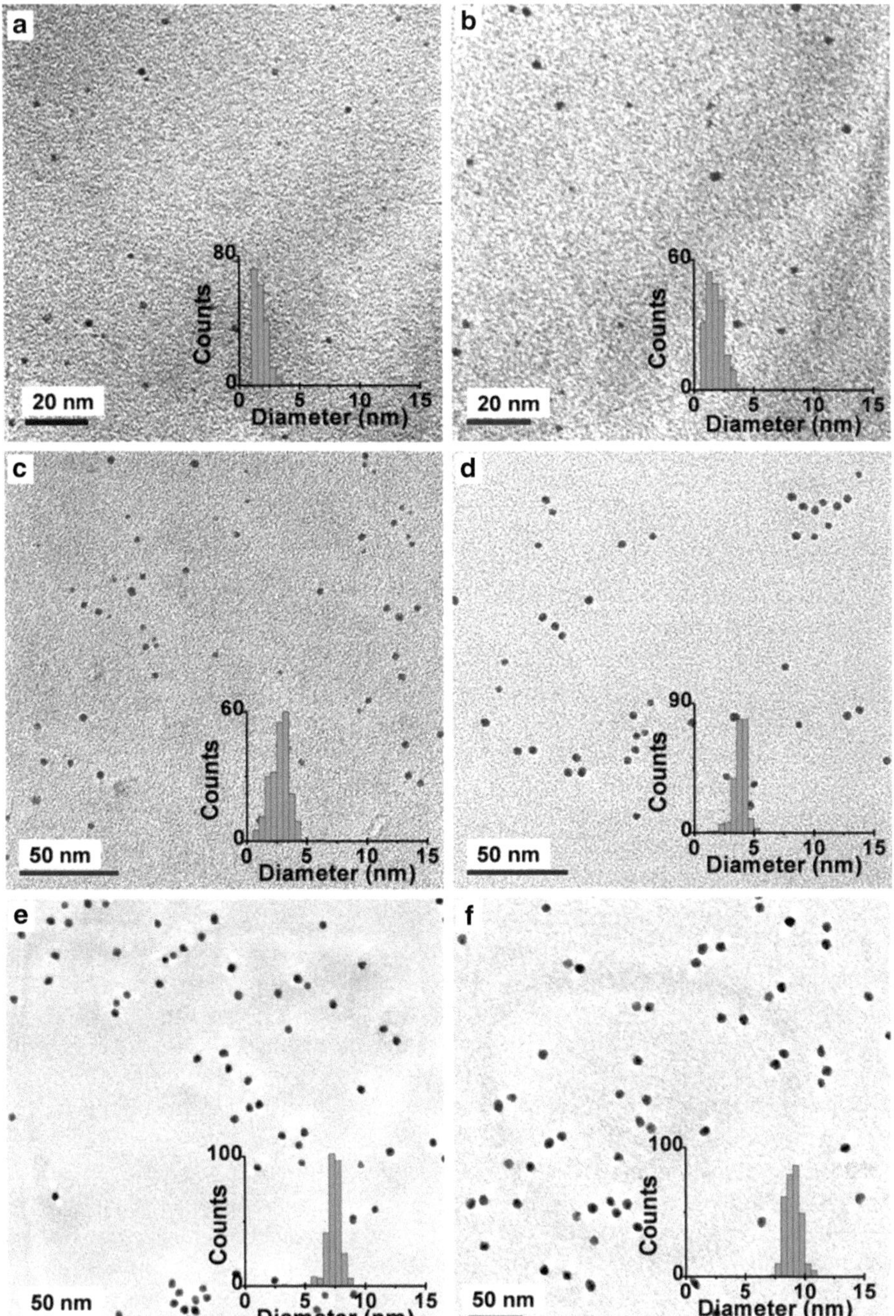
a
b
c
d
e
f
20 nm
20 nm
50 nm
50 nm
50 nm
50 nm
Counts
Diameter (nm)

4.5 Monodisperse Sub-10 nm Gold Nanoparticles by Reversing the Order of Addition in Citrate Method [27]

The ‘classical citrate’ method for synthesising gold nanoparticles, using sodium citrate as the reducing agent, is renowned for its ability to produce biocompatible colloids with a mean size >10 nm. Biocompatibility and ease of surface functionalisation have made the citrate method not only popular amongst researchers but also the method of choice for producing NIST reference standards for biological applications (NIST Reference Materials 8011, 8012 and 8013). Citrate-stabilised gold nanoparticles in the 1–10 nm size range are desirable for their ability to target the cell nucleus for gene/drug delivery and for the significant nanoscale size effects that occur in this size range.

Turkevich [12] showed that dicarboxy acetone, formed due to oxidation of citrate by chloroauric acid, facilitates nucleation of gold nanoparticles. Kumar et al. [28] have modelled the formation of gold nanoparticles by the citrate reduction method and elucidated the role of dicarboxy acetone as an ‘organiser’ for nucleation. Also, citrate exhibits pH-dependent ionisation and is an effective stabilising agent at alkaline pH. Based on our insights into the formation of gold nanoparticles by reduction with tannic acid, we hypothesised that smaller nanoparticles can be formed by reversing the order of addition, i.e. adding reactive chloroauric acid at acidic pH into alkaline solution of sodium citrate. 0.25 mL of 25.4 mM of chloroauric acid (measured pH of 1.6) was added to 24.75 mL of boiling citrate solution of required concentration. For comparison, we also carried out experiments wherein the same total amount (moles) of reagents was added in the standard configuration, i.e. adding 0.25 mL of sodium citrate solution of the desired concentration into 24.75 mL of boiling 0.256 mM chloroauric acid solution. Figure 11 shows some representative images and also compares the sizes of gold nanoparticles formed by these two alternative modes of addition.

At MR >5, addition of chloroauric acid into citrate solution, indeed, results in smaller nanoparticles than adding the alternative way. Figure 12a–b shows the digital images of the reaction mixture during the synthesis. The time required for the synthesis of nanoparticles was also found to be faster while adding chloroauric acid into citrate solution. Unlike classical citrate reduction, where citrate solution at room temperature is added to the reaction mixture, here, boiling citrate solution

Fig. 9 Representative TEM image of size-controlled nanoparticles formed by slow (dropwise) addition protocol. *Insets* show size distribution histograms. Nanoparticles were synthesised by slow (~1 mL/min) addition of (**a**) 0.2 mL, (**b**) 0.4 mL, (**c**) 2 mL, (**d**) 5 mL and (**e**) 40 mL of 0.64 mM chloroauric acid into 15 mL of 0.89 mM tannic acid, while maintaining reaction mixture pH above 6.4. The nanoparticle size distributions are (**a**) 1.7 ± 0.5 nm, (**b**) 1.8 ± 0.7, (**c**) 2.7 ± 0.8 nm, (**d**) 3.8 ± 0.6 nm and (**e**) 7.4 ± 0.6 nm. In (**f**), nanoparticles were synthesised by slow (dropwise) addition of 10 mL of 0.32 mM chloroauric acid solution into 15 mL of diluted gold colloid formed in experiment ‘I’. Diameter = 9.1 ± 0.7 nm (Reprinted from [26] under CCA terms)

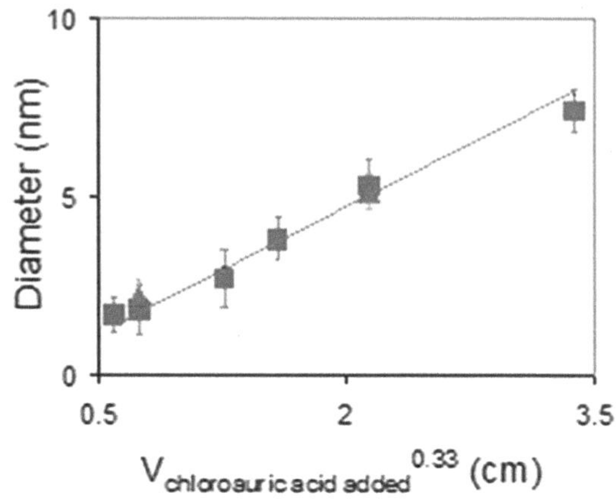

Fig. 10 A plot of expected and observed nanoparticle diameter as a function of the amount of the 1/3 power of the moles (equivalent to volume) of chloroauric acid added. The agreement between the predicted and observed diameters indicates that the slow (dropwise) addition process is akin to a one-shot nucleation-seeded growth process. *Triangle* and *square symbols* represent two different sets of experiment

is used. Thermal oxidation of citrate solution by dissolved oxygen can result in formation of acetone dicarboxylic acid and lead to faster reaction. To investigate this possibility, an experiment was performed by adding boiling citrate solution into boiling chloroauric acid. Figure 12c shows the digital images of the reaction mixture during the synthesis. The time taken for the formation of gold nanoparticles was similar to classical citrate method and it resulted in nanoparticles of size 14.0 ± 2.2 nm indicating that boiling of citrate solution does not significantly affect the synthesis.

Further experiments were performed by adding concentrated reactants into boiling water rather than diluting one of the reactant prior to reaction. Addition of citrate solution into boiling water followed by chloroauric acid resulted in faster reaction and smaller nanoparticles (8.1 ± 1.1 nm), while adding chloroauric acid into boiling water followed by citrate solution resulted in slower reaction and bigger nanoparticles (13.5 ± 2.3 nm). These experiments clearly prove that thermal degradation of citrate into acetone dicarboxylic acid, during the timescales involved in heating the reactants, does not account for the observed effects. To confirm the role of chloroauric acid reactivity, two further experiments were carried out: (1) the pH of the chloroauric acid solution was adjusted to be 3.2, similar to the classical citrate method, prior to addition into citrate, and (2) the pH of chloroauric acid solution was adjusted to 1.6 and citrate was added into it. Figure 12d–e shows that the kinetics is only affected by the initial pH of chloroauric acid solution and not by the mode of addition. However, the stability of the colloidal solution is enhanced when adding chloroauric acid at pH 1.6 into citrate solution.

Remarkably, the order of addition has a significant effect even when the observed induction periods were of the order of 100 s. An autocatalytic step in the oxidation of citrate by chloroauric acid is speculated to be responsible for this 'memory' effect, i.e. the presence of reactive chloroauric acid for short duration (5 s) during initial blending is sufficient to yield some acetone dicarboxylic acid, which then sustains the reaction.

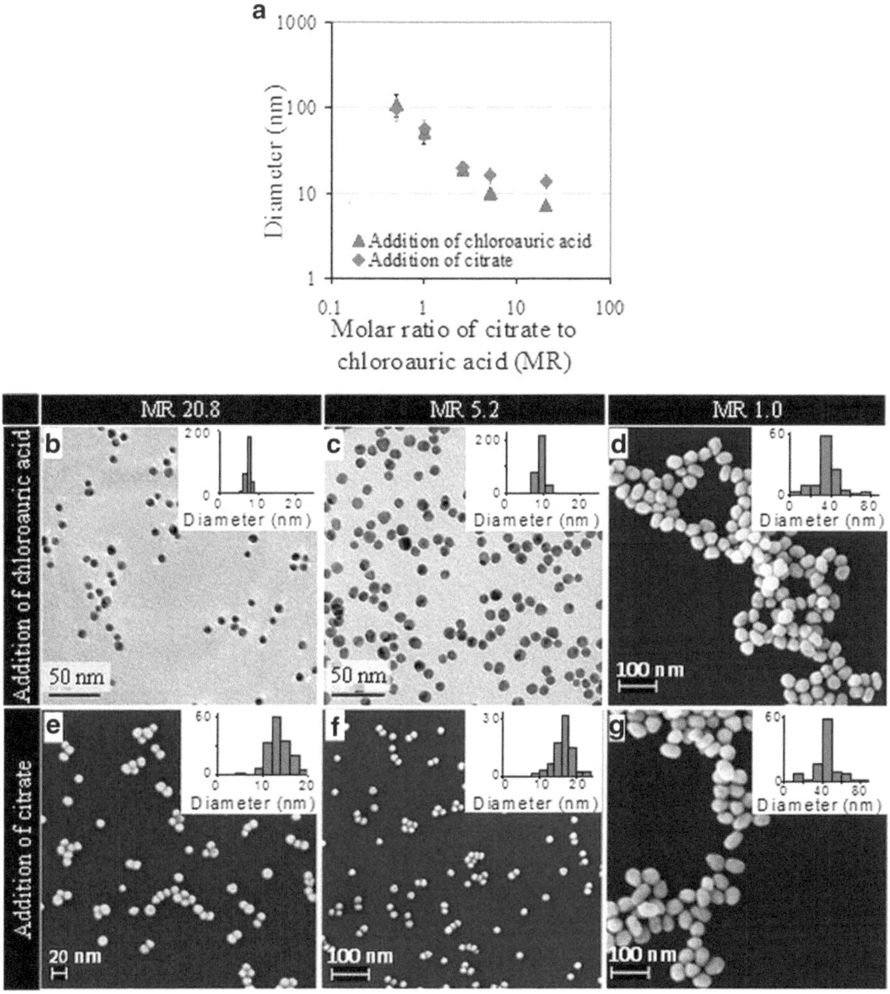

Fig. 11 Effect of reversing the order of addition of reactants. (**a**) Plot showing the variation of the mean particle size with the molar ratio for the two different modes of addition. The error bars represent one standard deviation. $HAuCl_4$ concentration was maintained constant at 25.4 mM. (**b–g**) Representative images with particle size histograms as *insets* are shown for colloids synthesised at three different molar ratios (MR) by the two different modes of addition of reactants. (**b**) MR = 20.8, gold chloride into citrate, diameter = 7.2 ± 0.8 nm. (**c**) MR = 5.2, gold chloride into citrate, diameter = 10.0 ± 1.0 nm. (**d**) MR = 1, gold chloride into citrate, diameter = 52.0 ± 15 nm. (**e**) MR = 20.8, citrate into gold chloride, diameter = 13.6 ± 2.7 nm. (**f**) MR = 5.2, citrate into gold chloride, diameter = 16.1 ± 3.3 nm. (**g**) MR = 1, citrate into gold chloride, diameter = 56.0 ± 15 nm (Reprinted from [27] with permission from Elsevier)

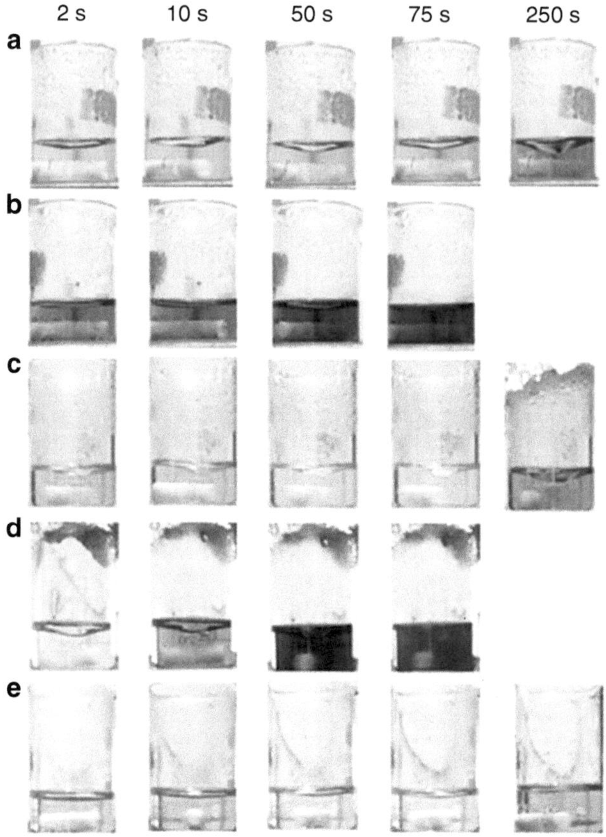

Fig. 12 Digital images of the reaction mixture taken during the course of nanoparticle synthesis. The MR value was 20.8 in all the above experiments. (**a**) Addition of citrate into boiling chloroauric acid (pH 3.2), standard Turkevich method. (**b**) Addition of chloroauric acid (pH 1.6) into boiling citrate. (**c**) Addition of citrate solution boiled for 10 min prior to addition into chloroauric acid (pH 3.2). (**d**) Addition of citrate into boiling chloroauric acid (pH 1.6). (**e**) Addition of chloroauric acid (pH 3.2) into boiling citrate (Color figure online) (Reprinted from [27] with permission from Elsevier)

5 Continuous-Flow Synthesis of Metal Nanoparticles

A primary objective of our project was to synthesise sub-10 nm metal nanoparticles in a continuous fashion using aqueous-phase processes. To achieve this, a simple coaxial flow microreactor setup was fabricated by moulding polydimethylsiloxane (PDMS) as shown in Fig. 13a. The metal salt was pumped through the inner tube, while tannic acid flowed around the core to avoid undesirable deposition and growth caused by adsorption of metal salt solutions on the reactor walls. The initial pH of the reagents was adjusted to ensure rapid reduction of metal salt by

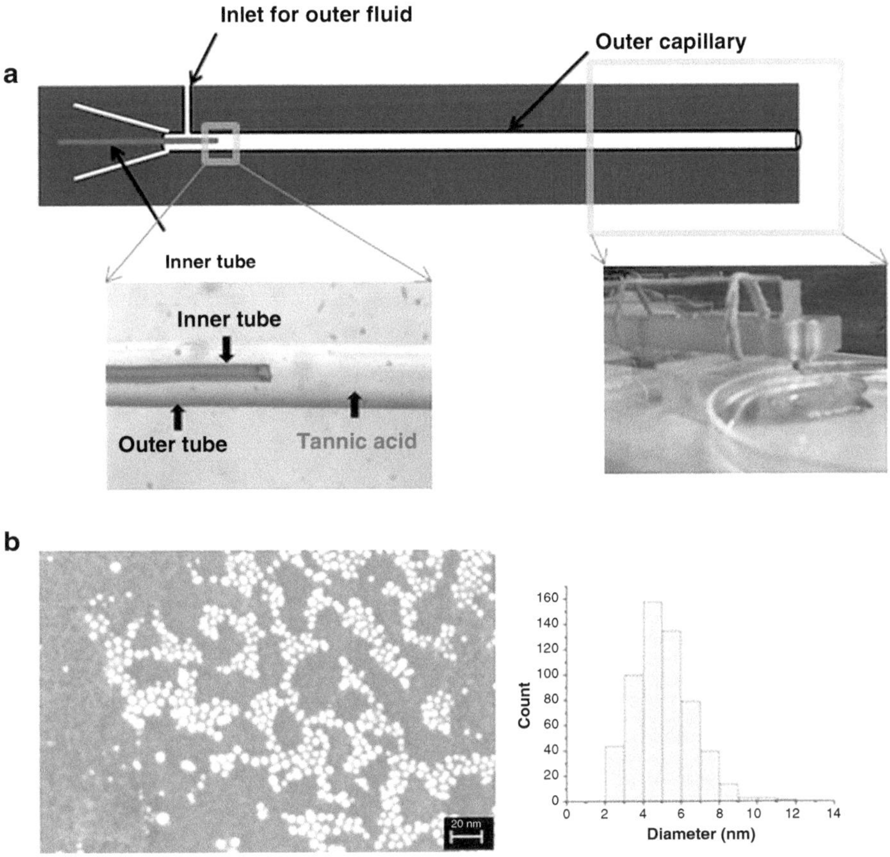

Fig. 13 (**a**) Schematic of the coaxial flow PDMS-based microchannel reactor along with digital images of the corresponding *insets*. The *yellow* colour of the collected solution reflects the formation of silver nanoparticles in this case. The inner diameter of the flow channel is 1 mm and it has a 150 μm capillary inserted coaxially at the entrance. Metal salt is introduced through the inner capillary, while tannic acid flows outside. (**b**) Representative SEM image of gold nanoparticles synthesised using the coaxial flow microchannel reactor and a histogram representing the corresponding size distribution (Color figure online)

tannic acid molecules. Complete conversion of the metal salt into silver or gold nanoparticles was found to occur for residence times of the order of 10 s, under the conditions of high reactivity in tune with earlier kinetic measurements. Interestingly, the nanoparticle size distribution was found to be similar to batch experiments at steady-state conditions (Fig. 13b) and was found to be insensitive to variations in flow velocities.

Microchannel flows can provide uniform and repeatable heat and mass transport conditions, which is purported to help in enhancing the reproducibility of nanoparticle synthesis; however, the widespread use and 'numbering-up' of microchannels

Fig. 14 Photograph of a novel flow reactor for metal nanoparticle synthesis. The channels were formed by cutting 2 mm wide × 2 mm deep slots into a PMMA sheet. The *blue colour* is due to the dye used for visualisation of the flow pattern in the reactor. During nanoparticle synthesis, tannic acid flows through the channels, while metal salt is added dropwise at selected points using syringes. A representative FESEM image of gold nanoparticles synthesised using the reactor is shown. The throughput in terms of gold was 1 g/h, and the average size of the gold nanoparticles was ~10 nm (Color figure online)

are hindered by large pressure drops due to small channel sizes, leading to very high power requirements for pumping the reagents. As our earlier results suggested that speed of mixing and flow conditions had very little effect on the nanoparticle size distribution, as long as their reactivity was tuned by choosing appropriate feed conditions, we designed a simple gravity-fed open channel reactor with multiple feed inlets to mimic the dropwise addition protocol for synthesising monodisperse gold nanoparticles (Fig. 14). Salient features of our process design are the very high throughput (g/h) vis-à-vis typical throughput values of microchannel reactors (mg/h) and the negligible power requirements for pumping liquids. The particle size distribution was found to be consistent with earlier batch experiments at steady-state conditions.

6 Reactive Inkjet Deposition of Metal Nanoparticles

Paper-based electronics is an emerging field with applications focused on low-cost electronic circuits for human as well as structural diagnostics, ubiquitous sensor networks, etc. A key process in the formation of such circuits is the fabrication of patterned, conductive features on paper, preferably using low-cost roll-to-roll compatible processes. In this context, inkjet printing is one of the foremost contenders for fabricating patterned features on paper. Metal nanoparticle or carbon-based inks are typically used to fabricate patterned conductive features on paper using inkjet printing tools. However, the complex rheology and very high particulate content required to form conductive features require the use of sophisticated inkjet deposition tools, which hinder their adaptation for low-cost device fabrication.

The core-flow geometry for microchannel reactors was adapted after preliminary experiments in a Y-channel micromixer led to undesirable deposition on reactor

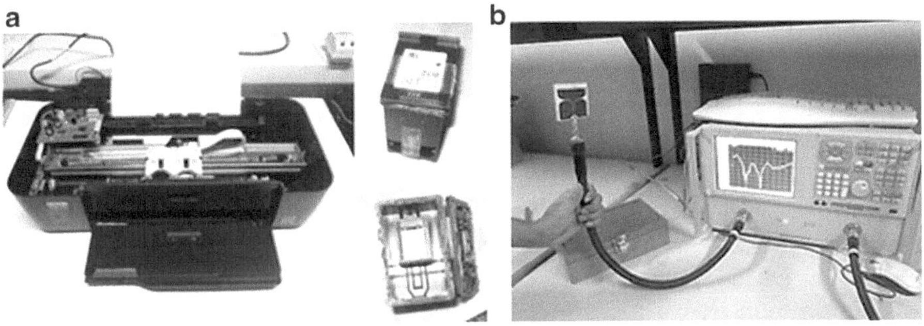

Fig. 15 (**a**) Photograph of the HP Deskjet 1000 printer used along with cartridges used for reactive inkjet printing. (**b**) Digital photographs of performance characterisation of UWB antenna fabricated on paper by using reactive inkjet printing and electroless deposition. The electrical performance of such paper-based antennas was found to be equivalent to that of antennas having the same design but fabricated using lithographic techniques on conventional RF substrates

walls by reduction of adsorbed metal salts. Such deposits were seen to form within seconds and also grow with extended exposure to the reagents. Although undesirable during bulk synthesis, the sturdy deposits formed rapidly, suggesting a simple route to fabricate conductive features on paper using office inkjet printers. We used an HP Deskjet 1000 printer and used reagent-filled ink cartridges to deposit silver nitrate and alkaline tannic acid solution sequentially on paper; desired patterns were generated using a standard computer interface. Silver nanoparticles formed *in situ* on the printed portions of the paper within seconds. This process represents a high-throughput method for synthesis and incorporation of metal nanoparticles into functional products. These silver nanoparticle patterns were further used as catalysts to generate conductive copper features on paper, using electroless plating. As a proof of concept, an ultrawideband (UWB) antenna was fabricated based on design from the literature (Fig. 15). The performance of the paper-based antenna was found to be equivalent to that reported for a similar antenna fabricated using conventional lithographic techniques [29].

7 Summary

A rapid, green, room temperature process for the synthesis of 2–20 nm gold and silver nanoparticles was developed. By simply altering the pH of the reagent solutions and modifying the mode of addition of reagents, monodisperse gold and silver nanoparticles were synthesised using tannic acid as the reducing and stabilising agent. An organiser-facilitated nucleation model was found to be suitable for describing the kinetics of nanoparticle formation. Similarly, reversing the order of addition in the widely used citrate method was found to result in highly monodisperse sub-10 nm gold nanoparticles. These results represent a significant advance in our knowledge of metal nanoparticle formation, especially considering the fact that these recipes have been widely studied for more than 100 years. Building upon this knowledge, high-throughput continuous-flow processes for the synthesis of gold and silver nanoparticles have been developed. Our results show that sub-10 nm gold and silver nanoparticles can be produced in continuous-flow reactors operated at room temperature. Furthermore, a low-cost process for the fabrication of conductive patterns on paper using an office inkjet printer was demonstrated. This process will be very useful for producing low-cost RFID tags and biodiagnostic devices on paper. Further investigations are required to understand the influence of flow patterns on nanoparticle size distribution for the continuous-flow reactor developed in this project.

Acknowledgements We gratefully acknowledge support from IRHPA scheme of DST. We also acknowledge the inputs of our students, project assistants and collaborators over the last 6 years. Excerpts have been reprinted with permission from Elsevier and Indian Academy of Sciences.

References

1. Pitkethly MJ (2004) Nanomaterials–the driving force. Mater Today 7:20–29
2. Ghosh P, Han G, De M, Kim CK, Rotello VM (2008) Gold nanoparticles in delivery applications. Adv Drug Deliv Rev 60:1307–1315
3. Muralidharan G, Bhat N, Santhanam V (2011) Scalable processes for fabricating non-volatile memory devices using self-assembled 2D arrays of gold nanoparticles as charge storage nodes. Nanoscale 3:4575–4579
4. Brown C, Bushell G, Whitehouse M, Agrawal DS, Tupe SG, Paknikar KM, Tiekink E (2007) Nanogoldpharmaceutics. Gold Bull 40:245–250
5. Wilson R (2008) The use of gold nanoparticles in diagnostics and detection. Chem Soc Rev 37:2028–2045
6. Santhanam V, Andres RP (2009) Metal nanoparticles: self-assembly into electronic nanostructures. In: Contescu CI, Putyera K (eds) Dekker encyclopedia of nanoscience and nanotechnology, 2nd edn. CRC Press, Boca Raton, pp 2079–2090
7. Feldheim DL, Colby AF Jr (2001) Metal nanoparticles: synthesis, characterization, and applications. CRC Press, New York
8. Thompson D (2007) Michael Faraday's recognition of ruby gold: the birth of modern nanotechnology. Gold Bull 40:267–269
9. Weiser HB (1933) Inorganic colloid chemistry. Wiley, New York
10. Slot JW, Geuze HJ (1985) A new method of preparing gold probes for multiple-labeling cytochemistry. Eur J Cell Biol 38:87–93
11. Frens G (1973) Controlled nucleation for the regulation of the particle size in monodisperse gold suspensions. Nat Phys Sci 241:20–22
12. Turkevich J (1951) A study of the nucleation and growth processes in the synthesis of colloidal gold. Discuss Faraday Soc 11:55–75
13. Perala SRK, Kumar S (2013) On the mechanism of nanoparticle synthesis in Brust-Schiffrin method. Langmuir 29:9863–9873
14. Hutchison JE (2008) Greener nanoscience: a proactive approach to advancing applications and reducing implications of nanotechnology. ACS Nano 2:395–402
15. Dykman LA, Bogatyrev VA (2007) Gold nanoparticles: preparation, functionalisation and applications in biochemistry and immunochemistry. Russ Chem Rev 76:181–194

16. Ostwald CWW (1917) An introduction to theoretical and applied colloid chemistry. Wiley, New York
17. Mühlpfordt H (1982) The preparation of colloidal gold particles using tannic acid as an additional reducing agent. Experientia 38:1127–1128
18. Bulut E, Özacar M (2009) Rapid, facile synthesis of silver nanostructure using hydrolyzable tannin. Ind Eng Chem Res 48:5686–5690
19. Tian X, Wang W, Cao G (2007) A facile aqueous-phase route for the synthesis of silver nanoplates. Mater Lett 61:130–133
20. Sivaraman SK, Elango I, Kumar S, Santhanam V (2009) A green protocol for room temperature synthesis of silver nanoparticles in seconds. Curr Sci India 97:1055–1059
21. Cruz BH, Díaz-Cruz JM, Ariño C, Esteban M (2000) Heavy metal binding by tannic acid: a voltammetric study. Electroanalysis 12:1130–1137
22. Bors W, Foo LY, Hertkorn N, Michel C, Stettmaier K (2001) Chemical studies of proanthocyanidins and hydrolyzable tannins. Antioxid Redox Signal 3:995–1008
23. Martinez-Castanon G, Nino-Martinez N, Martinez-Gutierrez F, Martinez-Mendoza J, Ruiz F (2008) Synthesis and antibacterial activity of silver nanoparticles with different sizes. J Nanopart Res 10:1343–1348
24. Liu J, Qin G, Raveendran P, Ikushima Y (2006) Facile "green" synthesis, characterization, and catalytic function of β-D-glucose-stabilized au nanocrystals. Chem Eur J 12:2131–2138
25. Belloni J (2006) Nucleation, growth and properties of nanoclusters studied by radiation chemistry: application to catalysis. Catal Today 113:141–156
26. Sivaraman SK, Kumar S, Santhanam V (2010) A room temperature synthesis of gold nanoparticles: size control by slow addition. Gold Bull 43:275–286
27. Sivaraman SK, Kumar S, Santhanam V (2011) Monodisperse sub-10 nm gold nanoparticles by reversing the order of addition in Turkevich method – the role of chloroauric acid. J Colloid Interface Sci 361:543–547
28. Kumar S, Gandhi KS, Kumar R (2007) Modeling of formation of gold nanoparticles by citrate method. Ind Eng Chem Res 46:3128–3136
29. Kumar S, Bhat V, Vinoy KJ, Santhanam V (2013) Using an office inkjet printer to define the formation of copper films on paper. IEEE Trans Nanotechnol 13:160–164

Carbon-Based Hierarchical Micro- and Nanostructures: From Synthesis to Applications

Chandra S. Sharma and Ashutosh Sharma

Abstract We review various methodologies as developed recently in our group for the production of carbon xerogel particles with a wide variety of morphologies (from spherical to fractal-like) in the size range of micro- to nanoscale. To name a few are sol–gel emulsification, electrospraying, electrospinning, and chemical vapor deposition. The role of various process parameters is studied in length to achieve a fine tuning and control on the size and morphologies of carbon structures. A large number of polymer precursors such as organic xerogel, photoresist materials, and polymers are employed as a source of carbon. Other than conventional photolithography, soft lithography and biomimicking approaches are used to fabricate micropatterned carbon surfaces which are further used to fabricate hierarchical carbon structures by combining top-down, bottom-up, and self-assembly processes. Thus, fabricated hierarchical carbon structures due to their unique properties such as controllable wettability, high surface area, and biocompatibility open up new possibilities in the area of carbon-based microelectrochemical systems, microfluidics, biosensors, and environmental pollution control. A more insight about some of these applications is presented in this work.

Keywords Carbon MEMS • Resorcinol–formaldehyde (RF) xerogel • Electrospun nanofibers • SU-8 photoresist • Hierarchical carbon structures

---

C.S. Sharma
Department of Chemical Engineering, Indian Institute of Technology, Hyderabad, Yeddumailaram, 502025 Telangana, India

A. Sharma (✉)
Department of Chemical Engineering and DST Unit on Soft Nanofabrication, Indian Institute of Technology, Kanpur, 208016 UP, India
e-mail: ashutos@iitk.ac.in

Y.M. Joshi, S. Khandekar (eds.), *Nanoscale and Microscale Phenomena*,
Springer Tracts in Mechanical Engineering, DOI 10.1007/978-81-322-2289-7_5

1 Introduction

Carbon, because of its unique properties of bonding chemically, is one of the most versatile elements in the periodic chart. There are a large number of allotropic forms of carbon that exist in the shape of naturally taking place diamond, graphite, nanotubes, fullerenes (C60), and so on. Furthermore, there is one more classification of the materials of carbon called "amorphous carbon" that does not reveal any order of the extensive range or crystallinity. Although, this amorphous carbon takes in two additional categories that include glassy carbon and diamond-like carbon derived from numerous percentage of the crystallites of diamond and graphite. In this report, the emphasis is more on the glassy carbon that is utilized in a number of applications such as in the carbon-based microelectromechanical systems (C-MEMS). Although most micro- and nanoelectromechanical (MEMS/NEMS) structures and devices are still currently based on silicon (Si) substrates as silicon processes are very mature compared to processes with any other material, however, glassy carbon can be a very important contender for miniaturization of future solid-state devices. We believe that C-MEMS and carbon nanoelectromechanical systems (C-NEMS), with sizes varying from the millimeter and micrometer to the nanometer, will offer solution, alone or in combination with Si and other organic, inorganic, and biological materials, in micro- and nanoelectronics, sensing devices, miniaturized power systems, etc.

The glassy carbon that has characteristics such as resistance to high temperature and a lower concentration is a hard (non-graphitizing) form of carbon [1–7]. In general, this glassy carbon is created in the form of bulk by heat treatment of polymers, also known as "polymeric predecessors" to carbon. These polymeric precursors include coal, wood, organic compounds, lignite, and so on. When these precursors are heated up in a static atmosphere in excess of 800 °C, the noncarbon substance of these precursors escapes in gaseous form and yields the pyrolytic carbon [7]. This procedure is known as carbonization or pyrolysis. Various polymers, for instance, phenolic resins, polyvinyl alcohol, polyacrylonitrile (PAN), epoxy resins, coal tar pitch, polyamides, and so forth, have been employed to manufacture the pyrolytic carbon by the help of the process of carbonization [7]. With the exception of the synthesis of bulk, thin film of the materials of carbon could be formulated by a number of other techniques, for example, the chemical deposition of vapor, the evaporation of electron beam, spluttering, and the polymeric thin films' pyrolysis.

Relying on the chemical configuration of the precursor of polymer and the conditions of heating, the materials of carbon with extensive variety of characteristics could be amalgamated. The carbon yields of numerous precursors of polymer have been reported by Jenkins et al. [7], whereas different attributes of a lot of materials of carbon were recapitulated by McCreery [4, 5], as these carbon materials were utilized as the materials of electrode. Because the glassy carbon takes in little resistance to electricity and is secure chemically and thermally as well, it is extensively utilized in electrochemistry [1–6].

In this work, we consider a large number of polymer precursors such as resorcinol–formaldehyde (RF)-based organic xerogel [8–10], epoxy-based negative photoresist, SU-8 [11–16], and polyacrylonitrile (PAN) along with employing a wide variety of synthesis methods including sol–gel emulsion [9, 17, 18], electrospinning and electrospraying [19–30], and chemical vapor deposition (CVD) [31–36] to yield carbon structures from micro- to nanoscale. Sol–gel polycondensation followed by inverse emulsion polymerization of RF gel was used to obtain the desired carbon particle's size and morphology, achieved by optimizing sol concentration, stirring time, type and amount of surfactants, and shearing conditions. A wide spectrum of morphologies from microspheres to highly folded, needlelike, branched, fractal-like carbon microstructures could be prepared [17, 18]. We further demonstrated the versatility of electrospinning process in synthesizing a wide variety of carbon nanostructures including the nanoparticles and nanofibers with tunable morphology [29, 30]. RF sol precursor chemistry was combined with electrospraying, a variation of electrospinning process, to produce nearly monodisperse carbon nanospheres. Similarly, a negative photoresist SU-8 that is commonly used in conventional photolithography was electrospun followed by pyrolysis to produce carbon nanostructures. Interestingly, the size and morphology of these carbon nanostructures could remarkably be changed by controlling the process variable, e.g., needle gauge opening, applied electric field, and polymer viscosity and its flow rate.

We developed some interesting ways to synthesize hierarchical carbon nanostructures which essentially possess higher surface area as compared to other distinct carbon nanostructures. In one of such attempts, CVD was combined with electrospinning to grow carbon nanofibers (average diameter ~30 nm) by CVD using metal nanoparticles as catalyst over the electrospun carbon nanofibers (average diameter ~150 nm) to fabricate hierarchical assemblies of carbon nanofibers. In another study, a facile way to obtain hierarchical fabric has been demonstrated by depositing PAN-based electrospun nanofibers on activated carbon fibers (ACFs) used as a support [37]. The wide variety of carbon micro- and nanostructures including the hierarchical assemblies synthesized in this work may have long-ranging potential applications, for example, filtration systems, composites, sensing devices, energy storage devices, and biomedical applications like bio-MEMS and tissue scaffolds. The development of photoresist-derived carbon nanostructures with assorted fiber morphologies can be particularly suitable in integration with photoresist-derived high-aspect-ratio C-MEMS to fabricate hierarchical structures that can be used as electrodes for energy storage devices.

Finally, we demonstrate that these carbon micro-, nano-, and multiscale structures have several unique properties such as controllable wettability [29, 30, 38], biocompatibility [39], and high specific surface area [37]. In the last section, we review the applications of these hierarchical carbon micro- and nanostructures by exploiting some of these properties.

2 Synthesis of Hierarchical Carbon Structures (From Micro- to Nano)

2.1 Synthesis of Carbon Xerogel Microstructures Using Sol–Gel Emulsion

As organic gels possess unique physical, chemical, and electrochemical properties, they are studied in detail from the past two decades [8–10, 17, 18, 40–44]. Depending on the drying methods used in synthesis, different kinds of organic gels (aerogel, cryogel, and xerogel) are produced. RF sol has been widely used as a precursor to synthesize carbon aerogels [8]; however, other precursors and solvents have also been used recently [45, 46]. The sol–gel process irrespective of precursor provides fine control over size, composition, and structural characteristics of the finishing materials. To choose the carbon gels for a specific application, surface morphology of these carbon gels plays an important role. A wide range of morphologies of these carbon gels, e.g., porous texture, microspheres, thin film, granule, honeycomb, etc., have been reported in literature [8–10, 17, 18, 40–46].

It is important to know that carbon aerogel and cryogel structures possess high porosity and are near spherical in shape as reported in literature [8, 9, 40]. However, organic xerogels [10, 17, 18, 43, 44] which are formed by evaporating the solvent using simple air oven drying have nearly nonporous structures. Further there are not many reports available in literature on carbon xerogels. During xerogel synthesis, the liquid inside the pore exhibits the capillary pressure that guides collapsing the gel network while drying in an oven resulting in the xerogel, a dense polymeric structure [9, 17, 18]. This xerogel is then heated at high temperature in the presence of inert environment (pyrolysis) to yield the carbogel. Recently, Job et al. [41, 42] and Matos et al. [46] have also described the synthesis of the porous carbon xerogels. Much of the work in this area is focused on synthesis of spherical particles; however, high external surface area carbon xerogels having dendritic and fractal-like morphologies have not been reported. A possible reason for that is because in almost all studies, low surfactant concentration was employed during the inverse emulsification before drying.

To produce RF xerogel-based carbon microparticles with different morphologies from spherical to high surface area fractal-like, we developed several ways for the tuning of particle shape and morphology. In particular, the effect of stirring conditions (time of stirring, methods of stirring), RF sol composition (catalyst ratio, dilution ratio), and types and amount of surfactants used during the inverse emulsion polymerization were investigated [17, 18].

To synthesize RF hydrogels, polycondensation of resorcinol (R) took place with formaldehyde (F) in aqueous medium (W), with the addition of a basic catalyst (C) [40]. The experimental details are given elsewhere [17, 18]. RF sol was added as spherical droplets in the cyclohexane in the presence of a nonionic surfactant to synthesize RF xerogel microspheres using inverse emulsion polymerization. The time of stirring and the amount of surfactant decided the size and shape of droplets

Table 1 Characteristic sizes of RF carbon xerogel particles synthesized at different *R*/*C* ratios with 5 h stirring time and 1 % (v/v) Span-80 concentration

R/*C* ratio	Mean particle diameter (μm)	Standard deviation
0.2	0.90	0.1
1	3.8	1.1
25	18.1	6.8
100	57.9	19.7

Reproduced with the permission by Sharma et al. [17]

in the emulsion. The colloidal solution of aqueous RF sol was formed by continuous stirring for a given time. The suspension of RF hydrogel particles on the substrate (silicon wafer) was quickly desiccated in order to arrest the movement of particles by partially evaporating the solvent and thereby measuring the particle size. Later, samples were dried in simple air oven by heating at 333 K for 12 h to yield the RF xerogel particles. Once dried, silicon wafer with the RF xerogel particles was laden in a quartz boat and then placed horizontally inside the high-temperature tubular furnace. Nitrogen gas was purged to ensure that there is no oxygen present inside the furnace. Furnace was then programmed to carry out the pyrolysis to yield finally carbon particles.

2.1.1 Effect of Catalyst Concentration

For the polymerization reaction between resorcinol and formaldehyde, potassium carbonate was chosen as a basic catalyst. Use of the catalyst guides the formation of small aggregates of cross-linked polymer. These small aggregates later aggregated and linked together during gelation. To examine the effect of catalyst ratio on final size distribution of particles, we varied the R/C molar ratio from 0.2 to 500. Table 1 summarizes the results of catalyst effect on the average particle diameter.

At very low *R*/*C* molar ratio (high catalyst concentration), small particles with average diameter varying from few hundred nanometers to few microns were obtained [17]. The mean size of the particles increased to ~58 μm with increase in the *R*/*C* ratio to 100. Further increase in *R*/*C* ratios ($R/C > 100$) distorted the spherical morphology of the particles. It was further noted that the particles' affinity to form aggregates increased with the increase in stirring time and with the *R*/*C* ratio as illustrated in Fig. 1 [17].

For $R/C = 10$, increase in the time of stirring from 5 to 24 h does not affect the aggregation significantly (Fig. 1a, b). However, at higher R/C ratios, it was clearly observed that initially the particles lose their identity followed by coalescence to form a fractal-like layered structure with an increase in stirring time to 24 h (Fig. 1c, d).

2.1.2 Effect of Dilution Ratio

Dilution ratio (*R*/*W*) is another variable in the sol–gel process that may influence the particle size distribution as well as shape. To examine the effect of this variable

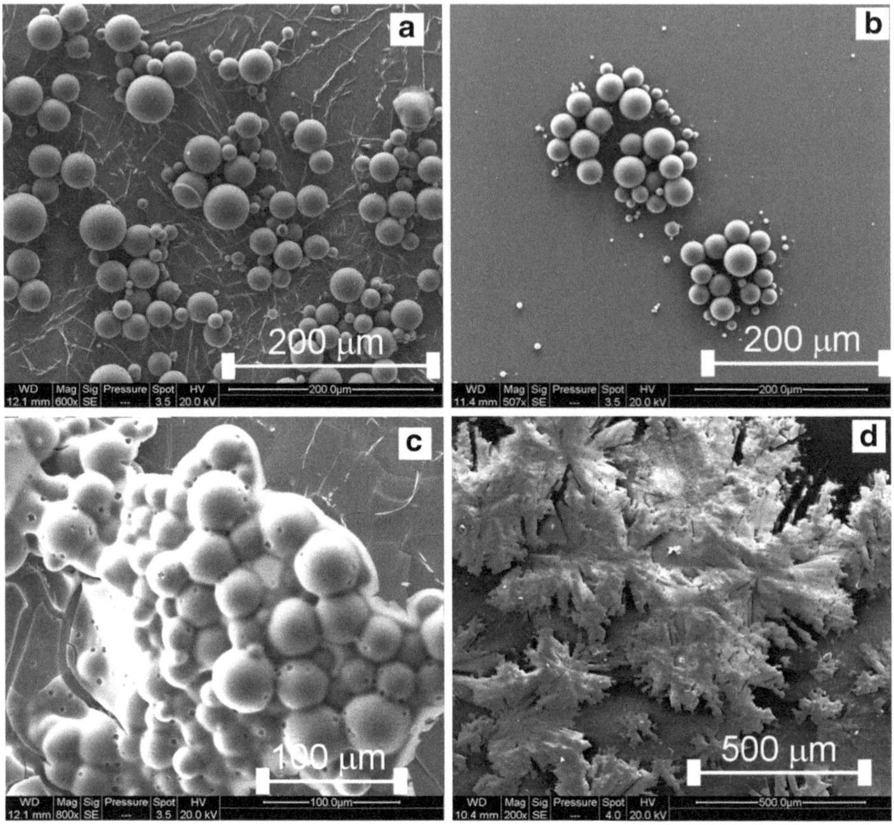

Fig. 1 SEM images showing aggregation and coalescence of RF-based carbon microspheres at Span-80 concentration 1 % (v/v): (**a**) $R/C = 10$, stirring time = 5 h; (**b**) $R/C = 10$, stirring time = 24 h; (**c**) $R/C = 25$, stirring time = 24 h; (**d**) $R/C = 500$, stirring time = 24 h (Reproduced with the permission by Sharma et al. [17])

on RF-based carbon xerogel particles, dilution ratio was varied from 0.0037 to 3.7 while preparing the RF sol. *R*/*C* molar ratio was fixed to be 25 for all the cases. RF sol thus prepared was then added in the cyclohexane to form microemulsion droplets which were continuously stirred for 5 h by magnetic stirrer.

It was found that with the increase in water concentration (which means decrease in *R*/*W* ratio from 3.7 to 0.0037), time for gelation of RF sol increased from about 4 h to 20 days. This clearly shows that the RF sol viscosity increases slowly which signifies that more time is required for sol to cross-link and thus to form a gel. With this increased gelation time or gradual increase in viscosity prior to the deferred onset of gelation time, a spherical particle deforms to elongated irregular shapes as shown in Fig. 2a, b. It is also consequently observed that particle shape does no longer remain spherical but rather distorted at such higher water concentration

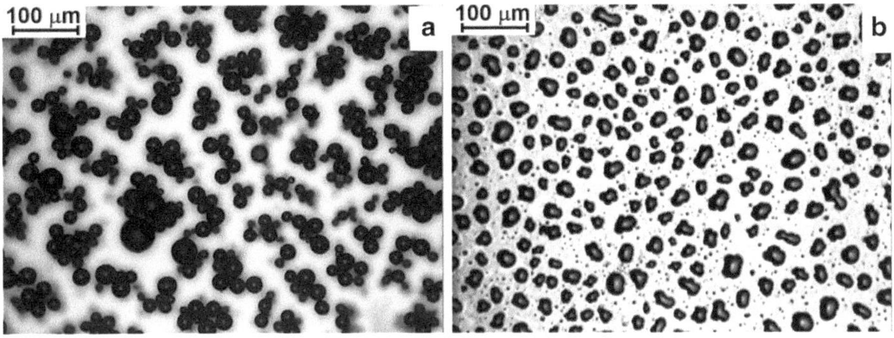

Fig. 2 Optical micrographs showing the transition to nonspherical shapes of RF-derived carbon xerogel microparticles with the change in dilution ratio: (**a**) $R/W = 0.037$, (**b**) $R/W = 0.0037$. Other conditions are $R/C = 25$, surfactant concentration = 1% (v/v), and stirring time = 5 h for both cases (Reproduced with the permission by Sharma et al. [18])

Table 2 Characteristic sizes of RF carbon xerogel particles synthesized at different *R/W* ratios with 5 h stirring time and 1 % (v/v) Span-80 concentration

R/W ratio	Mean particle diameter, d_{av} (μm)	Standard deviation (μm)
0.0037	27.9	11.3
0.037	17.2	4.9
0.37	11.8	4.5
3.7	1.4	0.6

Reproduced with the permission by Sharma et al. [18]

($R/W = 0.0037$). As a result of this, the change in dilution ratio finally affected the RF-derived carbon particle size. These results are summarized below in Table 2.

As a summary, by increasing the dilution ratio from 0.0037 to 3.7 (three orders of magnitude), average diameter of the RF-derived carbon particles (d_{av}) decreased from 28 μm to nearly 1 μm [18].

2.1.3 Effect of Surfactant Concentration

As discussed in the previous section, the average diameter and morphology of particles and their aggregates can be controlled by changing the RF sol composition and concentration. However, it was found that depending on the RF sol concentration, viscosity of the RF sol increases sharply at the start of gelation. For that reason, it is not easy to tailor the average size and surface morphologies of the final carbon structures simply by varying the RF sol composition and concentration. Therefore, a supplementary controlling mechanism was desired that would tender more flexibility in the engineering of carbon particles with different shape and size [17]. In this context, we discuss how modulating the interfacial tension by varying the composition of the continuous phase during the inverse emulsification affects

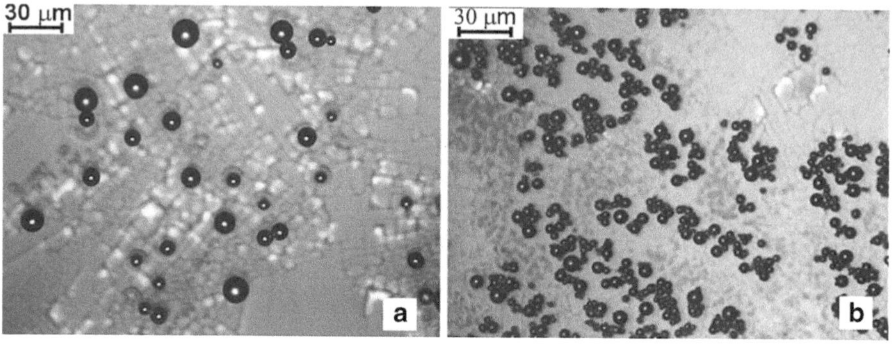

Fig. 3 Optical micrographs showing the effect of surfactant concentration on particle size at $R/C = 25$, stirring time 5 h in "low-surfactant-concentration region": (**a**) 1 % (v/v) Span-80; (**b**) 4% (v/v) Span-80 (Reproduced with the permission by Sharma et al. [17])

the size and morphology. This is in contrast to what has been discussed in the above Sects. 2.1.1 and 2.1.2 where properties of the dispersed phase were altered by varying the amount of catalyst and water concentration. The nonionic surfactant (Span-80) in varying concentration was employed to modify the properties of the oil–water interface and thereby of continuous phase [17]. For Span-80, HLB value is 4.3 and it is hydrophobic in nature.

The amount of surfactant added to cyclohexane was varied to change the concentration from 1 to 50 % (v/v). In low-surfactant-concentration regime, the average drop size shows a strong dependence on the surfactant concentration [17, 18]. For a given stirring time of 5 h in the "low-surfactant regime," the average size of RF sol-based particles was reduced from 10 to 5 μm with the increase in the amount of surfactant from 1 to 4 % (v/v) as shown in Fig. 3a, b. It could also be observed from Fig. 3b that the number density of the particles increased with increase in concentration, i.e., 4 % (v/v).

With further increase in the surfactant concentration inverted micelles aggregated considerably to decide the final shape of micelle. Beyond some minimum aggregation number, the morphology was distorted and no longer remains spherical. In case of large aggregation numbers, micellar shape may be envisage to be oblate, vesicles, and bilayers [47]. With further increase in aggregation number, thus distorted oval shapes finally transformed to fibrous, layered-like structures shown in Fig. 4a [17]. On further increasing the surfactant concentration up to 33–50 % (v/v) and by also varying the stirring time, high surface area carbon xerogel structures with features resembling sharp needles were obtained [17], as shown in Fig. 4b–l. At such a high micellar concentration, the lamellar crystalline phase was configured due to formation of ordered structures to arrangement of surfactant molecules [48]. Consequently the stirring-induced shear allowed the folding and subsequently breaking of the large bilayer sheets of the lamellar phase to yield the forked and folded fractal-like structures as shown here in Fig. 4b–l.

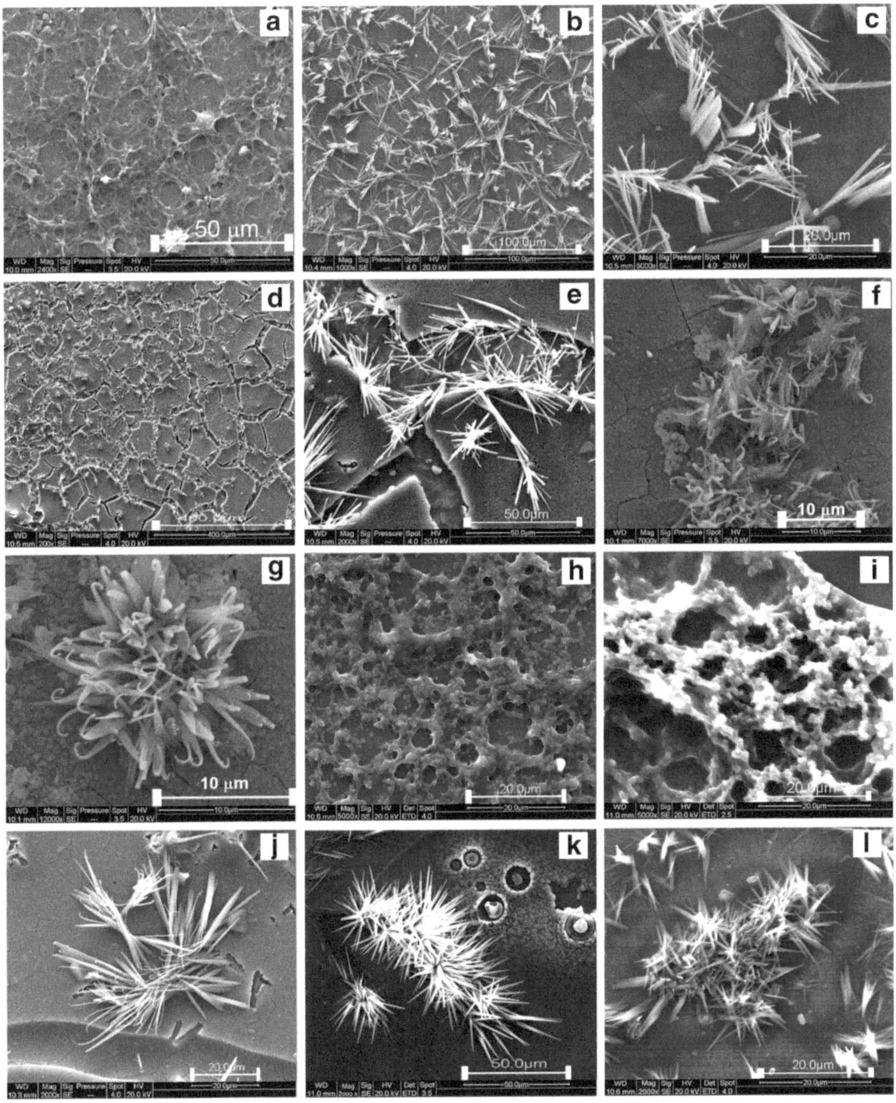

Fig. 4 SEM images of carbon "bush"- and "flower"-type fractal-like structures obtained at high surfactant concentrations: (**a**) 16 % Span-80, 2 h stirring; (**b**) 16 % Span-80, 5 h stirring; (**c**) 33 % Span-80, 1 h stirring; (**d**) 33 % Span-80, 2 h stirring; (**e**) high-magnification image of (**d**); (**f**) 50 % Span-80, 1 h stirring; (**g**) 50 % Span-80, 2 h stirring. Images (**a–g**) are obtained using the surfactant from Loba Chemie at $R/C = 25$. Images (**h**) and (**i**) are obtained at $R/C = 25$, 2 h stirring at 33% Span-80 using the surfactant from Sigma Aldrich and SD Fine Chemical, respectively. Images (**j–l**) are obtained at $R/C = 10$, 2 h stirring with 33 % Span-80 using the surfactant from Loba Chemie, Sigma Aldrich, and SD Fine Chemical, respectively (Reproduced with the permission by Sharma et al. [17])

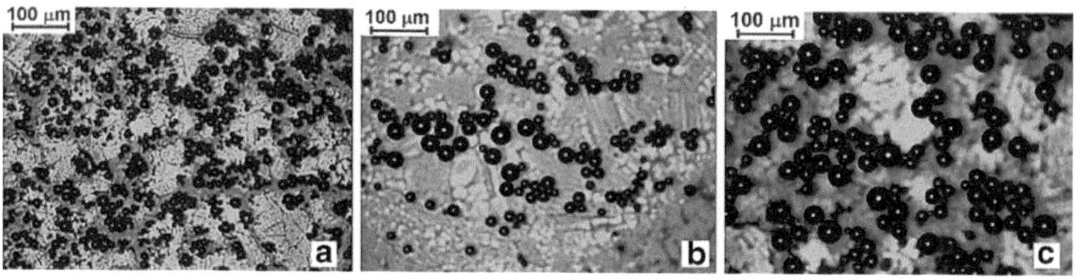

Fig. 5 Optical micrographs of the RF-based carbon microspheres obtained at $R/C = 25$, 1 % (v/v) Span-80 concentration with different stirring times: (**a**) 2 h, (**b**) 5 h, (**c**) 7 h (Reproduced with the permission by Sharma et al. [17])

Some of these carbon xerogel structures as shown in Fig. 4b–e are symmetrically branched over an order of magnitude length scale. Measuring the fractal dimension may be a mathematical way to quantify some of these structures. In this case, fractal dimension as measured was found to be in the range of 1.60 ± 0.10. Some of these aggregated needlelike structures (Fig. 4f, g) were highly folded and resembled such as high external surface area "carbon flower"-type structures.

2.1.4 Effect of Stirring Time

The optical micrographs in Fig. 5 show the dependency of RF-based carbon xerogel particle size with the stirring time of RF sol emulsion in the continuous phase. With the increase in the stirring time of emulsion from 2 to 7 h, the average diameter of the RF-based carbon xerogel particles increased from 5 to 46 μm [17]. However, the shape of the particles remains unchanged to spherical. For a stirring time of 2 h (Fig. 5a), almost all the particles were found to be having a mean diameter in the range of 5–15 μm and forming very dense clusters. As the stirring time increased, the size of the particles increased; however, the number density of the particles decreased as shown below in Fig. 5b, c. Figure 5b shows that the particle size ranges from 10 to 35 μm, while Fig. 5c shows particles with their diameter from 12 to 46 μm range.

2.1.5 Effect of Stirring Methods

The turbulence and shear flow caused by the stirring affects the dynamic equilibrium between the two opposing mechanisms of drop breakage and coalescence [49–51]. Stirring is one of the key aspects in determining the particle morphology as it modulates the flow field created in an emulsion. In this work, we employed three different stirring methods, (1) magnetic stirring, (2) probe ultrasonication, and (3) mechanical shaker.

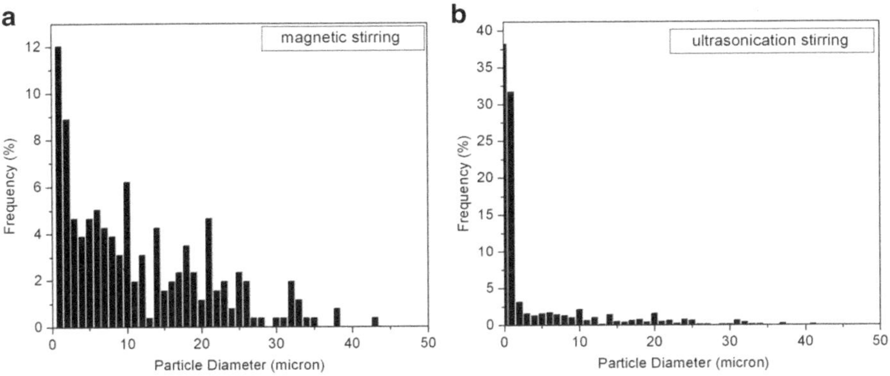

Fig. 6 Effect of method of stirring on particle size distributions of RF gel-derived carbon microparticles in low-surfactant region: (**a**) magnetic, (**b**) ultrasonication. Other conditions are $R/C = 25$, $R/W = 0.037$, surfactant concentration = 1% (v/v), and stirring time = 5 h for both the cases (Reproduced with the permission by Sharma et al. [18])

The size distributions of RF xerogel-based carbon particles produced using magnetic stirrer and ultrasonication for 5 h are shown in Fig. 6. The concentration of nonionic surfactant (Span-80) in each case was maintained to be 1 % (v/v). While comparing Fig. 6a, b, we observed that in case of ultrasonication stirring, smaller and more uniform particles with average diameter, $d_{av} = 4.79$ μm, were obtained as compared to magnetic stirring in which d_{av} was measured to be 2.31 μm. Thus, ultrasonication facilitates the size reduction due to its ability of intense stirring and thereby enhanced emulsifying effects.

Mechanical shaking did not provide sufficient shear and thus was not effective to have stabilized emulsion in comparison to ultrasonication and magnetic stirring. The unidirectional axial movement of the shaker plate was not sufficient to provide adequate shear which resulted in macroaggregation of the droplets in the emulsion at shaking speed as high as 180 strokes/min. The stroke length was 25 mm. With almost 3 h of stirring, emulsion destabilized forming a cluster of aggregated particles. As mechanical shaker was unable to agitate the emulsion sufficiently, it resulted in the suppression of droplets breakage. Thus, coalescence of RF sol droplets dominates to form the lumps and hence settling down in the solution.

As far as particle shape was concerned, there was no significant change for the magnetic and ultrasonication stirring in low-surfactant region. On the other hand, one could easily observe the considerable effect on particle morphology for higher surfactant concentration, i.e., 33 % (v/v) and 50 % (v/v) as summarized in Fig. 7. This is to mention that stirring time for all these cases was maintained constant as 2 h.

We observed that mechanical shaker was apparently not able to offer sufficient shear forces as was necessary to form the folded lamellas. Subsequently the coalescence of RF sol droplets produced irregularly shaped elongated carbon structures as aggregates (Fig. 7d).

Fig. 7 SEM images showing the effect of method of the stirring on RF gel-derived carbon microparticle morphology in the high-surfactant region: (**a**) magnetic stirring, 50 % Span-80; (**b**, **c**) ultrasonication stirring, 33 % (v/v) and 50 % (v/v) Span-80, respectively; (**d**) mechanical shaker, 50 % (v/v) Span-80. Other conditions are $R/C = 25$, $R/W = 0.037$, and stirring time = 2 h (Reproduced with the permission by Sharma et al. [18])

To summarize, it was observed that the mean particle size and morphology of the carbon xerogel microspheres could be easily tuned by controlling the rate of coalescence for the RF sol droplets during emulsification process. The firm control on the carbon xerogel particle morphology by simply manipulating the process variables and the easiness of synthesis route and uniform size distribution for spherical particles are the exclusive features of the methodology discussed in this section. Synthesis of larger external surface area, multifolded, fractal-like carbon xerogel structures may possibly be new prospects in the current study of xerogels. These carbon xerogel microparticles with wide range of morphologies may find potential scientific and technological applications. Some of these applications are discussed in a later section.

2.2 *Synthesis of Carbon Nanostructures Using Electrospinning*

Carbon nanostructures considered in this work include carbon nanoparticles and carbon nanofibers. There are a number of techniques for the synthesis of these carbon nanostructures [19–36, 52–57]. Some of them include electrospinning

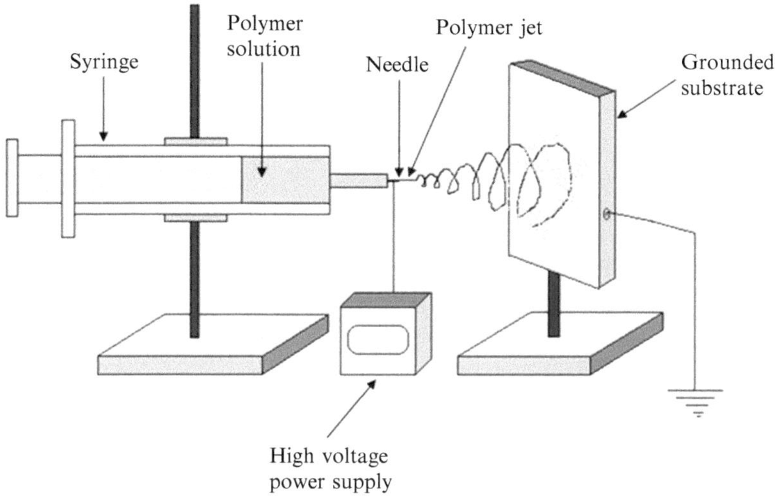

Fig. 8 Schematic of electrospinning setup (Reproduced with the permission by Sharma et al. [30])

[19–30], chemical vapor deposition [31–36], arc discharge [52, 53], pyrolysis [53], and pulsed laser vaporization [54–57]. In this work, we have primarily focused on electrospinning technique which is not only a simple and widely used method but also allows greater degree of control to produce wide spectrum of nanostructures.

The basic setup used for electrospraying/electrospinning is exactly the same as shown in Fig. 8. There are three major components [29, 30]: a high-voltage DC power supply (Gamma High Voltage Research, Inc., USA), a needle-mounted syringe pump (Longer Pump, China),, and a grounded collector plate. However, the operating conditions vary which controls the formation of charged-disrupted droplets rather than fibers.

A high-voltage source is used to generate a polymer jet which is electrically charged [19–30]. Beyond a threshold value of applied electric field, this charged polymer jet is ejected from the syringe needle forming a conical shape known as Taylor cone [30]. This charged polymer jet due to electrostatic repulsion experiences instabilities leading to a whipping motion. During this whipping motion, the charged polymer jet navigates a lengthy path enabling its thinning and at the same time evaporation of the solvent. As a result of this, a charged and dry ultrathin polymer fiber was collected on the grounded surface. The path of this charged polymer jet coming out from the syringe needle may be manipulated by varying the applied electric field [30]. Based on the similar principle but with different operating conditions, electrospinning process may be a promising method to synthesize nanoparticles as well. Historically in this case, the method is then termed as electrospraying [58–62]. In electrospraying, liquid coming out from a needle due to its surface tension is atomized in the presence of a high electric field. This elongates the liquid meniscus forming a jet which finally breaks intotiny charged droplets rather than fibers [29]. The main advantage of electrospraying as compared to other nanoparticle synthesis methods is its ability to form nanoparticles with a fine size distribution and higher yield [61, 62].

Among the precursors used for the synthesis of carbon nanostructures [19, 63–65], PAN is one of the most widely used. However, in this work we study electrospinning and/or electrospraying of two different polymer precursors to carbon. RF sol that has been employed to synthesize carbon xerogel microstructures using sol–gel emulsification has also been electrosprayed followed by pyrolysis to synthesize nearly monodisperse carbon nanospheres. Besides RF sol, we have electrospun SU-8, a negative photoresist to synthesize carbon nanostructures, primarily nanofibers. SU-8 photoresist is one of the key materials used in the fabrication of optoelectronic and lab-on-a-chip devices, including arrays of microfluidics and MEMS platforms [66]. Recently, C-MEMS fabrication based on SU-8 photoresist as a precursor has been demonstrated for various applications in microbatteries and biological platforms. [14–16, 66–68] This provides a key inspiration for SU-8 photoresist-derived nanostructures to enable their ease of incorporation and compatibility with the original devices and manufacturing techniques. Additionally, using negative photoresist as a polymer precursor to carbon nanofibers paves the way to fabricate different kinds of high-aspect-ratio patterned fibrous architecture using conventional photolithography which has several potential applications in C-MEMS and bio-MEMS [14–16, 68].

2.2.1 RF Sol-Derived Carbon Nanoparticles

The organic RF sol was electrosprayed just prior to gelation to yield RF-based nanospheres. The silicon wafer that was attached to a grounded copper screen was used as a substrate to collect the nanoparticle samples. Samples were then heated in air oven at 333 K for 12 h to evaporate the solvent completely. Various parameters were changed one by one maintaining the others constant to find out the best process variables for electrospraying. The different parameters tuned were as follows: needle diameter, applied electric field, and polymer flow rate. Following little understanding with a broad range of values that were apparently beyond the optimal ranges, the following ranges of various parameters were studied to achieve the tight control on droplet size, morphology, and monodispersity: (1) three different syringe needles with 18, 22, and 26 gauge corresponding to internal diameter of 1.02, 0.65, and 0.40 mm, respectively, were used; (2) the applied electric field between syringe needle and silicon wafer substrate was changed from 1.0 to 2.5 kV/cm in steps of 0.5 kV; and (3) the RF sol flow rate was increased from 0.2 to 2.0 ml/h in steps of 0.6 ml/h.

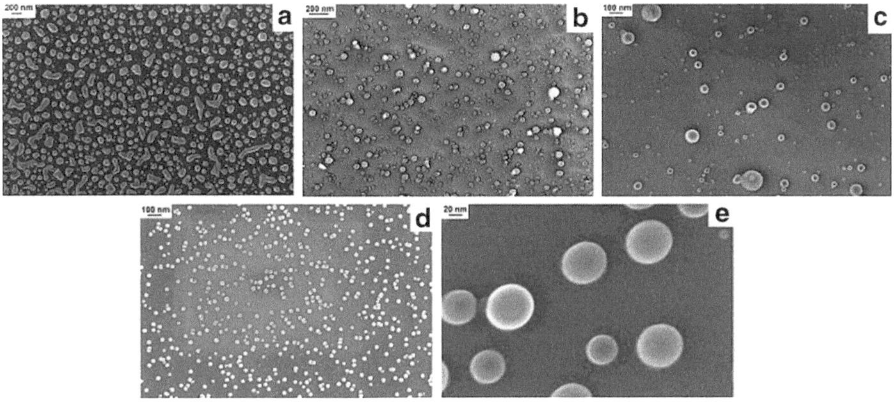

Fig. 9 SEM images showing the effect of needle gauge on the size and morphology of carbon nanospheres: needle gauge, (**a**) 18 g, (**b**) 22 g, (**c**) magnified view of (**b**), (**d**) 26 g, (**e**) magnified view of (**d**). Other parameters are electric potential, 2.0 kV/cm, and flow rate, 0.8 ml/h (Reproduced with the permission by Sharma et al. [29])

Effect of Needle Diameter

Three different syringe needles with their opening diameters 1.02, 0.65, and 0.40 mm, respectively (18, 22, and 26 g), were used to examine the effect of needle opening on particle size and shape as shown in Fig. 9. The electric field and polymer flow rate were kept constant as 2.0 kV/cm and 0.8 ml/h, respectively. In case of an 18 gauge needle diameter, a bimodal distribution of RF sol beads was observed (Fig. 9a). Upon pyrolysis, the mean diameter of primary carbon nanoparticles was found to be 156 ± 41.2 nm, while the secondary droplets which were of small fraction had a mean diameter of 32.1 ± 4.7 nm.

While using a needle of 22 g, polymer flow rate (0.8 ml/h) was suitably less to form a stable jet and thereby disrupted more uniformly into nanobeads with relatively less polydispersity as shown in Fig. 9b–c. The mean diameter of carbon nanoparticles in this case was measured to be 74.3 ± 14.2 nm. Next, the syringe needle of 26 g formed even finer jet as compared to two previous cases (18 and 22 g) which yielded nearly monodisperse carbon nanospheres with a mean diameter of 42.8 ± 6.3 nm (Fig. 9d–e). As needle of 26 g yielded spherical nearly monodisperse carbon nanoparticles with least diameter, it was used for further experiments.

Effect of Applied Electric Field

A pictorial summary of the effect of applied electric field on the mean size of carbon nanoparticles is shown in Fig. 10. While using the syringe needle of 26 g and fixing the polymer flow rate 0.8 ml/h, the electric potential was changed from 1.0 to 2.5 kV/cm with a linear increase of 0.5 kV/cm.

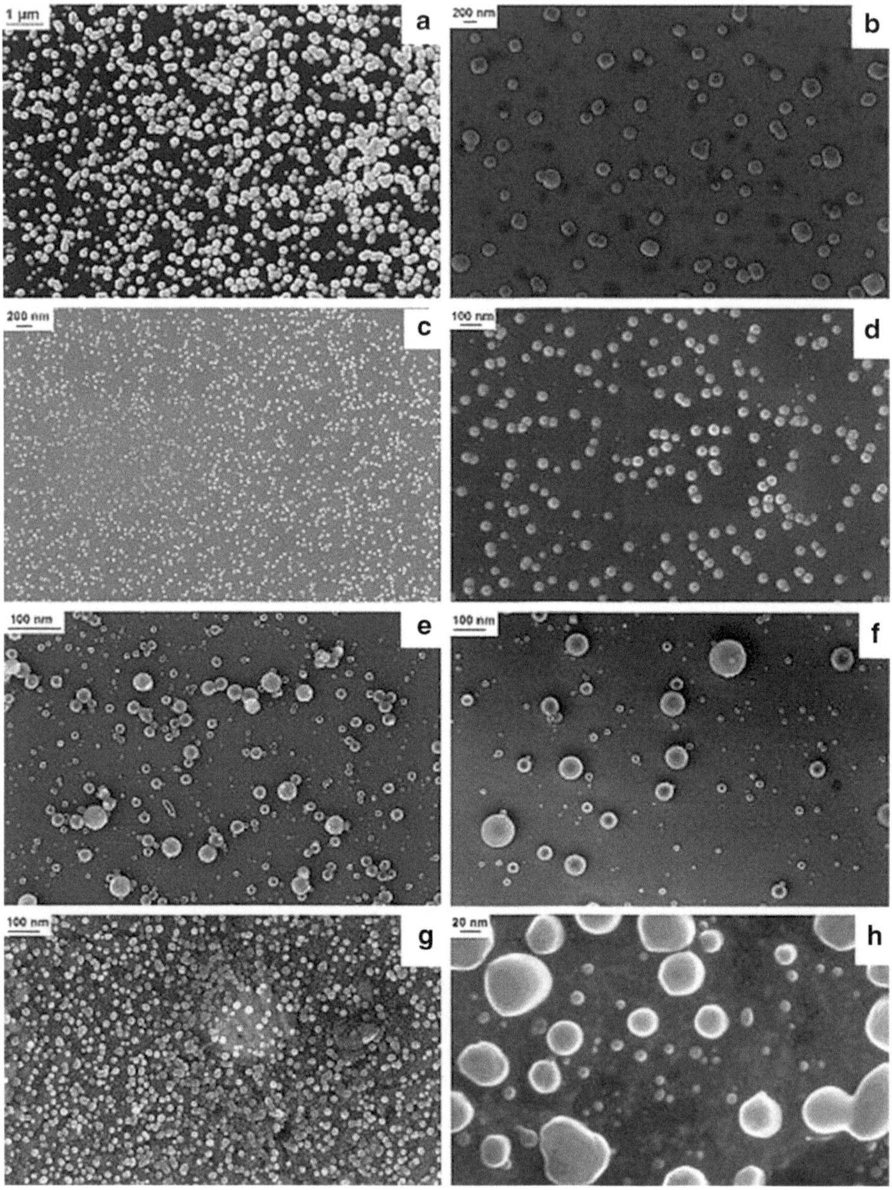

Fig. 10 SEM images showing the effect of applied electric potential on the size and morphology of carbon nanospheres: electric potential, (**a**) 1.0 kV/cm, (**c**) 1.5 kV/cm, (**e**) 2.0 kV/cm, (**g**) 2.5 kV/cm. Images (**b**, **d**, **f**, **h**) are magnified view of (**a**, **c**, **e**, **g**), respectively. Other parameters are needle gauge, 26 g, and flow rate, 0.8 ml/h (Reproduced with the permission by Sharma et al. [29])

At applied electric field 1.0 kV/cm, larger carbon particles with a mean diameter 206.8 ± 26.9 nm were produced as shown in Fig. 10a, b. On increasing the applied field to 1.5 kV/cm, there was a significant change in size range of carbon particles. The mean size of carbon nanospheres was reduced to 36.9 ± 8.1 nm (Fig. 10c, d). In both cases, particles were nearly monodisperse. However, further increase in the applied electric potential (2.0 kV/cm) resulted in carbon nanoparticles with a slightly reduced diameter (23.1 ± 16.7 nm), but with a wider size distribution as shown in Fig. 10e, f. At a very high electric potential (2.5 kV/cm), a prominent bimodal distribution of carbon nanospheres was observed. The mean diameter of primary carbon nanospheres in this case was 43.7 ± 14.9 nm, while the size of secondary particles was found to be 1.4 ± 4.0 nm (Fig. 10g, h). To summarize, carbon nanoparticles produced with applied electric field 2.0 kV/cm were found to have the smallest mean diameter (23.1 ± 16.7 nm); however, high polydispersity was observed. Thus, in the next set of experiments, applied electric field was fixed to be 1.5 kV/cm that resulted in nearly uniform-sized carbon nanospheres.

Effect of Polymer Flow Rate

RF sol was fed at a rate of 0.2–2.0 ml/h with a fixed increase of 0.6 ml/h to examine its effect on carbon nanoparticles' size and morphology. As shown in Fig. 11a, at 0.2 ml/h, mean size of carbon spheres was measured to be 17.1 ± 8.8 nm that was increased to 30.2 ± 7.1 nm with the increase in flow rate to 0.8 ml/h (Fig. 11b). With 0.8 ml/h RF sol feed rate, carbon nanospheres were nearly monodisperse. On further increasing the flow rate to 1.4 ml/h, mean size of carbon nanoparticles was found to be 88.4 ± 30.1 nm. Further increase in the flow rate to 2.0 ml/h leads the carbon particles agglomerated as shown in Fig. 11e. The mean diameter also in this case was increased to be 212 ± 37.9 nm.

The following study [29] suggested that during electrospraying, various process variables, e.g., needle diameter, applied electric field, and polymer (RF sol) flow rate, played a key role in determining the size and morphology of the carbon nanoparticles. The following set of parameters, needle of 26 g, applied electric field 1.5 kV/cm, and RF sol flow rate 0.8 ml/h, produced nearly monodisperse carbon nanoparticles with an average diameter of 30.2 ± 7.1 nm (Fig. 11b).

2.2.2 SU-8 Photoresist-Derived Carbon Nanofibers

SU-8 nanofibers were electrospun in the form of a nonwoven fiber mesh (fabric) on a Si wafer used as a substrate and attached to the grounded copper collector. Immediately after electrospinning, the samples were dried in a vacuum desiccator by ensuring the complete evaporation of the solvent. After drying, SU-8-based electrospun nanofibers collected on a Si wafer were cross-linked by exposing them to 365 nm UV light for 1 min. Finally these cross-linked photoresist nanofibers were

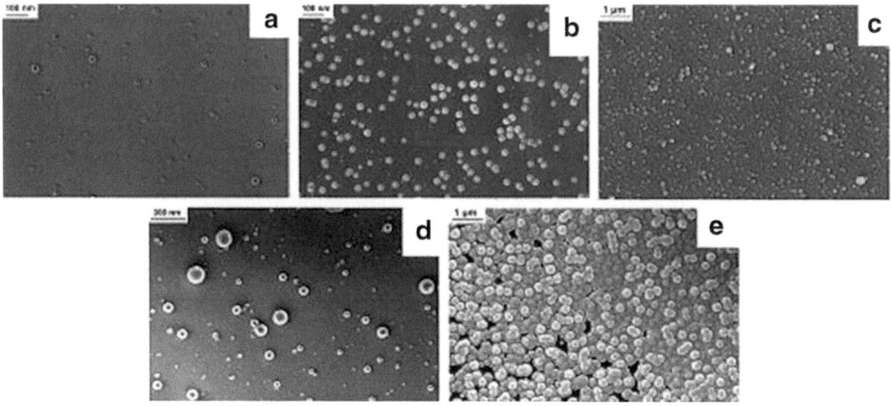

Fig. 11 SEM images showing the effect of volumetric flow rate on the size and morphology of carbon nanospheres: flow rate, (**a**) 0.2 ml/h, (**b**) 0.8 ml/h, (**c**) 1.4 ml/h, (**d**) magnified view of (**c**), (**e**) 2.0 ml/h. Other parameters are electric potential, 2.0 kV/cm, and flow rate, 0.8 ml/h (Reproduced with the permission by Sharma et al. [29])

heated in a high-temperature tubular furnace in the presence of nitrogen gas flow to yield carbon nanofibers.

Similar to electrospraying of RF sol into nanospheres, various parameters like electric field, distance between the syringe needle, and collector screen played a major role in modulating the average size of SU-8-derived electrospun nanofibers. However, the change in polymer flow rate did not alter the size and morphology to a significant level unlike electrospraying. The following set of parameters, applied electric field 2.0 kV/cm, 10 cm distance between needle and collector, and a flow rate of 0.3 ml/h, produced unbroken uniform fibers with a mean diameter of 190 ± 40 nm. Reduction of fiber diameter was possible by lowering the flow rate; however, the yield was less in that case. After studying the effect of these three process parameters (applied electric field, distance between source and collector, and flow rate) and optimizing the conditions to produce continuous nanofibers, the effect of SU-8 photoresist concentration and its viscosity on fiber morphology was examined as discussed below.

Effect of Viscosity of SU-8 Solution

Apart from the aforementioned process variables, we also studied the effect of viscosity of the photoresist on electrospun nanofiber morphology [30]. Experiments were carried out using two different solutions with viscosity 7 and 140 cSt keeping other process parameters similar to what were optimized in case of the highest viscosity SU-8 (applied electric field 2.0 kV/cm, distance with source and collector 10 cm, and a flow rate of 0.3 ml/h). The results revealed that in case of the lowest viscosity photoresist (SU-8 2002, 7 cSt), aggregated particles were produced with

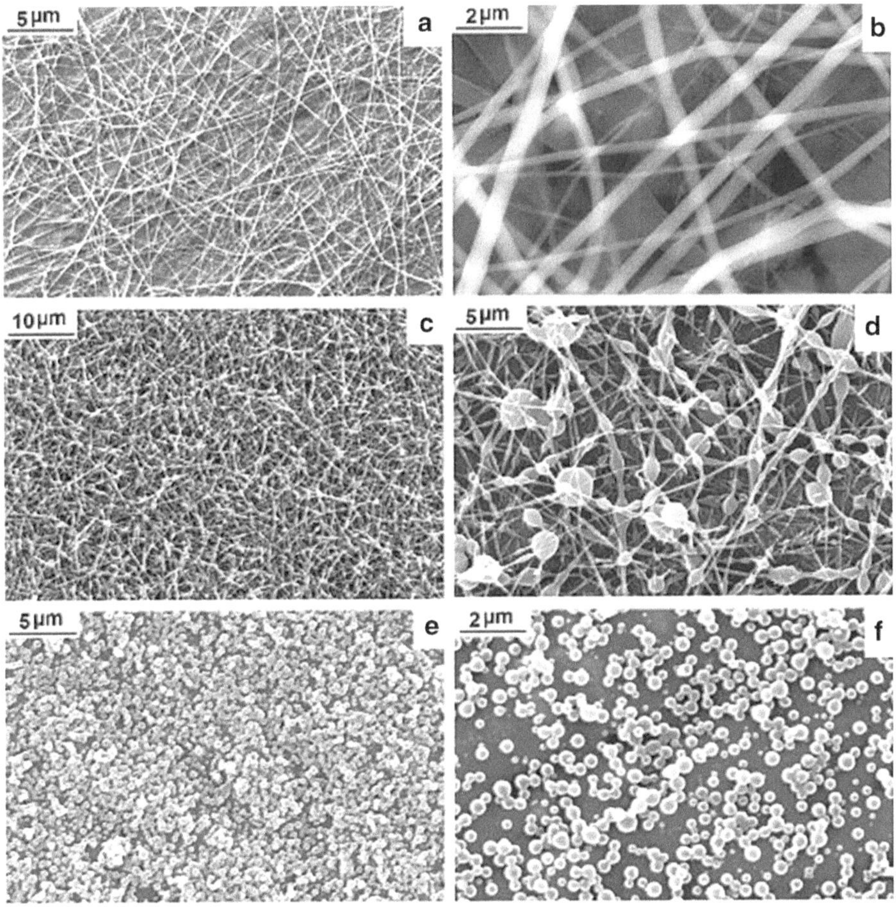

Fig. 12 SEM images showing the effect of viscosity on the morphology of carbon nanofibers: (**a**, **b**) highest viscosity, 1, 250 cSt; (**c**, **d**) medium viscosity, 140 cSt; and (**e**, **f**) lowest viscosity, 7 cSt. The applied electric potential, flow rate, and distance between source needle and collector were 2.0 kV/cm, 0.3 ml/h, and 10 cm, respectively, for (**a**–**e**). The applied electric potential was 1.0 kV/cm for f, with the other conditions remaining the same. (**b**, **d**) are high-magnification images of a and c, respectively (Reproduced with the permission by Sharma et al. [30])

a mean diameter varying from 180 to 620 nm. However, for medium viscosity photoresist (SU-8 2007, 140 cSt), long uniform carbon fibers connecting with beads were synthesized. These results signified that polymer solution viscosity (in this case, SU-8 photoresist) employed can significantly alter the fiber morphology [21, 27, 63]. These findings of the effect of viscosity on carbon nanofibers morphology are summarized in Fig. 12. Figure 12a, b shows the carbon nanofibers at different magnifications produced at optimized process conditions with mean fiber size 120 nm. Similarly Fig. 12c, d shows the SEM images of carbon fibers with beads present for medium viscous SU-8 photoresist. However, we observed small aggregates of uniform spherical carbon particles only for least viscous photoresist employed (SU-8 2002). Further by reducing the applied electric field to 1.0 kV/cm for SU-8 2002, we were able to produce uniform but discrete carbon beads as shown here in Fig. 12f.

2.3 Synthesis of Hierarchical Carbon Structures

2.3.1 RF Gel-Derived 3-D Hierarchical Carbon Structures

The conventional photolithography for SU-8 negative photoresist to fabricate 3-D microposts is discussed elsewhere [14, 15]. Since the RF gel is not photosensitive, we needed to come up with a new approach to pattern the material in order to work with 3-D configuration of RF gel-derived carbon. This approach is termed as replica molding [69–71], one of the widely used soft lithography techniques. In this method, RF sol was first poured in one of the SU-8-based masks (master pattern) to be replicated, followed by simple air oven drying in a controlled environment. The dried RF gel replica was then pyrolyzed to fabricated similar 3-D replica in carbon with reduced dimensions due to isotropic shrinkage [70, 71].

Further in order to fabricate multiscale carbon electrode, we deposited carbon micro- and nanospheres by electrospinning on to the RF-derived C-MEMS platform in the form of high-aspect-ratio pillars. As discussed in previous section, during electrospraying a polymer is ejected from a tiny spraying nozzle attached to a reservoir containing a solution of the polymer by applying a high voltage between a collector plate and an electrode in the polymer solution. Depending on the viscosity and the conductivity of the polymer solution, the nozzle, and the applied voltage, one obtains either a spray or a polymer nanowire [29, 30]. The integration of top-down (molding) and bottom-up (electrospraying/electrospinning) resulted in a hierarchical 3-D polymer structure that is composed of micrometer-sized 3-D posts covered with nanoscale beads/fibers [72]. One of such self-assembled hierarchical structures is shown in Fig. 13.

3 Applications

A wide range of carbonaceous materials for which synthesis and fabrication aspects were discussed in the previous section may find potential scientific and technological applications by exploring the different surface and bulk properties such as wettability, higher specific surface area, electrochemical behavior, and biocompatibility. Wettability is one of such properties which is greatly affected by surface morphology, textures, and chemistry of carbon structures. Wettability is an intrinsic property of a solid surface which is expressed by the water contact angle on that surface. Surfaces for which water contact angle is larger than

Fig. 13 SEM images of RF gel-derived 3-D carbon posts: (**a**) cross-shaped geometry, (**b**) cylindrical geometry, (**c**) SEM image showing suspended SU-8-derived CNF on cross geometry 3-D carbon posts, (**d**) higher-magnification image of (**c**), (**e**) SEM image showing suspended SU-8-derived CNF on cylindrical geometry 3-D carbon posts, (**f**) higher-magnification image of (**e**), (**g**) higher-magnification image of (**f**) showing an individual suspended CNF, (**h**) PAN-derived suspended CNF on RF gel-derived 3-D cross geometry carbon posts (Reproduced with the permission by Sharma et al. [72])

150° are known as non-wetting or superhydrophobic surfaces. Superhydrophobic surfaces have received a great interest in the last decade due to a wide variety of applications reported in several fields [73–84]. Some of these important areas and applications are optics [73, 74, 78, 80, 83] (transparent, non-reflective, and highly reflective surfaces), low-friction surfaces [73, 77, 78, 80, 81, 83, 84], anti-adhesion [73, 75, 76, 80–84] (architectural glasses, windshields of cars, non-wettable textiles), and anti-biofoulant [78, 80]. Further as carbon is electrically conductive and possesses high surface area [30, 85, 86], the superhydrophobic carbon surfaces might be used in microfluidics and bio-MEMS. Although there are various methods to fabricate superhydrophobic surfaces as discussed in literature such as etching [74, 77, 78, 80], lithography [77, 78, 80, 81], self-organization [80], layer-by-layer and colloidal assembly [78, 80], electrochemical deposition [78, 80], and electrospinning [78, 80, 85, 86], the most general way to produce superhydrophobic carbon surfaces is either growing, aligning, or rearranging carbon nanotubes/nanofibers [87–93] or annealing the carbon films [94, 95]. These processes are time consuming, expensive, and complicated. Here we describe few novel ways in which nearly superhydrophobic carbon surfaces can be fabricated by simply tuning the surface morphology and surface textures. Interestingly, we also show that role of surface roughness is not only limited to increase in hydrophobicity but also to achieve extreme hydrophilic behavior. The extremely hydrophilic carbon surfaces reported in this work may serve as cell scaffolds in tissue engineering, adsorbents, and electrode material for energy storage devices such as batteries, supercapacitors, and fuel cells.

3.1 *Controlled Wettability Surfaces*

3.1.1 Wetting Behavior of Carbon Surfaces Covered with RF Gel-Derived Carbon Nanospheres

To find a suitable application for the carbon nanospheres produced by electrospraying as discussed in Sect. 2.2.1, we require a thorough understanding of their wetting behavior. While measuring the water contact angle, it was found that surface morphologies significantly influence the surface properties.

Figure 14 summarizes the water contact angle measurement results on RF gel-based carbon thin film surface and surfaces covered by electrosprayed carbon nanospheres. The RF gel-based carbon thin films show very mild hydrophilic behavior (WCA 83.3° ± 1.6°). This may be due to the presence of hydroxyl (–OH) group present on its surface. However, the surface covered by electrosprayed carbon

Fig. 14 Water contact angles on (**a**) RF sol-derived smooth carbon thin film, (**b**) RF sol-derived carbon nanosphere film initially, and (**c**) RF-derived carbon nanosphere film after 8 min (Reproduced with the permission by Sharma et al. [29])

nanospheres shows the WCA as 10.1° ± 2.8°, which later decreased to 5° ± 1.8° after few minutes [29]. This observation essentially signifies the contribution of surface roughness and macroporosity of carbon nanospheres on the wetting behavior. Electrospraying of RF sol nanospheres increases the nanoscale surface roughness that governs the nearly superhydrophilic behavior of surface covered with carbon nanospheres. This situation in which wettability increases with increased roughness for a smooth hydrophilic material was first explained by Wenzel [96].

3.1.2 Wetting Behavior of SU-8-Derived Electrospun Carbon Nanofibers

The synthesis of SU-8-derived CNF with a wide range of morphologies such as fibers, beaded fibers, and only beads has been described in the previous section. To understand the wetting behavior of these fabrics, WCA and contact angle hysteresis (CAH) were measured.

Figure 15 shows the digital camera images of water droplet on various surfaces having different morphologies, i.e., only beads, fibers with beads, fibers, and thin film [30]. The first row of images (Fig. 15a–d) summarizes the water contact angle for the SU-8 surface with different morphologies, while the second row (Fig. 15e–h) represents SU-8-derived carbon surfaces. Similarly, water droplet images as shown in the third row (Fig. 15i–l) are for the UV-treated SU-8-derived carbon surfaces, while the fourth row (Fig. 15m–p) represents the plasma-treated SU-8-derived carbon surfaces with various morphologies. Different surface morphologies such

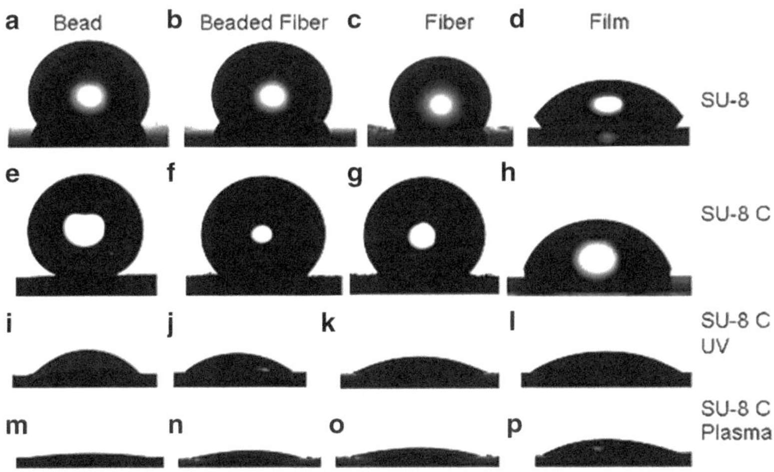

Fig. 15 Images of a water droplet on surfaces with various morphologies and chemical treatments: (**a–d**) SU-8, (**e–h**) SU-8-derived carbon, (**i–l**) SU-8-derived carbon after 30 min of UVO treatment, (**m–p**) SU-8-derived carbon after 5 min of oxygen plasma treatment. Morphology changes occurred in the following order (from *left* to *right*): beads, beaded fibers, fibers, and films. The drop volume was ~ 5 μL for all samples (Reproduced with the permission by Sharma et al. [30])

as only beads, beaded fibers, fibers, and thin film are represented by each column, respectively. For a given material (fixed row), the effect of different morphologies was observed by comparing the equilibrium WCA images column-wise. Similarly the effect of various surface treatments for a given morphology (column fixed) was observed by comparing the water droplet images row-wise.

SU-8 negative photoresist thin film prepared by spin coating was found to be mild hydrophilic (61.2° ± 1.3°) (Fig. 15d), while the WCA on SU-8-derived carbon film was measured to be slightly higher (70.8° ± 1.6°) (Fig. 15h). However, there was a considerable increase in the contact angle for SU-8 photoresist-derived carbon fibers which was measured to be 127.9° ± 3.1° (Fig. 15g). For beaded carbon fibers, there was further increase in contact angle, and it was found to be 138.4° ± 2.7° (Fig. 15f). We observe that formation of carbon beads imparts further to hydrophobicity, as revealed by measuring the WCA on carbon nanobead surfaces (142.7° ± 2.8°) (Fig. 15e).

To summarize, surface morphology and therefore roughness were observed to be playing a key role in determining the wettability of the electrospun fabrics [97–100]. Interestingly, only varying the carbon surface morphology with no surface chemical treatment could modulate the wettability characteristics to a wide range—from weakly hydrophilic (<90° contact angle) on a thin film surface to nearly superhydrophobic on a beaded fiber surface [30]. There are very few examples reported in literature where wettability transition across 90° contact angle has been demonstrated by changing the surface morphology [97, 99, 100].

As shown above, mildly hydrophilic behavior of thin film carbon surfaces was modulated to nearly superhydrophobic by simply varying the surface morphology alone. However, we also showed that the extent of hydrophilicity could also be enhanced largely by modifying the surface chemically [30]. Transition to extremely hydrophilic range may find potential applications for these fabric surfaces as bio-platforms and extending their compatibility in aqueous solutions. For SU-8-derived carbon nanofiber surface, it was observed that these can be made extremely hydrophilic by modifying their surface chemistry [30]. For example, the hydrophobicity of carbon nanofibers when treated by UVO for 30 min was changed to hydrophilicity (WCA 17.7° ± 0.6°). A similar behavior was observed when carbon fabric was treated with low-temperature oxygen plasma. WCA of plasma-treated SU-8 photoresist-derived carbon fibers mat was measured to be 10.3° (±0.9°). However, the surface modifications derived by these chemical treatments (UVO and plasma exposure) were found reversible. The hydrophobic nature of these chemically modified fabrics was examined after preserving the samples in ambient conditions for seven days. In case of SU-8-derived carbon nanofibers, contact angle was recovered to be 110.8° (±3.6°) after seven days. A relatively slow hydrophobic recovery for SU-8-based thin film after plasma treatment was also reported earlier [66].

3.1.3 Wettability on Micropatterned Electrospun Mat of SU-8-Derived Carbon Nanofibers

As described in the previous section, SU-8-derived electrospun CNF were deposited on Si wafer. As we observed, CNF show ultra-hydrophobic behavior (WCA ~ 128°) as compared to smooth SU-8-derived carbon film which is weakly hydrophilic (WCA ~ 60°). As SU-8 is a photosensitive material, it can be patterned by conventional photolithography. The ability to pattern SU-8 resist along with transforming the wettability characteristics by surface morphology provides a novel way for texturing the carbon surfaces as per the wettability behavior as shown in Fig. 16 [38].

Various patterned three-dimensional fabric structures are shown in Fig. 17a–f. The fabrics shown in Fig. 17a–d were electrospun directly on a Si wafer. For the rest of the samples (Fig. 17e, f), first a thin photoresist film was spin coated on Si wafer followed by pyrolysis. Thus, obtained carbon thin film was then exposed to plasma or UVO making it extremely hydrophilic before depositing the electrospun nanofibers. Figure 17a, b shows an array of microchannels fabricated by photopatterning the nonwoven fabric of electrospun SU-8 photoresist. Figure 17b is the high magnified view of the image shown in Fig. 17a. Each of these

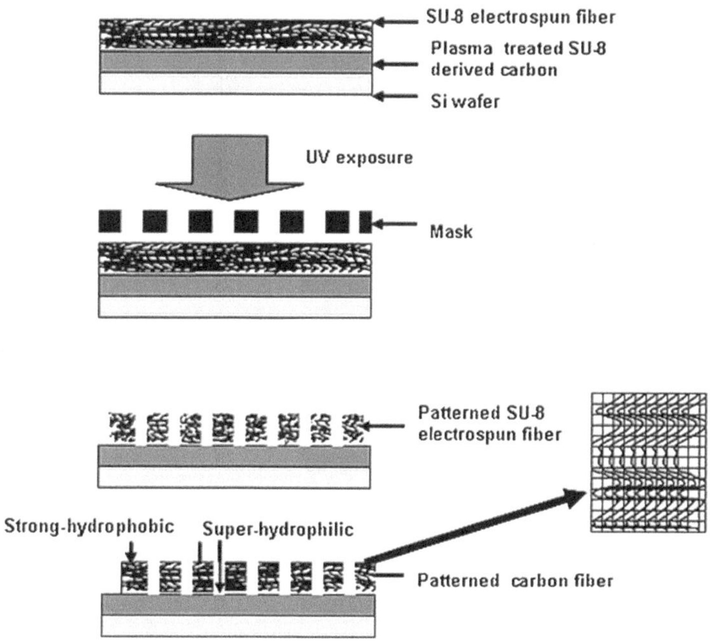

Fig. 16 Schematic of the fiber patterning process. Electrospun fibers are collected on very hydrophilic plasma or a UV-treated carbon surface on a Si wafer (Reproduced with the permission by Sharma et al. [38])

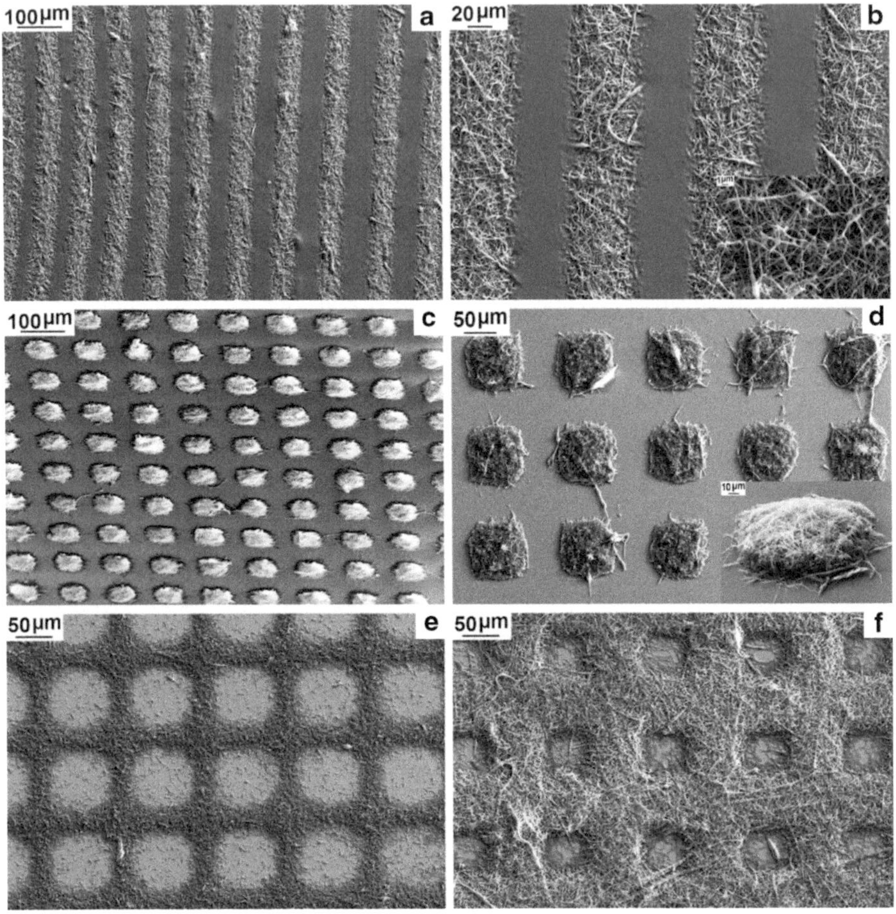

Fig. 17 SEM images of arrays of patterned fibrous carbon structures: (**a**, **b**) lines, (**c**, **d**) 3-D pillars. (**e**, **f**) Connecting squares. Images (**b**) and (**d**) are magnified view of images (**a**) and (**c**), respectively, and higher-magnification images of (**b**) and (**d**) are shown as *insets*. Images (**a–d**) are on Si wafer substrates, and images (**e**, **f**) are on a plasma-treated SU-8-derived carbon film on a Si wafer (Reproduced with the permission by Sharma et al. [38])

microchannels has porous carbon nanofibers walls and therefore may be utilized as a useful device for selective adsorption and filtration in microfluidics. Further, this photoresist-derived carbon which is one of the most widely used electrode materials shows electrical conductivity of the order of glassy carbon [72, 101, 102]. Thus, the electrospun nanofiber-based three-dimensional micropatterned carbon structures fabricated over a large area as shown in Fig. 17c, d may find potential use as an array of electrodes in microbatteries [14, 15, 38, 68, 103]. The micropatterned structures shown in Fig. 17e, f may also be utilized in selective immobilization of biomolecules and therefore developed as targeted reactors or as biosensors. Additionally, since

SU-8-derived carbon is shown as biocompatible [16], patterning nanofibrous carbon structures may pave the way for using them as a platform for cell culture and as tissue scaffolds.

Micropatterned electrospun carbon nanofibers demonstrate additional aspects of their applications as per their wettability behavior. As stated above, a thin film of SU-8 negative photoresist prepared by spin coating shows weakly hydrophilic (WCA $\sim 60°$) behavior, whereas the nano-textured surface by depositing electrospun CNF as shown in schematic (Fig. 16) is strongly hydrophobic with a WCA of about 130° and CAH 34°. Therefore, as shown in the schematic (Fig. 16), micropatterning hence engenders a large wettability contrast between the nanofiber domains with intervening micro-patches of chemically treated extremely hydrophilic carbon ($\sim$20°) or silicon substrates. Further the possibility of selective surface functionalization of carbon may present several innovative potential solutions in microfluidics, biosensing devices by selective adhesion, and the immobilization of biomaterials and also in electronic and photonic devices [104–106]. For example, this structural design could be applied to confine the flow of water-based solutions on hydrophilic regions of silicon or chemically treated carbon by screening hydrophobic solutes/oil phases from the hydrophobic porous carbon nanofiber membrane [38]. Namib desert beetle's wings are one such example found in nature with alternating hydrophilic and hydrophobic surfaces. Beetle's wings have an array of hydrophilic surface bumps with average diameter 100 μm on a hydrophobic background [38]. This particular structural arrangement in beetles helps them to collect and drink water from their fog-laden wing [104].

3.2 *Hierarchical Carbon Fabric for Environmental Remediation Applications*

Other properties of multiscale carbon structures are larger specific surface area, controllable porosity, and amenability to surface functionalization. We have presented a facile way of fabricating hierarchical carbon fabric by electrospinning PAN in the form of nanofibers on commercially available activated carbon microfiber (ACF) cloth and used this micro-nano hierarchical carbon web for adsorption of common atmospheric air pollutants, e.g., SO_x, NO_x, and volatile organic compounds (VOCs) [37]. SO_2, NO_x, and VOC are major atmospheric air pollutants emitted to the atmosphere largely from the combustion of fossil fuels. Adsorption is a preferred technique for the control of these pollutants at low concentration levels. Only ACF has been extensively used as adsorbent in air filtration [107–113]. Recent studies showed that the metal-impregnated ACF [114, 115] and/or surface functionalized with suitable reagent [116] had better efficiency in the catalytic oxidation of SO_x, NO_x, and VOCs. A micro-nano hierarchical carbon structure [117, 118] as fabricated by chemical vapor deposition (CVD)-grown CNF on ACF has also been used for controlling the NO and fluoride concentration from air and water, respectively, and was able to adsorb the contaminants more efficiently as compared to only ACF or ACF impregnated with metal [117, 118]. The carbon fiber mats have also been fabricated by electrophoretic deposition [119] or growing CNT by CVD [120, 121]. However, breakthrough curves for hierarchical carbon fabric prepared by depositing electrospun nanofibers on ACF suggest that so obtained hierarchical fabric has enhanced adsorption capacity as compared to only ACF or even with the hierarchical web of carbon micro-nanofibers fabricated by growing CNF on ACF impregnated with metals by the chemical vapor deposition technique. This study showed great potential for the hierarchical ACF prepared by electrospinning PAN nanofibers followed by their stabilization based in air pollution control [37].

Figure 18a describes the FESEM images of commercially available ACF cloth. The mean fiber diameter in this case was found to be 15.2 ± 1.4 μm. PAN polymer solution was prepared with DMF solvent followed by electrospinning on ACF to prepare hierarchical ACF-PAN fabric. These hierarchical fabrics were then kept in oven at 250 °C in the presence of air for thermal stabilization. The mean diameter of electrospun PAN nanofibers after stabilization (ACF-PANS) was measured to be 360 ± 38.1 nm (Fig. 18b) which is nearly two orders of magnitude smaller as compared to parent fibers (ACFs). Further the stabilized PAN fabric on ACF (ACF-PANS) was carbonized at 900 °C under continuous nitrogen gas flow to obtain hierarchical carbon fabric (ACF-PANC). Apparently, there was no change in the surface morphology for PAN nanofibers after pyrolysis (ACF-PANC) as shown in Fig. 18c.

3.2.1 Adsorption Behavior

Experiments were carried out on four different precursors, ACF, ACF-CNF, ACF-PANS, and ACF-PANC, to examine the adsorptive efficacy of the materials for SO_2, NO, and VOC (toluene). Figure 19 compares the experimental breakthrough curves of SO_2, NO, and VOC (toluene) for the prepared materials. The breakthrough time may be defined as the time when the outlet concentration reaches to 1 % of its inlet concentration [37]. Therefore, a large breakthrough time for a given adsorbent is desirable which means more efficient as the adsorbent is able to retain its adsorption capacity for larger duration with continuous flow of adsorbates through it. It was observed that the breakthrough time increased in the following order: ACF-CNF < ACF < ACF-PANC < ACF-PANS. The breakthrough time for the stabilized PAN fabric on ACF (ACF-PANS) was observed significantly larger as compared to three other adsorbents tested [37]. Further, time for complete saturation of the bed was also delayed significantly in case of ACF-PANS fabric sample compared to other three adsorbents used (only ACF, ACF-PAN, and ACF-PANC).

To summarize, the hierarchical ACF-PANS fabrics produced by electrospinning PAN nanofibers followed by its stabilization were found to be more efficient in

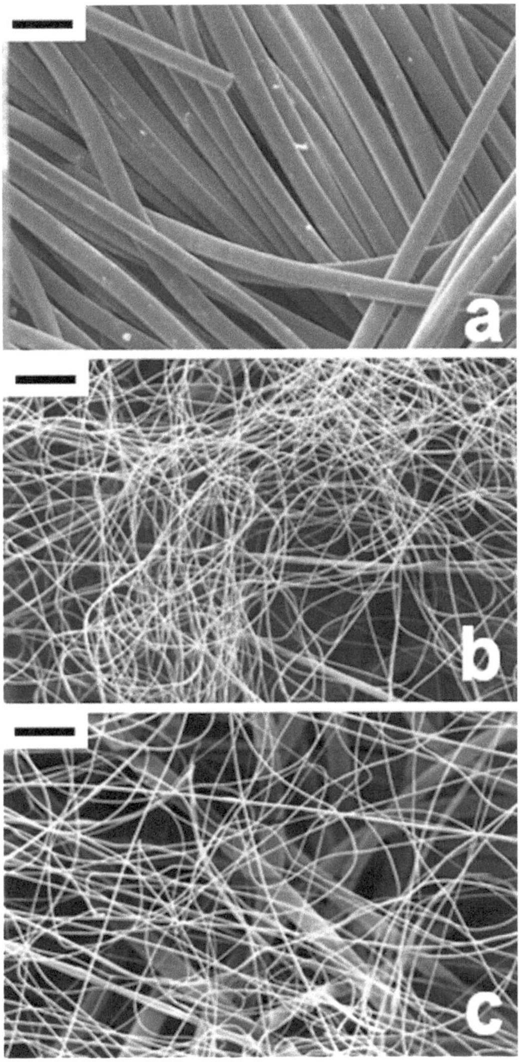

Fig. 18 SEM images of (**a**) the original ACF (average fiber diameter, 15 μm); (**b**) hierarchical web (ACF-PANS) derived by electrospinning PAN-based nanofibers (average fiber diameter, 360 nm), followed by thermal stabilization; (**c**) hierarchical carbon web (ACF-PANC) after pyrolysis. Scale bar equals to 20 μm in all cases (Reproduced with the permission by Katepalli et al. [37])

controlling the primary air pollutants such as SO_2, NO, and toluene [37]. The enhanced adsorption capacity for ACF-PANS may be attributed to the presence of nitrile and amine functional groups on their surface and higher specific surface area. The ACF-PANS-based hierarchical fabric may find sufficient potential in developing more efficient air filters for controlling the primary air pollutants. The ACF-PANS-based hierarchical fiber mat may also be utilized as a support to metal catalysts and also as an adsorbent in water purification.

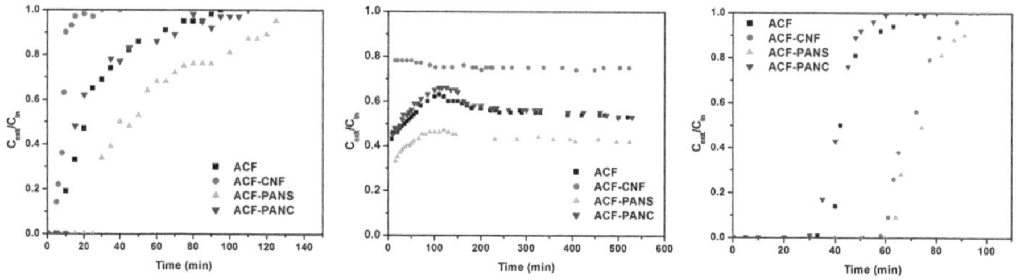

Fig. 19 Breakthrough profiles of SO_2, NO, and VOC (toluene) over different materials (ACF, ACF-CNF, ACF-PAN, ACF-PANS, and ACF-PANC) (Reproduced with the permission by Katepalli et al. [37])

3.3 Biocompatibility of Carbon Structures

3.3.1 Cellular Behavior

Besides wettability, higher surface area, and controlled porosity, we have studied another important property, biocompatibility, which adds the potential applications of carbon-based materials in a wide variety of bio-MEMS. Carbon is being used as biomaterial from the late 1960s [122]. However, while working with biosensors, microelectrode arrays, it is important to know the cell behavior with various patterned carbon structures that serve as platform to MEMS. One direct research application for the findings of cell behavior with patterned carbon structures may be the positioning and separation of cells selectively on a conductive artificial carbon platform. The ability to selectively grow and culture any specific cell line on a specific pattern would be of great value in developing biosensors. L929 fibroblast cell line was used to study the cytocompatibility of the micropatterned carbon surfaces through in vitro cellular biocompatibility using cell adhesion (Figs. 20 and 21) and cell viability (MTT) assessment (Fig. 22) [39]. Adhesion of cells on a substrate depends on a number of parameters such as surface morphology, porosity, and surface chemistry. It is therefore of great relevance to know the effects of patterned surface on the cellular behavior. Random and oriented patterned carbon surfaces, as a result of buckling [39], were used for studying the orientation of cells through cytoplasmic extension and attachment of extracellular matrix (ECM)-associated molecules during initial cell attachment [123]. Through the adhesion and viability studies on these buckling-based carbon patterns, we observed that cells preferred the randomly buckled patterns to grow which also provided cellular orientation supporting cell substrate and cell-to-cell communication (Figs. 21 and 22) [39].

Further it is worthwhile to mention here that the micropatterned carbon surfaces closely mimicked ECM in its structural design and therefore promoted the cell growth and viability, in comparison to the non-patterned surface. The cells initially attach to the substrate through integrins, a transmembrane protein present on the cell [39]. The micropatterned carbon surfaces influenced the organization

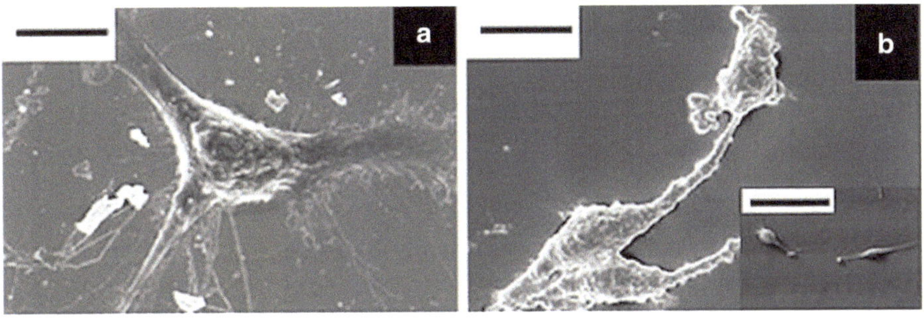

Fig. 20 Cell spreading on (**a**) control and (**b**) photoresist-derived patterned carbon surfaces (scale bar, 2 μm). *Inset* shows lower-magnification image of (**b**) (scale bar for inset image, 20 μm) (Reproduced with the permission by Kulkarni et al. [39])

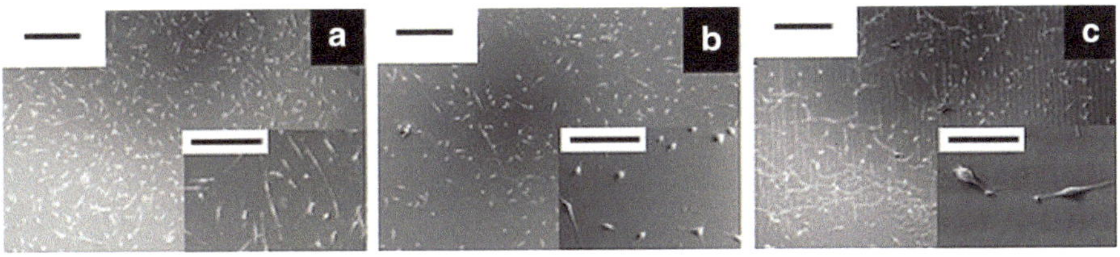

Fig. 21 SEM micrographs showing the cell (L929 mouse fibroblasts) adhesion after 24 h of seeding on various surfaces, (**a**) control, (**b**) smooth, and (**c**) patterned carbon (scale bar, 100 μm). Higher-magnification images are added as inset (scale bar, 20 μm) (Reproduced with the permission by Kulkarni et al. [39])

Fig. 22 Cell viability as shown by MTT assay results after 3 and 5 days of cell culture on control, smooth, and patterned carbon surface (Reproduced with the permission by Kulkarni et al. [39])

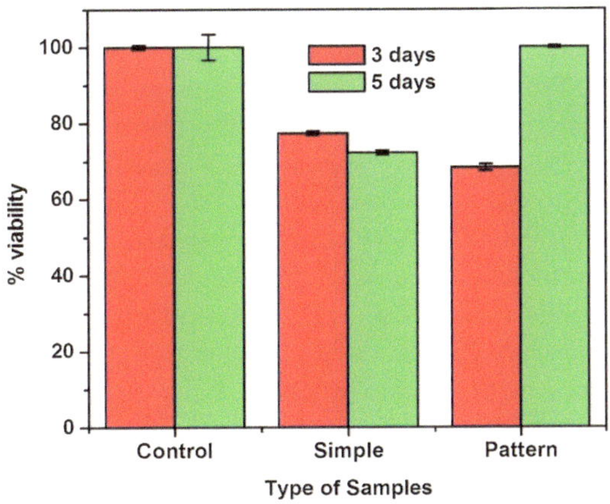

of integrin proteins explicitly. The cytoplasmic side of the integrins attaches to the cytoskeleton. As a result of this, the gene expression changes through up- and downregulation of specific gene and therefore ultimately regulates the ECM molecules [39, 124, 125]. As stated above, the ECM component of cell mimics closely with the patterned carbon surface with respect to its diameter. Moreover, the micropatterned surfaces adapt the cellular structure and its orientation, cytoskeleton reorganization, and ECM protein deposition on the modified surfaces [39, 126]. Extracellular components such as collagen and fibronectin facilitate the adhesion of cells during micropatterning at the initial phase of the cellular attachment [39]. Further, the substrate adhesion molecules and surface morphology affected the dynamics of cell–ECM interactions. These cell–ECM interactions further facilitate the cell migration, proliferation, division, differentiation, and programmed death, which are necessarily required for the successful and elongated cell biomaterial interactions [127]. Henceforth, the modified surface for the development of the connective tissue cell line like L929 fibroblast simply makes an interface with the micropatterned carbon surface [39]. This interface supports the cell growth and migration during reversible affection of cells on the surface [39, 128, 129].

This study confirmed that carbon substrates do not have any significant cytotoxic effect to L929 cells [39]. Further it was also revealed by this study that the surface morphology influences the cell viability significantly. The carefully synthesized micropatterned carbon surfaces were appraised to be the superior substrates to facilitate the enhanced cellular proliferation and extended cell viability [39].

4 Summary

In this report several methodologies to synthesize the hierarchical carbon micro- and nanostructures with a wide range of morphologies and functionalities are discussed. A variety of polymer precursors to carbon including organic gels and photoresist materials have been employed. Categorically, we have devised the several strategies in sol–gel processing and electrospinning in order to provide a firm control on the size and morphology of these carbon structures. Carbon xerogel microparticles with different morphologies such as microspheres and highly folded needlelike fractal structures were synthesized by RF sol polycondensation followed by inverse emulsification and subsequently pyrolysis. Several synthesis conditions such as sol concentration, stirring conditions, and type and amount of surfactants used were optimized. RF sol was also employed for electrospinning to synthesize nearly uniform-sized carbon nanospheres with mean diameter 30 nm. Similarly, SU-8, a negative photoresist that is a widely used material for C-MEMS, was electrospun to produce CNF with tunable morphology. Various electrospinning parameters in order to produce long, continuous CNF were optimized. Both RF sol and SU-8 photoresist are electrospun first ever to synthesize carbon nanostructures. For hierarchical carbon structures, a variety of micro-fabrication techniques (top-down approach) starting with conventional photolithography to soft lithographic techniques such as replica molding were combined with bottom-up approaches. In one of such attempts, a 3-D hierarchical carbon micropost array was fabricated by integrating electrosprayed carbon micro- and nanostructures (bottom-up approach) with lithographically (top-down approach) patterned carbon micropost arrays. Similarly, CVD, a well-established process to synthesize CNT, was also combined with electrospinning to assemble CNF in a unique hierarchical fashion over CNF.

Driven by the interest of using the carbon structures for various applications, several features of these multiscale carbon structures have been explored such as wettability, biocompatibility, and adsorption capacity. It was found that wettability of carbon surfaces could be tuned to a great extent from very hydrophilic to nearly superhydrophobic. The surfaces covered with RF gel-derived carbon nanospheres show extremely hydrophilic behavior. SU-8-derived electrospun nanofibers with various morphologies also show strong hydrophobic characteristics in contrast to SU-8-derived smooth film which is hydrophilic. Patterning of electrospun nanofibers thus allowed the fabrication of significant wettability contrast between the arrays of nanofibers with intervening micro-patches of hydrophilic carbon. This architecture similarly found in Namib desert beetle's wings may be useful for various applications including microfluidics. Another property that has been exploited is adsorption behavior which is associated with high surface area and specific surface chemistry of these hierarchical carbon assemblies. Hierarchical carbon fabric prepared by depositing electrospun PAN nanofibers on ACF fabric followed by thermal stabilization showed enhanced potential in mitigating the air pollutants such as SO_x, NO_x, and VOC. Furthermore, the cellular behavior of micropatterned carbon structures has been studied, and interestingly, the micropatterned carbon surfaces were found to be the superior substrates to facilitate the cellular proliferation and long-term cell viability. It is worth to mention here that the micropatterned carbon surfaces mimic the structural design of the ECM and promote cell growth and viability, when evaluated in contrast to the planar surface.

In conclusion, we have devised several control strategies to synthesize a wide spectrum of morphologies of multiscale carbon micro- and nanostructures. Various applications of these multiscale carbon micro- and nanostructures have been demonstrated based on their novel properties such as wettability, biocompatibility, and high surface area.

Acknowledgment We acknowledge the financial support from IRHPA project of DST to carry out this work. We also acknowledge the support from DST Unit on Soft Nanofabrication and Indo-US Centre of Excellence on Fabrionics.

References

1. Kinoshita K (1988) Carbon, electrochemical and physicochemical properties. Wiley, New York
2. Schueller OJA, Brittain ST, Marzolin C, Whitesides GM (1997) Fabrication and characterization of glassy carbon MEMS. Chem Mater 9:1399–1406
3. Schueller OJA, Brittain ST, Whitesides GM (1999) Fabrication of glassy carbon microstructures by soft lithography. Sensors Actuators A Phys 72:125–139
4. McCreery RL (1991) Carbon electrodes: structural effects on electron transfer kinetics. In: Bard AJ (ed) Electroanalytical chemistry, vol 17. Dekker, New York, p 221
5. McCreery RL (2008) Advanced carbon electrode materials for molecular electrochemistry. Chem Rev 108:2646–2687

6. Pocard NL, Alsmeyer DC, McCreery RL, Neenan TX, Callstrom MR (1992) Doped glassy carbon: a new material for electrocatalysis. J Mater Chem 2:771–784
7. Jenkins GM, Kawamura K (1976) Polymeric carbons–carbon fibre, glass and char. Cambridge University Press, Cambridge
8. Pekala RW (1989) Organic aerogels from the polycondensation of resorcinol with formaldehyde. J Mater Sci 24:3221–3227
9. Al-Muhtaseb SA, Ritter JA (2003) Preparation and properties of resorcinol–formaldehyde organic and carbon gels. Adv Mater 15:101–114
10. Lin C, Ritter JA (1997) Effect of synthesis pH on the structure of carbon xerogels. Carbon 35:1271–1278
11. Wang C, Madou M (2005) Short communication from MEMS to NEMS with carbon. Biosens Bioelectron 20:2181–2187
12. Singh A, Jayaram J, Madou M, Akbar S (2002) Pyrolysis of negative photoresists to fabricate carbon structures for microelectromechanical systems and electrochemical applications. J Electrochem Soc 149:E78–E83
13. Wang C, Taherabadi L, Jia G, Madou MJ (2004) Carbon-MEMS architectures for 3D microbatteries. Electrochem Solid-State Lett 7:A435–A438
14. Wang C, Jia G, Taherabadi LH, Madou MJ (2005) A novel method for the fabrication of high-aspect ratio C-MEMS structures. J Microelectromech Syst 14:348–358
15. Teixidor GT, Zaouk RB, Park BY, Madou MJ (2008) Fabrication and characterization of three-dimensional carbon electrodes for lithium-ion batteries. J Power Sources 183:730–740
16. Teixidor GT, Gorkin RA III, Tripathi PP, Bisht GS, Kulkarni M, Maiti TK, Battcharyya TK, Subramaniam JR, Sharma A, Park BY, Madou M (2008) Carbon microelectromechanical systems as a substratum for cell growth. Biomed Mater 3:034116
17. Sharma CS, Kulkarni MM, Sharma A, Madou M (2009) Synthesis of carbon xerogel particles and fractal-like structures. Chem Eng Sci 64:1536–1543
18. Sharma CS, Upadhyaya DK, Sharma A (2009) Controlling the morphology of resorcinol-formaldehyde based carbon xerogels by sol concentration, shearing and surfactants. Ind Eng Chem Res 48:8030–8036
19. Dzenis Y (2004) Spinning continuous fibers for nanotechnology. Science 304:1917–1919
20. Doshi J, Reneker HD (1995) Electrospinning process and application of electrospun fibers. J Electrostat 35:151–160
21. Renekar DH, Yarin AL, Fong H, Koombhongse SJ (2000) Bending instability of electrically charged liquid jets of polymer solutions in electrospinning. Appl Phys 87:4531–4547
22. Shin YM, Hohman MM, Brenner MP, Rutledge GC (2001) Electrospinning: a whipping fluid jet generates submicron polymer fibers. Appl Phys Lett 78:1149–1151
23. Hohman MM, Shin M, Rutledge G, Brenner MP (2001) Electrospinning and electrically forced jets. II. Applications. Phys Fluids 13:2221–2236
24. Fridrikh SV, Yu JH, Brenner MP, Rutledge GC (2003) Controlling the fiber diameter during electrospinning. Phys Rev Lett 90:144502
25. Theron SA, Zussman E, Yarin AL (2004) Experimental investigation of the governing parameters in the electrospinning of polymer solutions. Polymer 45:2017–2030
26. Tan S-H, Inai R, Kotaki M, Ramakrishna S (2005) Systematic parameter study for ultra-fine fiber fabrication via electrospinning process. Polymer 46:6128–6134
27. Thompson CJ, Chase GG, Yarin AL, Reneker DH (2007) Effects of parameters on nanofiber diameter determined from electrospinning model. Polymer 48:6913–6922
28. Tan S, Huang X, Wu B (2007) Some fascinating phenomena in electrospinning processes and applications of electrospun nanofibers. Polym Int 56:1330–1339
29. Sharma CS, Patil S, Saurabh S, Sharma A, Venkataraghavan R (2009) Resorcinol–formaldehyde based carbon nanospheres by electrospraying. Bull Mater Sci 32:239–246
30. Sharma CS, Vasita R, Upadhyay DK, Sharma A, Katti DS, Venkataraghavan R (2010) Photoresist derived electrospun carbon nanofibers with tunable morphology and surface properties. Ind Eng Chem Res 49:2731–2739

31. Iijima S (1991) Helical microtubules of graphitic carbon. Nature 354:56–58
32. Ebbesen TW, Ajayan PM (1992) Large-scale synthesis of carbon nanotubes. Nature 358: 220–222
33. Ando Y, Iijima S (1993) Preparation of carbon nanotubes by arc-discharge evaporation. Jpn J Appl Phys 32:L107–L109
34. Bower C, Zhou O, Zhu W, Werder J, Jin S (2000) Nucleation and growth of carbon nanotubes by microwave plasma chemical vapor deposition. Appl Phys Lett 77:2767–2769
35. Chhowalla M, Teo KBK, Ducati C, Rupesinghe NL, Amaratunga GAJ, Ferrari AC, Roy D, Robertson J, Milne WI (2001) Growth process conditions of vertically aligned carbon nanotubes using plasma enhanced chemical vapor deposition. J Appl Phys 90:5308–5317
36. Hofmann S, Ducati C, Kleinsorge B, Robertson J (2003) Direct growth of aligned carbon nanotube field emitter arrays onto plastic substrates. Appl Phys Lett 83:4661–4663
37. Katepalli H, Bikshapathi M, Sharma CS, Verma N, Sharma A (2011) Synthesis of hierarchical fabrics by electrospinning of PAN nanofibers on activated carbon microfibers for environmental remediation applications. Chem Eng J 171:1194–1200
38. Sharma CS, Sharma A, Madou M (2010) Multiscale carbon structures fabricated by direct micropatterning of electrospun mats of SU-8 photoresist nanofibers. Langmuir 26:2218–2222
39. Kulkarni MM, Sharma CS, Sharma A, Kalmodia S, Basu B (2012) Multiscale micro-patterned polymeric and carbon substrates derived from buckled photoresist films: fabrication and cytocompatibility. J Mater Sci 47:3867–3875
40. Horikawa T, Hayashi J, Muroyama K (2004) Size control and characterization of spherical carbon aerogel particles from resorcinol–formaldehyde resin. Carbon 42:169–175
41. Job N, Panariello F, Marien J, Crine M, Pirard J-P, Leonard A (2006) Synthesis optimization of organic xerogels produced from convective air-drying of resorcinol–formaldehyde gels. J Non-Cryst Solids 352:24–34
42. Job N, Pirard R, Marien J, Pirard JP (2004) Porous carbon xerogels with texture tailored by pH control during sol–gel process. Carbon 42:619–628
43. Kim SI, Yamamoto T, Endo A, Ohmori T, Nakaiwa M (2006) Influence of nonionic surfactant concentration on physical characteristics of resorcinol-formaldehyde carbon cryogel microspheres. J Ind Eng Chem 12(3):484–488
44. Matos I, Fernandes S, Guerreiro L, Barat S, Ramos AM, Vital J, Fonseca IM (2006) The effect of surfactants on the porosity of carbon xerogels. Microporous Mesoporous Mater 92:38–46
45. Shen J, Li J, Chen Q, Luo T, Yu W, Qian Y (2006) Synthesis of multi-shell carbon microspheres. Carbon 44:190–193
46. Xu L, Jianwei L, Jin D, Yiya P, Yitai Q (2005) A self-assembly template approach to form hollow hexapod-like, flower-like and tube-like carbon materials. Carbon 43:1560–1562
47. Bourret M, Schecter RS (1988) Microemulsions and related systems. In: Bourret M, Schecter RS (eds) Surfactant science series, vol 30. Marcel Dekker, New York
48. Miller CA, Neogi P (2007) Interfacial phenomena: equilibrium and dynamic effects, 2nd edn. CRC Press, Boca Raton
49. Velentas KJ, Bilous O, Amundson NR (1966) Analysis of breakage in dispersed phase systems. Ind Eng Chem Fundam 5:271–279
50. Tcholakova S, Denkov ND, Danner T (2004) Role of surfactant type and concentration for the mean drop size during emulsification in turbulent flow. Langmuir 20:7444–7458
51. Narsimhan G (2004) Model for drop coalescence in a locally isotropic turbulent flow field. J Colloid Interface Sci 272:197–209
52. Qiao WM, Song Y, Lim SY, Hong SH, Yoon SH, Mochida I, Imaoka T (2006) Carbon nanospheres produced in an arc-discharge process. Carbon 44:187–190
53. Govindraj A, Sen R, Nagaraju BV, Rao CNR (1997) Carbon nanospheres and tubules obtained by the pyrolysis of hydrocarbons. Philos Mag Lett 76:363–367
54. Journet C, Bernier P (1998) Production of carbon nanotubes. Appl Phys A 67:1–9
55. Yap YK, Yoshimura M, Mori Y, Sasaki T, Hanada T (2002) Formation of aligned carbon nanotubes by RF-plasma-assisted pulsed-laser deposition. Physica B 323:341–343

56. Thess A, Lee R, Nikolaev P, Dai H, Petit P, Robert J, Xu C, Lee YH et al (1996) Crystalline ropes of metallic carbon nanotubes. Science 273:483–487
57. Rinzler AG, Liu J, Dai H, Nikolaev P, Huffman CB, Rodrigues FG et al (1998) Large-scale purification of single-wall carbon nanotubes: process, product, and characterization. Appl Phys 67:29–37
58. Cloupeau M, Prunet-Foch BJ (1989) Electrostatic spraying of liquid in cone jet mode. J Electrostat 22:135–159
59. Cloupeau M, Prunet-Foch BJ (1990) Electrostatic spraying of liquids: main functioning modes. J Electrostat 25:165–184
60. Cloupeau M, Prunet-Foch BJ (1994) Electrohydrodynamic spraying functioning modes: a critical review. Aerosol Sci Technol 25:1021–1036
61. Jaworek A (2007) Micro- and nanoparticle production by electrospraying. Powder Technol 176:18–35
62. Jaworek A, Sobczyk AT (2008) Electrospraying route to nanotechnology: an overview. J Electrostat 66:197–219
63. Sutasinpromprae J, Jitjaicham S, Nithitanakul M, Meechaisue C, Supaphol P (2006) Preparation and characterization of ultrafine electrospun polyacrylonitrile fibers and their subsequent pyrolysis to carbon fibers. Polym Int 55:825–833
64. Wang Y, Serrano S, Santiago-Aviles JJ (2003) Raman characterization of carbon nanofibers prepared using electrospinning. Synth Met 138:423–427
65. Zussman E, Chen X, Ding W, Calabri L, Dikin DA, Quintana JP, Ruoff RS (2005) Mechanical and structural characterization of electrospun PAN-derived carbon nanofibers. Carbon 43:2175–2185
66. Walther F, Davidovskaya P, Zurcher S, Kaiser M, Herberg H, Gigler A, Stark RW (2007) Stability of the hydrophilic behavior of oxygen plasma activated SU-8. J Micromech Microeng 17:524–531
67. Ranganathan S, McCreery R, Majji SM, Madou M (2000) Photoresist-derived carbon for microelectromechanical systems and electrochemical applications. J Electrochem Soc 147:277–282
68. Park BY, Taherabadi L, Wang C, Zoval J, Madou MJ (2005) Electrical properties and shrinkage of carbonized photoresist films and the implications for carbon microelectromechanical systems devices in conductive media. J Electrochem Soc 152:J136–J143
69. Xia Y, Whitesides GM (1998) Soft lithography. Angew Chem Int Ed 37:550–575
70. Das A, Mukherjee R, Katiyer V, Kulkarni M, Ghatak A, Sharma A (2007) Generation of sub-micrometer-scale patterns by successive miniaturization using hydrogels. Adv Mater 19:1943–1946
71. Sharma CS, Verma A, Kulkarni MM, Upadhyay DK, Sharma A (2010) Microfabrication of carbon structures by pattern miniaturization in resorcinol-formaldehyde gel. ACS Appl Mater Interfaces 2:2193–2197
72. Sharma CS, Katepalli H, Sharma A, Madou M (2011) Fabrication and electrical conductivity of suspended carbon nanofiber arrays. Carbon 49:1727–1732
73. Blossey R (2003) Self-cleaning surfaces-virtual realities. Nat Mater 2:301–306
74. Cao M, Song X, Zhai J, Wang J, Wang Y (2006) Fabrication of highly antireflective silicon surfaces with superhydrophobicity. J Phys Chem B 110:13072–13075
75. Sun T, Feng L, Gao X, Jiang L (2005) Bioinspired surfaces with special wettability. Acc Chem Res 38:644–652
76. Truesdell R, Mammoli A, Vorobieff P, Swol FV, Brinker CJ (2006) Drag reduction on a patterned superhydrophobic surface. Phys Rev Lett 97:044504
77. Choi CH, Ulmanella U, Kim J, Ho CM, Kim CJ (2006) Effective slip and friction reduction in nanograted superhydrophobic microchannels. Phys Fluids 18:087105
78. Li X, Reinhoudt D, Calama MC (2007) What do we need for a superhydrophobic surface? A review on the recent progress in the preparation of superhydrophobic surfaces. Chem Soc Rev 36:1350–1368

79. Koch K, Bhushan B, Barthlott W (2008) Diversity of structure, morphology and wetting of plant surfaces. Soft Matter 4:1943–1963
80. Zhang X, Shi F, Niu J, Jiang Y, Wang Z (2008) Superhydrophobic surfaces: from structural control to functional application. J Mater Chem 18:621–633
81. Xiu Y, Zhu L, Hess DW, Wong CP (2008) Relationship between work of adhesion and contact angle hysteresis on superhydrophobic surfaces. J Phys Chem C 112:11403–11407
82. Bhushan B, Jung YC (2008) Wetting, adhesion and friction of superhydrophobic and hydrophilic leaves and fabricated micro/nanopatterned surfaces. J Phys Condens Matter 20:225010
83. Nosonovsky M, Bhushan B (2009) Superhydrophobic surfaces and emerging applications: non-adhesion, energy, green engineering. Curr Opin Colloid Interface Sci 14:270–280
84. Koch K, Bhushan B, Jung YC, Barthlott W (2009) Fabrication of artificial Lotus leaves and significance of hierarchical structure for superhydrophobicity and low adhesion. Soft Matter 5:1386–1393
85. Li Y, Huang XJ, Heo SH, Li CC, Choi YK, Cai WP, Cho SO (2007) Superhydrophobic bionic surfaces with hierarchical microsphere/SWCNT composite arrays. Langmuir 23:2169–2174
86. Banerjee D, Mukherjee S, Chattopadhyay KK (2010) Controlling the surface topology and hence the hydrophobicity of amorphous carbon thin films. Carbon 48:1025–1031
87. Lau KKS, Bico J, Teo KBK, Chhowalla M, Amaratunga GAJ, Milne WI, McKinley GH, Gleason KK (2003) Superhydrophobic carbon nanotube forests. Nano Lett 3:1701–1705
88. Li W, Wang X, Chen Z, Waje M, Yan Y (2005) Carbon nanotube film by filtration as cathode catalyst support for proton-exchange membrane fuel cell. Langmuir 21:9386–9389
89. Zou J, Chen H, Chunder A, Yu Y, Huo Q, Zhai L (2008) Preparation of a superhydrophobic and conductive nanocomposite coating from a carbon-nanotube-conjugated block copolymer dispersion. Adv Mater 20:3337–3341
90. Wang N, Xi J, Wang S, Liu H, Feng L, Jiang L (2008) Long-term and thermally stable superhydrophobic surfaces of carbon nanofibers. J Colloid Interface Sci 320:365–368
91. Lu SH, Tun MHN, Mei ZJ, Chia GH, Lim X, Sow C (2009) Improved hydrophobicity of carbon nanotube arrays with micropatterning. Langmuir 25:12806
92. Jung YC, Bhushan B (2009) Mechanically durable carbon nanotube-composite hierarchical structures with superhydrophobicity, self-cleaning, and low-drag. ACS Nano 3:4155–4163
93. Han ZJ, Tay BK, Shakerzadeh M, Ostrikov K (2009) Superhydrophobic amorphous carbon/carbon nanotube nanocomposites. Appl Phys Lett 94:223106
94. Feng L, Yang Z, Zhai J, Song Y, Liu B, Ma Y, Yang Z, Jiang L, Zhu D (2003) Superhydrophobicity of nanostructured carbon films in a wide range of pH values. Angew Chem Int Ed 42:4217–4220
95. Shakerzadeh M, Teo HE, Tan C, Tay BK (2009) Superhydrophobic carbon nanotube/amorphous carbon nanosphere hybrid film. Diamond Relat Mater 18:1235–1238
96. Wenzel RN (1936) Resistance of solid surfaces to wetting by water. Ind Eng Chem 28: 988–994
97. Ma M, Mao Y, Gupta M, Gleason KK, Rutledge GC (2005) Superhydrophobic fabrics produced by electrospinning and chemical vapor deposition. Macromolecules 38:9742–9748
98. Ma M, Hill RM (2006) Superhydrophobic Surfaces. Curr Opin Colloid Interface Sci 11: 193–202
99. Yoon Y, Moon HS, Lyoo WS, Lee TS, Park WH (2008) Superhydrophobicity of PHBV fibrous surface with bead-on-string structure. J Colloid Interface Sci 320:91–95
100. Feng L, Zhang Y, Xi J, Zhu Y, Wang N, Xia F, Jiang L (2008) Petal effect: a superhydrophobic state with high adhesive force. Langmuir 24:4114–4119
101. Maitra T, Sharma S, Srivastava A, Cho Y-K, Madou M, Sharma A (2012) Improved graphitization and electrical conductivity of suspended carbon nanofibers derived from carbon nanotube/polyacrylonitrile composites by directed electrospinning. Carbon 50:1753–1761
102. Sharma S, Sharma A, Cho Y-K, Madou M (2012) Increased graphitization in electrospun single suspended carbon nanowires integrated with carbon-MEMS and carbon-NEMS platforms. ACS Appl Mater Interfaces 4:34–39

103. Park BY, Zaouk R, Wang C, Madou MJ (2007) A case for fractal electrodes in electrochemical applications. J Electrochem Soc 154:P1
104. Zhai L, Berg MC, Cebeci F, Kim Y, Milwid JM, Rubner MF, Cohen RH (2006) Patterned superhydrophobic surfaces: toward a synthetic mimic of the Namib Desert beetle. Nano Lett 6:1213–1217
105. Manso-Silvan M, Valsesia A, Hasiwa M, Rodríguez-Navas C, Gilliland D, Ceccone G, Ruiz JPG, Rossi F (2007) Micro-spot, UV and wetting patterning pathways for applications of biofunctional aminosilane-titanate coatings. Biomed Microdevices 9:287–294
106. Ghosh H, Alves C, Tong Z, Tettey K, Konstantopoulos K, Stebe KJ (2008) Multifunctional surfaces with discrete functionalized regions for biological applications. Langmuir 24: 8134–8142
107. Bradley RH, Rand B (1995) On the physical adsorption of vapors by microporous carbons. J Colloid Interface Sci 169:168–176
108. Suzuki M (1994) Activated carbon fiber: fundamentals and applications. Carbon 32:577–586
109. Mochida I, Korai Y, Shirahama M, Kawano S, Hada T, Seo Y, Yoshikawa M, Yasutake A (2000) Removal of Sox and Nox over activated carbon fibers. Carbon 38:227–239
110. Huang ZH, Kang F, Zheng YP, Yang JB, Liang KM (2002) Adsorption of trace polar methyl-ethyl-ketone and non-polar benzene vapors on viscose rayon-based activated carbon fibers. Carbon 40:1363–1367
111. Adapa S, Gaur V, Verma N (2006) Catalytic oxidation of NO by activated carbon fiber (ACF). Chem Eng J 116:25–37
112. Gaur V, Sharma A, Verma N (2006) Preparation and characterization of ACF for the adsorption of BTX and SO2. Chem Eng Process 45:1–13
113. Gaur V, Asthana R, Verma N (2006) Removal of SO2 by activated carbon fibers in the presence of O2 and H2O. Carbon 44:46–60
114. Gaur V, Sharma A, Verma N (2005) Catalytic oxidation of toluene and m-xylene by activated carbon fiber impregnated with transition metals. Carbon 43:3041–3053
115. Gaur V, Sharma A, Verma N (2007) Removal of SO2 by activated carbon fibre impregnated with transition metals. Can J Chem Eng 85:188–198
116. Rathore R, Srivastava D, Agarwal A, Verma N (2010) Development of surface functionalized activated carbon fiber for control of NO and particulate matter. J Hazard Mater 173:211–222
117. Singhal RM, Sharma A, Verma N (2008) Micro—nano hierarchal web of activated carbon fibers for catalytic gas adsorption and reaction. Ind Eng Chem Res 47:3700–3707
118. Gupta AK, Deva D, Sharma A, Verma N (2009) Adsorptive removal of fluoride by micro-nanohierarchical web of activated carbon fibers. Ind Eng Chem Res 48:9697–9707
119. Rodriguez AJ, Guzman ME, Lim CS, Minaie B (2010) Synthesis of multiscale reinforcement fabric by electrophoretic deposition of amine-functionalized carbon nanofibers onto carbon fiber layers. Carbon 48:3256–3259
120. Hsieh CT, Chen WY (2010) Water/oil repellency and drop sliding behavior on carbon nanotubes/carbon paper composite surfaces. Carbon 48:612–619
121. Lim S, Yoon SH, Shimizu Y, Jung H, Mochida I (2004) Surface control of activated carbon fiber by growth of carbon nanofiber. Langmuir 20:5559–5563
122. Bokros JC (1997) Carbon biomedical devices. Carbon 15:355–371
123. Louise PC (1999) Role of actin-filament disassembly in lamellipodium protrusion in motile cells revealed using the drug jasplakinolide. Curr Biol 9:1095–1105
124. Anselme K, Davidson P, Popa AM, Giazzon M, Liley M, Ploux L (2010) The interaction of cells and bacteria with surfaces structured at the nanometre scale. Acta Biomater 6:3824–3846
125. Pennacchi M, Armentano I, Zeppetelli S, Fiorillo M, Guarnieri D, Kenny JM, Netti PA (2004) Influence of surface patterning on cell migration and spreading. Eur Cell Mater 7:77
126. Petreaca M, Martins-Green M (2008) Cell-ECM interactions in repair and regeneration. In: Atala A, Lanza R, Thomsan JA, Nerem R (eds) Principles of regenerative medicine, 5th edn. Academic Press, Burlington, pp 66–99

127. Albert B, Johnson A, Lewis J, Raff M, Roberts K, Walter P (2002) Molecule biology of the cell, 4th edn. Garland Science Group, Taylor and Francis, New York
128. Misra A, Pei ZLR, Wu Z, Thirumaran T (2007) N-WASP plays a critical role in fibroblast adhesion and spreading. Biomed Biophys Res Commun 364:908–912
129. Detrait E, Lhoest JB, Knoops B, Bertrand P, Aguilar PVB (1998) Orientation of cell adhesion and growth on patterned heterogeneous polystyrene surface. J Neurosci Methods 84:193–204

Part II
Wetting and Interfacial Dynamics

Dewetting and Hole Formation in Spin-Coated Films of Lipid Bilayers

S.K. Sundar and Mahesh S. Tirumkudulu

Abstract Solid-supported lipid bilayers (SLBs) are model systems used to study a number of aspects of biological membranes such as the structural organization of lipids in localization of lipid-anchored molecules for cell signaling, and interaction forces in biological membranes to name a few. One of the most common techniques to obtain SLBs is via the spin-coating technique where a lipid dispersed in volatile organic solvent is spin-coated on a substrate. The dried film though of uniform thickness is riddled with holes whose origins remain unclear. To gain a better understanding of the hole formation process, we spin-coated lipid films of four different lipids dispersed in ethanol and chloroform on glass substrates and investigated the role of the nature of lipid, solvent, and film thickness on the characteristic length scale of the holes and the number density of the holes. For a fixed solvent and rotation rate, the average size of the hole increased with dry film thickness while the number density decreased with the film thickness. However, the measured hole sizes are about an order of magnitude lower than that predicted by the spinodal dewetting theory. The length scale of the holes was greater in the case of ethanol compared to chloroform though the predicted trends are opposite. Our results indicate that despite the discrepancy, the spinodal dewetting process plays a role in the hole formation.

Keywords Supported lipid bilayer • Hole formation • Spinoidal dewetting • Hydration • Spin-coating

1 Introduction

Phospholipid membranes or bilayers are formed when lipid molecules self-assemble so that the unfavorable interaction between the hydrophobic lipid tails and a surrounding hydrophilic medium such as water or humid air is avoided. Because of their ability to mimic biological membranes, these lipid bilayers are extensively used

S.K. Sundar • M.S. Tirumkudulu (✉)
Indian Institute of Technology Bombay, Mumbai 400076, India
e-mail: sundar.bio@gmail.com; mahesh@che.iitb.ac.in

Y.M. Joshi, S. Khandekar (eds.), *Nanoscale and Microscale Phenomena*,
Springer Tracts in Mechanical Engineering, DOI 10.1007/978-81-322-2289-7_6

as model membranes to study biological processes such as binding or unbinding of signaling molecules from the surface of the bilayer, interaction forces in biological membranes, and changes in film properties resulting from changes in chemical potential, pH, etc. [3, 13]. The lipid bilayers when supported over a substrate such as glass or mica are called supported lipid bilayers (SLBs). In such systems, a thin layer (about 2 nm) of water between the bilayer and the substrate maintains the membrane fluidity [6, 23]. The coupling of the membrane to a solid surface allows for ease of handling of the samples and makes them amenable to sophisticated analytical techniques to investigate the molecular structure and the interaction forces of the bilayer [13]. It has also been proposed that the SLBs could form the base for many optical and electrical-based biosensors and lead to novel biomolecular materials [14].

Solid-supported lipid bilayers are typically produced via one of the following three methods [13]. In the *Langmuir-Schaefer* technique, the lipid molecules dissolved in an organic solvent are spread over the surface of water placed in a trough. After evaporation of the organic solvent, the surface is compressed using barriers to obtain a high-density packing. The lipid monolayer film is then transferred to a solid substrate with tails pointing outward when the substrate is dipped horizontally through the monolayer. A second set of monolayer is transferred to the substrate to form a bilayer when the process is repeated. This technique is time-consuming and one has to accurately maintain the surface pressure for proper packing of the molecules. The *lipid fusion* technique involves transfer of preformed liposomes onto the solid substrate. A suspension of small vesicles is dried on a hydrophilic substrate which leads to the vesicle adsorption to the substrate followed by rupture and spreading on the substrate. However, the dynamics of the bilayer formation depends on the lipid composition. Complete SLB formation may not occur in presence of cholesterol or charged lipids. SLB can also be formed using the *spin-coating* technique. In this method, a small amount of the lipid dissolved in an organic solution is spin-coated on the solid support. During the spin-coating process, the solvent evaporates, leaving behind a partially saturated film. The sample is next placed in a desiccator to completely remove the solvent and then rehydrated in humid air to yield highly aligned multilayer film. The thickness can be varied from a few tens of nanometers to microns by varying the concentration of lipid in the solution and rotation rate of the coater [3, 7, 8, 11, 20].

In this work, we shall focus on the process of obtaining thin films of lipids using the spin-coating process, given its versatility and ease of use. The spin-coating process has become a universal technique to produce thin, well-controlled, and homogeneous films on a substrate with thickness ranging from a few nanometers to tens of micrometer. In this study, lipids dissolved in a volatile organic solvent are spin-coated on a substrate to obtain dried films that exhibit holes similar to those observed in dewetting of thin films of polymer; the latter has been investigated extensively in the literature [4, 12, 15, 18, 22].

The spinodal dewetting process in thin films is controlled by a balance between the destabilizing long-range van der Waals force, the stabilizing surface tension force, and viscous dissipation. The destabilizing van der Waals forces occur when the attractive interaction between a film and surrounding medium is weaker than the substrate and the surrounding medium. The dewetting criteria depend on the effective interface potential, $\Phi(h)$, defined as the excess free energy needed to move two interfaces from infinity to the film thickness h and is given by

$$\Phi(h) = -\frac{A}{12\pi h^2}, \tag{1}$$

where A is the effective Hamaker constant that accounts for the interactions between the film, the substrate, and the surrounding medium and h is film thickness. The film stability is determined by the sign of the curvature of the effective interface potential ($\Phi''(h)$), and the film is deemed unstable if $\Phi''(h)$ is negative ($A > 0$). The surface fluctuation, $\delta h = \delta h_o \exp(t/\tau(q))$, is characterized by the wavelength of the instability, $\lambda = 2\pi/q$, and relaxation time $\tau(q)$, where q is the wave number. The onset of the instability is set by the critical wave number, $q_c = \sqrt{A/4\pi h^4 \gamma}$, corresponding to an increase in surface roughness, the latter given by $\sigma \sim \exp(t/\tau)$ with $\tau > 0$ for instability. The relaxation time of the instability depends on the initial film thickness; viscosity of solvent, μ; and surface tension, γ. The film is destroyed by the fastest-growing unstable capillary wave with a wave number $q_m = q_c/\sqrt{2}$. The expression for q_c in terms of h indicates that the predicted wavelength (λ) scales as h^2, whereas the number density of holes, N, scales as h^{-4} [9, 12, 16, 17, 19, 24, 25].

The first quantitative comparison of the above predictions with experiments was done by Reiter [12] for dewetting of nonvolatile polymer films. Thin films of polystyrene dissolved in an organic solvent were spin-coated on a silicon wafer. After complete drying, when the temperature of the film was raised above the glass transition temperature, the polymer film destabilized and dewetted from the substrate, resulting in an array of droplets. Optical microscopy was used to observe the breakup of polystyrene films on silicon substrates as a function of film thickness. The observed number density of the holes and the wavelength of the instability yielded power law exponents were close to the predicted values.

As mentioned previously, the spin-coating process offers a route to achieve thin films of lipid bilayers on solid substrates. Gallice et al. [3] were the first to perform a detailed study of the lipid bilayer formation using spin coating on solid substrates. Using 1,2-dimyristoyl-sn-glycero-3-phosphatidylcholine (DMPC) dissolved in isopropanol, they casted films on silicon wafers and observed the presence of holes ranging from 2 to 6 μm in the dry films. The observed unit thickness variation across the film corresponded to the multilamellar repeat distance of about 5.2 nm. The films were then exposed to 100 %RH environment at 45 °C for 4 h which led to a large-scale reorganization of the bilayers and resulted in an array of large holes and patches. The maximum thickness between the hole and the top of the film was of the order of 120 nm suggesting that films remained thin despite the rearrangement. They investigated theoretically the stability of the thin lipid bilayers supported on a solid substrate and found that the dry film is stable if the only long-range interactions (van der Waals) considered are that between the air and the lipid layer and that between the lipid layer and the substrate. In other words, the lipidic film completely wets the substrate. To explain the observed hole formation in the presence of hydration, they account for the van der Waals interactions between the bilayers, entropic confinement caused by neighboring layers, and hydration effects. Their analysis predicts $\lambda \approx 12\,\mu m$ which is of the same order as the observed wavelength of the dewetting patterns post hydration. Simonsen et al. [20] used atomic force microscopy (AFM) and fluorescence microscopy to study the detailed structure of spin-coated lipid films on mica before and after hydration. Their experiments were performed with 1-palmitoyl-2-oleoylphosphatidylcholine (POPC) and 1,2-dioleoyl-sn-glycero-3-phosphocholine (DOPC) lipids dissolved in chloroform. They observed holes in the dried film though their AFM measurements suggest that none of the holes extended all the way to the substrate. The characteristic dewetting pattern closely resembled those observed in thin polymer films, indicating that the thinning of the lipid film during spin coating led to the rupture of the outer lipid layers followed by hole formation and eventual film breakup. The dry films were found to be rigid and stable over time. The holes were present over the entire the dry film with a characteristic length scale of a few micrometers though they did not observe evidence of an undulated breakup pattern which is typically associated with the dewetting process. However, the authors did not perform a systematic study to determine the number density and size distribution of the holes as a function of film thickness (or equivalently, the rotation rate or lipid concentration) to confirm their hypothesis.

In this study, we focus on the hole formation observed after spin coating and drying but without the hydration step. The hole size is measured as a function of the rotation speed for four different lipids whose acyl chain length and gel-liquid crystal transition temperatures are different, for varying initial film thickness and two different solvents. For a fixed solvent and rotation rate, the average size of the hole increased with dry film thickness while the number density decreased with the film thickness. The behavior in both cases appear to follow power law, indicating that the hole formation is caused by the spinodal dewetting process. However, the measured hole sizes are about an order of magnitude lower than that predicted by the dewetting theory. The length scale of the holes was greater in the case of ethanol compared to chloroform though the predicted trends are opposite. Our results indicate that for the lipid concentrations used in this study, the hole formation is primarily controlled by spinodal dewetting though more work is needed to understand the hole formation mechanism.

2 Experimental Section

1,2-Dimyristoyl-sn-glycero-3-phosphocholine (DMPC), 1,2-dihexanoyl-sn-glycero-3-phosphocholine (DHPC), L-phosphatidylcholine (Soya-PC), and 1-myristoyl-2-6-[(7-nitro-2-1,3-benzoxadiazol-4-yl)amino]hexanoyl-sn-glycero-3-phosphocholine (NBD-PC) were obtained in solution form from Avanti Polar

Lipids®, Alabama. 1,2-Dioleoyl-sn-glycero-3 -phosphocholine (DOPC) was obtained in powder form from Sigma-Aldrich® and used without any further purification. Solvents (chloroform and ethanol) used to dissolve lipids were of high grade and obtained from Merck® Chemicals. Microscopic glass cover slips of dimensions 24 × 64 mm were used for all coating experiments.

Cover slips used for experiments were cleaned using ethanol and then rinsed with MilliQ water. The lipids were dissolved in chloroform/ethanol and 100 μl of the sample was placed at the center of the cover slip which in turn was placed on the spin coater. The substrate was rotated at constant angular velocities ranging from 1,000 to 5,000 rpm for 60 s. Liquid spreads out from the center toward the edge of the substrate due to the centrifugal force with simultaneous evaporation of the solvent. The final thickness depends on the concentration of the solution, evaporation rate of the solvent, rotation rate, and rotation time. Experiments were performed for lipid concentrations of 10, 25, and 50 mM at a constant rotation rate of 1,000 rpm while two other rotation rates, namely, 2,000 and 5,000 rpm, were applied to samples with a lipid concentration of 10 mM. The samples were kept under vacuum for 20 min after spin coating to allow evaporation of residual solvents. The samples were exposed to the ambient conditions during the microscope visualization. The lipid film is known to be stable in air as the polar head groups are concealed from the outside atmosphere and only the acyl chains point outward [20]. A picture of a coated cover slip is shown in Fig. 1 and the boundary of the coated region is demarcated. It is seen that the film covers almost the entire area of the glass substrate.

Optical microscopy (60X (1.42 NA) oil objective, IX71 Inverted Microscope, Olympus, Japan) and scanning electron microscopy (JEOL 7600F FESEM, Japan) were utilized for the observation of holes. The size of the holes were measured using ImageJ software. All experiments were done three times to ensure reproducibility and about 10–15 images were captured for each experiment. Majority of the holes

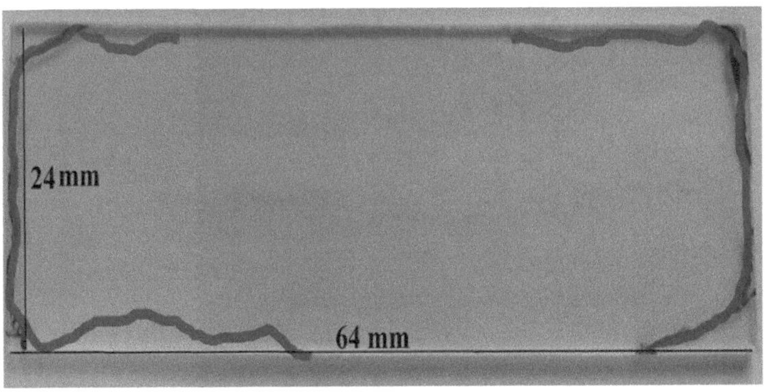

Fig. 1 Photographic image of a lipid coated slide. The boundaries of the coated section are demarcated

were greater than 500 nm as seen in scanning electron microscopy and atomic force microscopy, and so the size distribution of the holes was obtained using optical microscopy.

The thickness of the dry lipid film was measured by spectroscopic ellipsometry (M-2000V, J.A. Woollam Co, Inc.) [1, 10]. When a linearly polarized light irradiates a surface at a certain angle of incidence, the light undergoes reflection and refraction, and the light that reflects from the sample is elliptically polarized. Ellipsometry measures the polarization change of the incident light when it gets reflected after interacting with the sample. The polarization change is quantified by the change in amplitude ratio (Ψ) and the phase difference (Δ) to get information related to sample thickness. These changes are related to the Fresnel reflection coefficients, R_p and R_s, as $\tan(\Psi)\exp(i\,\Delta) = R_p/R_s$ for the p and s components of polarized light respectively. The Fresnel reflection coefficient is a function of refractive index and angle of incidence in air, film, and substrate. A suitable slab model is first constructed and then the values of amplitude and phase change provided by the model are compared with the experimental values by iterative procedures to get the film thickness. The measurement was done at wavelengths between 450 and 1,000 nm at angles of incidence of 65°, 70°, and 75°. The size of the laser spot was about 3 mm over the film surface. Data analysis and model fit were performed using a three-parameter Cauchy model [2]. The refractive index of the glass substrate was measured to be 1.47 and used as substrate reference to incorporate in the model for estimating the film thickness. The bottom side of the glass substrate was made opaque by attaching a tape. Thickness was measured twice and at three different spatial points per slide giving six thickness measurements per slide. Three such slides were prepared for each condition.

3 Results and Discussion

A thin liquid film will dewet if the effective Hamaker constant, $A \equiv A_{132}$, in the van der Waals interaction potential is greater than zero. Here, the subscripts $1, 3, 2$ represent the substrate, the solvent, and the air, respectively. The Hamaker constants for glass, chloroform, ethanol, and air were found to be $6.5 \times 10^{-20}, 5 \times 10^{-20}, 4.2 \times 10^{-20}$, and 14.8×10^{-20} J, respectively [5]. On substituting these values in the expression for the effective non-retarded Hamaker constant, $A_{132} = (\sqrt{A_{11}} - \sqrt{A_{33}})(\sqrt{A_{22}} - \sqrt{A_{33}})$, we find that $A_{132} \approx 5 \times 10^{-21}$ and 9×10^{-21} J for chloroform and ethanol, respectively, and the Hamaker constants are greater than zero in both cases. Assuming that the low concentration of lipid used in this study does not influence these values very much, the effective attraction between the two interfaces should lead to the breakup of the spin-coated thin film.

The surface morphology of the spin-coated lipid films was investigated using AFM and both optical and electron microscopy for varying lipid concentration, rotation rate, and solvent. Figure 2a, b presents the image of the same section of a thin dried film of DMPC observed using phase contrast and fluorescent microscopy,

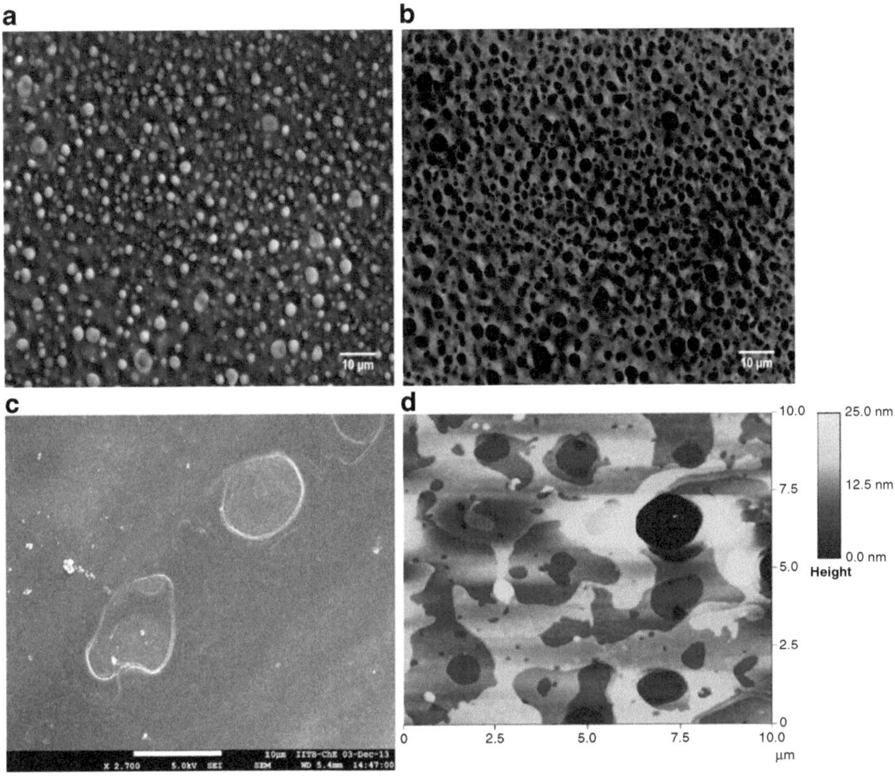

Fig. 2 Images showing holes in 10 mM DMPC lipid bilayer with chloroform as solvent and coated at 1,000 rpm. (**a**) Phase contrast microscopy, (**b**) fluorescent microscopy, (**c**) scanning electron microscopy, and (**d**) atomic force microscopy. Sub-images (**a**) and (**b**) pertain to identical sections of the same film

respectively. A small amount (1 % of the total lipid weight) of fluorescent lipid (NBD-PC) was added to the initial lipid solution so that the holes and patches could be easily distinguished. On comparing the two images, it is clear that the dark background in the phase contrast image corresponds to the patches of lipid bilayers while the holes correspond to the relatively lighter regions. The size of the holes varies from about 1 to about 5 μm. Figure 2c, d presents the SEM and AFM images of different regions on the film. While both the images confirm the morphology deduced from optical microscopy, the AFM image reveals the interconnecting patches of bilayer. Holes less than 500 nm in size were observed using AFM and SEM (image not shown), though the numbers were insignificant compared to holes greater than 500 nm. Hence, the size distribution of the holes was subsequently obtained solely using optical microscopy.

Figure 3 presents the size distribution of holes at three different concentrations (10, 25, and 50 mM) for the four different lipids. In all cases, the lowest concentra-

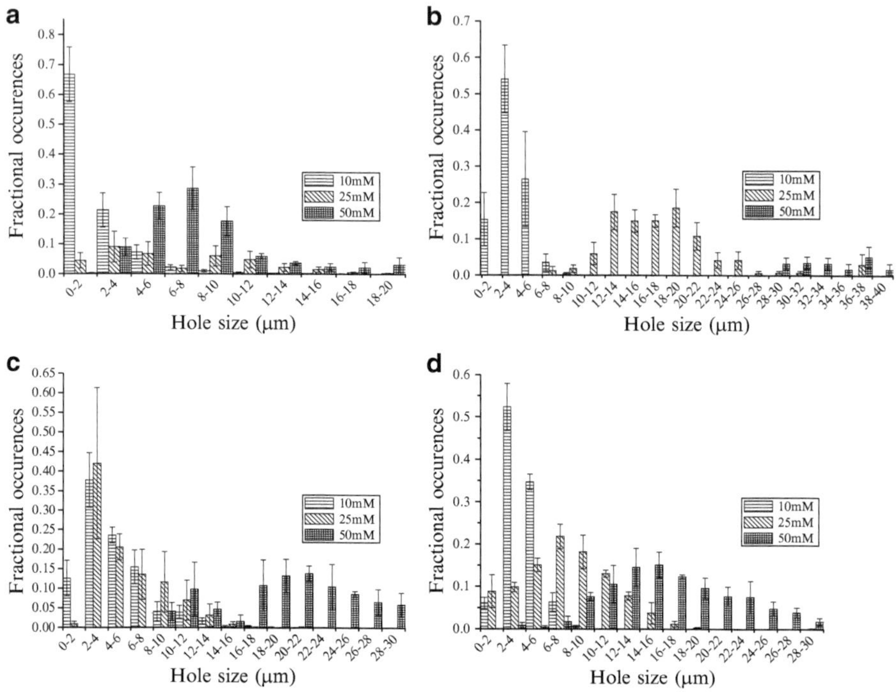

Fig. 3 Size distribution of holes obtained for the four different lipids in spin-coated films at a rotation rate of 1,000 rpm and at varying lipid concentrations for (**a**) DMPC (23 °C), (**b**) DOPC (−20 °C), (**c**) Soya-PC (0 °C), and (**d**) DHPC (43 °C). Here, the temperature in the bracket, the gel-liquid crystal transition temperature of each lipid

tion exhibits the smallest holes. Among the lipids, DMPC shows the smallest holes sizes over the three concentrations. In all cases, the size distribution is broad, leading to standard deviations of the order of 8–10 μm.

Figure 4 presents the size distribution for four different lipids for films spin-coated at 1,000 rpm in ethanol and chloroform. The hole sizes are much larger in the case of ethanol compared to those obtained with chloroform as the solvent although the effective Hamaker constant for a pure ethanol film is larger than that for a pure chloroform film on a glass substrate suggesting that the ethanol film should be more unstable and result in smaller holes. For the thin film instability caused by the destabilizing van der Waals attraction alone, the wavelength of the instability scales with square of the thickness,

$$\lambda = 4\pi \sqrt{\frac{\pi \gamma}{A_{132}}} h^2 .$$

On substituting the effective Hamaker constants for chloroform and ethanol and their respective surface tensions (27.5 and 22.1 mN/m), we find that the wavelength

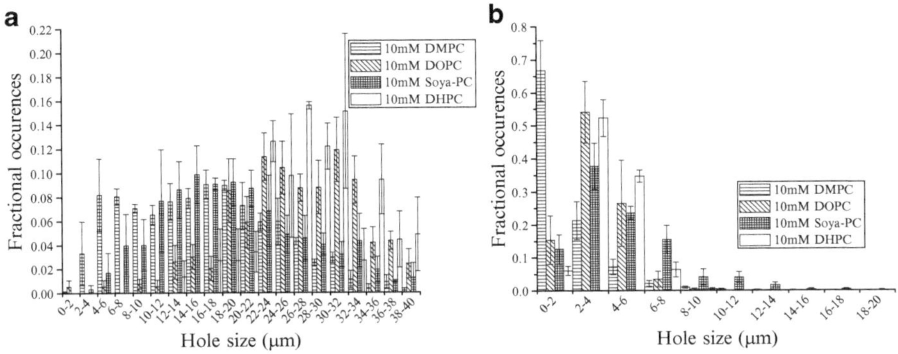

Fig. 4 Size distribution of holes obtained with different lipids at a rotation rate of 1,000 rpm with (**a**) ethanol as solvent and (**b**) chloroform as solvent

of the instability is about 350 and 520 μm for the ethanol and the chloroform films, respectively, for films of 100 nm thickness. These length scales are at least an order of magnitude larger than those of the measured hole size suggesting that effects other than those considered in this study may influence the hole formation. We also performed experiments with very high concentrations of lipids (~1 M) to determine if there existed a threshold thickness beyond which holes did not form. The film thicknesses were in the range of tens of micrometers and they too exhibited holes but of the order of 100 μm. However, multiple bilayers were observed inside the holes suggesting that the holes were not completely devoid of lipids. These films were not considered in the present analysis as the film thickness was not spatially uniform.

A possible alternate mechanism for hole formation in evaporating thin films may be due to condensation of water droplets, especially in the case of highly volatile solvents such as chloroform. Srinivasarao et al. [21] have shown that when a volatile liquid containing dispersed polymer, such as polystyrene, is evaporated in a humid environment, evaporative cooling at the liquid surface condenses water droplets over the film. If the solvent density is lower than that of water, the droplets sink into the film and form a 3D array of droplets surrounded by solvent containing the polymer. The polymer in the solvent acts to stabilize the droplets from coalescence. Complete evaporation of the solvent and water leaves behind a solid polymer film containing spherical air bubbles also known as breath figures. In the case of solvents denser than water, water droplets accumulate on the film surface resulting in a 2D array of spherical air bubbles. In our experiments with chloroform, the water droplets will remain on the solvent surface (chloroform being denser) and may influence the hole formation. However, in the case of ethanol, the condensing water drops will dissolve in the solvent and such effects will be absent, indicating that there should be no holes in the case of lipid films formed in ethanol. These observations suggest that formation of holes solely due to evaporative cooling and water condensation is unlikely.

The observed maxima in hole size distribution (λ) was plotted on logarithmic scale against the measured thickness (h) of dried lipid films obtained using ellipsometry for concentrations of 10, 25, and 50 mM (rotation rate of 1,000 rpm) and for rotation rates of 2,000 and 5,000 rpm (10 mM lipid concentration). By knowing the approximate area covered by the lipid film after spin coating and assuming head group area of lipid to be 0.6 nm^2 [5], calculations show the final dry film thickness to vary between 547 and 2,735 nm for lipid concentrations varying from 10 to 50 mM. The thickness measured using the ellipsometer was in the range of 40–430 nm which is lower than the estimate. This indicates that some of the fluid is thrown off the substrate during the spin-coating process.

Figure 5a presents the measured peak hole size as a function of the measured film thickness for all the experiments performed with both chloroform and ethanol. The average hole size and the peak hole size for all distributions are very close and so the latter is plotted for analysis. There is a large spread in the measured hole size and no clear trend can be ascertained. In order to eliminate the influence

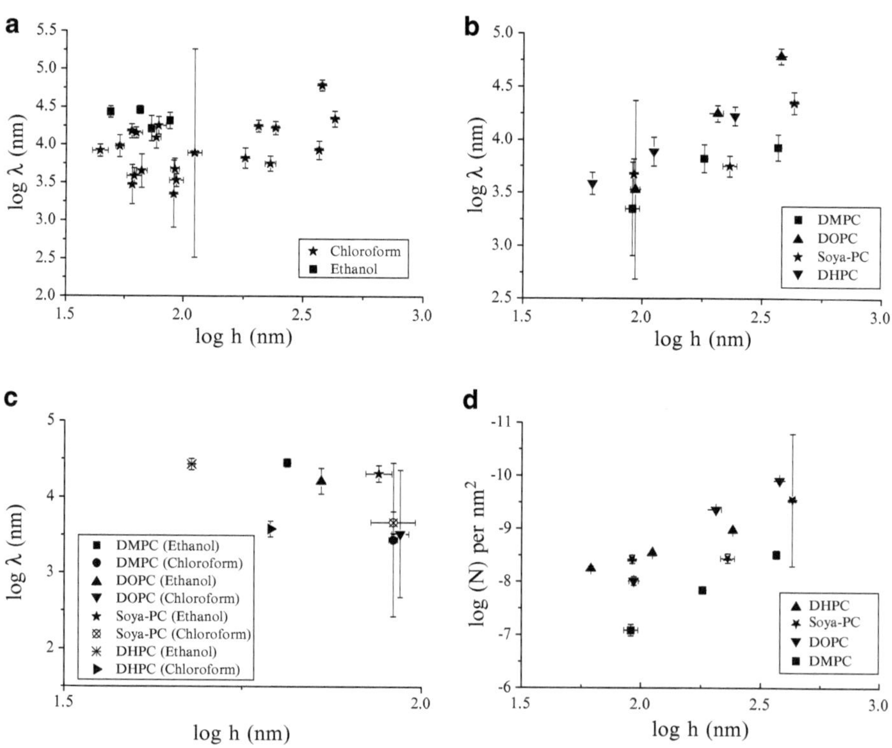

Fig. 5 Double logarithmic plot of hole size vs film thickness for (**a**) all lipid films with chloroform and ethanol as solvent, (**b**) all lipid films spin-coated at a rotation rate of 1,000 rpm in chloroform, and (**c**) all lipid films spin-coated at a rotation rate of 1,000 rpm in chloroform and ethanol at a lipid concentration of 10 mM. (**d**) Double logarithmic plot of number density of holes vs film thickness for all lipid films spin-coated at a rotation rate of 1,000 rpm in chloroform

of solvent and rotation rate, Fig. 5b plots the hole size measured for all lipids dispersed in chloroform and spin-coated at 1,000 rpm. Clearly, for a given lipid, the hole size increases with film thickness. Figure 5c plots the hole size for films spin-coated at 1,000 rpm for all lipids in both solvents at a lipid concentration of 10 mM, and it confirms that the hole sizes are larger in the case of ethanol than those with chloroform. An increasing hole size with film thickness suggests that the number density of the holes should concomitantly decrease with thickness, and this is confirmed in Fig. 5d for lipid films spin-coated in chloroform. The power law fit for λ vs h (Fig. 5b) yields exponents in the range of 0.9–2.1 and that for the number density vs h (Fig. 5d) gives an exponent of -1.2 to -3.2. While the exponents vary over a large range and are different from those predicted by the theory for spinodal dewetting, the observed trends do indicate that the unstable dewetting process of the solvent does play a role in the hole formation. Figure 5b, d indicate that the nature of lipid also influences the length scale of the hole and the number density of the holes. These results indicate that though the hole formation mechanism is controlled by the spinodal dewetting process, the Hamaker constant for the liquid film is influenced by the type of lipid dispersed in the solvent. We also varied the rotation rate from 1,000 to 5,000 rpm to obtain dry film thickness in the range of 40–100 nm. The hole size variation was of the order of the error bars and so no clear trend could be established.

The above results suggest that while the spinodal dewetting mechanism may play a key role in the hole formation, there are other effects that influence the hole formation. We have ignored the role of evaporation on the stability of thin films and it is possible that this may contribute to varying hole sizes. This may be one reason for the different hole sizes observed in chloroform and ethanol as the former has a much higher evaporation rate. We have also ignored the influence of lipid type on hole formation although experiments show that there is a distinct influence of the lipid on the hole size. These aspects will be investigated as part of a future investigation.

4 Conclusion

We have spin-coated lipid films of four different lipids dispersed in ethanol and chloroform on glass substrate and investigated the role of the nature of lipid, solvent, and film thickness on the characteristic length scale of the holes and the number density of the holes. For a fixed solvent and rotation rate, the average size of the hole increased with dry film thickness while the number density decreased with the film thickness. The behavior in both cases appears to follow a power law form, indicating that the hole formation is caused by the spinodal dewetting process. However, the measured hole sizes are about an order of magnitude lower than that predicted by the dewetting theory. The length scale of the holes was greater in the case of ethanol compared to chloroform though the predicted trends are opposite. Our results indicate that for the lipid concentrations used in this study, the dewetting process plays a role in the hole formation though more work is needed to fully understand the mechanism.

References

1. Benes M, Billy D, Benda A, Speijer H, Hof M, Hermens WT (2004) Surface-dependent transitions during self-assembly of phospholipid membranes on mica, silica and glass. Langmuir 20:10129–10137
2. Dufrent YF, Boland T, Schneider JW, Barger WR, Lee GU (1998) Characterization of the physical properties of model biomembranes at the nanometer scale with the atomic force microscope. Faraday Discuss 111:79–94
3. Gallice P, Fragneto G, Mennicke U, Salditt T, Rieutord F (2002) Dewetting of solid-supported multilamellar lipid layers. Eur Phys J E 8:275–282
4. Herminghaus S, Jacobs K, Mecke K, Bischof J, Fery A, Ibn-Elhaj M, Schlagowski S (1998) Spinodal dewetting in liquid crystal and liquid metal films. Science 282:916–919
5. Israelachvili JN (2011) Intermolecular and surface forces. Academic Press, Burlington
6. Johnson SJ, Bayerl TM, McDermott DC, Adam GW, Rennie AR, Thomas RK, Sackmann E (1991) Structure of an adsorbed dimyristoylphosphatidylcholine bilayer measured with specular reflection of neurons. Biophys J 59:289–294
7. Jurak M, Chibowski E (2007) Wettability and topography of phospholipid dppc multilayers deposited by spin-coating on glass, silicon, and mica slides. Langmuir 23:10156–10163
8. Mennicke U, Salditt T (2002) Preparation of solid-supported lipid bilayers by spin-coating. Langmuir 18:8172–8177
9. Mukherjee R, Das S, Das A, Sharma SK, Raychaudhuri AK, Sharma A (2010) Stability and dewetting of metal nanoparticle filled thin polymer films: control of instability length scale and dynamics. ACS Nano 4(7):3709–3724
10. Nielsen MMB, Simonsen AC (2013) Imaging ellipsometry of spin-coated membranes: mapping of multilamellar films, hydrated membranes, and fluid domains. Langmuir 29:1525–1532
11. Pompeo G, Girasole M, Cricenti A, Cattaruzza F, Flamini A, Prosperi T, Generosi J, Castellano AC (2005) AFM characterization of solid-supported lipid multilayers prepared by spin-coating. Biochimica et Biophysica Acta 1715(1):29–36
12. Reiter G (1992) Dewetting of thin polymer films. Phys Rev Lett 68:75–78
13. Richter R, Berat R, Brisson AR (2006) Formation of solid-supported lipid bilayers:? an integrated view. Langmuir 22(8):3497–3505
14. Sackmann E (1996) Supported membranes: scientific and practical applications. Science 271:43–48
15. Sferrazza M, Heppenstall-Butler M, Cubitt R, Bucknall D, Webster J, Jones R (1998) Interfacial instability driven by dispersive forces: the early stages of spinodal dewetting of a thin polymer film on a polymer substrate. Phys Rev Lett 81:5173–5176
16. Sharma A (2005) Self-organized structures in soft confined thin films. Pramana 65:601–614
17. Sharma A, Kargupta K (2003) Instability and dynamics of thin slipping films. Appl Phys Lett 83(17):3549–3551
18. Sharma A, Khanna R (1998) Pattern formation in unstable thin liquid films. Phys Rev Lett 81:3463–3466
19. Sharma A, Reiter G (1996) Instability of thin polymer films of coated substrates. rupture, dewetting, and drop formation. J Colloid Interface Sci 178:383–399
20. Simonsen AC, Bagatolli LA (2004) Structure of spin-coated lipid films and domain formation in supported membranes formed by hydration. Langmuir 20:9720–9728
21. Srinivasarao M, Collings D, PHilips A, Patel S (2001) Three dimensionally ordered array of air bubbles in a polymer film. Science 292:79–83
22. Stange T, Evans D (1997) Nucleation and growth of defects leading to dewetting of thin polymer films. Langmuir 13:4459–4465
23. Tamm LK, McConnell HM (1985) Supported phospholipid bilayers. Biophys J 47:105–113

24. Verma A, Sharma A (2010) Enhanced self-organized dewetting of ultrathin polymer films under water-organic solutions: fabrication of sub-micrometer spherical lens arrays. Adv Mater 22:5306–5309
25. Wyart FB, Daillant J (1990) Drying of solids wetted by thin liquid films. Can J Phys 68(9):1084–1088

Understanding Wetting Transitions Using Molecular Simulation

Tarak K. Patra, Sandip Khan, Rajat Srivastava, and Jayant K. Singh

Abstract The thermophysical and transport properties of fluid at nanoscale, adjacent to the solid surface, are quite different from that of macroscopic level. It has been realized that the fluid behavior can be tuned via modification of surfaces, which can unfold the underlying physics or mechanism of many biochemical processes or phenomena that exist in nature. In this chapter, we mainly emphasize the characteristic equilibrium properties of complex fluids near surfaces. In particular, we investigate different types of surface phase transitions such as layering transition, prewetting transition, and 2D vapor-liquid transition for associating (hydrogen-bond-forming) fluids using molecular simulations. These phases can be altered or controlled using chemically modified surfaces to tune the structure and the stability of the adsorbed layers. In addition, the wetting behavior of different amphiphilic molecules such as water and ethanol and their mixture on smooth and rough surfaces is examined based on the contact angle of the liquid droplet on the surface. Different types of wetting modes such as Cassie-Baxter, Wenzel, and impregnation of a droplet are observed as a function of roughness factor, surface fraction, and composition of the binary mixture. The effect of electric field on the phase transition of water under nanoconfinement is also presented. Further, we also discuss the phase transition and transport properties of 2D thin films.

Keywords Wetting transition • Chemically and physically modified surface • Phase equilibria • Transport properties • Line tension • Confinement • Electrical field • Thin film

1 Introduction

Wetting transitions of fluids on surfaces have recently received increasing research interests due to their potential applications in fields such as coating [1], micro-/nano-fluidics [2], chemical or biological sensor [3], and nano-tribology [4]. Wettability of

T.K. Patra • S. Khan • R. Srivastava • J.K. Singh (✉)
Department of Chemical Engineering, Indian Institute of Technology, Kanpur, 208016 UP, India
e-mail: jayantks@iitk.ac.in

Y.M. Joshi, S. Khandekar (eds.), *Nanoscale and Microscale Phenomena*,
Springer Tracts in Mechanical Engineering, DOI 10.1007/978-81-322-2289-7_7

a surface is generally characterized by the contact angle of a liquid droplet on the surface. A surface with contact angle of liquid droplet more than 150° is known as the superhydrophobic surface. Superhydrophobic surfaces are frequently observed in nature, especially in plants and insects such as lotus leaves, rice leaves, fisher spiders, and water strider feet. Usually, the superhydrophobicity of a surface is caused due to the presence of micro-/nanoscale textures on it. Textured surfaces are expected to have properties for anti-frosting, antifouling, and super-oleophobicity. The advent of micro-fabrication techniques has made it possible to control the chemical and topographical patterns on a substrate of micrometer length scales, and thus those surfaces become a viable option for micro- or nanoscale devices. Hence, understanding of fluid behavior on rough surfaces at micro-/nanoscale is required for efficient design of these devices.

Wettability and phase transitions of fluids under confinements, as in slit pores or cylindrical pores, have different characteristics in comparison to that on flat substrates. Water molecules, for example, bind themselves under confinement and form small clusters, whose hydrogen bonding network is strongly altered by the confining conditions [5], and evaporate spontaneously if confined in sufficiently narrow hydrophobic nanopore [6]. Hence, an understanding of the structural and dynamical properties of confined fluids continues to be a subject of increasing research interests in recent years. In addition to scientific interests, a fundamental understanding of the wetting phenomena inside pore geometries is necessary for many industrial and geophysical operations, as micro- and meso-porous materials are widely used for selective adsorption, pollution control and catalyst support for chemical reactions [7], designing nanosensors [8], artificial biological membrane, and drug delivery system [9]. In the case of slit-shape pores, the three-dimensional fluid behavior reduces to two-dimensional fluid-like behavior with decrease in the slit width. At a limiting condition, when slit width is of the order of fluid size, a monolayer or very thin film of fluid forms which has a dramatically different phase coexistence curve and critical properties from the bulk state. In case of incomplete spreading of liquid on a substrate, a film adsorbs on the solid surface, from the vapor phase, and may control the extent of incomplete spreading. Therefore, the thermophysical properties of thin films, vapor or liquid, play an important role in surface phase transition.

In this chapter, we present our recent efforts using molecular simulations to understand the surface phase transitions on smooth and rough substrates, effect of confinement on phase transition, and phase transition and dynamical properties of monolayers.

2 Thermodynamics of Wetting Transitions

Wettability of a surface is generally characterized by the behavior of fluid molecules near the surface. Therefore, a liquid drop on a surface (as shown in Fig. 1) can be used to quantify the wettability of the surface. The liquid drop on a substrate

Fig. 1 A liquid drop on a flat substrate: (**a**) partial wetting state; (**b**) complete wetting state; (**c**) complete drying state

under mechanical equilibrium yields a contact angle due to the interfacial tensions experienced by the drop at the three-phase contact line. The relation between contact angle and these interfacial tensions at the three-phase contact line is given by Young's equation:

$$\gamma_{\mathrm{SV}} = \gamma_{\mathrm{SL}} + \gamma_{\mathrm{LV}} \cos\ \theta, \tag{1}$$

where γ is the interfacial tension between two phases as indicated by the subscripts. The subscript S, L, and V refer to solid, liquid, and vapor, respectively, and θ is the contact angle of the droplet on the surface. The contact angle of a droplet can be found in between 0° and 180° depending on the spreading of the droplet on the surface. The spreading of the droplet or the length of the three-phase contact line is controlled by the interfacial tensions. For example, the length of the three-phase contact line becomes infinite ($\theta = 0°$) when the sum of the solid-liquid and liquid-vapor interfacial tensions is equal to the solid-vapor interfacial tension. As a result, the surface is completely covered by a liquid film as illustrated from Fig. 1b. On the other hand, the three-phase contact line reduces to a point when the sum of solid-vapor and the liquid-vapor interfacial tensions is equal to the solid-liquid interfacial tension ($\theta = 180°$). This situation is known as complete drying state and is shown in Fig. 1c. An intermediate state of the water droplet appeared when there is a finite contact line among the three phases. This state is known as partial wetting and is recognized by the contact angle in between 0° and 180° (see Fig. 1a). A surface is called hydrophilic when the contact angle is less than 90°. In case of hydrophobic surfaces, the contact angle is greater than 90°.

Young's equation is not applicable for micron-size liquid droplet due to the absence of three-phase interaction across the three-phase contact line. To account for the three-phase contact line, the line tension, τ, is introduced in Young's equation as follows:

$$\gamma_{\mathrm{SV}} = \gamma_{\mathrm{SL}} + \gamma_{\mathrm{LV}} \cos\ \theta + \left(\frac{\tau}{r_{\mathrm{B}}}\right), \tag{2}$$

where r_{B} and θ are the base radius of the droplet and the microscopic contact angle of the droplet, respectively. The line tension is defined as the interfacial energy per unit length across the three-phase contact line. Hence, this equation is useful for the

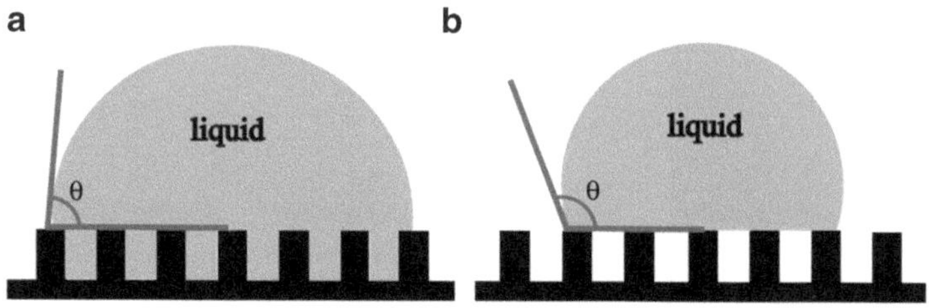

Fig. 2 Wetting states of a liquid drop on a textured surface: (**a**) Wenzel state; (**b**) Cassie-Baxter state

analysis of micron (or lesser)-size droplet [10]. Macroscopic contact angle can be evaluated from Eq. 2 by extrapolating $1/r_B \rightarrow 0$, which yields Young's equation:

$$\cos\,\theta_{\infty} = \left(\frac{\gamma_{SV} - \gamma_{SL}}{\gamma_{LV}}\right). \tag{3}$$

where θ_{∞} is the contact angle of an infinitely large droplet on a surface. Therefore, microscopic and macroscopic contact angle on a smooth surface can be correlated through Eqs. 2 and 3 as given below:

$$\cos\,\theta = \cos\,\theta_{\infty} - \left(\frac{\tau}{\gamma_{LV}}\right)\frac{1}{r_B}. \tag{4}$$

The contact angle of a liquid drop on a rough surface, on the other hand, depends on the roughness of the surface. Wenzel [11] introduced a factor in Young's equation to include the effect of roughness. Therefore, Young's equation for a rough surface becomes:

$$\cos\,\theta' = \frac{r\,(\gamma_{LV} - \gamma_{SL})}{\gamma_{LV}} = r\cos\,\theta \tag{5}$$

where θ' and r are the contact angle of the droplet on the rough surface and the roughness factor, respectively. The roughness factor is defined as the ratio of the actual area of a rough surface to the projected one. In this formalism, the wetting interface is assumed to be homogenous, i.e., there is no air trapped in between the droplet and the texture of the surface. On the other hand, Cassie-Baxter assumed that the droplet sits on the tip of the texture of the surface and proposed an equation based on such heterogeneous system as follows:

$$\cos\,\theta'' = rf\cos\,\theta' + f - 1 \tag{6}$$

where f is the surface fraction of a rough surface which is measured as the fraction of a rough surface wetted by the liquid.

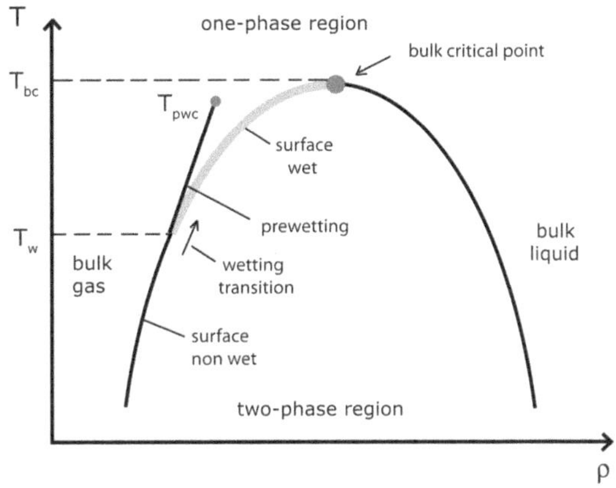

Fig. 3 Typical phase diagram of fluid near a surface

3 Phase Transitions on Surfaces

The adsorption of fluid molecules on solid surfaces at temperatures below their bulk critical temperature results in different phases such as percolation, layering, and prewetting. The adsorbed phase depends on the nature of surfaces. Figure 3 relates the surface phase transitions with the bulk phase diagram of a fluid. At the saturation condition, a fluid may wet the surface above a particular temperature called the wetting temperature (T_w), which is shown as a green curve in the figure. The wetting temperature for a system is defined as a temperature where a microscopic thin film discontinuously transforms to a macroscopically thick film. Prior to wetting the surface, i.e., at a subsaturation condition, a first-order transition might be observed between two surface phases of the film differed by its thickness. This transition is known as the prewetting transition. This prewetting transition stems from the saturation line at the wetting temperature and goes away from the saturation line with decrease in the density difference of thin and thick films and becomes indistinguishable at prewetting critical temperature, T_{pwc} (see Fig. 3). To observe such phase transition, precise control over pressure or density of the system is necessary, as it is very close to the bulk saturation line. Perhaps due to this fact, though it was theoretical prediction in 1977 [12, 13], the experimental evidence came much later in 1992. Such phase transition is observed for inert gases on alkali metal surfaces [14–16] and acetone and water on graphite surface [17, 18].

3.1 Prewetting Transitions

The prewetting transition is a first-order transition which occurs at a temperature in between the wetting temperature and the prewetting critical temperature of a system. Therefore, the length of prewetting transition is generally measured by the difference between the wetting temperature and the prewetting critical temperature. However, it is very difficult to locate the accurate wetting temperature and prewetting temperature from adsorption isotherms. As the prewetting transition stems from bulk saturation line at the wetting temperature, the bulk chemical potential and prewetting chemical potential should be same at the wetting temperature. Therefore, the difference between the bulk saturation chemical potential and prewetting chemical potential, $\triangle\mu$, can be used to evaluate the wetting temperature. It has been found that locating the locus of the prewetting line close to the bulk saturation line is very difficult from usual methods [13, 19–24]. For example, prediction of the wetting temperature (T_w) for the argon gas on a solid carbon dioxide substrate is found to be quite diverse varying from 0.58 to 0.96, depending on the methods and the temperature range considered for evaluating T_w. However, the grand canonical transition matrix Monte Carlo (GC-TMMC) method is quite effective to capture the coexistence of thin and thick films close to the bulk saturation line and found $\Delta\mu$ to follow the power law $(T\text{–}T_\mathrm{w})^{3/2}$ as also predicted by the theory [25]. On the other hand, prewetting critical temperature can be obtained from extrapolating the series of true boundary tension as a function of temperature to zero. We have applied the above methods to investigate the prewetting transition of different multiple-site associating fluids, which is discussed in the next section.

3.1.1 Prewetting Transitions on Smooth Surfaces

In our recent works [26, 27], we have investigated the prewetting transition of one-site, two-site, and four-site associating fluids on smooth surface for various associating strengths. These sites mimic the strong and short-range directional attraction of real associating fluids such as alcohol, hydrogen fluoride, and water. In case of two-site associating fluid, sites are located opposite to each other. On the other hand, sites are arranged in tetrahedral form for four-site associating fluid, similar in spirit of water models. The complete potential model [28] for associating fluids is

$$\begin{aligned} u_{\mathrm{ff}}\left(r_{ij},\theta_i,\theta_j\right) &= u_{\mathrm{LJ}-tr}\left(r_{ij}\right) + u_{\mathrm{af}}\left(r_{ij},\theta_i,\theta_j\right), \\ u_{\mathrm{af}}\left(r_{ij},\theta_i,\theta_j\right) &= \begin{cases} -\varepsilon_{\mathrm{af}} & \text{if } \sigma < r_{ij} < r_c, \theta_i < \theta_c \text{ and } \theta_j < \theta_c, \\ 0 & \text{Otherwise} \end{cases} \\ u_{\mathrm{LJ}-tr}\left(r_{ij}\right) &= \begin{cases} 4\varepsilon\left[\left(\frac{\sigma}{r_{ij}}\right)^{12} - \left(\frac{\sigma}{r_{ij}}\right)^{6}\right] & \text{if } r_{ij} < r_{\mathrm{cut}}, \\ 0 & \text{Otherwise} \end{cases} \end{aligned} \tag{7}$$

where θ_i and θ_j are angles between the center-center and the center-site vectors of molecules i and j, respectively. $\varepsilon_{\mathrm{af}}$ is the association energy and r_c is the range of site-site interaction. σ and ε are the model parameters for the Lennard-Jones (LJ) potential. r_{cut} is the cutoff diameter for the LJ potential. We adopt units such that σ and ε are united. In this study, θ_c, r_c, and r_{cut} are fixed at 27°, 1.00, and 2.5, respectively. Based on the number of sites and the association strength, this model

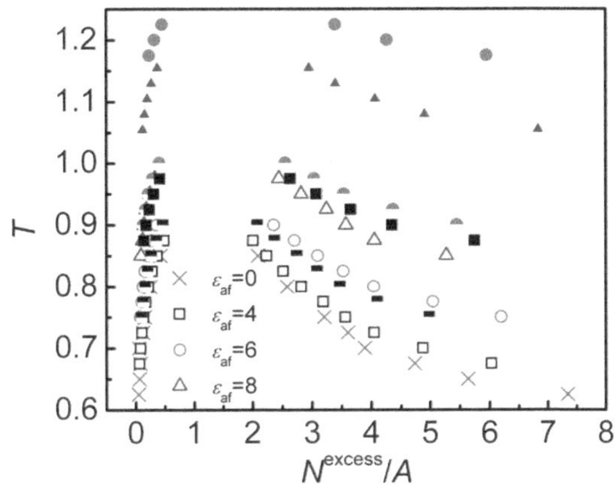

Fig. 4 Surface phase coexistence curves for multiple associating fluids. *Cross*, *open*, *half close*, and *close symbols* represent zero-site, one-site, two-site, and four-site associating fluids, respectively (Reprinted (adapted) with permission from Ref. [27]. Copyright (2012) Taylor & Francis)

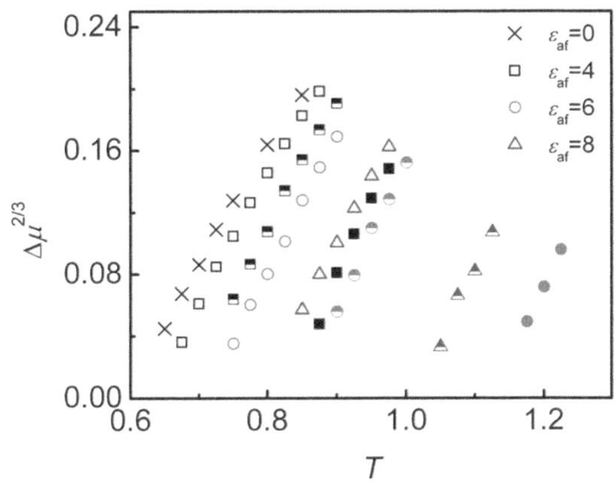

Fig. 5 Difference between bulk saturation chemical potential and chemical potential for surface phase transition vs. temperature for multiple-site associating fluids (Reprinted (adapted) with permission from Ref. [27]. Copyright (2012) Taylor & Francis)

can be used for real associating fluids. For example, Tsangaris and de Pablo [29] had extensively studied this model to parameterize the acetic acid with single-site associating fluid and found $\varepsilon_{af} = 20$ as the optimized value for acetic acid.

We observe a first-order transition between thick and thin films for different temperature ranges depending on the number of sites and site strength of associating fluids. Figure 4 presents the phase diagram of thin and thick films of different associating fluids. In general, the prewetting critical temperature increases with increase in associating strength as well as with number of sites. However, the prewetting span decreases with increase in associating strength as well as number of sites. Thick film density is also found to increase with increasing associating strength and number of sites. This is due to the dimer formation in case of one-site associating fluid, chain formation in case of two-site associating fluid, and three-dimensional complex structures in case of four-site associating fluid.

Figure 5 shows the span of prewetting transition for multiple-site associating fluid with various associating strength. The prewetting regime for all associating fluids is found to fall within a triangle. Hence, the triangle can provide a complete picture of the region of prewetting transition for all sets of associating fluids with variable sites (0–4) and associating energies (0–8). For example, the prewetting transition for one-site associating fluid can be observed for large values of associating strength, whereas for two- and four-site associating fluids, the prewetting transition is absent beyond associating strength 8 and 6, respectively. This suggests that the prewetting transition of multiple-site associating fluids with high associating strength such as hydrogen fluoride and water is very difficult to observe, which is evident from a very few number of experimental observations for polar molecules [17].

3.1.2 Prewetting Transitions on Functionalized Surfaces

The presence of surface sites can also influence the growth of adsorbed film through network-like bridging across the adjacent layers. Hence, the stability of the adsorbed film can be improved through proper arrangement of surface sites as well as their orientations. These surface sites can even lead to different kinds of surface phase transitions. For example, hydroxyl group density on the silica surface significantly influences the water adsorption on the surface. The hydrogen bonding across hydroxyl group and interfacial water molecules compensates the hydrogen bond loss in the bulk and increases the stability of the adsorbed layers [28, 30–32].

In this work, functional surface is modeled with a structured surface with active sites uniformly distributed as shown in Fig. 6. The site density ρ_s is defined as the number of sites per unit of area. Surface site interaction is equal to the fluid site interaction, but the interaction range r_c is used as 1.2.

The growth of a thick film is substantially suppressed in the presence of surface sites for low associating fluids with decrease in temperature. This is a characteristic feature of the quasi 2D vapor-liquid transition, where coexistence chemical potential shifts from the bulk saturation line with decrease in temperature and finally splits in layering transition as shown in Fig. 7. In the presence of surface sites, an additional fluid layer adjacent to the surface is observed, which represents the bonded molecules with surface sites. The number of molecules in that layer increases with decrease in temperature and significantly affects the subsequent

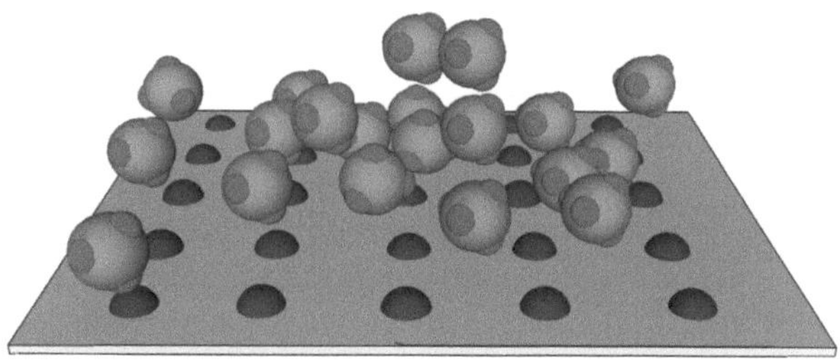

Fig. 6 Schematic diagram of associating molecules on a surface with associating sites

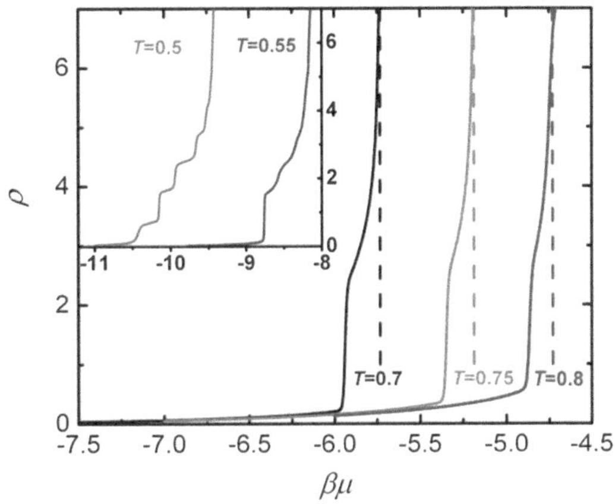

Fig. 7 Adsorption isotherm for $\varepsilon_{af} = 4.0$ and $\rho_s = 1.0$ at different temperatures. The *inset* represents adsorption isotherms for lower temperature range, which is not shown in the phase diagram. *Dash lines* represent the bulk saturation chemical potential (Reprinted (adapted) with permission from Ref. [26]. Copyright (2011) American Chemical Society)

layers, which is responsible for the shift in nature of phase transition. However, at higher associating strength, prewetting transition though intact shifts slightly toward lower temperature range.

3.2 *Wetting Transitions*

The knowledge of the surface properties of various materials and their temperature dependence allows one to utilize them in a more efficient way. One of the methods frequently used to obtain information about surface properties is via the contact angle, which a drop makes with a solid surface. Various methods can be employed to calculate the contact angle [33–36]. Contact angle of a nanodroplet on a surface can also be obtained directly by molecular simulations [37, 38]. We have conducted extensive molecular dynamics simulations to estimate the contact angle of liquids on smooth and textured surfaces.

3.2.1 Wetting on Smooth Surfaces

In our recent work [39], we have studied the wetting temperature and the line tension for water-graphite system using molecular dynamics simulations. The water-water interaction is described by the SPC/E [40] model, and the surface-water interaction is represented by the Lennard-Jones potential. Finite-size scaling approach has been used to obtain the macroscopic contact angle. In this approach, we have performed series of simulations using different system sizes, viz., 4,000, 5,000, 6,000, and 7,000 water molecules, at different temperatures. The microscopic contact angle for each system size is evaluated as per the method described by Dutta et al. [39]. These microscopic contact angles are then extrapolated to get the macroscopic contact angle as shown in Fig. 8.

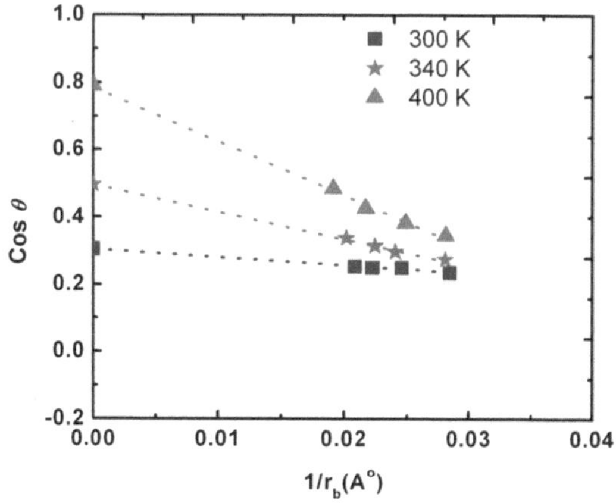

Fig. 8 Dependence of the size of a water droplet on its contact angle on a BN surface at different temperatures. Symbols, from *right* to *left*, represent droplet with 2,000, 3,000, 4,000, and 5,000 water molecules (Reprinted (adapted) with permission from Ref. [39]. Copyright (2011) Elsevier)

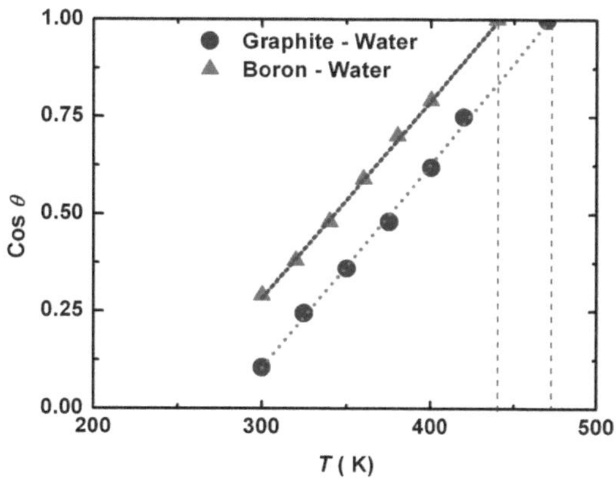

Fig. 9 Dependency of cosine of macroscopic contact angle θ against temperature for graphite-water and BN-water systems. *Dashed line* along the data points is a guide to the eyes. *Dashed vertical lines* represent the estimated wetting temperatures (Reprinted (adapted) with permission from Ref. [39]. Copyright (2011) Elsevier)

Wetting temperature, T_w, is evaluated from a series of contact angles as a function of temperature. The data is extrapolated linearly to obtain the temperature at which contact angle becomes zero as described in Fig. 9. Using the above approach, we have calculated T_w for the graphite-water (470 ± 5 K) and boron-nitride (438 ± 5 K) systems. Our estimate of wetting temperature of water on the graphite surface is in agreement with that obtained from the grand canonical Monte Carlo (GCMC) simulations of Zhao [41].

To calculate the line tension, we have taken the advantage of Eq. 4. System size analysis data can be used to obtain the slope, $-\tau/\gamma_{LV}$. The numerical values for vapor-liquid surface tension, γ_{LV}, were taken from the study of Vega and de Miguel [42] for SPC/E water at different temperatures. The calculated values of line tensions are in the order of 10^{-11} N and are in good agreement with the line tension values reported in the literature [43, 44]. We have obtained 3.06×10^{-11} N as the line tension value at 300 K [39], which is in good agreement with the previously reported values on graphite-water system at the same temperature by Werder et al. [45]. In a recent work, Zangi and Berne [46] assumed a constant line tension for different

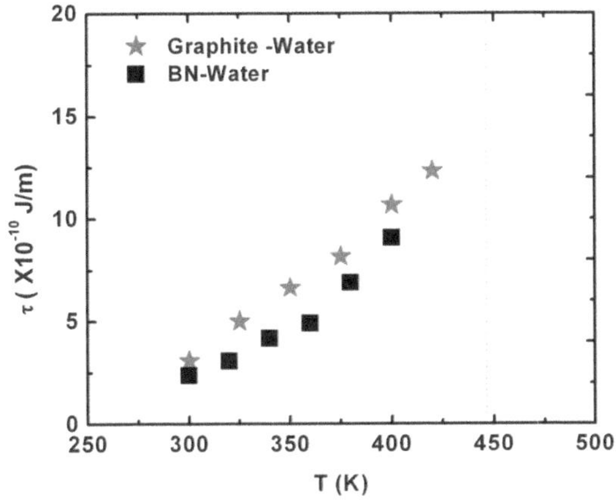

Fig. 10 Line tension vs. temperature for graphite-water and BN-water systems (Reprinted (adapted) with permission from Ref. [39]. Copyright (2011) Elsevier)

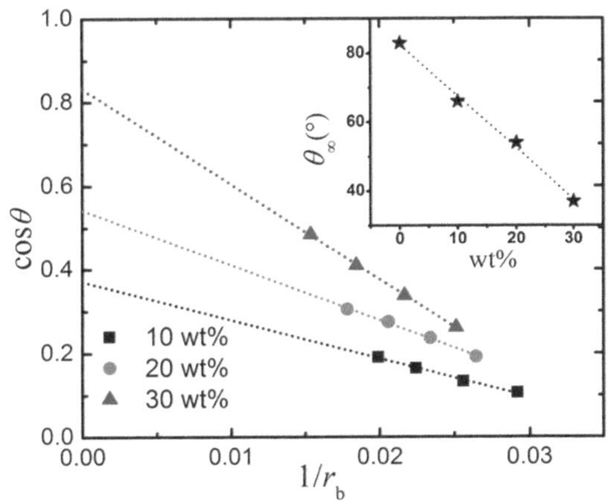

Fig. 11 Dependence of the size of an ethanol-water droplet on its contact angle on a smooth graphite surface, at $T = 300$ K, for different ethanol concentrations. The inset shows the variation of contact angle of an infinitely large droplet as a function of weight percentage of ethanol

temperatures for their study of water on a hydrophobic plate. On the contrary, we have found that the line tension substantially increases with increasing temperature. The variation of the line tension for water on graphite and BN surfaces is shown in Fig. 10.

We have also computed the line tension for water-ethanol solution for different ethanol concentrations at 300 K from the system size analysis of the contact angle of water-ethanol solution as shown in Fig. 11. The line tension is found to increase with increase in ethanol concentration. On the other hand, the contact angle of water-ethanol solution is observed to decrease with increase in ethanol concentration. Therefore, it also suggests that the line tension increases as the system approaches the complete wetting, i.e., with increasing ethanol concentration.

In general, there is a consensus of positive line tension near the first-order wetting transition [47]. At the complete wetting, short-ranged interactions are characterized by a finite τ with finite slope, while the retarded van der Waals interactions exhibit finite τ with diverging slope, and the long range interactions (e.g., non-retarded van der Waals) exhibit divergence in τ. However, which of the two forces (long or short ranged) dominate cannot be ascertained from the calculated line tension data.

3.2.2 Wetting on Textured Surfaces

It is well known that roughness plays an important role over wetting phenomena. Both experiments and simulations find significant effect of surface roughness on the wetting properties, which can be explained by two main hypotheses attributed to Wenzel and Cassie [11, 48]. Numerous fractal structures have been reported experimentally, which are able to significantly improve the surface property of a solid surface [49–55]. Along with experiments, numerous simulation works have been reported on the effects of the surface texture on wetting transition [56–59]. We have studied the wetting behavior of pure water and as well as water-ethanol mixture on textured surfaces.

Water Droplet on Textured Surfaces

We have studied the effect of roughness on wetting behavior of grooved patterned graphite surfaces [60] by examining the change in contact angle of a water droplet. In this work, we have investigated different types of wetting states for various pillar heights, h(1–8 atomic layers), and surface fractions (of pillars), f(0.25–0.51). We have observed that the equilibrium state of the water droplet on the rough surfaces significantly depends on the pillar height, surface fraction, and the way the droplet is placed on the surfaces. The typical snapshots of the equilibrium state of a water droplet at lower pillar height $h=2$ and at higher pillar height $h=4$ for various surface fractions are shown in Fig. 12. In general, for higher surface fraction and for higher pillar height, the equilibrium state of the water droplet is found to be in the Cassie-Baxter state. For example, for a surface fraction $f=0.51$, a water droplet is always in the Cassie-Baxter state for all pillar heights irrespective of its initial position. On the other hand, for a surface fraction $f=0.39$, there is a critical pillar height ($h=2$) below which the droplet is found to be in the Wenzel state. Above the critical pillar height ($h>2$), the droplet is always in the Cassie-Baxter state irrespective of its initial position. With further decrease in surface fraction, $f=0.25$, the critical pillar height increases to $h=3$ (corresponding to a roughness factor $r=1.95$). However, at this surface fraction, the equilibrium state of the droplet depends on its initial position. If the initial state of a droplet is close to the Wenzel state, then the equilibrium state of the droplet is found to be in the Wenzel state and vice versa. This observation clearly affirms the coexistence of the Wenzel and the Cassie-Baxter state on textured surfaces. It is obvious that there is a free energy barrier between two states for such systems. We have estimated the free energy for those systems using phantom wall method, which is described in detail in the next section. The crossover behavior of water droplet due to pillar height is also observed in many experimental studies, and the critical roughness factor is found to

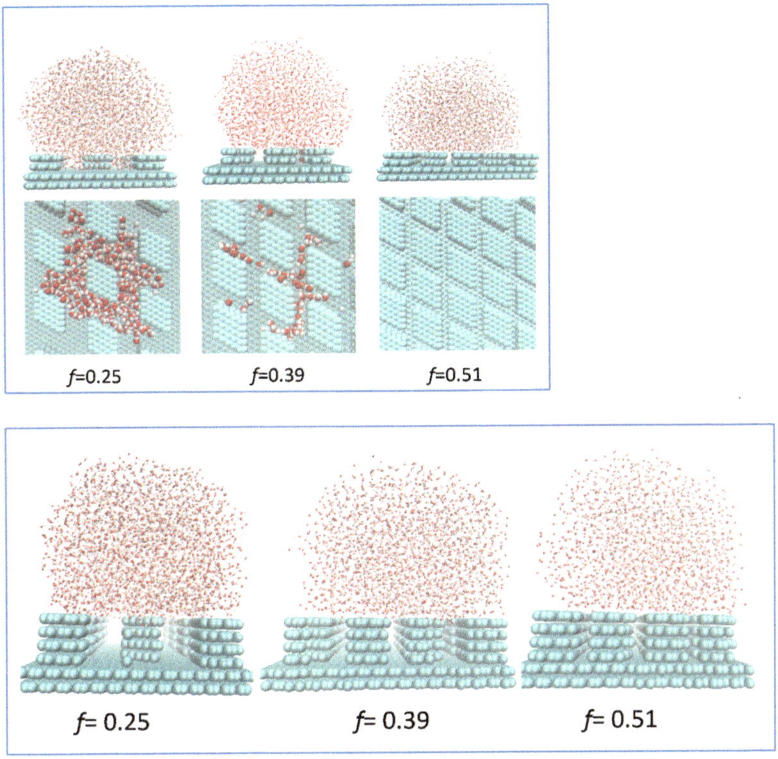

Fig. 12 Configuration snapshots of a water droplet on pillared surfaces for different surface fractions at $h=2$ (*top panel*) and at $h=4$ (*bottom panel*)

Table 1 Different wetting states of a water droplet as a function of surface fraction and pillar height

	$f=0.25$	$f=0.39$	$f=0.51$
$h=1$	Wenzel	Partial Wenzel	Cassie
$h=2$	Wenzel	Partial Wenzel	Cassie
$h=3$	Wenzel	Cassie	Cassie
$h=4$	Wenzel/Cassie	Cassie	Cassie
$h=6$	Wenzel/Cassie	Cassie	Cassie

be in between 1.2 and 1.35 [49, 55, 61, 62]. However, the theoretical prediction for the critical roughness factor at this range of surface fractions is in between 1.9 and 2.3. Recent studies of molecular dynamic simulations [63, 64] are also in line with our observation and find the transition roughness factor to be around 1.8–2.0. The equilibrium states of a water droplet for different pillar heights and different surface fractions are summarized in the Table 1.

The wettability of the rough surfaces is also examined by the contact angle of an equilibrated water droplet. It is found that the wetting behavior of a water droplet is significantly affected by the sharp edges of pillars. As a result, the contact angle

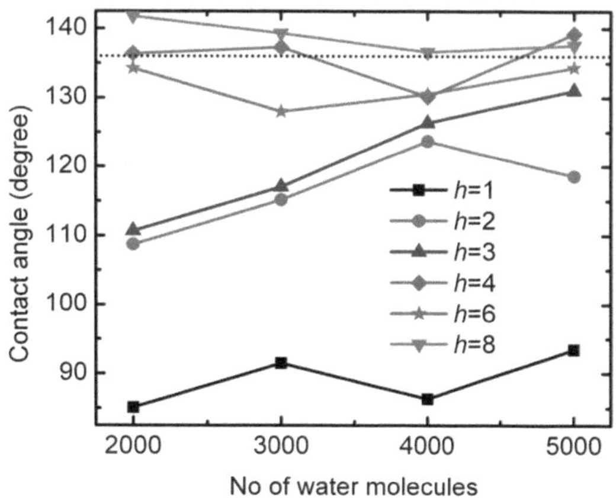

Fig. 13 Contact angle variation as a function of number of water molecules, at $f = 0.25$, for different pillar heights, $h = 1$–8. The *dotted line* represents the contact angle value predicted by the Cassie-Baxter equation. Errors associated with the contact angles are within 1°

of a water drop depends on the system size and on the initial configuration as well. We have rigorously investigated the system size effect on the contact angle of a water droplet. We measure the contact angle of a droplet as a function of its size for different pillar heights at a surface fraction $f = 0.25$. The contact angle of a droplet varies with its size and with the pillar height as shown in Fig. 13. The droplet is in the Wenzel and Cassie-Baxter states for $h \leq 3$ and $h > 3$, respectively, as we have discussed in the previous section for $f = 0.25$. Thus, the fluctuation of the contact angle in the Wenzel state with the change in system size is more pronounced in comparison to that for the Cassie-Baxter state, as clearly visible in Fig. 13. It is also observed that the contact angle of the droplet on the Cassie state is very close to the value predicted by the Cassie-Baxter equation (around 136°). On the other hand, the contact angle of the water drop in the Wenzel regime is significantly off from the value predicted by the Wenzel equation. We have also observed that the base layer of a droplet dose not spread as its size increases in the Wenzel state. The spreading is blocked in between the grooves of the surface. This results in fluctuation in the contact angle values with increase in the size of a droplet. We have extensively studied this issue for one atomic pillar height surface ($h = 1$). We have found that a droplet should contain at least 5,000 molecules to minimize the effect of the surface texture for $h = 1$. This also suggests that when droplet size is comparable with pillar dimensions, there could be many possible localized stable structures of Wenzel drops. In the next section, we describe the different localized stable structures that coexist on a rough surface and the procedure to estimate the free energy difference between those structures.

Earlier, we have discussed that there is an intermediate region for surface fraction (in that case, $f = 0.25$) for which both the Cassie and Wenzel states can coexist beyond the critical pillar height ($h > 3$). In recent years, some methods are developed to calculate the free energy difference between two wetting states [64–69]. For example, Koshi et al. [64] applied an external force to visit both the wetting states and calculated the work done during the transition between the two states. Kumar et al. [68] also demonstrated a method to evaluate the free energies of different wetting states for the Lennard-Jones system using the grand canonical Monte Carlo (GCMC) simulation. Further, Leroy and Muller-Plathe [69] developed a phantom wall method to create a path between specific wetting states and that of a droplet on a smooth surface. Thus, the free energies of different wetting states were calculated with respect to that of a droplet on a smooth surface. Recently, we have also created an artificial path to connect the Wenzel state and the Cassie-Baxter state with a droplet on a smooth surface through solid-fluid interactions [60]. The solid-fluid interactions is modulated thorough a coupling parameter λ, where $\lambda = 1$ represents the full interaction between the pillars and the water droplet and $\lambda = 0$ represents no interaction between them. We have started with two initial configurations, one is the Cassie state and the other is the Wenzel state, which are coexisted configurations for the system $f = 0.25$ and $h = 4$ as discussed earlier. The initial configurations are represented by A and D at $\lambda = 1$ as shown in Fig. 14. We have varied the λ value from 1 to 0 to reach the target system (a droplet on the smooth surface). The transformations are shown by the solid red (a droplet at the Cassie-Baxter to the target system) and blue (a droplet at the Wenzel state to the target system) lines in the figure. The derivative of free energy with respect to λ value is computed along the routes. The solid red line is quite smooth, whereas the solid blue line changes abruptly with change in λ value. The close examination of the structures, corresponding to the λ values, reveals some metastable structures that are designated by B and C in Fig. 14. These metastable structures associated with the thermodynamic route make it difficult to compute the free energy difference between the wetting states on a rough surface and on a smooth surface. In this work, we have found four possible equilibrium structures for the surface fraction $f = 0.25$

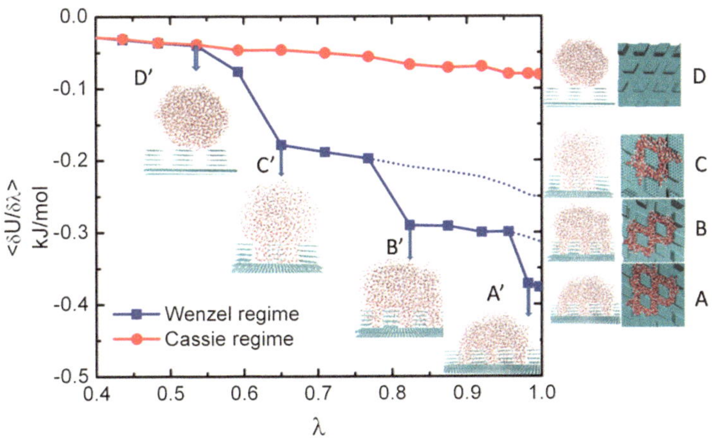

Fig. 14 The free energy change during transformation between the pillared surface and the smooth surface at $f = 0.25$ and $h = 4$. States A, B, C, and D represent the different equilibrium structures during the transformation. *Solid line* represents the reverse path ($\lambda = 1$–0) and *dotted line* represents the forward path. States A', B', C', and D' represent intermediate structures during the transformation

at $h = 4$ (one Cassie state and three Wenzel states). We further extend the phantom wall method to estimate the free energy difference between those stable structures, which is discussed below.

The phantom wall method is basically a thermodynamic integration technique. In this method, a phantom wall carries a water droplet and places it on the top of a texture surface or vice versa, which is completely reversible and computes the free energy change during the whole transformation. Hence, it basically calculates the free energy of a droplet on textured surfaces with respect to a droplet on the phantom wall. In our study, a smooth graphite surface is used as the phantom wall. An equilibrated droplet on the phantom wall is placed above a pillared surface beyond the water-pillar interaction cutoff distance. Subsequently, the phantom wall is then gradually shifted toward the pillared surface, in increment of 1 Å. As there is no interaction between phantom wall and pillared surface, the water droplet only interacts with pillars, and it goes through the pillars and finally sits on the base layer of the pillared surface while the phantom wall crosses the pillared surface. The free energy difference between the droplet on the pillared surface and the droplet on the phantom wall is computed along the path and is shown on Fig. 15 as a function of distance between the base layer of pillared surface and smooth surface. The reversibility of the transformation is also checked by shifting the phantom wall with an equilibrated water droplet toward the pillared surface and is represented by a solid blue line. We have observed a sharp change in free energy at the locations where the phantom wall crosses the top of the pillars and the base layers of the pillared surface. In between, the change in free energy is smooth. Therefore, the whole transformation consists of three distinct regimes: (1) the droplet at the Cassie-Baxter state transforms into the Wenzel state, (2) the droplet is carried to the base of

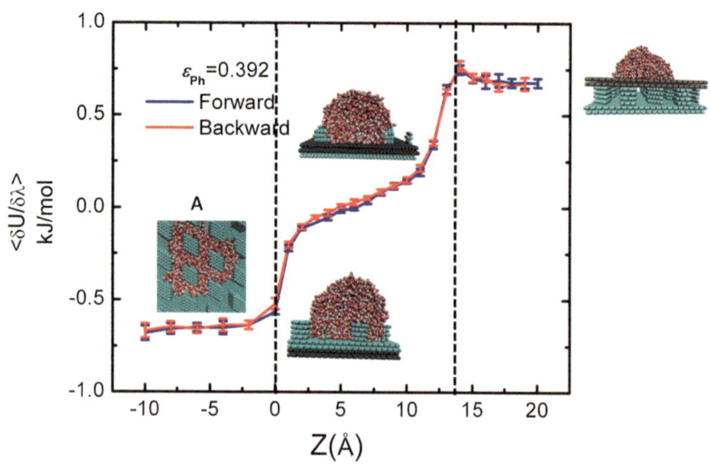

Fig. 15 Free energy change for state *A* during the transformation between the pillared surface and the smooth surface at $f = 0.25$ and $h = 4$. Strength of the phantom wall, $\varepsilon_{ph} = 0.392$, is equivalent to that of the smooth graphite-water system

the pillared surface, and (3) finally the droplet gets into a specific Wenzel state (here, it is the state A as defined in the earlier section). The strength of the phantom wall is found to be the key parameter for determining the final wetting state. Therefore, the final equilibrated wetting states B, C, and D are also achieved by changing the strength of the phantom wall, 0.30, 0.20, and 0.10 kJ/mol, respectively. Overall, the Wenzel states are found to be more stable compared to the Cassie state. In general, this method can be employed to other complex systems but needs to find the appropriate phantom wall strength with respect to the fluid.

Water-Ethanol Mixture on Textured Surfaces

The fundamental understanding of wetting behavior of binary mixtures is of great importance for many biological and chemical processes. One of such mixtures is ethanol-water, which has many practical applications in chemical, medical, and pharmaceutical industries. Hence, we have investigated the wetting behavior for water-ethanol mixture and used the same rough surfaces for the sake of comparison with the wetting behavior of pure water droplets.

The wetting behavior of an ethanol-water mixture is investigated on the rough surface with $f = 0.25$ and $h = 4$ at different ethanol compositions. As seen earlier, the equilibrated water droplet on the top of the pillars is in the Cassie-Baxter state. However, the droplet can change its wetting state to the Wenzel state upon application of an external force. The external force helps to overcome the free energy barrier between the Cassie-Baxter and the Wenzel states. In this study, we have started with a droplet in the Cassie-Baxter state with variable ethanol concentration and found different wetting states. Figure 16 represents MD snapshots of those different wetting states with increase in ethanol concentration. Particularly, the equilibrated droplet is found to be in the Cassie-Baxter state up to ethanol concentration of 5 wt-%. At 10 wt-%, some water and ethanol molecules are found to penetrate into groves of the textured surface leading to a state in the partially wetting regime, referred as the partial Wenzel state. Beyond ethanol concentration

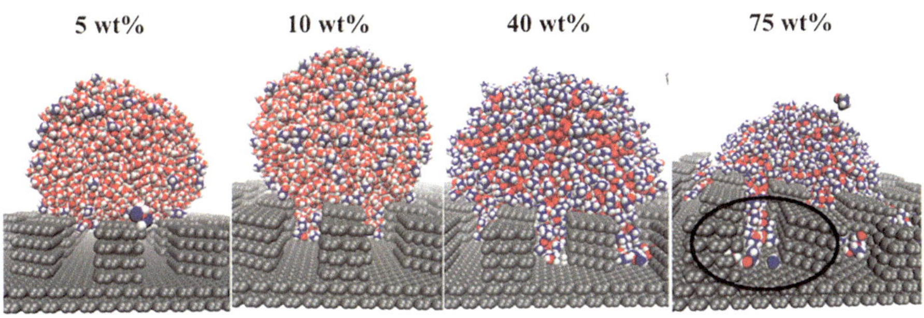

Fig. 16 Configuration snapshots of ethanol-water droplet on a rough surface ($h = 4$ and $f = 0.25$) for different ethanol composition 5, 10, 40, and 75 wt-%, respectively

of 20 wt-%, the ethanol-water droplet is found in the complete Wenzel state. The impregnation of the ethanol-water droplet (indicated by a black circle in the corresponding snapshot) is also found at an ethanol concentration of 75 wt-%. In general, the addition of ethanol molecules has helped to overcome the free energy difference between different wetting states. The analysis of the density and hydrogen bond profile (not shown) has revealed that the change in wetting state with increasing ethanol concentration is mainly due to the distribution of ethanol molecules in the droplet. It is clearly seen that the ethanol molecules are preferentially adsorbed at the vapor-liquid interface and close to the solid surface. Therefore, the ethanol molecules close to the groves assist in changing its wetting states from Cassie-Baxter → partial Wenzel → Wenzel → impregnation state with increase in ethanol concentration.

The presence of ethanol molecules enhances the wetting behavior of water-ethanol solutions, and as a result, the wetting behavior of water-ethanol solution shows significant difference from that of pure of water with increase in ethanol concentration. For example, at a surface fraction $f = 0.25$, the critical pillar height for pure water droplet is $h > 3$ (as seen from Table 1); but it increases to $h > 4$ for 10 wt-% ethanol solution and even increases to $h > 6$ for 75 wt-% ethanol solution. Similarly, for a surface fraction $f = 0.39$, the water droplet is in the partial Wenzel state, whereas for 10 wt-% ethanol solution, it is found in the complete Wenzel state. On the other hand, for a surface fraction $f = 0.51$, we could not find the water droplet in the Wenzel state even in the lower pillar heights. However, for 50 wt-% ethanol solutions, we have observed complete Wenzel state up to $h = 2$. In general, the ethanol concentration can be used to modulate the different wetting states on a given rough surface. We have also computed the work of adhesion for 10 wt-% water-ethanol solution at different rough surfaces. It is observed that the work of adhesion is linearly proportion to the surface fraction and surface heterogeneity factor for both the Cassie and Wenzel states.

4 Phase Transitions Inside Nanopores

Wettability can alter the flow of confined water inside nanopores in response to external stimuli [70] such as pressure, temperature, chemical patterning, and external field. Hence, it is very important for various applications to analyze the phase diagram and dynamical properties of water in nanopores in order to understand the effects of thermophysical conditions on wetting and dewetting behavior of confined water. We recently investigated the phase behavior of water confined inside graphite and mica slit pores of variable size using all-atom molecular dynamics simulations [71]. In Fig. 17a, b, we plot vapor-liquid phase diagrams for water confined inside graphite and mica slit pores with the width ranging from 10 to 60 Å.

Our findings reveal that the presence of confining surfaces causes a reduction in the critical temperature of water inside pores with respect to its bulk value.

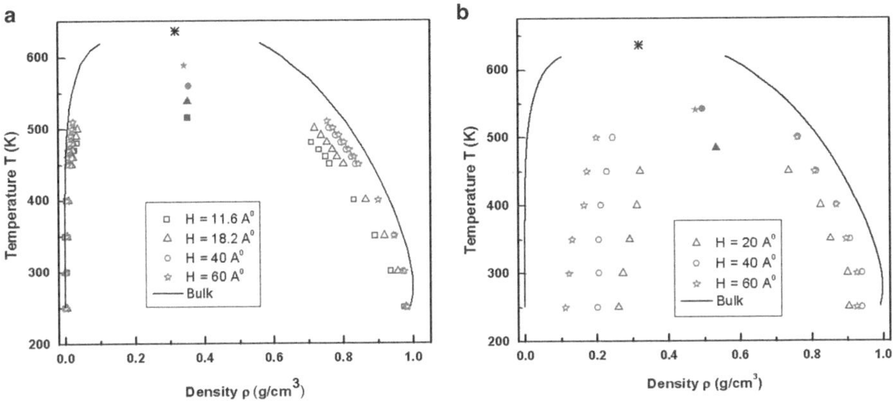

Fig. 17 Vapor-liquid phase coexistence curve of water confined inside (**a**) graphite and (**b**) mica pores, respectively (Reprinted (adapted) with permission from Ref [71]. Copyright (2011) American Chemical Society)

This is because as the separation between the two surfaces decreases, the surface effects become increasingly important. The nature of confining surfaces can alter the capillarity at nanoscale. Capillarity is closely related to wetting of liquid in the confined space of a narrow tube or between at least two surfaces [72]. If the water-surface interaction is strong (as in the case of hydrophilic pores), the condensation occurs at a lower pressure than that in the case of bulk water, and this phenomenon is called capillary condensation. In the case of hydrophobic pores, condensation occurs at higher pressure. At the bulk saturated pressure, the hydrophobic pore is filled with only vapor, and this phenomenon is called capillary evaporation. The phase transition of the confined fluid, corresponding to capillary condensation and capillary evaporation, causes the narrowing of the vapor-liquid coexistence curve, which in turn leads to the lowering of the critical temperature. It is also observed that the presence of hydrophilic surfaces, such as mica, causes a shift in the critical density of the confined water and a greater reduction in the critical temperature as compared to hydrophobic surfaces such as graphite. This is mainly due to the hydrophilic nature of the mica surface, which allows water to wet the surface leading to a considerable increase in vapor phase density, in comparison to the bulk value.

As reported above, confining surfaces can profoundly affect the wetting and phase behavior of fluids. In addition to effects of confining surfaces, there is a need to understand the behavior of confined water in the presence of external fields such as electric and magnetic fields. Recent studies have suggested that the wetting/dewetting of water in nanopores can be switched by applying an electric field across a nanopore [73] and is a key mechanism for the artificial biological channels. In our recent study, we have investigated the effect of electric field on thermophysical behavior of liquids confined in hydrophobic and hydrophilic nanopores [74, 75]. Our observations reveal that the application of an electric field perpendicular to the confining surface decreases the liquid phase density of the confined water film and increases the evaporation rate of the confined water, which in turn results in a further decrease in the pore critical temperature of the confined water. This is in line with the observation that the application of electric field reduces the free energy barrier for capillary evaporation of water confined between the surfaces [76].

5 Phase Transitions and Transport Properties of Thin Films

The phase behavior of liquids is strongly influenced by the effects of confinement. At a very strong confinement, the behavior of confined fluids is very similar to that of 2D systems. For example, the scaling of critical volume and radius of gyration of a polymeric molecule with its size in a slit pore of width three molecular diameters is identical to that in a strictly 2D environment [77]. Thus, a 2D system represents essential features of strongly confined systems. In this section, we discuss the coexistence of vapor and liquid phases in a two-dimensional thin film and its dynamical properties. The thin film is modeled as a 2D monolayer of Lennard-Jones (LJ) fluid or short LJ chain molecules (oligomers). The van der Waals interaction between any pair of particles is represented by the truncated and shifted Lennard-Jones (LJ) potential:

$$u^{\mathrm{LJ}}\left(r_{ij}\right) = 4\varepsilon\left[\left(\frac{\sigma}{r_{ij}}\right)^{12} - \left(\frac{\sigma}{r_{ij}}\right)^{6}\right] - 4\varepsilon\left[\left(\frac{\sigma}{r_{\mathrm{c}}}\right)^{12} - \left(\frac{\sigma}{r_{\mathrm{c}}}\right)^{6}\right]. \tag{8}$$

Here, ε and σ are the characteristic energy and length scales, respectively. All quantities are reduced with respect to ε, σ, and M (the mass of a particle). The cutoff distance, r_{c}, is taken as 4σ. In case of oligomers, m numbers of beads are connected tangentially; m varies from 1 to 8. Adjacent bead separations are fixed to 1σ. There are no angular or torsional constrains in chain molecules. Therefore, chains are fully flexible. The grand canonical transition matrix Monte Carlo (GC-TMMC) simulations are performed to calculate the vapor-liquid phase diagrams of oligomers. The details of the simulations can be found in Ref. [78].

The temperature-density phase diagrams of thin films are shown in the Fig. 18 for varying m. As the chain length increases, the phase envelope shifted to higher temperature regions. At a particular temperature, the vapor density of a longer chain is lower than that of a shorter chain. However, the liquid density of a longer chain is higher than that of a shorter chain at a fixed temperature. The critical temperature, calculated from the scaling law [78], of oligomers are represented by filled symbols in Fig. 18. The critical temperature increases as the chain length increases, which is also the case for 3D chain molecules (bulk) [79]. However, the critical temperature is substantially suppressed in 2D in comparison to that in 3D. For example, the critical temperature (T_{C}) of the 8-mer is 2.65 ± 0.05 and 0.835 ± 0.002 in 3D and 2D, respectively. On the other hand, the critical density of a chain is significantly

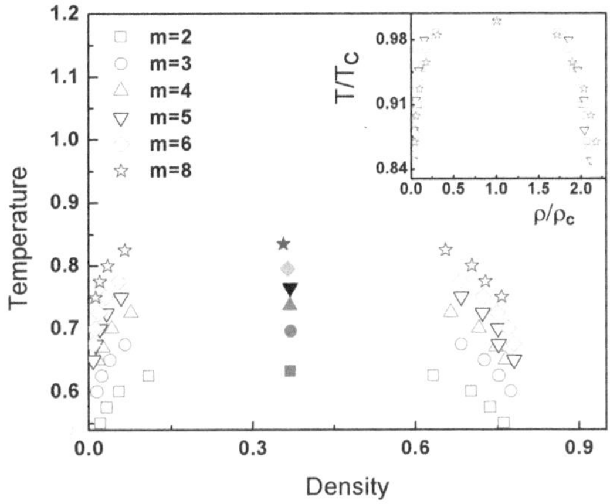

Fig. 18 Temperature-density vapor-liquid coexistence of 2D oligomers. The critical points are represented by filled symbols. The corresponding state plot is shown in the inset (Reprinted (adapted) with permission from Ref. [78]. Copyright (2012) American Institute of Physics)

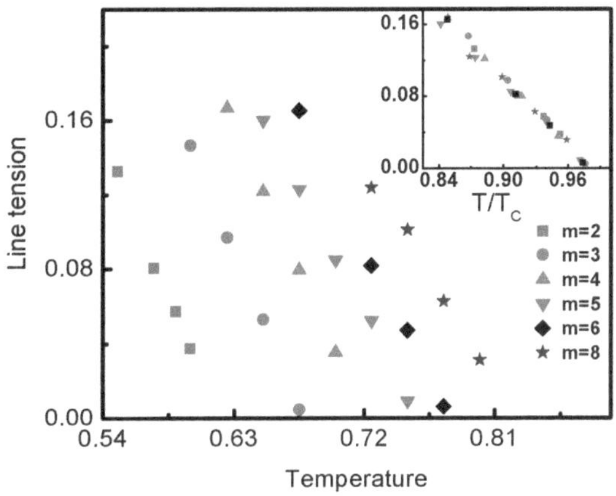

Fig. 19 The line tension of oligomers is shown as a function of temperature. In the *inset*, line tension is shown as a function of reduced temperature (Reprinted (adapted) with permission from Ref. [78]. Copyright (2012) American Institute of Physics)

higher in 2D in comparison to its 3D value. The critical density (ρ_C) of 3D 8-mer is 0.234 ± 0.007, and it is 0.355 ± 0.001 in case of 2D. Further, the critical densities of 2D polymeric fluids do not change significantly with the chain length, in contrast with 3D polymeric fluids. The corresponding state (CS) plot is shown in the inset of Fig. 18. In the CS plot, temperatures and densities of a system are reduced by its critical temperature and critical density, respectively. The phase envelopes of all the systems fall onto a single master curve, indicating that 2D polymers follow the CS principles, similar to 3D polymers [79].

The line tension between the vapor and liquid regions, calculated as per the Binder's formalism [80], is shown in Fig. 19 as a function of temperature. The line tension decreases with increasing temperature. The line tension as a function of reduced temperature is shown in the inset of Fig. 19. The line tension of various oligomers at the reduced temperature falls on a master curve. This implies that the line tension is insensitive to the chain length for the systems considered in our work. Blas et al. [79] has reported that, in case of 3D systems, the surface tension of different chain molecules as a function of reduced temperature falls on a single curve when the chain length is greater than eight. Thus, the interfacial tension of 3D oligomers with chain length m <8, at a reduced temperature, depends on their size. On the other hand, the interfacial tension of 2D oligomers is found to be independent of their size.

Thin films are characterized by their strong correlations in structural and dynamical properties. The velocity autocorrelation function (VACF) and stress autocorrelation function (SACF) are very important to understand the behavior of a thin film. Further, the VACF and the SACF of a system can be integrated to obtain its diffusion coefficient and viscosity, respectively, which are known as the Green-Kubo relations [81]. Molecular dynamics of a monolayer of oligomers is simulated in the canonical ensemble to study their correlation functions and estimated their diffusion coefficient and viscosity. The VACF is calculated as

$$\mathrm{VACF}(t) = \langle v(t)v(0)\rangle\,. \tag{9}$$

Here $v(t)$ is the velocity of a particle at time t. The VACF is integrated to study its convergence. The integration also gives the diffusion coefficient of the system as

$$D = \frac{1}{d}\int\limits_{t=0}^{t=\infty} \langle v(t)v(0)\rangle \mathrm{d}t. \tag{10}$$

Here, d represents the dimension of the system. The convergence of autocorrelation functions, calculated from phase space coordinates, is a long-standing issue [82]. Capturing the important fluctuation and essential decay may suffer from numerical errors. Alder and Wainright predicted a long-time power law decay of the VACF of moderately dense hard-sphere fluid using MD simulations [83]. They determined that the VACF decays as $t^{-d/2}$, where d being the dimension of the system. It indicates a logarithmic divergence of the VACF in 2D. In a recent study, Isobe [84] used a large sample population and long correlation time to measure more accurate VACF. The author demonstrated that the VACF of moderately dense hard-sphere fluid decays a little faster than t^{-1} in 2D [84]. In our work, we have further investigated the convergence of autocorrelation functions for a monolayer of LJ fluid. The *VACF* and diffusion coefficient are calculated using Eqs. 9 and 10, respectively, at a reduced density $\rho = 0.7$ and a reduced temperature $T = 3.0$.

Figure 20 (top panel) shows the VACFs of various molecular systems for one unit of time. VACFs show oscillatory behavior in the long-time range as shown in the insets of the top panel. The integrations of VACFs, shown in the bottom panel of Fig. 20, converge within reasonable time. However, as the chain length increases, VACF converges very slowly. The diffusion coefficient of a system has been calculated by taking the average of the data points in the plateau region of its VACF, as shown in the bottom panel of Fig. 20. Now, we discuss the SACF and viscosity of oligomers. The SACF of a system is written as

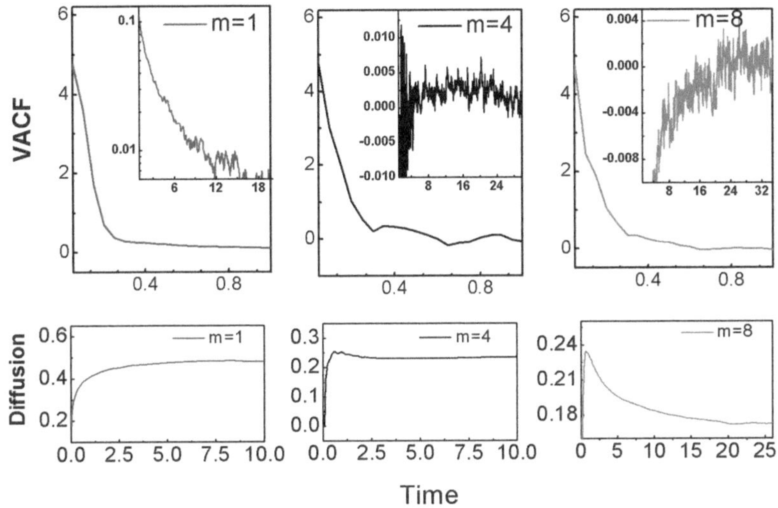

Fig. 20 Velocity autocorrelation functions (*VACF*) and diffusion coefficients (time integration of VACF) of oligomers (Reprinted (adapted) with permission from Ref. [78]. Copyright (2012) American Institute of Physics)

$$\mathrm{SACF}(t) = \langle \sigma_{xy}(t)\sigma_{xy}(0)\rangle, \tag{11}$$

where a component of the stress matrix is calculated as

$$\sigma_{xy} = \frac{1}{V}\left(\sum_{i=1}^{N} M_i v_{xi} v_{yi} + \sum_{i=1}^{N}\sum_{j=i+1}^{N} \mathrm{d}x_{ij} f_{yij}\right). \tag{12}$$

Here, v and M represent the velocity and mass of a particle, respectively; f_{ij} is the total force exerted by the ith particle due to the jth particle, and V is the volume of the system. The viscosity of the system can be estimated as

$$\eta = \frac{V}{k_B T}\int_{t=0}^{t=\infty} \langle \sigma_{xy}(t)\sigma_{xy}(0)\rangle \,\mathrm{d}t. \tag{13}$$

Here, k_B and T are the Boltzmann constant and temperature of the system, respectively. The SACFs and their time integrations for oligomers are shown in Fig. 21, at a reduced density $\rho = 0.7$ and a reduced temperature $T = 3.0$. The long-time oscillation is seen in SACFs, similar to that seen for VACFs. The integration of a SACF, which is the viscosity, shows a plateau region indicating the convergence of the function. As the chain length increases, the SACF takes longer time to converge, which is also seen for the VACF.

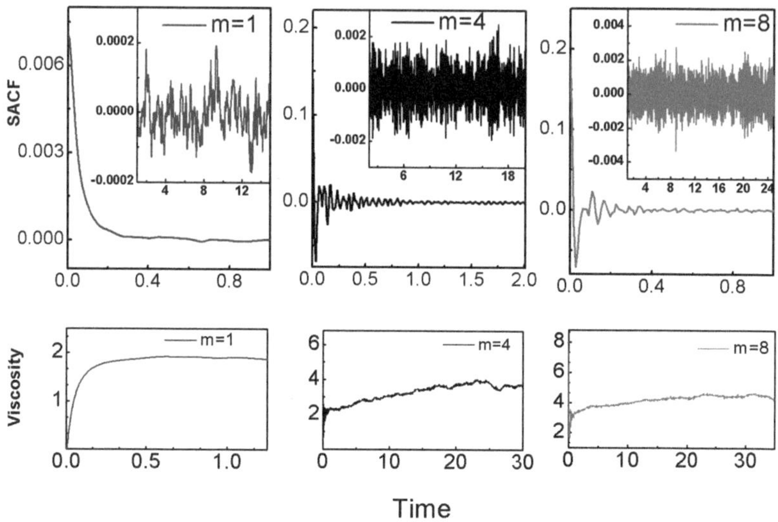

Fig. 21 Stress autocorrelation function (*SACF*) and viscosity (time integration of SACF) (Reprinted (adapted) with permission from Ref. [78]. Copyright (2012) American Institute of Physics)

Table 2 The ratio of the diffusion coefficient and viscosity of oligomers between 2D and 3D at different corresponding state (CS) points

m	$T/T_c = 0.9$	$T/T_c = 0.95$	$T/T_c = 1.5$	$T/T_c = 2.0$
		D_{3D}/D_{2D}		
1	3.871 (5)	3.138 (7)	2.415 (5)	2.464 (5)
4	7.924 (2)	7.313 (7)	5.031 (1)	4.838 (4)
8	6.992 (2)	8.682 (4)	7.221 (2)	5.231 (2)
		η_{2D}/η_{3D}		
1	2.18 (3)	2.14 (4)	1.88 (3)	1.63 (3)
4	3.81 (1)	3.78 (2)	4.04 (7)	3.43 (5)
8	5.57 (3)	4.72 (5)	3.68 (2)	3.01 (2)

Reprinted (adapted) with permission from Ref. [78]. Copyright (2012) American Institute of Physics

The viscosity of a system is calculated from the average over data points in the plateau region of its SACF, as shown in the bottom panel of Fig. 21. As the thin film dynamics differ from that of 3D bulk fluid, the difference in the transport properties of fluids between 2D and 3D is very important to study. Relative measurements of the diffusion coefficient and the viscosity of oligomers in 2D with respect to their 3D bulk properties are shown in Table 2. For the sake of comparison, the calculations of the diffusion coefficient and the viscosity of oligomers are done at the same 2D and 3D corresponding state (CS) points. The diffusion coefficient decreases, and the viscosity increases by many folds in 2D with respect to their bulk 3D values, as shown in Table 2. The amount of change in the diffusion coefficient and in the viscosity of an oligomer, while moving from 3D to 2D, depends on its size.

Finally, a monolayer (2D) has a higher critical density and lower critical temperature in comparison to its 3D counterpart. The stress correlation and velocity correlation in a thin film of molecular width converge slowly. Therefore, the strong dynamical correlation in a thin film makes it highly viscous and less diffusive. We believe that this has a great impact on the wetting behavior of fluids on substrates (either confined or monolayer).

6 Conclusions

We have presented some of our recent studies on surface phase transitions using molecular simulations. Percolation, layering, prewetting, and vapor-liquid phase coexistence are observed at solid-liquid interface, depending on the structure and interaction of fluids and solid. The wettability strongly depends on the surface geometry and the chemical nature of substrate and fluid. Roughness of a substrate can be tuned to obtain desired contact angle of a droplet on substrate. Further, associating fluid exhibits a crossover from vapor-liquid phase coexistence to prewetting state on functionalized surface, depending on the strength of association. The vapor-liquid phase diagram shifted as the dimensionality decreases as seen in case of water under mica and graphite confinements and monolayer of LJ fluids. There is a reduction in critical temperature and enhancement in critical density, which affect the wetting phenomena in lower dimension. Wetting phenomena are also affected by the enhanced structural and dynamical correlations, which cause high shear viscosity, slow relaxation, and nonliner responses to external perturbations in lower dimensions. Therefore, this chapter illustrates the wetting phenomena at different substrate geometry and its dependence on fluid-fluid and fluid-solid interaction. The molecular level understanding of the wetting behavior of fluids, which we have presented here, will be useful for designing efficient micro- and nano-fluidic devices.

References

1. Reisch A, Voegel JC, Gonthier E, Decher G, Senger B, Schaaf P, Mesini PJ (2009) Polyelectrolyte multilayers capped with polyelectrolytes bearing phosphorylcholine and triethylene glycol groups: parameters influencing antifouling properties. Langmuir 25:3610
2. Sung KE, Vanapalli SA, Mukhija D, McKay HA, Millunchick JM, Burns MA, Solomon MJ (2008) Programmable fluidic production of microparticles with configurable anisotropy. J Am Chem Soc 130:1335
3. Wu C, Chen M, Xing C (2010) Molecular understanding of conformational dynamics of a fibronectin module on rutile (110) surface. Langmuir 26:15972
4. Yoon E-S, Yang SH, Kong H, Kim K-H (2003) The effect of topography on water wetting and micro/nano tribological characteristics of polymeric surfaces. Tribol Lett 15:145
5. Guse C, Hentschke R (2012) Simulation study of structural, transport, and thermodynamic properties of TIP4P/2005 water in single-walled carbon nanotubes. J Phys Chem B 116:751

6. Lum K, Chandler D, Weeks JD (1999) Hydrophobicity at small and large length scales. J Phys Chem B 103:4570
7. Gelb LD, Gubbins KE, Radhakrishnan R, Sliwinska-Bartkowiak M (1999) Phase separation in confined systems. Rep Prog Phys 62:1537
8. Wirtz M, Martin CR (2002) Nanotube membrane sensors: resistive sensing and ion channel mimetics. Sensor Technol 11:35
9. Chaban VV, Prezhdo OV (2011) Water boiling inside carbon nanotubes: toward efficient drug release. ACS Nano 5:5647
10. Heine DR, Grest GS, Webb EB (2005) Surface wetting of liquid nanodroplets: droplet-size effects. Phys Rev Lett 95:107801
11. Wenzel RN (1936) Resistance of solid surfaces to wetting by water. Ind Eng Chem 28:988
12. Cahn JW (1977) Critical point wetting. J Chem Phys 66:3667
13. Ebner C, Saam WF (1977) New phase-transition phenomena in thin argon films. Phys Rev Lett 38:1486
14. Hallock RB (1995) Review of some of the experimental evidence for the novel wetting of helium on alkali metals. J Low Temp Phys 101:31
15. Phillips J, Ross D, Taborek P, Rutledge J (1998) Superfluid onset and prewetting of 4He on rubidium. Phys Rev B 58:3361
16. Cheng E, Mistura G, Lee HC, Chan MHW, Cole MW, Carraro C, Saam WF, Toigo F (1993) Wetting transitions of liquid hydrogen films. Phys Rev Lett 70:1854
17. Kruchten F, Knorr K (2003) Multilayer adsorption and wetting of acetone on graphite. Phys Rev Lett 91:0855021
18. Zhao X (2007) Wetting transition of water on graphite: Monte Carlo simulations. Phys Rev B 76:0414021
19. Evans R, Tarazona P (1983) Wetting and thick-thin film transition in a model of argon at a solid CO2 substrate. Phys Rev A 28:1864
20. Meister TF, Kroll DM (1985) Density functional theory for inhomogeneous fluids: application to wetting. Phys Rev A 31:4055
21. Finn JE, Monson PA (1989) Prewetting at a fluid–solid interface via Monte Carlo simulation. Phys Rev A 39:6402
22. Shi W, Zhao X, Johnson JK (2002) Phase transitions of adsorbed fluids computed from multiple histogram reweighting. Mol Phys 100:2139
23. Bojan MJ, Stan G, Curtarolo S, Steele WA, Cole MW (1999) Wetting transitions of Ne. Phys Rev E 59:864
24. Errington JR (2004) Prewetting transitions for a model argon on solid carbon dioxide system. Langmuir 20:3798
25. Ancilotto F, Toigo F (1999) Prewetting transitions of Ar and Ne on alkali-metal surfaces. Phys Rev B 60:9019
26. Khan S, Singh JK (2011) Surface phase transition of associating fluids on functionalized surfaces. J Phys Chem C 115:17861
27. Khan S, Bhandary D, Singh JK (2012) Surface phase transition of multiple sites associating fluids. Mol Phys 110:1241
28. Plooster MN, Gitlin SN (1971) Phase transitions in water adsorbed on silica surfaces. J Phys Chem 75:3322
29. Tsangaris DM, de Pablo JJ (1994) Bond bias simulation of phase equilibria for strongly associating fluids. J Chem Phys 101:1477
30. Takei T, Konishi T, Fuji M, Watanabe T, Chikazawa M (1995) Phase transition of capillary condensed liquids in porous silica: effect of surface hydroxyl groups. Thermochim Acta 267:159
31. Bogdan A, Kulmala M, Avramenko N (1998) Reduction of enthalpy of fusion and anomalies during phase transitions in finely divided water. Phys Rev Lett 81:1042
32. Argyris D, Tummala NR, Striolo A, Cole DR (2008) Molecular structure and dynamics in thin water films at the silica and graphite surfaces. J Phys Chem C 112:13587

33. Suzuki A, Koboki Y (1999) Static contact angle of sessile air bubbles on polymer gel surfaces in water. J Appl Phys 38:2910
34. Jones JB, Adamson AW (1968) Temperature dependence of contact angle and of interfacial free energies in the naphthalene-water-air system. J Phys Chem 72:646–650
35. Bernardin JD, Mudawar I (1997) Film boiling heat transfer of droplet streams and sprays. Int J Heat Mass Transfer 40:2579–2593
36. Klier PSJ, Wyat AFG (1995) Heat capacity measurements of ^{3}He-^{4}He mixtures in aerogel. Phys Rev Lett 75:3705
37. De Ruijter MJ, Blake TD, Coninck JD (1999) Dynamic wetting studied by molecular modeling simulations of droplet spreading. Langmuir 15:7836
38. Giovambattista N, Debenedetti PG, Rossky PJ (2007) Effect of surface polarity on water contact angle and interfacial hydration structure. J Phys Chem B 111:9581
39. Dutta RC, Khan S, Singh JK (2011) Wetting transition of water on graphite and boron-nitride surfaces: a molecular dynamics study. Fluid Phase Equilib 302:310–315
40. Berendsen HJC, Grigera JR, Straatsma TP (1987) The missing term in effective pair potentials. J Phys Chem 91:6269–6271
41. Zhao X (2007) Wetting transition of water on graphite: Monte Carlo simulations. Phys Rev B 76:041402(R)
42. Vega C, Miguel ED (2007) Surface tension of the most popular models of water by using the test-area simulation method. J Chem Phys 126:154707
43. Mugele F, Becker T, Nikopoulos R, Kohonen M, Herminghaus S (2002) Capillarity at the nanoscale: an AFM view. J Adhes Sci Technol 16:951
44. Rowlinson JS, Widom BW (1982) Molecular theory of capillarity. Clarendon Press, Oxford
45. Werder T, Walther JH, Jaffe RL, Halicioglu T, Koumoutsakos P (2003) On the water-carbon interactions for use in MD simulations of graphite and carbon nanotubes. J Phys Chem B 107:1345
46. Zangi R, Berne BR (2008) Temperature dependence of dimerization and dewetting of large scale hydrophobes: a molecular dynamics study. J Phys Chem B 112:8634
47. Indekeu JO (1994) Line tension at wetting. Int J Mod Phys B 8:309
48. Cassie ABD, Baxter S (1944) Wettability of porous surfaces. Trans Faraday Soc 40:546
49. Yoshimitsu Z, Nakajima A, Watanabe T, Hashimoto K (2002) Effects of surface structure on the hydrophobicity and sliding behavior of water droplets. Langmuir 18:5818
50. Extrand CW (2002) Model for contact angles and hysteresis on rough and ultraphobic surfaces. Langmuir 18:7991
51. Lenz P, Lipowsky R (1998) Morphological transitions of wetting layers on structured surfaces. Phys Rev Lett 80:1920
52. He B, Patankar NA, Lee J (2003) Multiple equilibrium droplet shapes and design criterion for rough hydrophobic surfaces. Langmuir 19:4999
53. Patankar NA (2004) Transition between superhydrophobic states on rough surfaces. Langmuir 20:7097
54. Shirtcliffe NJ, McHale G, Newton MI, Perry CC (2005) Wetting and wetting transitions on copper-based super-hydrophobic surfaces. Langmuir 21:937
55. Ran C, Ding G, Liu W, Deng Y, Hou W (2008) Wetting on nanoporous alumina surface: transition between Wenzel and Cassie states controlled by surface Structure. Langmuir 24:9952
56. Lundgren M, Allan NL, Cosgrove T (2007) Modeling of wetting: a study of nanowetting at rough and heterogeneous surfaces. Langmuir 23:1187
57. Hirvi JT, Pakkanen TA (2008) Nanodroplet impact and sliding on structured polymer surfaces. Surf Sci 602:1810
58. Yang C, Tartaglinoa U, Persson BNJ (2008) Nanodroplets on rough hydrophilic and hydrophobic surfaces. Eur Phys J E Soft Matter 25:139
59. Hirvi JT, Pakkanen TA (2007) Wetting of nano grooved polymer surfaces. Langmuir 23:7724
60. Khan S, Singh JK (2014) Wetting transition of nanodroplets of water on textured surfaces: a molecular dynamics study. Mol Sim 40:458

61. Yeh K-Y, Chen L-J, Chang JY (2008) Contact angle hysteresis on regular pillar-like hydrophobic surfaces. Langmuir 24:245
62. Yeh K-Y, Cho K-H, Chen L-J (2009) Preparation of superhydrophobic surfaces of hierarchical structure of hybrid from nanoparticles and regular pillar-like pattern. Langmuir 25:14187
63. Jeong W-J, Ha MY, Yoon HS, Ambrosia M (2012) Dynamic behavior of water droplets on solid surfaces with pillar-type nanostructures. Langmuir 28:5360
64. Koishi T, Yasuoka K, Fujikawa S, Ebisuzaki T, Zeng XC (2009) Coexistence and transition between Cassie and Wenzel state on pillared hydrophobic surface. Proc Natl Acad Sci 106:8435
65. Cui XS, Li W (2010) On the possibility of superhydrophobic behavior for hydrophilic materials. J Colloid Interface Sci 347:156
66. Gross M, Varnik F, Raabe D, Steinbach I (2010) Small droplets on superhydrophobic substrates. Phys Rev E 81:051606
67. Liu HH, Zhang HY, Li W (2011) Thermodynamic analysis on wetting behavior of hierarchical structured superhydrophobic surfaces. Langmuir 27:6260
68. Kumar V, Sridhar S, Errington JR (2011) Monte Carlo simulation strategies for computing the wetting properties of fluids at geometrically rough surfaces. J Chem Phys 135:184702
69. Leroy F, Muller-Plathe F (2011) Rationalization of the behavior of solid–liquid surface free energy of water in Cassie and Wenzel wetting states on rugged solid surfaces at the nanometer scale. Langmuir 27:637
70. Smirnov S, Vlassiouk I, Takmakov P, Rios F (2010) Water confinement in hydrophobic nanopores. Pressure-induced wetting and drying. ACS Nano 4:5069
71. Srivastava R, Docherty H, Singh JK, Cummings PT (2011) Phase transitions of water in graphite and mica pores. J Phys Chem C 115:12448
72. Honschoten JW, Brunets NB, Tas NR (2010) Capillarity at the nanoscale. Chem Soc Rev 39:1096
73. Powell MR, Cleary L, Davenport M, Shea KJ, Siwy ZS (2011) Electric-field-induced wetting and dewetting in single hydrophobic nanopores. Nat Nanotech 6:798
74. Srivastava R, Singh JK, Cummings PT (2012) Effect of electric field on water confined in graphite and mica pores. J Phys Chem C 116:17594
75. Bhandary D, Srivastava K, Srivastava R, Singh JK (2014) Effects of electric field on the vapor–liquid equilibria of nanoconfined methanol and ethanol. J Chem Eng Data 59:3090
76. Vaitheeswaran S, Yin H, Rasaiah JC (2005) Water between plates in the presence of an electric field in an open system. J Phys Chem B 109:6629
77. Hehmeyer OJ, Arya G, Panagiotopoulos AZ (2004) Phase transitions of confined lattice homopolymers. J Phys Chem B 108:6809
78. Patra TK, Hens A, Singh JK (2012) Vapor–liquid phase coexistence and transport properties of two dimensional oligomers. J Chem Phys 137:084701
79. Blas FJ, MacDowell LG, Miguel ED, Jackson G (2008) Vapor–liquid interfacial properties of fully flexible Lennard-Jones chains. J Chem Phys 129:144703
80. Binder K (1982) Monte Carlo calculation of the surface tension for two- and three-dimensional lattice-gas models. Phys Rev A 25:1699
81. Allen MP, Tildesley DJ (1987) Computer simulations of liquids. Clarendon, Oxford
82. Chen T, Smit B, Bella AT (2009) Are pressure fluctuation-based equilibrium methods really worse than nonequilibrium methods for calculating viscosities? J Chem Phys 131:246101
83. Alder BJ, Wainwright TE (1977) Decay of the velocity auto correlation function. Phys Rev A 1:18
84. Isobe M (2008) Long-time tail of the velocity autocorrelation function in a two-dimensional moderately dense hard-disk fluid. Phys Rev E 77:021201

Effect of Electric Field on Planar Fluid-Fluid Interfaces

Priya Gambhire and Rochish Thaokar

Abstract Electrohydrodynamic instabilities at fluid-fluid interfaces are being extensively explored for potential technological applications such as soft lithography. Although extensive work has been done in this field experimentally, the state of the art in the theory is the leaky dielectric model which might not be valid in certain practically relevant systems. In the present work, various scenarios where the leaky dielectric model is found lacking are identified and alternate theories have been proposed. Experimental validation is also provided to show the practical relevance of these parametric regimes, and a big picture of the interfacial instability under electric fields is obtained.

Keywords Electrohydrodynamic instabilities • Pattern formation • Alternating fields • Leaky dielectric model • Linear stability analysis

1 Introduction

Electrohydrodynamics (EHD) is the study of fluid flow under the action of an applied electric field. In the context of a planar interface between two fluids, an applied normal electric field destabilizes the interface to form columns of one fluid growing into the other [1–5]. This instability termed 'electrohydrodynamic (EHD) instability' carries immense technological importance. The wide applicability of this phenomenon stems from the fact that it can be modelled using the simple fundamental laws of electrostatics (Gauss' law) and fluid dynamics (Navier-Stokes' equations) (hence the name *electrohydrodynamics*) providing a robust tool for making realistic predictions about the features of the instability.

P. Gambhire (✉)
Indian Institute of Technology Bombay, Mumbai 400076, India
e-mail: priya.gambhire@gmail.com

R. Thaokar
Department of Chemical Engineering, Indian Institute of Technology Bombay, Mumbai 400076, India
e-mail: rochish@che.iitb.ac.in

Y.M. Joshi, S. Khandekar (eds.), *Nanoscale and Microscale Phenomena*, Springer Tracts in Mechanical Engineering, DOI 10.1007/978-81-322-2289-7_8

$$\int n.\boldsymbol{E}\,dS = \frac{\rho_f}{\epsilon_0} \tag{1}$$

$$\nabla \cdot \boldsymbol{V} = 0 \tag{2}$$

$$\rho(\partial_t \boldsymbol{V} + \boldsymbol{V} \cdot \nabla \boldsymbol{V}) = -\nabla p + \mu \nabla^2 \boldsymbol{V} + \rho_f E \tag{3}$$

where $\boldsymbol{E}$ is the electric field, S is the area of the interface, ρ_f is the free charge, ϵ_0 is the permittivity of vacuum, $\boldsymbol{V}$ is the velocity of the fluid, p is the hydrostatic pressure and μ is the viscosity.

Electrohydrodynamics (EHD) is a term unique to the study of systems in which the effects due to current are minimal. Unlike the related field of electrokinetics, fluids that show EHD behaviour are generally dielectric materials. Pure or perfect dielectric (PD) materials have dipoles in them which align along the electric field to create a polarization field within the fluid. The Gauss' law for such a fluid is modified to consider the contribution of the polarization to the field. The differential form of the Gauss' law is given as

$$\nabla^2 \phi = \frac{\rho_f}{\epsilon_0 \epsilon} \tag{4}$$

where ϕ is the electric potential and ϵ is the dielectric constant of the fluid.

To describe the electric field in the present fluid-fluid system, a few boundary conditions across the interface are made use of. One of the condition being that the electric field vector is continuous across the interface. In the normal direction,

$$[\epsilon_0 \epsilon_i (-\nabla \phi_i \cdot \hat{\boldsymbol{n}})] = 0 \tag{5}$$

where the operator $[X_i]$ denotes the jump $X_1 - X_2$ across the fluid-fluid interface $h(x,t)$. $i = 1, 2$ denotes the upper and the lower fluid, respectively. A balance of the tangential component of the field gives the continuity of the potentials across the interface:

$$\phi_1 = \phi_2 \tag{6}$$

An EHD model for a real dielectric material considers both the alignment of dipoles within the fluid and the flow of a negligible yet finite free charge present in the fluid, under the action of an electric field. The term 'leaky dielectric (LD)' describes such a material. The assumption of an infinitesimal free charge present in an LD fluid makes this model more powerful in terms of explaining phenomena which cannot be explained by a simple PD or a perfect conductor theory. The free charge accumulates at the interface and can be estimated by a material balance where the net current across the interface is equivalent to the accumulation.

$$\partial_t q + \boldsymbol{V_1}.\nabla_s q - q\hat{\boldsymbol{n}}.(\hat{\boldsymbol{n}}.\nabla)\boldsymbol{V_1} = [\sigma_i \nabla \phi_i \cdot \hat{\boldsymbol{n}}] \tag{7}$$

where q is the free charge density at the interface and σ is the conductivity of the fluid. The last two terms on the left-hand side of the equation physically signify

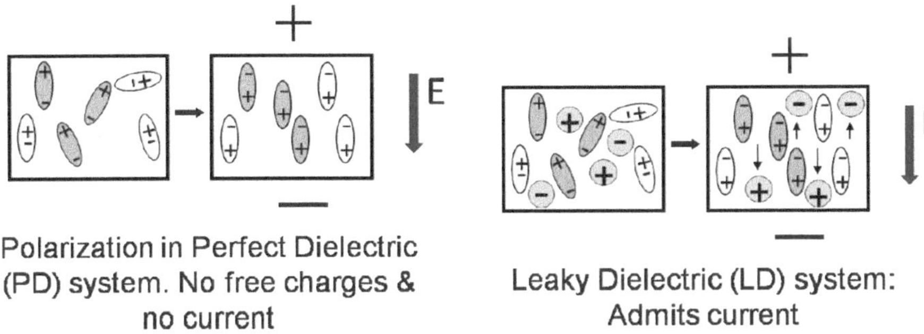

Fig. 1 Schematic of a PD-PD interface with dipoles aligned under an electric field

the surface convection of the charge and the change in the surface charge due to stretching of the interface. The right-hand side denotes the net current due to conduction. Equation (5) is modified for a leaky dielectric as (Fig. 1)

$$[\epsilon_0\epsilon_i(-\nabla\phi_i \cdot \hat{\boldsymbol{n}})] = q \tag{8}$$

The coupling of the electrostatics and the hydrodynamics occurs through the stress balance at the interface. The total stress which has contributions from hydrodynamic stress tensor and the Maxwell's stress tensor is balanced by the surface tension.

$$[\hat{\boldsymbol{n}} \cdot \boldsymbol{\tau}_i \cdot \hat{\boldsymbol{n}}] = \gamma C \tag{9}$$

where C is the curvature given by $C = \nabla \cdot \hat{\boldsymbol{n}} = -\partial_x^2 h/(1 + \partial_x h^2)^{3/2}$. The term on the right-hand side in the above equation represents the stabilizing force due to surface tension γ. The balance of stresses in the tangential direction is given by

$$[\hat{\boldsymbol{t}} \cdot \boldsymbol{\tau}_i \cdot \hat{\boldsymbol{n}}] = 0 \tag{10}$$

$\boldsymbol{\tau}_i$, the total stress, i.e. the sum of the hydrodynamic and the electrical stresses, for the fluid i, is given by

$$\boldsymbol{\tau}_i = -p_i I + \mu_i(\nabla \boldsymbol{V}_i + \nabla \boldsymbol{V}_i{}^T) + \boldsymbol{\tau}_e \tag{11}$$

The superscript T indicates transpose, $i = 1, 2$ denotes the two fluid layers, and $\boldsymbol{\tau}_e$ is Maxwell's stress tensor $\tau_e = \epsilon_0\epsilon\left(\boldsymbol{EE} - \frac{1}{2}E^2 I\right)$.

Although popular for its simple and elegant mathematical formulation, the LD theory has certain limitations. With the tremendous evolution of technology, it is vital to have equivalently powerful tools for modelling and simulation. Hence, in the present chapter, the limitations of the LD theory in four scenarios which arise in contemporary technological applications of EHD are discussed and appropriate alternate theories proposed.

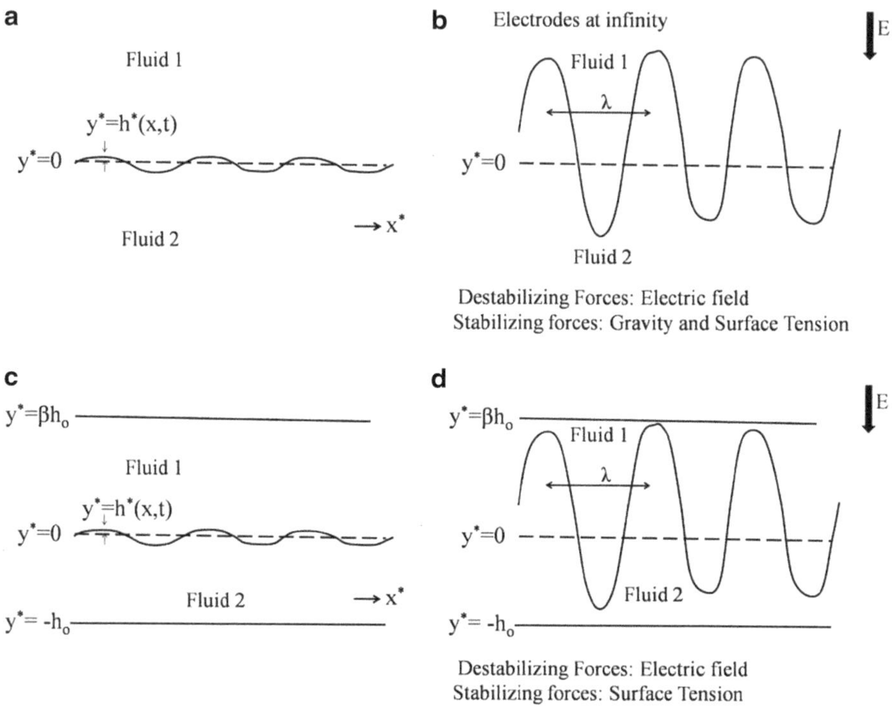

Fig. 2 Schematic of an EHD instability at an interface when gravity is significant (**a**, **b**) and when it is negligible (**c**, **d**)

Early research in the field of EHD interfacial instabilities dealt with inviscid fluid interfaces [3, 4, 6], destabilizing under the combined action of gravity, surface tension and high electric fields (refer Fig. 2a, b). It is found that a critical field is required to induce an instability which consists of a typical undulation, crests of which grow to form droplet ejecting jets. The wavelength of this instability is referred to as the 'Taylor wavelength' and can be obtained considering the fluids to be either PD or PC materials, as water, surfactant solutions and organic solvents are the commonly used fluids in these studies.

A mismatch of experimental observations with either of the PD or PC theory occurs when a dielectric fluid interface is subjected to a DC field. This is attributed to interfacial shear stresses that arise as a result of accumulation of the negligible yet finite charge, present in a real dielectric fluid, at the interface [7]. This is explained by the LD model [8–13]. Most of the fluids used practically have low conductivities which indicate a slow albeit finite charge relaxation time. This implies that true perfect dielectric behaviour can be realized, in practice, only at frequencies high enough that the rapid alteration of the field doesn't allow the inherent free charge to migrate to the interface.

2 The Effect of AC Fields

This poses quite interesting questions as to what would happen if the frequency of the applied field is not high enough for the charges to remain fixed at their position but transits from this high value to a value which is almost constant when compared to the experimental timescale. How does the conductivity of the leaky dielectric material compete with the alternating effect of the electric field and affect the final instability? The leaky dielectric model which assumes time-invariant electric forcing cannot be used to answer the aforementioned questions, and hence, the 'Floquet theory' is made use of. A systematic linear stability analysis of a planar fluid interface subjected to an alternating field shows that an AC field indeed can be used to probe new regimes of system behaviour with the LD and PD behaviour being the limiting cases at a low and high frequency, respectively (refer Fig. 3). The transition from a PD behaviour to an LD behaviour occurs when the frequency equals the inverse of the charge relaxation time. It is found that this trend is same across the three orders of conductivity values studied except that the transition frequency ($\Omega_{\text{trans}} = \tau_c$ where Ω is the scaled frequency and τ_c is the charge relaxation time) shifts to lower frequencies with decrease in the conductivity.

Recent electrohydrodynamic studies at planar interfaces deal more with thin films of viscous fluids as opposed to reservoirs of inviscid liquids considered in the previous studies owing to their application in producing micro-patterned devices. This thin fluid film system (parts c, d in Fig. 2), in contrast to the earlier ones, has insignificant effects of gravity due to the low length scales involved. The leaky dielectric theory for viscous systems involves the Stokes' equation (inertial terms

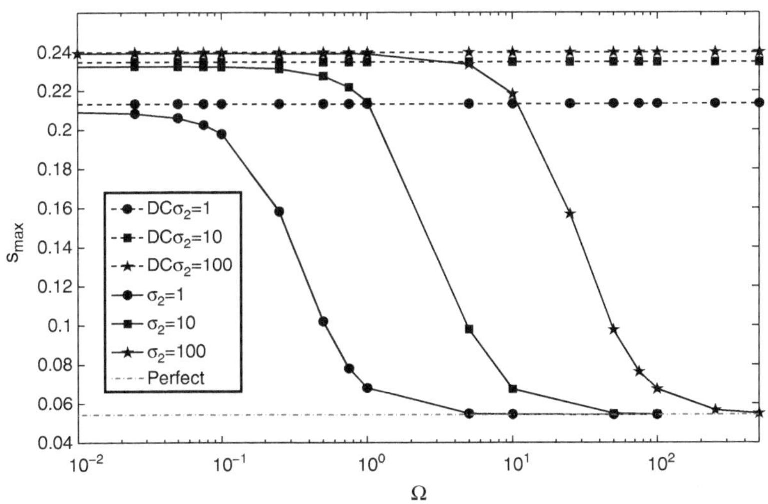

Fig. 3 Dependence of the maximum growth rate on frequency for a perfect dielectric-leaky dielectric system. The parameters used are $\epsilon_1 = 1$, $\epsilon_2 = 2.5$, $\lambda = 10^{-6}$, $B = 0.01$ and $\sigma_1 = 0$

neglected), governing the hydrodynamics. Due to the small length scales involved in the system, the so-called thin film approximation (TFA) is found to be a useful tool in studying this instability. This approximation yields simplified evolution equations of the position of the interface (h) and the charge at the interface (q) with time.

3 The Effect of Conductivity

It is, therefore, interesting, not only to study the effect of alternating fields on the instability but also an in-depth study of how the low conductivities of these viscous fluids effect the wavelength of the instability. The key assumption in the leaky dielectric model is that the charge relaxation is instantaneous. Although quite powerful, this assumption cannot be used without a knowledge of the conductivity of the system which for deionized water itself is $\sim 10^{-6}$ S/m and can go to as low as 10^{-9}–10^{-12} S/m for viscous polymeric fluids. The need of the hour is a modelling tool which can predict the wavelength of the instability depending upon the conductivity.

To overcome the limitation of the LD model in the context of low conductivity fluids, non-linear analysis is used. The evolution equations of the interface position and interfacial charge, obtained after the TFA, are numerically integrated in time from an initial condition of a random disturbance of infinitesimal amplitude. The result from the simulation shows the formation of columns, the mean spacing between which has been compared to experiments carried out at the same conditions (a few representative images are given in Fig. 4). It is observed from the results

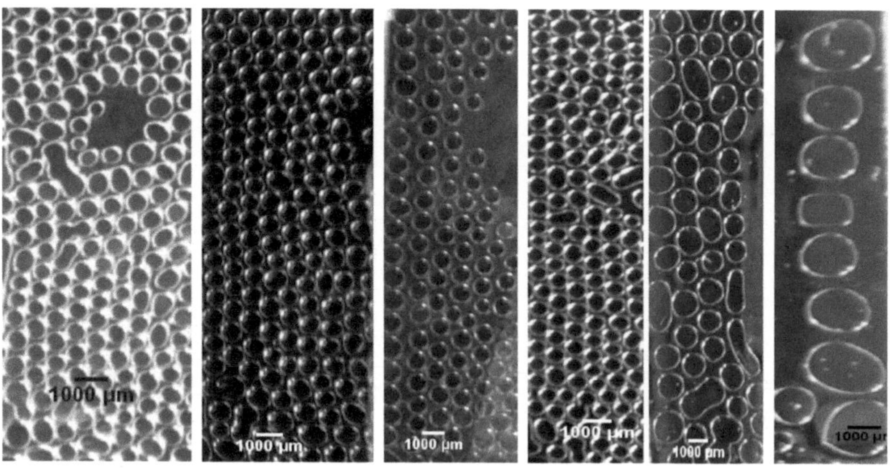

Fig. 4 Images of the instability in the form of columns rising from the pdms film and seen from the top view at different voltages. Starting from the left, the images are at 140 V & 1 mHz, 70 V & 1 mHz, 35 V & DC, 140 V & 0.3 Hz, 70 V & 10 Hz and 35 V & 1 Hz, respectively

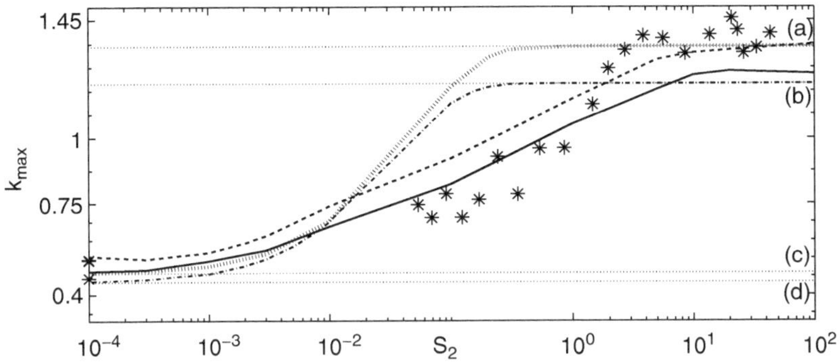

Fig. 5 A comparison of $k_{\max}$ as a function of S_2 between results obtained from experiments and non-linear analysis. The values of the parameters used in the theoretical analysis are $\epsilon_1 = 1, \epsilon_2 = 3, \mu_r = 10^{-7}$ and $S_1 = 0$. The $(\cdots)$ lines (a) and (b) indicate the linear LD-PD theory, and the $(\cdots)$ lines (c) and (d) indicate the PD-PD theory for $\beta = 0.65$ and $\beta = 0.7$, respectively. The $(\cdots)$ line curve and the $(- \cdot -\cdot)$ line curve represent the results from the LD modified theory for $\beta = 0.65$ and $\beta = 0.7$, respectively, while the $(- - -)$ curve and the *solid line* curve represent the results from the non-linear analysis for $\beta = 0.65$ and $\beta = 0.7$, respectively. The $*$ symbols indicate experimental results

of the non-linear analysis (refer Fig. 5) that the wavelength ($\lambda = 1/k_{\max}$) of the instability increases with a decrease in the conductivity of the fluid under DC fields. At very low conductivity, the wavelength predicted by the non-linear analysis is equivalent to that predicted by a perfect dielectric theory. This dependence of the instability wavelength can be explained in terms of the timescales involved in the system, namely, the charge relaxation time (τ_c) and time for the growth (τ_s) of the instability. The system would behave like a PD material or an LD material depending on whether $\tau_c > \tau_s$ or $\tau_c < \tau_s$. If $\tau_c \approx \tau_s$, then it would show a transition behaviour which can be predicted by the non-linear analysis alone.

It is interesting to study the effect of using alternating fields in the present systems. In the presence of AC fields, an additional timescale, that of the time period (τ_f) of the AC field, becomes relevant. The interplay of all the three timescales gives a total of nine regimes which have been studied in the present work (Fig. 6). It is seen that at very low values of τ_f, in spite of the magnitude of the other two timescales (τ_c and τ_S), the system always shows the signature of a perfect dielectric material. At very high values of τ_f, the relative values of τ_c, τ_f and the type of waveform used play an important role in deciding the wavelength of the instability. At very large τ_f (low Ω), the Floquet theory always predicts a limiting value equivalent to that of an LD fluid under a DC voltage equivalent to the r.m.s voltage of the applied AC field. It is only through the non-linear analysis that one can show that this limiting case is not always an LD behaviour but depends on the conductivity of the system. To elaborate, for low conductivity fluids (i.e. $\tau_c > \tau_s$), the system behaves like a perfect dielectric material. But at such low frequencies, as the field is constant within the timescale τ_s, the system experiences the peak voltage and not the r.m.s voltage as

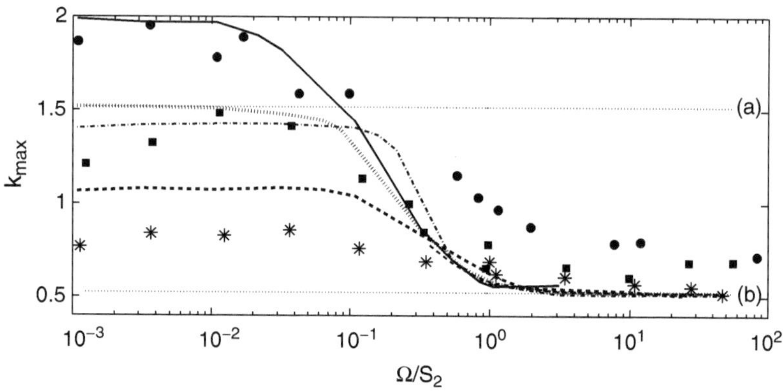

Fig. 6 A comparison of $k_{\max}$ as a function of Ω/S_2 between results obtained from experiments and non-linear analysis. The values of parameters used are $\epsilon_1 = 0, \epsilon_2 = 3, \mu_r = 10^{-7}, \beta = 0.6$ and $S_1 = 0$. The $(\cdots)$ lines (a) and (b) indicate results from the linear LD-PD and PD-PD theories, respectively, under DC fields. The $(\cdots)$ curve indicates collapse of results from the Floquet theory for the three conductivity ranges 0.073–0.16, 1.17–2.56 and 18.73–41.08. The *solid line* curve, the $(- \cdot -)$ line curve and the $(- - -)$ curve indicate results from the non-linear analysis for the conductivity ranges 18.73–41.08, 1.17–2.56 and 0.073–0.16, respectively, while the corresponding experimental results are indicated by the symbols $\bullet$, ■ and $*$

assumed by the linear theory. It is shown that such regimes are easily realizable in experiments. A similar deviation from the linear theory is seen when $\tau_c < \tau_s$ and $\tau_c \approx \tau_s$ at large τ_f (low Ω). In the former case, the system is similar to a leaky dielectric material subjected to the equivalent of the peak voltage, while in the later case, one would obtain a limiting value as predicted by the non-linear analysis under DC fields (refer Fig. 5). At intermediate values of τ_f, a smooth transition between these limiting values is seen.

This study is very important as EHD instabilities in viscous thin films find direct application in the field of soft lithography. The wavelength of the instability decides the pitch of the features that can be produced on a polymeric soft film, and hence, it is vital to be able to predict this value accurately. With the previous work, it is clear that this wavelength depends critically on the conductivity of the system. It is also demonstrated that the frequency of AC fields can be used to tune the wavelength of the instability to any value between those obtained in an LD or a PD material.

4 In the 'Small Feature Size' Limit

With the growing popularity of nanofluidics, the demand for producing sub-micron-sized features on substrates, inexpensively, is increasing. In this regard, the EHD-based soft lithography can prove to be useful as the size of the features produced using this technique can easily be controlled by varying the system

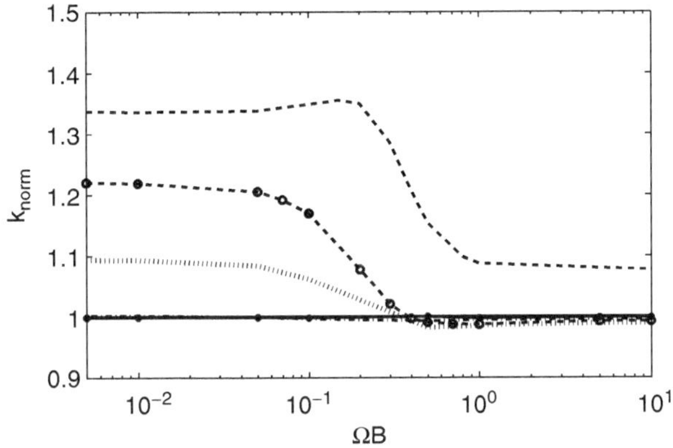

Fig. 7 Comparison between the model with and without the thin film approximation for the case of a PD-LD system under AC fields. The values of the parameters are $\epsilon_1 = 1, \epsilon_2 = 3, \mu_r = 0, \beta = 1, \sigma_1 = 0$ and $\sigma_2 = 1$. The *solid line* represents the TFA, (— — —) denotes $B = 0.1$, (— — —) with ○ denotes $B = 0.3$, (· · ·) denotes $B = 0.5$, (· — · — ·) denotes $B = 2$, and (· — · — ·) with the $*$ symbol denotes $B = 15$. Note that the $B = 2$ and $B = 15$ curves merge with the TFA and hence are indistinguishable

parameters such as thickness of thin film, surface tension and electric field. For a given air-polymer system, there is a practical limit to which the thickness of the thin film can be decreased to obtain smaller pattern sizes. Further decrease in the pattern sizes can be obtained by increasing the electric field applied to the system. This situation violates the TFA which has been used as a simplification tool. To overcome this limitation, the LD theory without the TFA, referred to as the 'generalized model' (GM) in the present work, is used. A detailed study is carried out with the GM under both DC and AC fields. A comparison of the wavelength and the growth rate of the instability as obtained from the GM and the TFA is carried out. It is seen to be dependent on the ratio of the length scales in the normal and the transverse directions, denoted by δ, with a deviation in the two theories occurring when $\delta \sim O(1)$. A parametric plot of the exact values of the validity of the theories is provided (Fig. 7).

5 In the Presence of Diffuse Layers of Charge

A second outcome of working with sub-micron thick fluid layers is that of an overlap of diffuse layers (DLs) of charge formed at the electrode surface and at the fluid-fluid interface. The assumption of electroneutrality made in the LD model leads to an unrealistic picture of charge being confined to the interfacial plane alone. Although valid for perfect conductors which form equipotential surfaces under electric field, this assumption is especially violated when very thin films of fluids are used. A

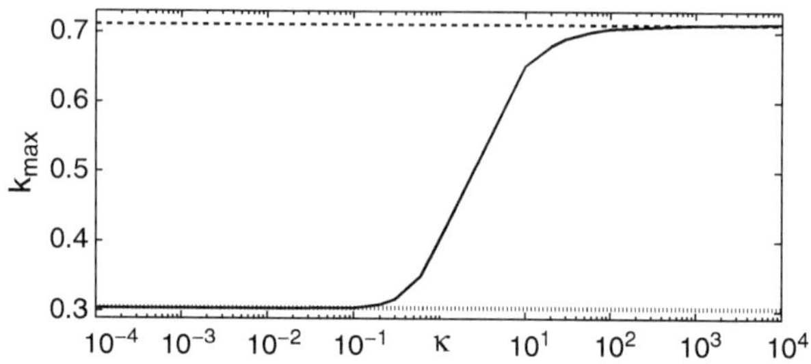

Fig. 8 Variation of $k_{\max}$ as a function of the inverse Debye length κ. The *dotted line* indicates results obtained from the PD-PD theory, the *dashed line* indicated those from the PD-PC, and the *solid line* indicates results from the electrokinetic theory. The values of parameters used are $\epsilon_1 = 1, \epsilon_2 = 3, \sigma_1 = 0, \sigma_2 = 10^{10}, \mu_r = 0, \beta = 1$ and $B = 1$

typical DL thickness for the commonly used polymeric liquid polydimethylsiloxane is around 10–100 nm, while the thin films used in the experiments are 100–1,000 nm thick. This estimate indicates that the thickness of the DLs of charge is comparable to the length scales in the system and in certain cases the DL from the electrode surface might overlap with that formed at the interface giving rise to a different field within the fluid. In the present work, an electrokinetic model is proposed as an alternative to the EHD LD model. In this model, the Debye-Huckel approximation is used to get a linear equation for Gauss' law of electrostatics. That is, $\nabla^2\phi = \kappa^2(\phi - \phi_b)$ where ϕ is the potential, κ is the inverse Debye length, and ϕ_b is a reference potential. The wavelength of the instability is studied as a function of the scaled inverse Debye length κ which is a ratio of the thickness of the fluid to the Debye length. The key result of this work is that as $\kappa \to 0$, i.e. the Debye-layer thickness is larger than the fluid thickness, the system approaches the perfect dielectric behaviour. While in the other extreme of $\kappa \gg 1$, the system approaches the perfect conductor behaviour. A transition from PD to PC is seen with changing κ (Fig. 8).

6 Summary

To summarize, through the present work, a complete understanding of the instabilities at planar fluid interfaces under electric fields is obtained. The key outcomes are listed as follows:

- The effect of alternating fields on the instability is studied in both the cases, i.e. where gravity effects are comparable to surface tension and where they are negligible.
- Using the three different timescales inherent to the present system, a comprehensive picture of the various possible wavelengths of the instability has been studied.

- Non-linear analysis which was hitherto used for qualitative understanding of the instability has been shown to possess powerful capabilities in predicting quantitatively the wavelength of the instability, in regimes where the LD theory fails.
- A parametric study of when the thin film approximation can be used and under which conditions it is violated is carried out and the alternative generalized model is used.
- The other violation of the EHD theory which occurs when one approaches a thin film limit is that of the overlap of diffuse layers of charge. This is taken into account using an electrokinetic theory.
- To conclude, three ways of obtaining a transition behaviour with the limiting extreme cases of a PD material and an LD material have been studied:
  - By changing the frequency of the applied alternating field
  - By changing the system timescales via changing the conductivity of the fluids or the applied field
  - By changing the thickness of the fluid layers

This study furthers the understanding of the interfacial EHD phenomenon while suggesting new tools for analysis and in context to applicability to technologies like soft lithography.

References

1. Melcher JR (1961) Electrohydrodynamic and magnetohydrodynamic surface waves and instabilities. Phys Fluids 4:1348
2. Atten P, Malraison B, Ali Kani S (1982) Electrohydrodynamic stability of dielectric liquids subjected to ac fields. J Electrost 12:477–488
3. Melcher JR, Smith CV (1969) Electrohydrodynamic charge relaxation and interfacial perpendicular field instability. Phys Fluids 12(4):778–790
4. Terasawa H, Mori YH, Komotori K (1983) Instability of horizontal fluid interface in a perpendicular electric field: Observational study. Chem Eng Sci 38:567–573
5. Schaffer E, Thurn-Albrecht T, Russell TP, Steiner U (2000) Electrically induced structure formation and pattern transfer. Nature 403:874–877
6. Dong J, de Almeida VF, Tsouris C (2001) Formation of liquid columns on liquid-liquid interfaces under applied electric fields. J Colloid Interface Sci 242:327–336
7. Saville DA (1997) Electrohydrodynamics: the taylor-melcher leaky dielectric model. Ann Rev Fluid Mech 29:27–64
8. Pease LF, Russel WB (2002) Linear stability analysis of thin leaky dielectric films. J Non-Newtonian Fluid Mech 102:233–250
9. Pease LF, Russel WB (2003) Electrostatically induced submicron patterning of thin perfect and leaky dielectric films: A generalized linear stability analysis. J Chem Phys 118:3790–3803
10. Pease LF, Russel WB (2006) Charge driven, electrohydrodynamic patterning of thin films. J Chem Phys 125:184716
11. Shankar V, Sharma A (2004) Instability of the interface between thin fluid films subjected to electric fields. J Colloid Interface Sci 274:294–308
12. Thaokar RM, Kumaran V (2005) Electrohydrodynamic instability of the interface between two fluids confined in a channel. Phys Fluids 17:084104
13. Roberts SA, Kumar S (2009) Ac electrohydrodynamic instabilities in thin liquid films. J Fluid Mech 631:255–279

Suppression of Interfacial Instabilities using Soft, Deformable Solid Coatings

V. Shankar and Gaurav Sharma

Abstract Interfacial instabilities in multilayer flows of immiscible liquid layers are often driven by viscosity contrasts in Newtonian liquids and/or elasticity contrasts in the case of polymeric liquids. Such instabilities, when present, can often be detrimental to many applications such as coating applications and polymer extrusion processes. In this review, we summarize our research which has explored the possibility of using a soft solid layer coating to manipulate interfacial instabilities. We have explored a variety of geometric configurations, as well as fluid constitutive behavior. We show that in general, it is possible to suppress interfacial instabilities in both Newtonian and polymeric liquids by lining the rigid wall with a deformable solid layer. We show that there is a window of parameters (shear modulus, layer thickness, ratio of solid to fluid viscosity) where the solid layer suppresses the interfacial instabilities at all wavenumbers without triggering new instabilities. We also show that solid layer deformability can further be used to promote interfacial instabilities in core–annular flows. This could be potentially used in microfluidic applications which are used to prepare monodisperse droplets and emulsions.

Keywords Interfacial instability • Viscoelastic flows • Deformable solids • Manipulation and control of interfacial instability • Core–annular flows

1 Introduction

Multilayer flows involving flow of two or more immiscible liquid layers are frequently encountered in a wide variety of physical settings and technological applications such as manufacturing of photographic films and coating processes [68], polymer co-extrusion [16, 44, 62], and lubricating pipelines [36, 50] and more recently in microfluidic devices [2] to produce monodisperse emulsions. These

V. Shankar (✉)
Department of Chemical Engineering, Indian Institute of Technology, Kanpur, 208016 UP, India
e-mail: vshankar@iitk.ac.in

G. Sharma
Department of Chemical Engineering, Indian Institute of Technology, 247667, Roorkee, India
e-mail: goravfch@iitr.ac.in

Y.M. Joshi, S. Khandekar (eds.), *Nanoscale and Microscale Phenomena*,
Springer Tracts in Mechanical Engineering, DOI 10.1007/978-81-322-2289-7_9

flows of multiple liquid layers are susceptible to various interfacial instabilities at the interface between two fluid layers owing to jump in viscosity and/or elasticity across the fluid–fluid interface. It is frequently desired to control and manipulate (suppress or enhance) these interfacial instabilities depending on the end-product requirements. For example, instabilities are detrimental to product quality in coating processes, where they result in nonuniform film thickness. On the other hand, these instabilities are desirable to enhance mixing properties and in several patterning applications or in creation of nanoliter-sized droplets in microfluidic devices involving flow of two or more liquids. An important aspect in multilayer flows is the ability to control and manipulate the interfacial instabilities depending on the end-product requirements. In the present review, we demonstrate the capability of using a "passive" deformable solid coating to inhibit or enhance the instabilities present between one or more fluid–fluid interfaces.

There have also been several other attempts to explore the possibility of controlling interfacial instabilities by different methodologies. The effect of in-plane horizontal oscillations imposed in the direction parallel to the flow has been investigated for free-surface instability of single Newtonian liquid film falling down an inclined plane for both two- and three-dimensional disturbances by using Floquet theory [45, 46]. It was shown that by choosing appropriate amplitudes and frequencies of forced oscillations, it is possible to completely suppress the instability of falling liquid film. A similar conclusion for instability suppression was achieved by Jiang and Lin [33] for the case of two-layer Newtonian liquid film flow. Huang and Khomami [29, 30] explored the role of imposing such dynamic modulation on single and two-layer viscoelastic liquid film flow, respectively, and demonstrated the possibility of using in-plane oscillations to either stabilize or destabilize flow of Newtonian and viscoelastic fluids down an inclined plane. Demekhin [12] investigated the possibility of imposing a temperature gradient along the wall to control free-surface instabilities present for a Newtonian falling film. A downstream linearly increasing temperature distribution was found to be stabilizing for small Prandtl numbers, while for large Prandtl numbers, increasing the temperature gradient first destabilizes the flow and then stabilizes the falling film configuration. Several studies have also investigated the effect of insoluble surfactants present at liquid–liquid or gas–liquid interface in context of altering the stability behavior for different flow configurations [3, 4, 64–66].

While these studies have suggested "active" methods (i.e., externally imposed oscillations or heating of the plate) toward suppressing the fluid–fluid instabilities, here, we examine the feasibility of a "passive" method, where a soft solid coating is proposed for suppressing the interfacial instabilities in different flow situations. When the fluid flows past a soft, deformable solid with relatively low shear modulus ($\sim O(10)$ kPa -1 MPa), the solid can be deformed by moderate fluid stresses and these deformations can in turn alter the adjacent fluid flow. Such "elastohydrodynamic coupling" makes the deformable solid a potential candidate to (1) trigger new instabilities which remain absent in flow past rigid surfaces and/or (2) to inhibit the flow instabilities that are already present in flow past rigid surfaces. The role of this coupling in inducing new interfacial instabilities between fluid–solid interface is well known both at zero and nonzero Reynolds number [18–21, 39, 41, 42]. We are interested in presenting the exactly opposite behavior of deformable wall, i.e., the suppression and controlling the instabilities that remain present for flow past rigid surfaces.

Kramer [38] was the first to propose the role of compliant coatings in the postponement of laminar-turbulent transition and related the suppression of Tollmien-Schlichting waves and the subsequent delay of laminar-turbulent transition to the efficiency of swimming of dolphins. Following this, there have been extensive studies to explore the role of wall compliance or deformability to delay the Tollmien-Schlichting instability for boundary layer and pressure-driven flows [5–7, 11, 15, 25, 31, 49, 54]. The Tollmien–Schlichting mode is present for a single fluid layer and becomes unstable for high Reynolds numbers. In this work, we explore the role of deformable solid wall in manipulating the interfacial instabilities present for multilayer flows at lower values of Reynolds number ($Re \sim O(1)$). The flow configuration, in general, comprises of one or more fluid layers flowing past a soft, deformable solid layer. The soft solid layer is assumed to be strongly adhered to a rigid surface. Some of the typical flow configurations that are frequently encountered in a wide variety of natural and industrial settings are shown in Fig. 1. For example, falling liquid film(s) past a rigid surface is a fundamental configuration in different coating applications such as manufacturing of photographic films or wire coating applications. Similarly, flow of two or more fluid layers often occur in different polymer processing applications and in lubricated pipeline transport. A common feature to all these flow configurations is the presence of one or more fluid–fluid interfaces. The interface could be a liquid–liquid (LL) interface or a gas–liquid (GL) free surface which are prone to interfacial instabilities caused due to discontinuity of viscosity and elasticity across the interface [34, 35, 44]. Further, the stability behavior depends strongly on the physical configuration as well as the rheological properties of the fluids. For example, plane-Couette flow of two Newtonian liquid layers or single liquid film falling down an incline cannot become unstable in the absence of inertia [69, 70], while multiple Newtonian liquid films falling down an incline or plane-Couette flow of two viscoelastic liquid layers could become unstable even in creeping flow limit [9, 53, 62, 67]. Similarly, the presence of a cylindrical fluid–fluid interface causes capillary instability in addition to the interfacial instabilities caused due to viscosity or elasticity contrasts [27, 50, 63].

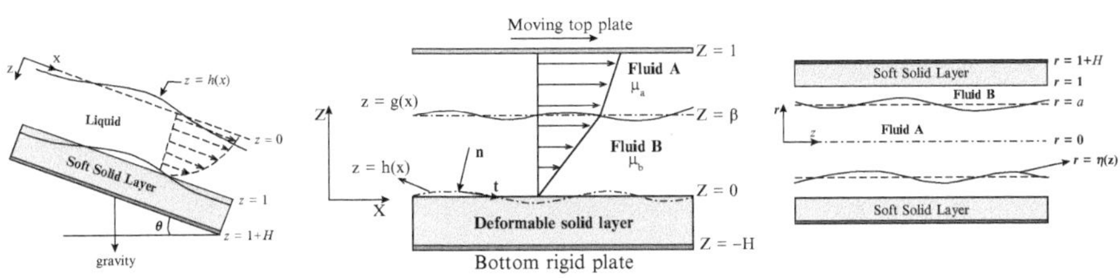

Fig. 1 Different configurations depicting flow of one or more fluid layers past a soft, deformable solid surface

As shown in Fig. 1, in this work, a soft solid layer is attached to the rigid surface in order to demonstrate the possibility of using a soft solid coating to control such qualitatively different interfacial instabilities in different physical settings.

2 Problem Formulation

2.1 Governing Equations

In general, the physical system of interest involves a fluid–solid interface in addition to an LL and/or a GL interface. A linear stability analysis is performed for different flow situations considering both Newtonian and viscoelastic fluids in order to illustrate the role of fluid–solid interface in manipulating the LL/GL interfacial instabilities. In the following, we describe the mathematical formulation of the linear stability problem by taking an example of a Newtonian liquid film falling down an inclined plane which is coated with a soft, deformable solid layer (refer to Fig. 2). A similar procedure is followed for any other flow configuration in the respective coordinate system using the appropriate constitutive relation for both fluid and solid layers.

2.1.1 Newtonian Liquid Layer

Figure 2 shows the sketch of gravity-driven laminar flow of an incompressible Newtonian liquid film flowing past an impermeable and incompressible soft solid layer. The solid layer (of thickness HR, shear modulus G, density ρ, and viscosity μ_s) is strongly bonded to a rigid inclined plane at $z^* = (1 + H)R$ which makes

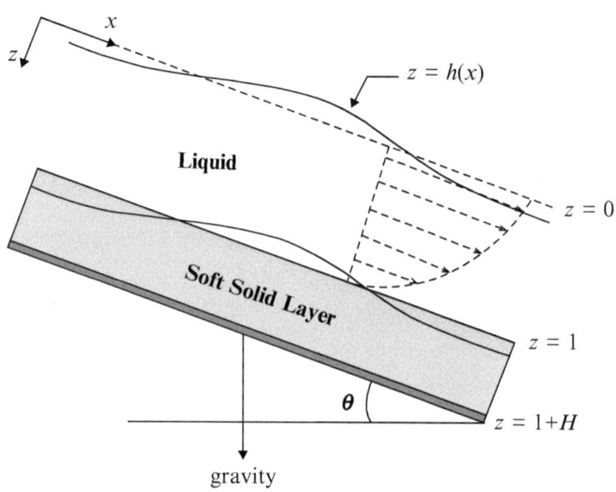

Fig. 2 Schematic diagram showing the configuration and nondimensional coordinate system: gravity-driven flow of a Newtonian liquid layer falling down an incline which is coated with a deformable solid layer

an angle θ with the horizontal surface. The liquid layer (of viscosity μ and density ρ) occupies a region $0 \leq z^* \leq R$ and is in contact with a passive gas at $z^* = 0$. We use the following scales to nondimensionalize various physical quantities: fluid thickness R for lengths and displacements, free-surface/maximum velocity of the liquid layer in unperturbed steady base-state $V = \frac{\rho g R^2 \sin\theta}{2\mu}$ for velocities, and $\mu V/R$ for stresses and pressure. The resulting nondimensional governing equations for the liquid layer are

$$\nabla \cdot \mathbf{v} = 0\,, \tag{1}$$

$$Re[\frac{\partial \mathbf{v}}{\partial t} + \mathbf{v} \cdot \nabla \mathbf{v}] = \nabla \cdot \mathbf{T} + \frac{2}{\sin\theta}\hat{\mathbf{g}}\,, \tag{2}$$

where $\mathbf{v}$ represents the velocity field in the liquid, $\hat{\mathbf{g}}$ is the unit vector pointing in the direction of gravity, and $Re = \rho V R/\mu$ is the Reynolds number. The total stress tensor $\mathbf{T}$ for liquid is

$$\mathbf{T} = -p\mathbf{I} + \boldsymbol{\tau}\,. \tag{3}$$

where $\boldsymbol{\tau} = [\nabla \mathbf{v} + (\nabla \mathbf{v})^T]$ is the nondimensional Cauchy stress tensor for Newtonian liquid.

2.1.2 Deformable Solid Layer

Since, our goal is to demonstrate the potential of deformable solid wall in suppressing interfacial instabilities, it is crucial to choose an appropriate solid model which would accurately describe the behavior of soft, elastic solids in real systems. The dynamics of the deformable solid layer is described by using both a simple linear viscoelastic model and a nonlinear neo-Hookean constitutive relation. While the linear viscoelastic solid model is quickly amenable to mathematical analysis, it has a drawback that it is valid only when the deformation gradients in the solid layer are small compared to unity. The neo-Hookean solid model is one of the simplest nonlinear rheological solid model for elastic solids which capture the behavior of real soft, deformable materials like rubber reasonably well [47]. It will be demonstrated by using a long-wave asymptotic analysis and numerical results that the deformation gradients that are required to be developed within the solid layer to suppress interfacial instabilities remain small compared to unity. Hence, the phenomena of instability suppression is independent of the choice of constitutive relation used to model the soft solid surface. However, several earlier studies related to flow past flexible surfaces have shown that the fluid–solid interface becomes unstable when the strain in the solid layer increases above a critical $O(1)$ value [20–22, 39, 42]. This is outside the validity domain of linear theory of elasticity and it is expected that the linear solid model will not be sufficiently accurate to predict the instability of fluid–solid mode. Indeed, it has been noted by Gkanis and Kumar

[20, 21] and Gaurav and Shankar [18, 19] that it is important to include constitutive nonlinearities in the solid model to accurately capture the stability behavior of fluid–solid interface. In order to realize instability suppression in a given physical setting, it is necessary that both fluid–fluid and fluid–solid interfaces remain stable. Thus, we use both linear elastic and neo-Hookean solid model to represent the dynamics of the deformable solid layer.

Linear Viscoelastic Solid

Following the earlier works on flow past linear elastic solid [42, 56, 59], the governing equations for the fluid written above (Eqs. 1 and 2) and linear elastic solid are written in terms of spatial (Eulerian) coordinates ($\mathbf{x} = x, y, z$). In the Eulerian description, the nondimensional equations governing the deformation in linear viscoelastic solid are conveniently written in terms of the displacement vector:

$$\nabla \cdot \mathbf{u} = 0\,, \tag{4}$$

$$Re\left[\frac{\partial^2 \mathbf{u}}{\partial t^2}\right] = \nabla \cdot \boldsymbol{\sigma} + \frac{2}{\sin\theta}\hat{\mathbf{g}}\,, \tag{5}$$

where $\mathbf{u}(\mathbf{x}) \equiv (u_x, u_y, u_z)$ is the displacement of a particle from the initial unstressed (reference) configuration $\mathbf{X} = (X, Y, Z)$ to a current configuration $\boldsymbol{w} \equiv (w_x, w_y, w_z)$. The initial reference and current configuration are related through displacement $\mathbf{u}(\mathbf{x})$ as $\boldsymbol{w} = \mathbf{X} + \mathbf{u}(\mathbf{x}, t)$. In the momentum equation (5), $\boldsymbol{\sigma}$ is the nondimensional Cauchy stress tensor similar to the one used in previous studies [42, 56]:

$$\boldsymbol{\sigma} = -p_s\mathbf{I} + \left(\frac{1}{\Gamma} + \mu_{rs}\partial_t\right)[\nabla\mathbf{u} + (\nabla\mathbf{u})^T]\,. \tag{6}$$

where p_s is the pressure in the solid layer and $\mu_{rs} = \mu_s/\mu$ is the ratio of solid to fluid viscosity. In the above equation, $\Gamma = \mu V/GR$ is the nondimensional parameter characterizing the deformability of the solid layer and can be interpreted as ratio of viscous shear stresses in liquid to elastic stresses in the solid layer. Higher values of Γ indicate more soft solid or increasing deformability of the solid layer and $\Gamma \to 0$ implies the limit of a rigid solid layer.

Neo-Hookean Viscoelastic Solid

In considering flow past neo-Hookean solid, the governing equations for fluid, as usual, are written in a Eulerian framework, and thus, Eqs. 1 and 2 are used as mass and momentum conservation equations for the liquid layer. However, we find it convenient to express the conservation equations for neo-Hookean solid in

Lagrangian description, where the deformations in the solid layer are expressed in terms of the spatial positions in a reference (unstressed) configuration denoted by $\mathbf{X} = (X, Y, Z)$. Thus, $\mathbf{X} = (X, Y, Z)$ denotes the position vector of material particle in unstressed configuration and is used as the independent variable for all the dynamical variables in neo-Hookean deformable solid medium. In the deformed state of the solid, the current position of the material particle is denoted by $\boldsymbol{w}(\boldsymbol{X}) = (w_X, w_Y, w_Z)$. Thus, the displacement vector $\boldsymbol{u}(\boldsymbol{X})$ of a particle in current deformed state which was originally at spatial location $\mathbf{X} = (X, Y, Z)$ in the reference configuration is $\boldsymbol{u}(\boldsymbol{X}) = \boldsymbol{w}(\boldsymbol{X}) - \mathbf{X}$. The past and present states are related by deformation gradient tensor $\mathbf{F} = \nabla_{\mathbf{X}}\boldsymbol{w}$, where subscript $\mathbf{X}$ denotes the gradient with respect to reference configuration. The mass and momentum conservation equations for an incompressible neo-Hookean viscoelastic solid are [28, 48]:

$$\det(\mathbf{F}) = 1\,, \tag{7}$$

$$Re\left[\frac{\partial^2 \boldsymbol{w}}{\partial t^2}\right]_{\mathbf{X}} = \nabla_X \cdot \mathbf{P} + \frac{2}{\sin\theta}\hat{\mathbf{g}}\,. \tag{8}$$

where $\mathbf{P}$ is the first Piola-Kirchhoff tensor and is related to Cauchy stress tensor as $\mathbf{P} = \mathbf{F}^{-1} \cdot \boldsymbol{\sigma}$. The Cauchy stress tensor for a neo-Hookean viscoelastic solid comprises of an elastic part $\boldsymbol{\sigma}_e$ and a dissipative part, $\boldsymbol{\sigma}_d$ [13, 14, 26]:

$$\boldsymbol{\sigma} = \boldsymbol{\sigma}^e + \boldsymbol{\sigma}^d\,, \tag{9}$$

$$\boldsymbol{\sigma}^e = -p_s\mathbf{I} + 2G\mathbf{E}\,, \quad \boldsymbol{\sigma}^d = \mu_{rs}\left(\boldsymbol{L} + \boldsymbol{L}^T\right)\,. \tag{10}$$

Here, p_s is the pressure in the solid medium, G is the shear modulus of the elastic solid, and $\mathbf{L} = \dot{\mathbf{F}} \cdot \mathbf{F}^{-1}$ is the spatial velocity gradient with overdot representing material time derivative. $\mathbf{E}$ is the general nonlinear strain tensor given as [48]:

$$\mathbf{E} = \frac{1}{2}\left(\mathbf{F} \cdot \mathbf{F}^{\mathrm{T}} - \mathbf{I}\right) \text{ or} \tag{11}$$

$$E_{ij} = \frac{1}{2}\left(\frac{\partial u_i}{\partial X_j} + \frac{\partial u_j}{\partial X_i} + \frac{\partial u_i}{\partial X_k}\frac{\partial u_j}{\partial X_k}\right)\,. \tag{12}$$

Note that when the deformation gradients in solid medium are small compared to unity, the nonlinear quadratic terms in strain tensor can be neglected in comparison to linear terms and we obtain the linearized strain tensor $(E_{ij})_{\mathrm{lin}} = \frac{1}{2}\left(\frac{\partial u_i}{\partial X_j} + \frac{\partial u_j}{\partial X_i}\right)$. When this linearized strain tensor $(E_{ij})_{\mathrm{lin}}$ is used in $\boldsymbol{\sigma}_e$, we obtain Cauchy stress tensor for a linear elastic solid. It is important to mention here that these nonlinear terms give rise to a nonzero first normal stress difference in the base state and several additional coupling terms in governing stability equations of solid which in turn significantly affects the stability behavior of fluid–solid mode. Within molecular framework, the constitutive relation for neo-Hookean solid given above can be derived by considering the soft elastomer as a three-dimensional network of polymer chains connected together at random sites. The elasticity of soft material is a result of net entropic restoring force when individual chains are stretched from its random coil conformations. Further, the change in entropy of the network caused due to deformation is assumed to be the sum of change of entropy for all polymer chains.

2.1.3 Interfacial and Boundary Conditions

For the configuration shown in Fig. 2, a GL free surface and a liquid-solid (LS) interface are present. The gas adjacent to the liquid layer is assumed to be passive, and hence, zero tangential and normal stress conditions are used along with the kinematic condition for the evolution of free surface:

$$\partial_t h + v_x \partial_x h = v_z \,, \tag{13}$$

$$\mathbf{n} \cdot \mathbf{T} \cdot \mathbf{t} = 0 \,, \tag{14}$$

$$\mathbf{n} \cdot \mathbf{T} \cdot \mathbf{n} + \Sigma_{\text{gl}} (\boldsymbol{\nabla} \cdot \mathbf{n}) = 0 \,, \tag{15}$$

where $z = h(x)$ is the gas–liquid interface position in the perturbed state, $\mathbf{n}$ and $\mathbf{t}$ are the normal and tangential unit vectors to the GL interface, respectively, while $\Sigma_{\text{gl}} = \gamma_{\text{gl}}/\mu V$ is the nondimensional interfacial tension parameter, with γ_{gl} being the dimensional GL interfacial tension.

In general, a liquid–liquid interface can also be present in the flow configuration of interest. For example, Fig. 1b, c show the presence of an LL interface in addition to an LS interface in the case of plane-Couette flow and core–annular flow of two immiscible liquids past a flexible surface. Similarly, gravity-driven flow of two liquid layers down an incline coated with a deformable solid layer involves the presence of all three LL, GL, and LS interfaces. In such cases where an LL interface is present, the interfacial conditions at liquid–liquid interface are the continuity of tangential and normal velocities and stresses along with kinematic condition relating the velocity in two liquid layers to the location of the perturbed LL interface (denoted by $z = g(x)$):

$$\partial_t g + v_x^a \partial_x g = v_z^a \,, \tag{16}$$

$$\mathbf{v}^a = \mathbf{v}^b \,, \tag{17}$$

$$\mathbf{n} \cdot \mathbf{T}^a \cdot \mathbf{t} = \mathbf{n} \cdot \mathbf{T}^b \cdot \mathbf{t} \,, \tag{18}$$

$$\mathbf{n} \cdot \mathbf{T}^a \cdot \mathbf{n} - \mathbf{n} \cdot \mathbf{T}^b \cdot \mathbf{n} = \Sigma_{\text{ll}} (\boldsymbol{\nabla} \cdot \mathbf{n}) \,, \tag{19}$$

where superscripts a and b are used to denote two fluid layers and $\Sigma_{\text{ll}} = \gamma_{\text{ll}}/\mu V$ is the nondimensional liquid–liquid interfacial tension parameter, with γ_{ll} being the dimensional LL interfacial tension. Here, V is the characteristic velocity for the flow system under consideration and μ will be the viscosity of the liquid layer adjacent to the deformable solid medium.

At the LS interface, the velocities and stresses in liquid and solid layers are continuous. For linear elastic solid, the condition of velocity continuity is given by

$$\mathbf{v} = \frac{\partial \mathbf{u}}{\partial t}, \tag{20}$$

and for neo-Hookean solid

$$\mathbf{v} = \frac{\partial \mathbf{w}}{\partial t}. \tag{21}$$

The normal and tangential stress balance are given as

$$\mathbf{n} \cdot \mathbf{T} \cdot \mathbf{t} = \mathbf{n} \cdot \boldsymbol{\sigma} \cdot \mathbf{t}, \tag{22}$$

$$\mathbf{n} \cdot \mathbf{T} \cdot \mathbf{n} - \mathbf{n} \cdot \boldsymbol{\sigma} \cdot \mathbf{n} = \Sigma_{\mathrm{ls}}(\nabla \cdot \mathbf{n}). \tag{23}$$

Here, $\mathbf{n}$ and $\mathbf{t}$ are the normal and tangential unit vectors for LS interface. $\Sigma_{\mathrm{ls}} = \gamma_{\mathrm{ls}}/\mu V$ is the nondimensional LS interfacial tension parameter, with γ_{ls} being the dimensional LS interfacial tension. The condition at the rigid surface ($z = 1 + H$) is the zero-displacement condition, which is given for linear elastic solid as

$$\mathbf{u} = 0, \tag{24}$$

and for neo–Hookean solid as

$$\mathbf{w} = \mathbf{X}. \tag{25}$$

2.1.4 Base State

The laminar steady-state solution of the liquid layer whose stability is of interest is obtained by solving the governing equations for liquid layer (Eqs. 1 and 2) under the assumption of unidirectional steady-state flow in x-direction along with the interfacial conditions at GL interface (Eqs. 14 and 15) and LS interface (Eqs. 20, 22, and 23 for linear elastic solid and 21, Eqs. 22, and 23 for neo-Hookean solid). Both GL and LS interfaces remain flat in the base state. The solid layer remains at rest with a nonzero displacement in x-direction which can be obtained by solving appropriate governing equation for linear or nonlinear solid (Eqs. 4 and 5 for linear viscoelastic solid or Eqs. 7 and 8 for neo-Hookean viscoelastic solid) along with the boundary conditions at LS interface and at rigid surface. The nondimensional velocity profile and pressure distribution in the liquid layer are given as

$$\overline{v}_x = (1 - z^2)\,, \quad \overline{v}_z = 0\,, \quad \overline{p} = (2\cot\theta)z\,. \tag{26}$$

The displacement and pressure field for the linear elastic solid are given by:

$$\overline{u}_x(z) = \Gamma[(1 + H)^2 - z^2]\,, \quad \overline{u}_z = 0\,, \quad \overline{p}_s = (2\cot\theta)z\,. \tag{27}$$

and the deformation and pressure field for the neo-Hookean solid is given as:

$$\overline{w}_X = X + \Gamma[(1 + H)^2 - Z^2]\,, \quad \overline{w}_Z = Z\,, \quad \overline{\hat{p}}_s = \frac{1}{\Gamma} + (2\cot\theta)Z\,. \tag{28}$$

where $\hat{p}_s$ is a pressure-like function related to actual pressure in solid as $\hat{p}_s = p_s + \frac{1}{\Gamma}$.

2.1.5 Linear Stability Analysis

A standard temporal linear stability analysis is performed in order to determine the fate of small perturbations imposed on base-state solution: $f = \overline{f} + f'$. Here, f denotes any variable like velocity, deformation, or pressure in a fluid or solid layer; $\overline{f}$ denotes the variable in the base state; and f' is the small perturbation imposed on $\overline{f}$. This perturbed variable f is substituted in the governing equations and interfacial conditions and the resulting equations are then linearized to obtain a set of equations in terms of small perturbation variable f'. The perturbations are assumed to be two dimensional (in x and z directions), which are expanded in the form of Fourier modes, $f' = \tilde{f}(z)\exp[\mathrm{i}k(x - ct)]$, where $\tilde{f}(z)$ is the complex amplitude function of the disturbance, k is the wavenumber of perturbation, and $c = c_r + \mathrm{i}c_i$ is the complex wavespeed. If $c_i > 0$ (or $c_i < 0$), the flow will be unstable (or stable). For the case of flow past neo-Hookean solid, x and z are replaced by X and Z.

The linearized governing equations for the liquid layer are

$$\frac{\mathrm{d}\tilde{v}_z}{\mathrm{d}z} + \mathrm{i}k\tilde{v}_x = 0\,, \tag{29}$$

$$Re\,[\mathrm{i}k(\overline{v}_x - c)\tilde{v}_x + (\mathrm{d}_z\overline{v}_x)\tilde{v}_z] = -\mathrm{i}k\tilde{p} + \left[\frac{\mathrm{d}^2}{\mathrm{d}z^2} - k^2\right]\tilde{v}_x\,, \tag{30}$$

$$Re\,[\mathrm{i}k(\overline{v}_x - c)\tilde{v}_z] = -\frac{\mathrm{d}\tilde{p}}{\mathrm{d}z} + \left[\frac{\mathrm{d}^2}{\mathrm{d}z^2} - k^2\right]\tilde{v}_z\,. \tag{31}$$

The governing equations for linear elastic solid are

$$\frac{\mathrm{d}\tilde{u}_z}{\mathrm{d}z} + \mathrm{i}k\tilde{u}_x = 0\,, \tag{32}$$

$$-Re\, k^2 c^2 \tilde{u}_x = -\mathrm{i}k\, \tilde{p}_s + \left(\frac{1}{\Gamma} - \mathrm{i}kc\mu_{rs}\right)\left[\frac{\mathrm{d}^2}{\mathrm{d}z^2} - k^2\right]\tilde{u}_x\,, \qquad (33)$$

$$-Re\, k^2 c^2 \tilde{u}_z = -\frac{\mathrm{d}\tilde{p}_s}{\mathrm{d}z} + \left(\frac{1}{\Gamma} - \mathrm{i}kc\mu_{rs}\right)\left[\frac{\mathrm{d}^2}{\mathrm{d}z^2} - k^2\right]\tilde{u}_z\,, \qquad (34)$$

and for neo-Hookean viscoelastic solid are

$$\frac{\mathrm{d}\tilde{w}_Z}{\mathrm{d}Z} + \mathrm{i}k\tilde{w}_X - \underline{\left(\frac{\mathrm{d}\overline{w}_X}{\mathrm{d}Z}\right)\mathrm{i}k\tilde{w}_Z} = 0\,, \qquad (35)$$

$$\begin{aligned} -\mathrm{i}k\,\tilde{p}_s + \underline{(2\cot\theta)\mathrm{i}k\tilde{w}_Z} + \left(\frac{1}{\Gamma} - \mathrm{i}kc\eta_r\right)\left[-k^2 + \frac{\mathrm{d}^2}{\mathrm{d}Z^2}\right]\tilde{w}_X \\ \underline{-\mathrm{i}kc\eta_r\left[-\left(\frac{\mathrm{d}\overline{w}_X}{\mathrm{d}Z}\right)2\mathrm{i}k\frac{\mathrm{d}\tilde{w}_X}{\mathrm{d}Z} + \mathrm{i}k\frac{\mathrm{d}\tilde{w}_Z}{\mathrm{d}Z}\right]} \\ \underline{+\mathrm{i}kc\eta_r\left[\left(\frac{\mathrm{d}\overline{w}_X}{\mathrm{d}Z}\right)^2 k^2\tilde{w}_X + \left(\frac{\mathrm{d}^2\overline{w}_X}{\mathrm{d}Z^2}\right)\mathrm{i}k\tilde{w}_X\right]} + k^2c^2 Re\,\tilde{w}_X = 0\,, \end{aligned} \qquad (36)$$

$$\begin{aligned} -\underline{(2\cot\theta)\mathrm{i}k\,\tilde{w}_X} - \frac{\mathrm{d}\tilde{p}_s}{\mathrm{d}Z} + \left(\frac{1}{\Gamma} - \mathrm{i}kc\eta_r\right)\left[-k^2 + \frac{\mathrm{d}^2}{\mathrm{d}Z^2}\right]\tilde{w}_Z \\ \underline{\left(\frac{\mathrm{d}\overline{w}_X}{\mathrm{d}Z}\right)\mathrm{i}k\,\tilde{p}_s + \mathrm{i}kc\eta_r\left[\left(\frac{\mathrm{d}\overline{w}_X}{\mathrm{d}Z}\right)(-k^2\tilde{w}_X + 3\mathrm{i}k\frac{\mathrm{d}\tilde{w}_Z}{\mathrm{d}Z})\right]} \\ \underline{+\mathrm{i}kc\eta_r\left[\mathrm{i}k\left(\frac{\mathrm{d}^2\overline{w}_X}{\mathrm{d}Z^2}\right)\tilde{w}_Z + 2\left(\frac{\mathrm{d}\overline{w}_X}{\mathrm{d}Z}\right)^2 k^2\tilde{w}_Z\right]} + k^2c^2 Re\,\tilde{w}_Z = 0\,. \end{aligned} \qquad (37)$$

2.1.6 Linearized Interfacial Conditions

In this section, we briefly discuss the linearization of interfacial conditions at LS and GL interfaces. At LS interface, the velocity and stresses in liquid and solid layers are continuous. However, the linearization procedure depends upon whether Eulerian or Lagrangian approach is used to express variables in a fluid or solid layer. As customary in fluid dynamics, we always express the fluid variables in a Eulerian framework, while Lagrangian approach is used when the solid is modeled as a neo-Hookean viscoelastic solid and Eulerian approach is used for linear viscoelastic solid. Thus, a Eulerian–Eulerian framework is used for flow past a linear solid model and a Eulerian–Lagrangian approach is used for coupled fluid–neo-Hookean solid system. Denoting F and S to be solid dynamical variables, the continuity conditions at perturbed LS interface (see Fig. 3) can be written as

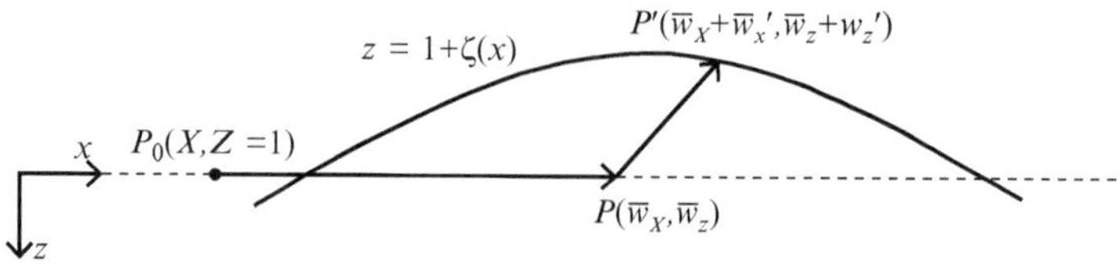

Fig. 3 Schematic diagram illustrating the application of boundary conditions over the perturbed fluid–solid interface

$$F(x,z,t)|_{P'} = S(X,Z,t)|_{P'}. \tag{38}$$

for liquid–neo-Hookean solid and

$$F(x,z,t)|_{P'} = S(x,z,t)|_{P'}. \tag{39}$$

for liquid–linear elastic solid interface. Here, point P' denotes the current position of a material particle on perturbed interface which was originally at point P_0 in unstressed configuration and at P in base-state configuration. The current position of the point P' is unknown a priori, and since Eulerian coordinates are used for the liquid layer, any fluid dynamical quantity on the left side of equation (38) or (39) is expressed by Taylor expansion about the known position of the unperturbed flat interface ($z = 1$). Thus, the fluid variable is expressed as

$$F(x,z,t)|_{P'} = F|_{P(z=1)} + \zeta(x)\frac{\partial F}{\partial z}|_{P(z=1)}. \tag{40}$$

where $\zeta(x) = w_Z'(X, Z = 1)$. Note that the only difference in Eq. (38) or (39) lies in identifying the material particle in the solid layer by using the spatial positions in reference unstressed configuration (X, Z) for neo-Hookean solid or using the current spatial location (x, z) for linear elastic solid. Thus, the labeling of material particle remains identical in unstressed, or base-state, or perturbed configuration and the LS interface is always identified by $X, Z = 1$ in Lagrangian framework. Because of these reasons, a Taylor series expansion is not required to evaluate any solid variable on perturbed interface in the Lagrangian description of solid. On the other hand, if a Eulerian approach is used for solid (as in the case of linear elastic model), Taylor expansion about flat interface position $z = 1$ must be used for the solid layer in a manner similar to the liquid layer (see Eq. 40). The detailed explanation of the above argument is given in Gaurav and Shankar [19]. Following it, the linearized conditions at LS interface for flow past a neo-Hookean solid are

$$\tilde{v}_z = -ikc\,\tilde{w}_Z\,, \tag{41}$$

$$\tilde{v}_x + \tilde{w}_Z\,(\mathrm{d}_z\overline{v}_x)_{z=1} = -ikc\,\tilde{w}_X\,, \tag{42}$$

$$
\begin{aligned}
&\left(\frac{1}{\Gamma} - ikc\eta_r\right)\left(\frac{d\tilde{w}_X}{dZ} + ik\tilde{w}_Z\right) \\
&\underline{\frac{1}{\Gamma}\left(\frac{d\overline{w}_X}{dZ}\right)\frac{d\tilde{w}_Z}{dZ} - \frac{1}{\Gamma}\left(\frac{d\overline{w}_X}{dZ}\right)^2 ik\tilde{w}_Z} \\
&\underline{-k^2 c\eta_r \left(\frac{d\overline{w}_X}{dZ}\right)\tilde{w}_X - \tilde{w}_Z\,(d_z^2\overline{v}_x)} = \left(\frac{d\tilde{v}_x}{dz} + ik\tilde{v}_z\right), \qquad (43)
\end{aligned}
$$

$$
\begin{aligned}
&-\widetilde{p_s} + 2\left(\frac{1}{\Gamma} - ikc\eta_r\right)\frac{d\tilde{w}_Z}{dZ} - k^2\Sigma_{ls}\tilde{w}_Z \\
&\underline{+\tilde{w}_Z\frac{d\overline{p}}{dz} - 2k^2 c\eta_r\left(\frac{d\overline{w}_X}{dZ}\right)\tilde{w}_Z} = -\tilde{p} + 2\frac{d\tilde{v}_z}{dz}. \qquad (44)
\end{aligned}
$$

The underlined terms present in the above interfacial conditions and governing equations for neo-Hookean solid represent the additional coupling terms that are present due to constitutive nonlinearities present in neo-Hookean solid model. As discussed above, the linearized conditions at LS interface with linear viscoelastic solid are obtained by Taylor expanding both fluid and solid variables about $z = 1$:

$$\tilde{v}_z = -ikc\,\tilde{u}_z\,, \qquad (45)$$

$$\tilde{v}_x + \tilde{u}_z\,(d_z\overline{v}_x)_{z=1} = -ikc\,\tilde{u}_x\,, \qquad (46)$$

$$\frac{d\tilde{v}_x}{dz} + ik\tilde{v}_z = \frac{1}{\Gamma}\left(\frac{d\tilde{u}_x}{dz} + ik\tilde{u}_z\right), \qquad (47)$$

$$-\tilde{p} + 2\frac{d\tilde{v}_z}{dz} + k^2\Sigma_{ls}\tilde{u}_z \quad -\widetilde{p_s} + \frac{2}{\Gamma}\frac{d\tilde{u}_z}{dz}, \qquad (48)$$

Similarly, the gas–liquid interface ($z = 0$) also involves only Eulerian variables, and hence, the interfacial conditions at free surface are linearized by Taylor expanding the conditions about $z = 0$. The linearized kinematic and boundary conditions at unperturbed free surface ($z = 0$) are

$$ik[\overline{v}_x(z = 0) - c]\tilde{h} = \tilde{v}_z(z = 0)\,, \qquad (49)$$

$$-2\tilde{h} + \left(\frac{d\tilde{v}_x}{dz} + ik\tilde{v}_z\right) = 0\,, \qquad (50)$$

$$-\tilde{p} - (2\cot\theta)\tilde{h} + 2\frac{d\tilde{v}_z}{dz} - k^2\Sigma_{gl}\,\tilde{h} = 0\,. \qquad (51)$$

Finally, the boundary conditions at rigid surface ($z = 1 + H$) are no displacement condition $\tilde{\mathbf{u}} = 0$, for linear elastic solid, and no deformation condition $\tilde{\boldsymbol{w}} = 0$, for neo-Hookean solid. This completes the description of governing stability equations for the configuration shown in Fig. 2. Recall that we used this falling film configuration as an example to illustrate the general procedure that is followed for any other flow setting involving one or more fluid layers past a deformable solid surface. Several other flow configurations (e.g., see Fig. 1) were analyzed to investigate the role of wall deformability in controlling interfacial instabilities. Examples include gravity-driven flow of single polymeric liquid or two Newtonian liquid layers past a flexible surface, plane-Couette flow of two immiscible viscoelastic or Newtonian fluid layers past a soft surface, core–annular flows in flexible tubes, and so on. The stability behavior for each of these configurations in the absence of deformable wall is different. For example, gravity-driven flow of two Newtonian liquid layers involves an interaction of LL interface and free surface which renders flow unstable for any parameter regime [9] or elasticity in the liquid layer causes falling liquid film to become unstable even in creeping flow limit [43, 60]. Similarly, CAFs involve capillary instability in addition to long-wave instability due to viscosity stratification. Note that the above equations were developed considering Newtonian fluids, while it is evident from the literature that viscoelastic effects in fluids significantly affect the stability behavior. Here, we evaluate the capability of soft solid coatings to suppress such interfacial instabilities in different flow settings and for different types of fluid rheology. While we do not attempt to give a complete set of stability equations involving a polymeric viscoelastic fluid, it is instructive to discuss briefly the two commonly used constitutive relation for viscoelastic fluids.

2.1.7 Oldroyd-B Model

One of the simplest constitutive relations useful for describing the behavior of polymeric solutions is the Oldroyd-B fluid model. The total stress tensor $\mathbf{T} = -p\mathbf{I} + \boldsymbol{\tau}$, where $\boldsymbol{\tau}$ is the extra stress tensor which is composed of a polymeric contribution and a Newtonian solvent contribution:

$$\boldsymbol{\tau} = \boldsymbol{\tau}^p + \boldsymbol{\tau}^s\,, \tag{52}$$

The solvent contribution to is simply $\boldsymbol{\tau}^s = \mu_{\text{sol}}\left[\nabla\mathbf{v} + (\nabla\mathbf{v})^T\right]$, while the polymeric contribution is given by upper-convected Maxwell equation:

$$\tau_R\left[\frac{\partial\boldsymbol{\tau}^p}{\partial t} + \mathbf{v}\cdot\nabla\boldsymbol{\tau}^p - (\nabla\mathbf{v})^T\cdot\boldsymbol{\tau}^p - \boldsymbol{\tau}^p\cdot\nabla\mathbf{v}\right] + \boldsymbol{\tau}^p = \mu_p\left[\nabla\mathbf{v} + (\nabla\mathbf{v})^T\right], \tag{53}$$

where μ_p is the polymer contribution to the shear viscosity and τ_R is the polymer relaxation time. If solvent contribution is set to zero, the model reduces to upper-convected Maxwell model which describes the rheological behavior of polymer melts. The constitutive equation for the Oldroyd-B model can be derived by idealizing the polymer molecule consisting of two Brownian beads (representing the source of hydrodynamic resistance) connected by an infinitely extensible spring which mimics the entropic elasticity of a single polymer molecule. This model retains essential physics, like nonzero first normal difference in simple shear flows, which is necessary to capture interfacial instabilities in multilayer flows of polymeric liquids. If we assume a characteristic shear rate like V/R and a characteristic length scale for the flow, then, two dimensionless parameters arise out of the Oldroyd-B equation. These are Weissenberg number $W = \tau_R V/R$ which is a measure of elasticity or relaxation time of fluid and a ratio of solvent to total viscosity $S = \mu_{\text{sol}}/(\mu_{\text{sol}} + \mu_p)$.

2.1.8 FENE-p Model

The Oldroyd-B model presented above predicts the shear-independent viscosity and first normal stress difference. It also predicts the zero second normal stress difference and unbounded growth of extensional viscosity, while viscosity saturates in real polymeric fluids because of finite extensibility of the polymer molecule. Similar to the Oldroyd-B fluid model, FENE-p constitutive relation can also be derived by considering two Brownian beads connected by a finitely extensible spring. This model predicts a shear-dependent viscosity and shear-dependent first normal stress difference for simple shear flows and, hence, overcomes the limitation of the Oldroyd-B model. The polymeric contribution to the stress tensor is given as

$$\boldsymbol{\tau}^p = \frac{f(\boldsymbol{Q})\boldsymbol{c} - \boldsymbol{I}}{W} \tag{54}$$

where $\boldsymbol{c}$ is the nondimensional conformation tensor characterizing the average second moment of end-to-end vector $\boldsymbol{Q}$ of the polymer molecule, and $\boldsymbol{c}$ reduces to the identity tensor $\boldsymbol{I}$ at equilibrium. The conformation tensor is nondimensionalized by group $k_B T/H_{\text{spring}}$, where H_{spring} is the spring constant, k_B is the Boltzmann constant, and T is the absolute temperature. The equation governing the time-evolution of conformation tensor is

$$\frac{\partial \boldsymbol{c}}{\partial t} + \mathbf{v} \cdot \nabla \boldsymbol{c} - \left[\boldsymbol{c} \cdot \nabla \mathbf{v} + (\nabla \mathbf{v})^T \cdot \boldsymbol{c}\right] = -\left(\frac{f\boldsymbol{c} - \boldsymbol{I}}{W}\right) \tag{55}$$

The term $f(\boldsymbol{Q})$ in Eq. 54 is the Peterlin function given as

$$f = \frac{b - 3}{b - \text{trace}(\boldsymbol{c})} \tag{56}$$

where b is the square of the nondimensional maximum extensibility of the spring (nondimensionalized by $k_B T/H_{\text{spring}}$) and $\text{trace}(\boldsymbol{c}) \leq b$. This parameter b characterizes the finite extensibility and the limit $b \to \infty$ represents the limit of an infinitely extensible polymer molecule chain and hence, the model reduces to the Oldroyd-B model.

3 Suppression of Long-Wave Interfacial Instabilities

It is well known that the discontinuity of viscosity or elasticity across the interface causes a long wavelength instability in the flows involving a GL or LL interface in the absence of deformable solid wall. Thus, in this section, we carry out an analysis similar to Yih's original long-wave analysis [69, 70] but include a deformable solid layer adjacent to the fluid layer in order to demonstrate the role of soft solid coatings in manipulating these long-wave interfacial instabilities. The configuration shown in Fig. 2 is used as an illustration to show the procedure and key features of long-wave or low-k asymptotic analysis. Subsequently, we discuss long-wave results for other flow configurations.

3.1 Suppression of Falling Film Instabilities with Free Surface

3.1.1 Single Newtonian Falling Film

Considering a Newtonian liquid film and linear viscoelastic solid, the equations that govern the stability of the present system are Eqs. 29–31 (liquid layer), (32)–(34) (solid layer), free- surface conditions (49)–(51), interfacial conditions at LS interface (45)–(47), and no displacement conditions at rigid surface. For dimensionless solid thickness $H \sim O(1)$, $k \ll 1$ implies low-k or long-wave limit and parameters Re and Γ are treated as independent parameters which do not show any particular scaling with wavenumber. The complex wavespeed c is expanded in an asymptotic series in k: $c = c^{(0)} + kc^{(1)} + \cdots$. The order of magnitude of different variables in liquid and solid layers (e.g., $\tilde{v}_z$, $\tilde{v}_x$, $\tilde{p}$, $\tilde{u}_z$, etc.) with respect to k is determined by using the respective governing equations in the fluid or solid layer. For example, if we assume $\tilde{v}_z \sim O(1)$, then the continuity equation and x-momentum equation implies $\tilde{v}_x \sim O(k^{-1})$ and $\tilde{p} \sim O(k^{-2})$, respectively. Thus, liquid variables are expanded as

$$\tilde{v}_z = \tilde{v}_z^{(0)} + k\tilde{v}_z^{(1)} + \cdots, \tag{57}$$

$$\tilde{v}_x = k^{-1}\tilde{v}_x^{(0)} + \tilde{v}_x^{(1)} + \cdots, \tag{58}$$

$$\tilde{p} = k^{-2}\tilde{p}^{(0)} + k^{-1}\tilde{p}^{(1)} + \cdots, \tag{59}$$

In a similar fashion, setting $\tilde{u}_z \sim O(1)$ and determining scalings of $\tilde{u}_x$ and $\widetilde{p_s}$, these variables are expanded as:

$$\tilde{u}_z = \tilde{u}_z^{(0)} + k\tilde{u}_z^{(1)} + \cdots, \tag{60}$$

$$\tilde{u}_x = k^{-1}\tilde{u}_x^{(0)} + \tilde{u}_x^{(1)} + \cdots, \tag{61}$$

$$\tilde{p}_s = k^{-2}\tilde{p}_s{}^{(0)} + k^{-1}\tilde{p}_s{}^{(1)} + \cdots, \tag{62}$$

The kinematic condition at GL interface (Eq. 49) shows that $h \sim O(1/k)$ and hence $\tilde{h} = k^{-1}\left(\tilde{h}^{(0)} + k\tilde{h}^{(1)} + \cdots\right)$. All the above expansions are substituted in the governing equations and boundary and interfacial conditions to obtain sets of equations governing the problem at different orders in k. The analysis shows that it is sufficient to calculate the leading order and first correction to velocities in the fluid layer, while only the leading-order displacement field is necessary to reveal the role of wall deformability for the present configuration. The equations for liquid (at $O(1)$ and $O(k)$) and solid (leading-order) layers are

$$\mathrm{d}_z^4 \tilde{v}_z^{(0)} = 0\,, \tag{63}$$

$$\mathrm{d}_z^4 \tilde{v}_z^{(1)} = \mathrm{i}Re\left[\left(\overline{v}_x - c^{(0)}\right)\mathrm{d}_z^2 \tilde{v}_z^{(0)} - \left(\mathrm{d}_z^2 \overline{v}_x\right)\tilde{v}_z^{(0)}\right], \tag{64}$$

$$\mathrm{d}_z^4 \tilde{u}_z^{(0)} = 0\,. \tag{65}$$

The leading-order conditions at $z = 1 + H$ are $\tilde{u}_x^{(0)} = 0, \tilde{u}_z^{(0)} = 0$ and at free surface ($z = 0$) are

$$-2\tilde{h}^{(0)} + \mathrm{d}_z \tilde{v}_x^{(0)} = 0\,; \quad -\tilde{p}^{(0)} = 0 \tag{66}$$

The leading-order LS interfacial conditions show that the leading order fluid velocities ($\tilde{v}_z^0$ and $\tilde{v}_x^0$) vanish at LS interface similar to the case of flow past a rigid boundary.

$$\tilde{v}_z^{(0)} = 0\,, \tag{67}$$

$$\tilde{v}_x^{(0)} = 0\,, \tag{68}$$

$$\mathrm{d}_z \tilde{v}_x^{(0)} = \frac{1}{\Gamma}\mathrm{d}_z \tilde{u}_x^{(0)}\,, \tag{69}$$

$$\tilde{p}^{(0)} = \widetilde{p_s}^{(0)}\,. \tag{70}$$

The leading-order governing equation for the liquid layer (Eq. 63), two free-surface conditions, and two velocity continuity conditions do not involve solid variables and are sufficient to solve for leading-order velocity field in the liquid layer. Therefore, the liquid layer is decoupled from the soft solid layer and the velocity field at this order remains identical to the velocity field for flow past rigid incline. Further, it is expected that the leading-order wavespeed $c^{(0)}$ will be identical to that of Yih's [69] analysis for flow past rigid incline. However, an important feature that emerges out of these interfacial conditions is that the liquid layer exerts a tangential stress on the deformable solid medium (see Eq. 70) which in turn sets up the deformation field in the solid layer at leading order. Subsequent analysis shows that this leading-order deformations in the solid layer affects the first correction to the fluid velocity field via velocity continuity conditions. At leading order, the fluid governing equation

(Eq. 63), the boundary conditions (Eqs. 66–68), and the leading order linearized kinematic condition (Eq. 49) can be solved to obtain leading-order wavespeed:

$$c^{(0)} = 2 \,. \tag{71}$$

As expected, this is exactly identical to the result for flow past rigid inclined plane [69], and the value of $c^{(0)}$ shows that the configuration is neutrally stable at leading order and it is necessary to perform the analysis at the next higher order in k to determine the stability of the system.

The leading-order displacement field ($\tilde{u}_z^{(0)}$ and $\tilde{u}_x^{(0)}$) in the solid layer which is required to calculate $c^{(1)}$ can be obtained by solving Eqs. 65 along with zero-displacement conditions and leading-order stress continuity conditions (refer to Eqs. 69–70):

$$\tilde{u}_z^{(0)} = \Gamma[z-(1+H)]^2 \,; \quad \tilde{u}_x^{(0)} = 2\mathrm{i}\Gamma[z-(1+H)] \,. \tag{72}$$

This leading-order deformation field developed in the solid layer affects the first correction to velocity field in the liquid layer via velocity continuity conditions at $z=1$.

$$\tilde{v}_z^{(1)} = -\mathrm{i}c^{(0)}\tilde{u}_z^{(0)} \,, \tag{73}$$

$$\tilde{v}_x^{(1)} + \mathrm{d}_z\overline{v}_x|_{z=1}\tilde{u}_z^{(0)} = -\mathrm{i}c^{(0)}\tilde{u}_x^{(0)} \,. \tag{74}$$

The first correction to free-surface conditions are

$$-3\tilde{h}^{(1)} + \mathrm{d}_z\tilde{v}_x^{(1)} = 0 \,; \quad -\tilde{p}^{(1)} - 3\cot\beta\tilde{h}^{(0)} = 0 \,. \tag{75}$$

The final steps involved in calculation of $c^{(1)}$ are as follows. The first correction to the velocity field is determined from Eq. 64 and the constants of integration are calculated using LS interfacial conditions (Eqs. 73–74) and the normal stress balance at free surface. (Eq. 75). The first correction to wavespeed $c^{(1)}$ is then obtained using tangential stress balance at free surface and the first correction to linearized kinematic condition (Eq. 49) [59].

$$c^{(1)} = \frac{2\mathrm{i}}{15}\left[(4Re - 5\cot\theta) - 30\Gamma H\right] \,. \tag{76}$$

This expression of $c^{(1)}$ is purely imaginary and hence determines the stability of the composite fluid–solid system. Qualitatively, the expression for $c^{(1)}$ consists of two contributions. The first contribution consists of terms proportional to Re and $\cot\theta$ which are identical to the expression of $c^{(1)}$ for flow past a rigid incline [69]. The term proportional to ΓH represents the contribution of the deformable solid layer and it occurs with a negative sign which implies that the effect of the solid layer is always stabilizing for this flow configuration. The rigid contribution clearly shows that flow becomes unstable to low-k disturbances when $Re > 5/4 \cot\theta$ in the absence of deformable solid layer ($\Gamma \to 0$ or $H \to 0$). In such cases, nonzero values of Γ and H, i.e., the presence of a soft solid coating, can suppress the free-surface instability in a low-wavenumber limit.

The low-k analysis presented above was performed for the gravity-driven flow of a Newtonian liquid past an inclined plane coated with a soft solid layer. However, there are several qualitative salient features that are common in the long-wave analysis of the problems involving GL and/or LL interface in the presence of a deformable LS interface. For example, the leading-order velocity field in fluid layers always gets decoupled from the solid layer (refer to Eqs. 67 and 68), and hence, the leading-order wavespeed (with deformable wall) remains identical to the leading-order wavespeed in the absence of a deformable solid layer regardless of the flow configuration under consideration. In several cases, this leading-order wavespeed is real and hence it is necessary to determine the first correction in order to determine the stability of the system. The expression for first correction to wavespeed $c^{(1)}$ generally consists of two qualitatively different Contributions: first, the rigid contribution which consists of terms that are present for a particular configuration when the flow occurs past a rigid surface. For example, in the expression of $c^{(1)}$ above (Eq. 76), the terms proportional to Re and $\cot\theta$ are present identically in the absence of a deformable solid layer as well and are termed as rigid contribution in the present discussion. The second contribution arises due to soft solid layer. This soft solid contribution could be stabilizing or destabilizing depending on the flow configuration and parameter regimes under consideration. Thus, rigid contribution remains unchanged in the presence of deformable wall, while an additional soft solid contribution arises and both rigid and soft solid contributions appear at $O(k)$. Therefore, wall deformability could affect the stability of flow system at this order. For example, the destabilizing inertial contribution, i.e., the term proportional to Re in the expression of $c^{(1)}$ given above, appears only at the first correction to the fluid velocity field. The leading-order deformations in the solid layer induce a perturbation flow field at $O(k)$ in fluid layers. Thus, leading-order deformations in solid affect the first correction to fluid velocity field which in turn affects the first correction to wavespeed $c^{(1)}$. In the above problem, $c^{(1)}$ can be related to $\tilde{v}_x^{(1)}$ as: $c^{(1)} = \frac{3i}{4} \mathrm{d}_z \tilde{v}_x^{(1)}|_{z=0}$. It is observed that the quantity $\mathrm{d}_z \tilde{v}_x^{(1)}|_{z=0}$ is negative in the presence of a soft solid layer, while it is positive for rigid surface. These arguments lead to the conclusion that the $O(k)$ flow field created by leading-order deformations in solid are responsible for altering the nature of free-surface instability in the above example and interfacial instabilities in general. However, it is important to mention here that the effect of the deformable solid layer could be either stabilizing or destabilizing depending on the flow configuration and parameter regimes, in contrast to the results presented above, where the effect of deformability was always found to be stabilizing.

3.1.2 Single Newtonian Falling Film: Neo-Hookean Solid

In the preceding section, we presented the long-wave asymptotic analysis considering the deformable solid layer as a linear viscoelastic medium. In this section, we represent the dynamics of deformable solid medium by a more accurate nonlinear neo-Hookean model. The governing equations are reported in Sect. 2.1.5 (Eqs. 35–37) and conditions at LS interface are Eqs. 41–44. The underlined terms in these equations are the additional coupling terms between the base state and perturbations which are present because of nonlinear neo-Hookean constitutive relation. We demonstrate that these additional terms remain absent in the long-wave limit and hence, the instability suppression predicted by the linear model in low-k limit holds for a more accurate neo-Hookean model as well. The scalings for fluid variables, governing equations for the liquid layer and conditions at GL interface, remain identical as presented in the previous section. Thus, $\tilde{v}_z \sim O(1)$, $\tilde{v}_x \sim O(k^{-1})$ and $\tilde{p} \sim O(k^{-2})$ and expansions 57–59 holds. Similarly, we set $\tilde{w}_Z \sim O(1)$, and this implies that $\tilde{w}_X \sim O(1/k)$ and $\widetilde{p_s} \sim O(1/k^2)$ according to solid continuity (Eq. (35)) and x–momentum balance (Eq. 36), respectively. If we expand the solid variables according to the scalings mentioned above, the following equations govern the leading-order continuity and momentum balance for neo-Hookean viscoelastic solid:

$$\frac{\mathrm{d}\tilde{w}_Z^{(0)}}{\mathrm{d}Z} + \mathrm{i}\tilde{w}_X^{(0)} = 0\,, \tag{77}$$

$$-\mathrm{i}\widetilde{p_s}^{(0)} + \frac{1}{\Gamma}\frac{\mathrm{d}^2\tilde{w}_X^{(0)}}{\mathrm{d}Z^2} = 0\,, \tag{78}$$

$$\frac{\mathrm{d}\widetilde{p_s}^{(0)}}{\mathrm{d}Z} = 0\,. \tag{79}$$

Note that all the additional coupling terms remain absent in the leading-order equations for the neo-Hookean solid layer. Similarly, it is ensured that the additional underlined terms in tangential and normal stress balance at LS interface (Eqs. 43 and 44) remain at least $O(k)$ smaller than the other terms present in these equations. This implies that at leading order, the equations for both linear and neo-Hookean model are identical. Also, the low-k analysis reveals that only leading-order deformations are required to determine the stability of system in long-wave limit. Thus, the leading-order and first correction to wavespeed with neo-Hookean solid is identical to the wavespeed obtained for linear elastic model in Shankar and Sahu [59] or in the preceding section:

$$c^{(0)} = 2\,, \tag{80}$$

$$c^{(1)} = \frac{2\mathrm{i}}{15}\left[(4Re - 5\cot\theta) - 30\Gamma H\right]\,. \tag{81}$$

Hence, the conclusion of free-surface instability suppression holds for a nonlinear neo-Hookean viscoelastic model as well.

3.1.3 Viscoelastic Falling Film Past a Deformable Solid

Here, we discuss the low-k asymptotic results for the gravity-driven flow of single liquid film as shown in Fig. 2 but considering the liquid layer to be a viscoelastic fluid modeled as the Oldroyd-B fluid. Recall that the Oldroyd-B model predicts a nonzero first normal stress difference in the base state for simple shear flows. Two dimensionless parameter appear: Weissenberg number $W = \tau_R V/R$ which is a measure of elasticity of fluid and a ratio of solvent to total viscosity $S = \mu_{\rm sol}/(\mu_{\rm sol} + \mu_p)$. The limit $S = 0$ corresponds to upper-convected Maxwell model. The original long-wave analysis for a viscoelastic liquid film falling down a rigid incline predicts an additional destabilizing contribution due to elasticity of the fluid [24, 43]:

$$c^{(0)} = 2\,; \quad c^{(1)} = \frac{2i}{15}\left[(4Re - 5\cot\theta) + 10W(1-S)\right]. \tag{82}$$

Note that the terms proportional to Re and $\cot\theta$ are present for Newtonian film as well, while the term proportional to W represents the additional destabilizing contribution due to the viscoelastic nature of fluid. A low-k analysis with the Oldroyd-B fluid in the presence of a deformable solid layer reveals that while $c^{(0)}$ is identical to the case of flow past rigid surface, the expression for $c^{(1)}$ is modified and wall deformability affects the stability behavior of viscoelastic film falling down an inclined plane [32]:

$$c^{(1)} = \frac{2i}{15}\{[4Re + 10W(1-S) - 5\cot\theta] - 30\Gamma H\}. \tag{83}$$

The above equation for $c^{(1)}$ shows that the rigid contribution is not altered and is identical to as given in Eq. 82. However, similar to the asymptotic analysis presented above for Newtonian liquid, a stabilizing contribution proportional to ΓH appears due to the presence of soft solid coating. Thus, for the viscoelastic liquid layer as well, it is possible to suppress free-surface instability by appropriately choosing wall deformability (or Γ) and solid thickness H. A more striking feature on comparing the above Eq. 83 with Eq. 81 is that the contribution of the soft solid layer remains identical to the earlier result for Newtonian fluids past a neo-Hookean or linear elastic solid surface. This implies that the contribution due to wall deformability is independent of both fluid (Newtonian vs. viscoelastic) and solid (neo-Hookean vs. linear model) rheology. This can be understood if we recall the general comments mentioned in the last of the preceding section for long-wave analysis for any flow configuration. The leading-order velocity field and the wavespeed for the Oldroyd-B model is identical to those in the Newtonian case, and hence, the leading-order deformation field set up in the solid layer also remains unaltered for the case of viscoelastic fluids. It is this leading-order deformations in solid which affects the first correction to velocity field and wavespeed $c^{(1)}$, and hence, the solid contribution remains identical to the Newtonian case.

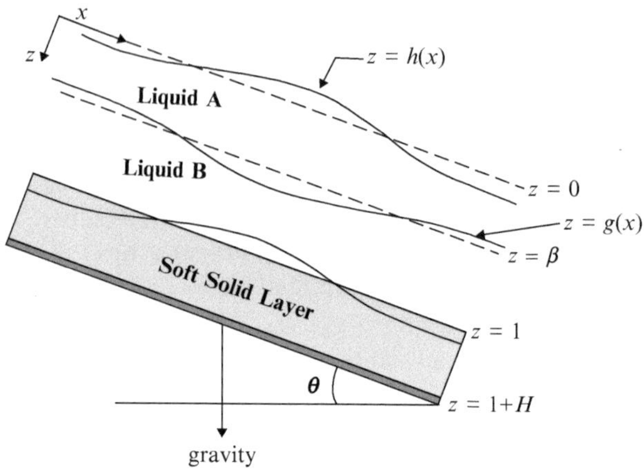

Fig. 4 Schematic diagram showing the configuration and (nondimensional) coordinate system: two Newtonian liquid films flowing past an inclined plane lined with a viscoelastic neo-Hookean solid layer

3.1.4 Gravity-Driven Two-Layer Flow with Free Surface

In this section, we extend the analysis given in the previous Sect. 3.1.1 to include one more liquid layer. The configuration under consideration consists of two liquid layers falling down a flexible surface which is inclined at an angle θ to the horizontal (see Fig. 4). It was shown in Sect. 3.1.1 that the GL free-surface instability could be completely suppressed for long-wavelength disturbances by manipulating the deformability of the wall. Here, we address the question whether the instability suppression by the deformable solid layer holds if an LL interface is also present in addition to a GL free surface. For such a two-liquid-layer flow configuration in the absence of soft solid coating, i.e., gravity-driven flow of two liquid layers past a rigid incline, a key nondimensional parameter, apart from Re and θ, which govern the stability behavior, is viscosity ratio $\mu_r = \mu_a/\mu_b$. Here, μ_a is the viscosity of the top liquid layer and μ_b is the viscosity of the layer adjacent to the (rigid or deformable) wall. This implies that $\mu_{\rm r} < 1(\mu_{\rm r} > 1)$ represents the case when more (less) viscous fluid is adjacent to the wall. The dimensionless mean position of the LL interface is represented as $z = \beta$ and the undisturbed mean position of the GL interface is $z = 0$. The qualitative nature of rigid and soft solid contribution in low-k limit depends on whether more or less viscous fluid is adjacent to the wall. We first discuss the case of $\mu_{\rm r} < 1$ and then proceed to the results pertaining to $\mu_{\rm r} > 1$.

As a representative example of results in low-wavenumber limit for $\mu_{\rm r} < 1$, we consider $\mu_{\rm r} = 0.5$ and $\beta = 0.5$. In the expressions for $c^{(1)}$ given below, the underlined terms represent the contributions that are identically present for flow past rigid surface [37], while the term proportional to ΓH represents the contribution due to wall elasticity:

$$c_{\rm gl}^{(0)} = 2.42539\,, \tag{84}$$

$$c_{\rm gl}^{(1)} = {\rm i}\,[\underline{0.904(Re) - 0.720(\cot\theta)} - 4.319(\Gamma H)]\,, \tag{85}$$

$$c_{\rm ll}^{(0)} = 0.824609\,, \tag{86}$$

$$c_{\rm ll}^{(1)} = {\rm i}[\underline{3.86 \times 10^{-3}(Re) - 3.011 \times 10^{-2}(\cot\theta)} - 1.807 \times 10^{-1}(\Gamma H)]. \tag{87}$$

Before making any comments related to the role of wall deformability, it is first instructive to discuss briefly the results in rigid limit. For flow past rigid incline, two interfaces, i.e., GL and LL interface, are present, and hence, we obtain two expressions for the $c^{(0)}$ and first correction to wavespeed $c^{(1)}$ corresponding to the GL and LL interface, respectively. The underlined terms, i.e., the rigid contributions, for both GL and LL mode show that inertia is destabilizing and the term proportional to $\cot\theta$ is stabilizing. Thus, both LL and free surface become unstable above a critical Re for rigid inclined plane and for any nonzero Re for vertical incline. The third term, proportional to ΓH in expressions of $c_{\rm gl}^{(1)}$ and $c_{\rm ll}^{(1)}$, shows that the deformability of the solid layer is stabilizing for both GL and LL mode, respectively. Thus, for $\mu_{\rm r} = 0.5, \beta = 0.5$, the wall elasticity can be chosen appropriately to suppress the long-wave perturbations for the present two-layered configuration. We have verified that the qualitative nature of different terms in the expression of $c^{(1)}$ for both interfaces remains independent of the values of mean LL interface position β and $\mu_{\rm r} < 1$. This implies that when more viscous liquid is adjacent to the deformable wall, the solid contribution is stabilizing and it is possible to simultaneously stabilize both free- surface and liquid–liquid interfacial instability.

The qualitative nature of different terms in the expression of $c^{(1)}$ changes when the less viscous liquid flows next to the deformable (or rigid) wall. We consider $\mu_{\rm r} = 2$ and $\beta = 0.5$ to illustrate results for $\mu_{\rm r} > 1$. The leading order and first correction to wavespeed for both interfaces are

$$c_{\rm gl}^{(0)} = 1.80902\,, \tag{88}$$

$$c_{\rm gl}^{(1)} = {\rm i}\,[\underline{0.406(Re) - 0.638(\cot\theta)} - 3.831(\Gamma H)]\,, \tag{89}$$

$$c_{\rm ll}^{(0)} = 0.690983\,, \tag{90}$$

$$c_{\rm ll}^{(1)} = {\rm i}[\underline{-1.20 \times 10^{-3}(Re) + 1.36 \times 10^{-2}(\cot\theta)} + 8.156 \times 10^{-2}(\Gamma H)]. \tag{91}$$

The qualitative nature of each term in the expression of $c_{\rm gl}^{(1)}$ remains similar to the $\mu_{\rm r} < 1$ case. The solid contribution is stabilizing, inertia is destabilizing, and term proportional to $\cot\theta$ is stabilizing. Therefore, ΓH can be altered to stabilize the GL interface. However, the behavior of both rigid and deformable solid contribution is reversed as compared to the $\mu_{\rm r} < 1$ case. For example, the inertia is stabilizing, while the term proportional to $\cot\theta$ and ΓH is destabilizing. A careful analysis of relative magnitudes of each term in the expressions of $c_{\rm gl}^{(1)}$ and $c_{\rm ll}^{(1)}$ reveals that it is not possible to simultaneously stabilize both GL and LL interface for $\mu_{\rm r} > 1$. For example, if $\mu_{\rm r} = 2, \beta = 0.5, \theta = \pi/2$, then $(\Gamma H) < 0.0147 Re$ for LL mode to remain stable, while $(\Gamma H) > 0.106 Re$ for stabilizing the GL-mode low-k perturbations. Thus, it is not possible to obtain stable flow configuration by manipulating wall elasticity.

We performed the low-k analysis and presented the results for this three-layered composite system considering Newtonian fluid layers and a neo-Hookean or linear viscoelastic solid. We have already shown in Sect. 3.1.3 that the solid layer contribution remains identical, both in magnitude and in terms of stabilizing nature, for both the Newtonian and Oldroyd-B fluid models in the case of a liquid film falling down a flexible inclined surface. Following the same arguments given in Sect. 3.1.3, we expect that the role of the deformable solid layer and its contribution will not be altered even if we consider the flow of two viscoelastic Oldroyd-B fluid films past a flexible surface. To summarize the low-k results, the effect of the soft solid layer could be stabilizing or destabilizing depending on the viscosity ratio μ_r, and it is possible to obtain a stable flow configuration by using a deformable solid layer only for $\mu_r < 1$, when the configuration is otherwise unstable in the rigid limit.

3.2 *Suppression of Fluid–Fluid Interfacial Instabilities in Plane-Couette Flow*

In the previous Sect. 3.1.1, we demonstrated the ability of deformable solid surface in suppressing the low-k free surface instabilities for single liquid (Newtonian or Oldroyd-B) falling film or simultaneous suppression of free surface and liquid–liquid interface in the case of two liquid films flowing down a plane under the influence of gravity. Here, we consider Couette flow of two fluid layers past a flexible surface as shown in Fig. 5. Thus, liquid–liquid and liquid–solid interfaces are present, while gas–liquid free surface is absent in the present configuration. As mentioned earlier, the LL interface undergoes a long-wave interfacial instability due to viscosity stratification in the case of two-layer Newtonian fluid flow [70] and due to a jump in elasticity (and/or viscosity) in the case of two viscoelastic liquid layers [8]. A long-wave asymptotic analysis similar to the analysis presented above in Sect. 3.1.1 is carried out for the system shown in Fig. 5 to explore the role of deformable solid surface on fluid–fluid interfacial instabilities caused by a

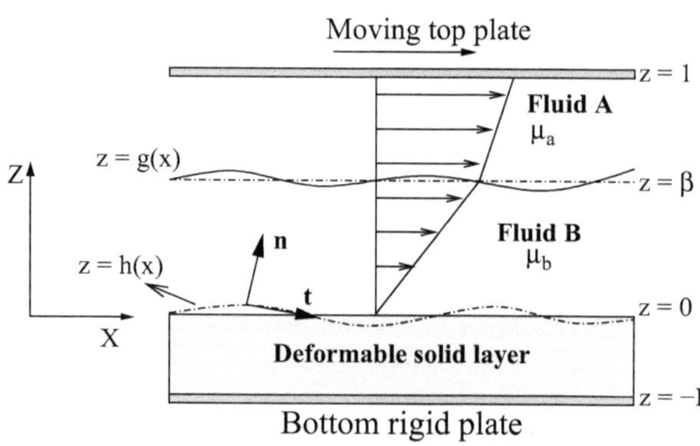

Fig. 5 Schematic diagram showing the configuration and (nondimensional) coordinate system: two fluid layers flowing past a deformable solid

viscosity/elasticity jump across the interface. A linear viscoelastic model is used as it is sufficient to accurately capture the effect of the soft solid layer in the long-wave limit (refer to Sect. 3.1.2 or [17]). The two fluid layers are first considered as Newtonian fluid layers in order to focus on instability caused by viscosity stratification and finally as two upper-convected Maxwell (UCM) fluids to consider the effect of the jump in fluid elasticity in addition to a viscosity jump across the interface. In either case (Newtonian or UCM fluids), the asymptotic results for c can be written as

$$c = c^{(0)} + k(c^{(1)}_{\text{visc}} + c^{(1)}_{\text{elas}} + c^{(1)}_{\text{solid}}) \tag{92}$$

As observed in earlier low-k results, the leading-order wavespeed $c^{(0)}$ is real and is identical to the leading-order wavespeed for plane-Couette flow of two Newtonian liquid layers past a rigid surface [70]. Thus, neither solid deformability nor fluid elasticity affects the stability problem at leading order. The first correction to wavespeed $c^{(1)}$ has three contributions and all three are imaginary. The first term $c^{(1)}_{\text{visc}}$ is proportional to Re and is present because of discontinuity of viscosity across the interface. This $c^{(1)}_{\text{visc}}$ remains identical to the first correction to wavespeed obtained by Yih [70] for the two-layer flow of Newtonian liquids. The second term $c^{(1)}_{\text{elas}}$ arises due to elasticity stratification and is proportional to difference in Weissenberg number of two fluids. This is identical to Chen's result [8] for plane-Couette flow of two UCM fluids. Finally, $c^{(1)}_{\text{solid}}$ is the contribution from the soft solid layer which represents the effect of deformable solid surface on two-fluid interfacial mode which could undergo instability due to viscosity/elasticity stratification. All these three terms could have a stabilizing/destabilizing effect depending on μ_r, W_a, W_b, and β. In the following, we first discuss the result for two Newtonian liquid layers flowing past a deformable solid surface.

3.2.1 Two-Layer Newtonian Plane-Couette Flow

In the case of two-layer Newtonian plane-Couette flow past a deformable solid surface, the key nondimensional parameters that determine the stability behavior are viscosity ratio μ_r, mean dimensionless LL interface position β, solid deformability parameter Γ, and thickness of the solid layer H. The analytical expression for $c^{(1)}$ (refer to Eq. 92) consists of only the viscous contribution $c^{(1)}_{\text{visc}}$ and soft solid contribution $c^{(1)}_{\text{solid}}$, where $c^{(1)}_{\text{visc}}$ is exactly identical to Yih's [70] result for two-layer Newtonian plane-Couette flow past a rigid surface. For flow past a rigid surface, the fluid–fluid interface becomes unstable to long-wave disturbances when less viscous fluid occupies more space. In terms of dimensionless parameters introduced, the above statement translates as $c^{(1)}_{\text{visc}} = C_1 Re$, where $C_1 > 0$, i.e., inertia is destabilizing, for $\mu_r < 1, \beta < 0.5$, and $C_1 < 0$ which implies that the contribution is stabilizing for $\mu_r > 1, \beta > 0.5$. It is also important to remark here that the placement of more or less viscous fluid with respect to moving plate does not affect the stability behavior. For example, the above-written conditions clearly show that if the less viscous fluid occupies less space, the two-fluid configuration is stable irrespective of whether the less viscous fluid is adjacent to moving or stationary plate.

Let us consider a case when $\mu_r = 0.5, \beta = 0.4$:

$$c^{(1)} = \mathrm{i}[0.00012Re - 0.034\Gamma H(H^2 + 2.24H + 1.5)] \tag{93}$$

For $\mu_r = 0.5, \beta = 0.4$, the more viscous liquid occupies less Space, and the term proportional to Re is destabilizing, and hence, the configuration remains unstable in the absence of solid wall. However, the contribution proportional to ΓH, i.e., $c^{(1)}_{\text{solid}}$ occurs with a negative sign and is stabilizing. Therefore, in the presence of the deformable solid layer, i.e., for nonzero values of Γ and H, it is possible to suppress the long-wave fluid–fluid interfacial mode instability originally present in the rigid configuration due to a jump in viscosity. Next, we increase β from 0.4 to 0.7 keeping $\mu_r = 0.5$. This implies that the less viscous fluid now occupies less space and for such cases, the $c^{(1)}_{\text{visc}}$ contribution becomes stabilizing. Thus, such a configuration is stable in the rigid limit. The asymptotic result with the solid layer included, for $\mu_r = 0.5, \beta = 0.7$, is given as

$$c^{(1)} = \mathrm{i}[-0.000152Re + 0.026\Gamma H(H - 0.286)(H + 0.8)] \tag{94}$$

The above expression for $c^{(1)}$, with the deformable solid layer included, brings out an interesting feature that the role of the solid layer could be stabilizing (if $H < 0.286$) or destabilizing (if $H > 0.286$) depending on the solid layer thickness. A careful analysis of long-wave results for several values of μ_r and β reveals the general qualitative trend for the role of deformable LS interface. The low-k asymptotic result for the present two-layer Newtonian configuration can, in general, be expressed as

$$c^{(1)} = \mathrm{i}[C_1 Re + \Gamma H f(H)] \tag{95}$$

where the qualitative nature of rigid contribution $c^{(1)}_{\text{visc}} = C_1 Re$ has already been described in the above discussion. The soft solid contribution $c^{(1)}_{\text{solid}} = \Gamma H f(H)$ could be stabilizing (negative) or destabilizing (positive) depending on μ_r, β, and H. A few representative values of C_1 and $f(H)$ are shown in Tables 1 and 2 for $\mu_r = 0.5$ and $\mu_r = 2$, respectively, at different values of β. The results presented in these tables illustrate that the solid layer usually has a stabilizing effect when the two fluid layers are so arranged such that the system undergoes a long-wave fluid–fluid interfacial instability due to a viscosity jump across the interface. On the other hand, when the configuration remains stable in the rigid limit, the solid contribution is usually destabilizing, but for certain values of H, for which the solid layer could further provide an additional stabilizing effect. The results presented in Tables 1 and 2 also show that the stabilizing/destabilizing behavior of the soft

Table 1 Qualitative summary of long-wave results for $\mu_r = 0.5$ at different values of β

β	$f(H)$	$C_1 Re$
0.1	Stable for all H	$6.811 \times 10^{-5} Re$
0.2	Stable for all H	$1.781 \times 10^{-4} Re$
0.3	Stable for all H	$2.006 \times 10^{-4} Re$
0.4	Stable for all H	$1.2 \times 10^{-4} Re$
0.5	Stable for all H	$2.860 \times 10^{-3} Re$
0.6	Stable for $H < 7.92$ Unstable for $H > 7.92$	$-9.6 \times 10^{-5} Re$
0.7	Stable for $H < 0.286$ Unstable for $H > 0.286$	$-1.527 \times 10^{-4} Re$
0.8	Unstable for all H	$-1.553 \times 10^{-4} Re$
0.9	Unstable for all H	$-9.504 \times 10^{-5} Re$

Table 2 Qualitative summary of long-wave results for $\mu_r = 2$ at different values of β

β	$f(H)$	$C_1 Re$
0.1	Unstable for all H	$-4.75 \times 10^{-4} Re$
0.2	Unstable for all H	$-7.76 \times 10^{-5} Re$
0.3	Unstable for all H	$-7.63 \times 10^{-5} Re$
0.4	Unstable for all H	$-4.80 \times 10^{-5} Re$
0.5	Unstable for $H < 0.865$ Stable for $H > 0.865$	$1.43 \times 10^{-6} Re$
0.6	Stable for all H	$6.002 \times 10^{-4} Re$
0.7	Stable for all H	$1.003 \times 10^{-4} Re$
0.8	Stable for all H	$8.906 \times 10^{-5} Re$
0.9	Stable for all H	$3.406 \times 10^{-5} Re$

solid layer crucially depends on the placement of two fluid layers with respect to the deformable solid layer. For example, when $\mu_r = 0.5, \beta = 0.6$ (i.e., more viscous fluid near deformable surface), Table 1 shows that the solid contribution is stabilizing for $H < 7.92$ and destabilizing for $H > 7.92$. On the other hand, for $\mu_r = 2, \beta = 0.4$, i.e., when the two fluid layers are interchanged and less viscous fluid is adjacent to soft solid, the solid contribution is always stabilizing. This is in contrast to the rigid results where the relative placement of two fluid layers with respect to moving/stationary plate was not important in determining the sign of C_1.

3.2.2 Two-Layer Viscoelastic Plane-Couette Flow

While the preceding section demonstrated the control of LL interfacial mode instability caused due to viscosity contrast, in the following, we discuss the effect of the deformable solid layer in the presence of both elasticity and viscosity jump across the LL interface. The two fluid layers are represented by an upper-convected Maxwell model and the expression for $c^{(1)}$ contains all three contributions due to viscosity jump ($c^{(1)}_{\text{visc}}$), elastic stratification ($c^{(1)}_{\text{elas}}$), and soft solid contribution ($c^{(1)}_{\text{solid}}$)

(refer to Eq. 92). Thus, apart from μ_r, β, Γ, and H, the stability also depends on the difference of Weissenberg number of two fluids. The elastic part in the expression of $c^{(1)}$ could be stabilizing or destabilizing depending on the difference in Weissenberg number of two fluids. The viscous part $c^{(1)}_{\text{visc}}$ remains identical to the case of flow of two Newtonian fluid layers and is stabilizing if less viscous liquid occupies less space. The third part, i.e., the soft solid contribution, depends on μ_r and β. As an illustration, consider low-k results for $\mu_r = 0.5$ and $\beta = 0.4$:

$$c^{(1)} = \mathrm{i}\times 10^{-2}[\underline{0.012Re + 1.8(W_b - 1.13W_a)} - 3.4\Gamma H(H^2 + 2.24H + 1.5)] \quad (96)$$

The underlined terms in the imaginary part of the above expression are present identically in the absence of the deformable solid layer, i.e., for flow past rigid surface. Thus, the term proportional to Re is identical to the Yih's result [70] for two Newtonian fluid layers, and the term proportional to difference in W_a and W_b is identical to Chen's result [8] for flow of two UCM fluids in rigid channel. The contribution due to viscosity stratification is destabilizing; the second term in $c^{(1)}$, i.e., the elastic part, is destabilizing for $W_b > 1.13W_a$; and the third term representing the effect of solid layer deformability occurs with a negative sign and is therefore stabilizing. More importantly, the soft solid contribution for Couette flow of two UCM fluids remains identical to the soft solid contribution for plane-Couette flow of two Newtonian fluid layers (refer to Eq. 93). A similar behavior was also observed for the case of falling film past a deformable solid surface where the contribution due to solid deformability remains independent of whether the liquid film is Newtonian or Oldroyd-B viscoelastic film (refer to Eqs. 81 and 83, respectively). This essentially indicates that the effect of solid deformability does not depend on the fluid elasticity and is dictated mainly by viscosity jump across the interface. This feature is more clearly illustrated in the low-k results presented in Tables 3 and 4 which reports the three distinct contributions to $c^{(1)}$ for two different values of mean fluid–fluid interfacial position $\beta = 0.4$ and $\beta = 0.6$ for various values of μ_r. The results for $\beta = 0.4$ show that the solid deformability is

Table 3 Long-wave results for $\beta = 0.4$ at different values of μ_r

μ_r	$c^{(1)}_{\text{elas}}$	$c^{(1)}_{\text{visc}}$	$c^{(1)}_{\text{solid}}$
0.2	$0.012(W_b - 1.75W_a)$	$7.37 \times 10^{-4}Re$	$-0.048\Gamma H(H^2 + 2.08H + 1.36)$
0.4	$0.018(W_b - 1.21W_a)$	$1.98 \times 10^{-4}Re$	$-0.042\Gamma H(H^2 + 2.18H + 1.44)$
0.6	$0.017(W_b - 1.07W_a)$	$7.49 \times 10^{-5}Re$	$-0.026\Gamma H(H^2 + 2.32H + 1.57)$
0.8	$0.015(W_b - 1.02W_a)$	$2.59 \times 10^{-5}Re$	$-0.011\Gamma H(H^2 + 2.5H + 1.75)$
1	$0.011(W_b - W_a)$	0	0
1.1	$0.0098(W_b - 1.005W_a)$	$-8.96 \times 10^{-6}Re$	$0.00413\Gamma H(H^2 + 2.9H + 2.16)$
2	$0.0008(W_a - 3.94W_b)$	$-4.8 \times 10^{-5}Re$	$0.007\Gamma H(H + 1.03)(H + 8.92)$
4	$0.008(W_a - 2.1W_b)$	$-7.06 \times 10^{-5}Re$	$-0.0963\Gamma H(H - 1.22)(H + 0.87)$
6	$0.0086(W_a - 2.73W_b)$	$-7.2 \times 10^{-5}Re$	$-0.247\Gamma H(H - 0.481.03)(H + 0.8)$

Table 4 Long-wave results for $\beta = 0.6$ at different values of μ_r

μ_r	$c^{(1)}_{\text{elas}}$	$c^{(1)}_{\text{visc}}$	$c^{(1)}_{\text{solid}}$
0.2	$0.0088(W_b - 2.37W_a)$	$-3.64 \times 10^{-4} Re$	$-0.034\Gamma H(H^2 + 2.89H + 2.18)$
0.4	$0.0041(W_b - 2.018W_a)$	$-8.37 \times 10^{-3} Re$	$-0.037\Gamma H(H + 10.02)(H + 1.03)$
0.6	$0.0093(W_b - 2.43W_a)$	$-6.4 \times 10^{-5} Re$	$0.007\Gamma H(H - 2.74)(H + 0.93)$
0.8	$0.00727(W_b - 1.035W_a)$	$-2.43 \times 10^{-5} Re$	$0.0069\Gamma H(H - 1.08)(H + 0.88)$
1	$0.011(W_a - W_b)$	0	0
1.1	$0.013(W_a - 1.0036W_b)$	$9.21 \times 10^{-6} Re$	$-0.005\Gamma H(H - 0.456)(H + 0.828)$
2	$0.018(W_a - 1.13W_b)$	$6 \times 10^{-5} Re$	$-0.0685\Gamma H(H + 0.019)(H + 0.738)$
4	$0.014(W_a - 1.52W_b)$	$1.2 \times 10^{-4} Re$	$-0.197\Gamma H(H + 0.267)(H + 0.629)$
6	$0.0098(W_a - 2W_b)$	$1.65 \times 10^{-4} Re$	$-0.271\Gamma H(H + 0.378)(H + 0.548)$

stabilizing for $\mu_r < 1$ and destabilizing/stabilizing for $\mu_r > 1$ depending on H. Similarly, the solid layer is stabilizing for $\mu_r > 1$ and destabilizing/stabilizing for $\mu_r < 1$ when $\beta = 0.6$. In both cases, the solid layer has a stabilizing contribution whenever the term $c^{(1)}_{\text{visc}}$ is destabilizing. Further, we have verified that the $c^{(1)}_{\text{solid}}$ for two UCM fluid flow is always identical to the $c^{(1)}_{\text{solid}}$ for flow of two Newtonian fluid layers. The Weissenberg number or alternatively elasticity of fluid layers does not affect the soft solid contribution, and the stabilizing/destabilizing nature of the solid layer is controlled only by viscosity ratio of the two fluid layers. Thus, as in the case of two Newtonian fluid layers, it is possible to suppress fluid–fluid interfacial mode whenever the two fluids (viscoelastic or Newtonian) are placed in such a manner that viscosity stratification is destabilizing. For other configurations, the solid deformability could be stabilizing or destabilizing depending on β and H.

Another feature worth noting is that the solid layer contribution vanishes for $\mu_r = 1$. Shankar [55] investigated the effect of solid layer deformability on two-fluid interfacial instability with discontinuity in elasticity across the interface but with matched viscosities. Their results demonstrated that the dimensionless parameter $\Gamma \propto k^{-1}$ for $k \ll 1$ in order for the solid layer to have an effect on two-fluid interfacial instability. On the other hand, we have assumed $\Gamma \sim O(1)$ in the present analysis, which implies that Γ does not scale with the wavenumber in any particular fashion. For the case of unmatched viscosities ($\mu_r \neq 1$), this assumption $\Gamma \sim O(1)$ is sufficient in order for the solid deformability to have an effect on fluid–fluid interfacial mode. However, when $\mu_r = 1$, Γ must increase as $1/k$ in $k \ll 1$ limits in order for solid deformability to effect the fluid–fluid perturbations. This limit of $\Gamma \sim 1/k$ physically corresponds to an ultra-soft solid layer. We have verified that if we let $\Gamma \sim O(1/k)$ in the low-k analysis, then the asymptotic results matches with earlier results of Shankar [55] for the special case of matched viscosities. In real situations, viscosity difference would always be present and hence the conclusions presented for $\mu_r \neq 1$ are more relevant for any practical situation.

3.3 Simultaneous Suppression of Capillary and Interfacial Instabilities

Thus far, we have demonstrated the efficacy of the deformable solid layer in suppressing free-surface and/or two-fluid interfacial instabilities. However, due to the planar nature of flow configurations, the surface tension forces remain subdominant in low-k limit. We now consider interfacial flows in cylindrical geometry where capillary forces destabilize fluid–fluid interface in addition to existing free-surface or two-fluid interfacial instability. Specifically, we consider Newtonian liquid film flow outside wires and inside tubes where a free-surface instability is present along with the presence of capillary instability. Following it, we consider the pressure-driven flow of two immiscible Newtonian liquids in a core–annular arrangement in a tube where two-fluid interfacial instability (due to a viscosity jump) and capillary instability is present.

The configurations under consideration for liquid film flow outside the wire and inside the tube are shown in Fig. 6. For both cases, r_0, L_0, and h_0 are dimensional tube radius, solid thickness, and unperturbed liquid film thickness, respectively. The lengths are nondimensionalized using unperturbed liquid film thickness, and the dimensionless wire/tube radius, solid thickness, and film thickness are, respectively, S, H, and unity. For the flow inside the tube, the geometry imposes the constraint $S - H - 1 > 0$ for liquid film to exist inside the tube. For both cases in the presence of the deformable solid layer, the value of leading order wavespeed is $c^{(0)} = 2$ which is identical to the case of liquid film flow past rigid cylindrical surface. The first correction to wavespeed $c^{(1)}$ is purely imaginary and consists of three different contributions:

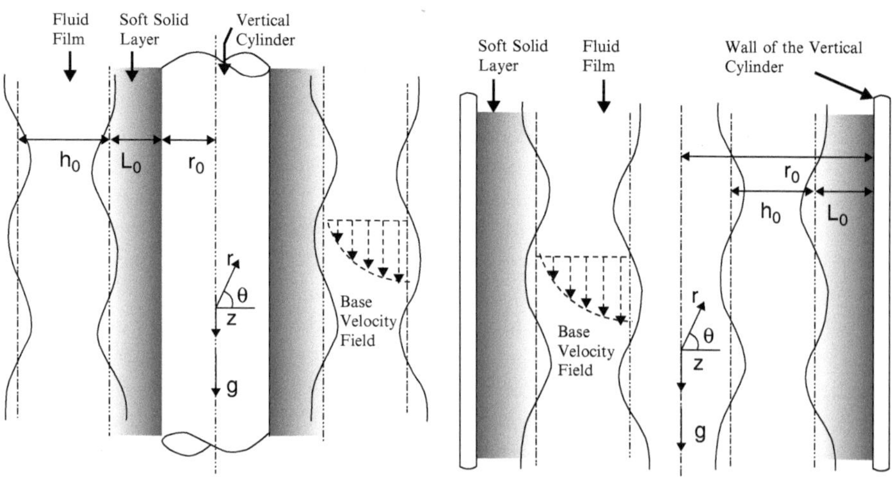

Fig. 6 Schematic diagram showing the configuration and coordinate system when there is liquid film flow outside the wire and inside the tube

$$c^{(1)} = (\underline{c^{(1)}_{\text{free}} + c^{(1)}_{\text{cap}}} + c^{(1)}_{\text{solid}}) \tag{97}$$

The underlined terms represent rigid contributions. The first term $c^{(1)}_{\text{free}}$ is proportional to Re and represents free-surface instability. The second term $c^{(1)}_{\text{cap}}$ is proportional to Σ_{gl} and represents contribution due to capillary forces, while the third term represents the effect of the deformable solid layer. The values of different terms in the expression of $c^{(1)}$ for the flow outside the cylinder coated with a soft solid layer are shown in Table 5 at different values of S and $H = 1$. These results show that both $c^{(1)}_{\text{free}}$ and $c^{(1)}_{\text{cap}}$ are always destabilizing and hence it is not possible to obtain stable film flows in the rigid limit. Again, the contribution proportional to Γ always occurs with a negative sign, and hence, the effect of the deformable solid layer is stabilizing. We have also observed a similar qualitative nature of the three contributions in the expression of $c^{(1)}$ for the case of the flow inside the tube lined with a deformable solid coating. Thus, for both cases, the wall deformability effect is stabilizing and it is possible to choose solid thickness and elasticity in such a manner that both capillary and free-surface instabilities are suppressed for long-wave perturbations. Note that in the limit of very high radius of curvature ($S \gg 1$), the present result reduces to the case of planar liquid film flow past a vertical incline coated with deformable solid layer (refer to Eq. 81).

We now consider the effect of wall deformability on capillary and two-fluid interfacial mode instability present for core–annular flow of two immiscible, Newtonian liquids in a flexible tube. The configuration is shown in Fig. 7. This core–annular fluid arrangement in the rigid tube is susceptible to long-wave capillary instability which is independent of the viscosity ratio of two liquids. The interface between core and annular liquids also becomes unstable to low-wavenumber perturbations

Table 5 Effect of the soft solid layer for the flow outside the cylinder coated with the deformable solid layer of thickness $H = 1$

S	$c^{(1)}$
1	$-5.234\ \Gamma + 0.6114\ Re + 0.037895\ \Sigma_{\text{gl}}$
5	$-4.269\ \Gamma + 0.561\ Re + 0.0068\ \Sigma_{\text{gl}}$
10	$-4.1344\ \Gamma + 0.549\ Re + 0.0232\ \Sigma_{\text{gl}}$
50	$-4.026\ \Gamma + 0.537\ Re + 0.000123\ \Sigma_{\text{gl}}$
100	$-4.0131\ \Gamma + 0.535\ Re + 0.000032\ \Sigma_{\text{gl}}$

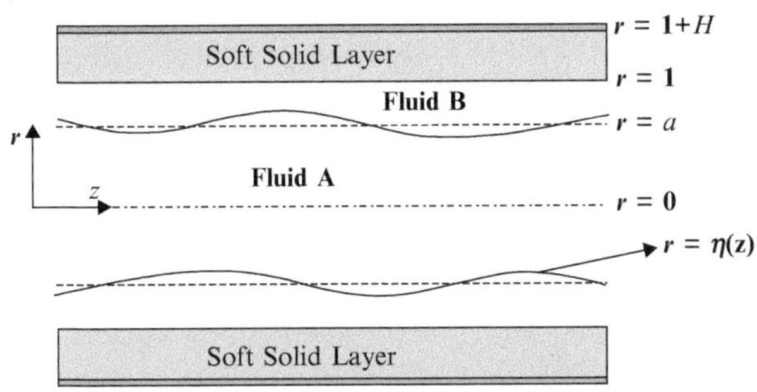

Fig. 7 Schematic diagram showing the configuration and (nondimensional) coordinate system: two immiscible fluids flowing in core–annular arrangement in a tube coated with a deformable solid layer

Table 6 Long-wave results for $\mu_r = 0.5$ and $H = 2$ at different values of a

a	$c^{(0)}$	$c^{(1)}$
0.1	0.98990	$i(-1.3602 \times 10^{-6} Re + 4.5067 \times 10^{-2} \Sigma - 0.00784 \Gamma)$
0.2	0.95847	$i(-2.0210 \times 10^{-5} Re + 5.5552 \times 10^{-2} \Sigma - 0.11740 \Gamma)$
0.3	0.90269	$i(-8.6712 \times 10^{-5} Re + 5.3401 \times 10^{-2} \Sigma - 0.52378 \Gamma)$
0.4	0.81903	$i(-1.9466 \times 10^{-4} Re + 4.4125 \times 10^{-2} \Sigma - 1.33954 \Gamma)$
0.5	0.70588	$i(-2.2769 \times 10^{-4} Re + 3.1496 \times 10^{-2} \Sigma - 2.34480 \Gamma)$
0.6	0.56657	$i(-5.9478 \times 10^{-6} Re + 1.8834 \times 10^{-2} \Sigma - 2.94633 \Gamma)$
0.7	0.41125	$i(4.1462 \times 10^{-6} Re + 8.8005 \times 10^{-3} \Sigma - 2.61970 \Gamma)$
0.8	0.25539	$i(5.9698 \times 10^{-6} Re + 2.7445 \times 10^{-3} \Sigma - 1.51624 \Gamma)$
0.9	0.11472	$i(2.9115 \times 10^{-6} Re + 3.4476 \times 10^{-4} \Sigma - 0.41716 \Gamma)$

Table 7 Long-wave results for $\mu_r = 2$ and $H = 2$ at different values of a

a	$c^{(0)}$	$c^{(1)}$
a	$c^{(0)}$	$c^{(1)}$
0.1	0.99005	$i(6.5539 \times 10^{-7} Re + 3.6062 \times 10^{-2} \Sigma + 0.00392 \Gamma)$
0.2	0.96077	$i(8.7759 \times 10^{-6} Re + 3.9639 \times 10^{-2} \Sigma + 0.05912 \Gamma)$
0.3	0.91370	$i(3.2523 \times 10^{-5} Re + 3.4191 \times 10^{-2} \Sigma + 0.27159 \Gamma)$
0.4	0.85089	$i(6.3829 \times 10^{-5} Re + 2.5685 \times 10^{-2} \Sigma + 0.75101 \Gamma)$
0.5	0.77419	$i(7.5613 \times 10^{-5} Re + 1.7087 \times 10^{-2} \Sigma + 1.54678 \Gamma)$
0.6	0.68434	$i(4.1090 \times 10^{-5} Re + 9.9002 \times 10^{-3} \Sigma + 2.59604 \Gamma)$
0.7	0.57957	$i(-4.1336 \times 10^{-5} Re + 4.7319 \times 10^{-3} \Sigma + 3.66622 \Gamma)$
0.8	0.45271	$i(-1.3441 \times 10^{-4} Re + 1.6207 \times 10^{-3} \Sigma + 4.22275 \Gamma)$
0.9	0.28275	$i(-1.4953 \times 10^{-4} Re + 2.4594 \times 10^{-4} \Sigma + 3.12267 \Gamma)$

due to discontinuity in viscosity when more viscous liquid is placed adjacent to the rigid wall ($\mu_r > 1$). Stable core–annular flow configuration could be obtained in the rigid limit when less viscous liquid is near the rigid wall and occupies much less space than the core viscous liquid [50]. We performed a long-wave asymptotic analysis in order to determine the effect of the deformable solid layer on these low-k capillary and fluid–fluid interfacial instabilities. Again, the stability in low-k limit is determined by the first correction to wavespeed and the leading-order wavespeed is real and remains unaffected by the presence of a deformable solid layer. The result for $c^{(1)}$ comprises three different contribution, and the expression for $c^{(1)}$ is qualitatively similar to the expression of $c^{(1)}$ given above for the case of liquid film flow inside a tube or outside a cylinder (see Eq. 97). Tables 6 and 7 reports the expression of $c^{(0)}$ and $c^{(1)}$ for $\mu_r = 0.5$, i.e., when more viscous fluid is adjacent to the wall, and for $\mu_r = 2$ when the less viscous liquid is placed near the tube wall. These results clearly show that the effect of the deformable solid layer is stabilizing for $\mu_r = 0.5$ and destabilizing for $\mu_r = 2$. We have verified that the qualitative nature of soft solid contribution remains similar to $\mu_r = 0.5$ for all $\mu_r < 1$ and to $\mu_r = 2$ for all $\mu_r > 1$. Therefore, for $\mu_r < 1$ when the more viscous liquid is near the wall, it is possible to suppress the fluid–fluid interfacial mode perturbations caused due to either capillary forces or viscosity contrast by using a deformable solid coating on the inside of tube wall. On the other hand, when $\mu_r > 1$, the solid layer provides a destabilizing contribution and the flow can be rendered unstable by altering wall properties when the flow can be otherwise stable in the rigid limit.

4 Suppression of Interfacial Instabilities for Arbitrary Disturbances

The central conclusion of the foregoing discussion on the effect of the deformable solid layer on fluid–fluid interfacial instabilities in long-wave limit is that the wall elasticity could be tuned to control gas–liquid and/or liquid–liquid interfacial instabilities for a wide variety of flow configurations and for different fluid rheological behaviors. The solid deformability parameter Γ and dimensionless solid thickness H can be chosen appropriately to suppress interfacial instabilities in low-k limit for different flow settings when the corresponding flow configuration was otherwise unstable in the rigid limit. While for parameter regimes where flow is already stable in the rigid limit, the deformable solid layer could have a stabilizing/destabilizing effect. However, as noted, the above asymptotic results for the control of interfacial instabilities by the soft solid layer are valid only in the limit of very long-wave perturbations. An arbitrary disturbance is expected to be composed of modes with all wavelengths. Thus, if the predicted suppression is to be realized in real systems, it is necessary to ensure that the instabilities are suppressed for disturbances corresponding to all wavelengths. Therefore, the first objective of this section is to extend the low-k asymptotic results to finite and arbitrary values of wavenumber by solving the governing stability equations using numerical methods. This will ensure whether the predicted long-wave suppression holds for any arbitrary disturbance or not. Second, there have been several studies exclusively focusing on the stability of deformable fluid–solid interface, and these studies demonstrated that the fluid–solid interface could become unstable even in creeping flow limit when the solid deformability parameter Γ is increased above a critical value [20, 21, 39, 40, 42]. Further, several additional liquid–solid unstable modes proliferate in the presence of inertia when the wall deformability is sufficiently increased [10, 18, 19, 57, 58]. Thus, when Γ increases beyond a critical value, the fluid–fluid interfacial instabilities wherever possible are suppressed. With further increase in $\Gamma \sim O(1)$, the fluid–solid interface also becomes unstable. Consequently, the question arises whether there exist sufficient gap in between the two values of Γ where both fluid–fluid and fluid–solid interfaces remain stable. If we envisage the use of deformable solid coating in suppressing instabilities, we must ensure that the fluid–solid interface also remains stable. The results are presented in terms of neutral stability diagrams which demarcate the stable and unstable regions. These neutral curves are constructed by solving governing stability equations using a numerical shooting technique and Chebyshev spectral collocation method. In the following, we first present results for free-surface problems and then proceed to discuss results for other flow configurations.

However, before proceeding to discuss results for arbitrary wavenumbers, it is important to remark here the procedure that we follow in identifying and labeling an interfacial mode. In general, the flow configuration under consideration will consist of fluid–fluid (gas–liquid and/or liquid–liquid) interface and a deformable fluid–solid interface. All of these interfaces could potentially become unstable as parameters are varied. As clear from the previous section, for $Re \sim O(1)$, the fluid–fluid interface undergoes a low-wavenumber instability caused due to a jump in viscosity or elasticity across the interface. For $Re \sim O(1)$ and in the absence of deformable wall, the only eigenmode which becomes unstable is this low-k fluid–fluid interfacial mode. This long-wave interfacial mode will be referred as fluid–fluid interfacial mode or gas–liquid (GL) mode when a gas–liquid interface is present or liquid–liquid (LL) mode if a liquid–liquid interface is present. The other potential interface which could become unstable is liquid–solid (LS) interface. This LS interfacial mode could become unstable via different mechanisms depending on the presence/absence of inertial forces. In the absence of inertia, the flow past a linear elastic or neo-Hookean solid becomes unstable for finite wavenumber perturbations when Γ is increased over a critical value [39]. The short-wave perturbations also become unstable when $\Gamma \sim O(1)$ is sufficiently increased as predicted first by Gkanis and Kumar [20] for a more accurate neo-Hookean solid model. In addition to this creeping flow interfacial unstable mode, the coupled fluid–solid problem also admits multiple solutions which correspond to shear waves in a solid medium which exist only for finite inertia in the solid [1]. These free shear waves in solid could also be destabilized by fluid flow in the presence of inertia when the solid deformability parameter increases beyond a critical value [18, 19, 57, 58]. Thus, there are two broad categories of LS unstable modes: first, the creeping flow finite wavenumber and short-wave instability which continue to finite Re and second, the free shear waves in solid destabilized by fluid flow for which the presence of inertia, however small, is necessary. Here, since we are more interested in the question that whether a deformable solid layer could be used to suppress interfacial instabilities or not, therefore, we refer both these unstable LS modes as simply LS modes even though a more precise term for a second class of LS modes would be inertial LS modes.

4.1 Suppression of Falling Film Instabilities with Free Surface

Following the sequence presented in Sect. 3, we first present the neutral stability diagrams for Newtonian liquid film falling down an inclined plane (refer to Fig. 2 for schematic). The low-k analysis in the previous Sect. 3.1.1 demonstrated that it is possible to suppress the long-wave free- surface instability by appropriately choosing solid deformability parameter Γ and dimensionless solid thickness H. Figure 8 presents the neutral stability curves for GL and LS modes for both neo-Hookean and linear elastic solid model. In the limit of $\Gamma \rightarrow 0$, i.e., the rigid limit, the GL free surface remains unstable. As Γ is increased above the lower neutral

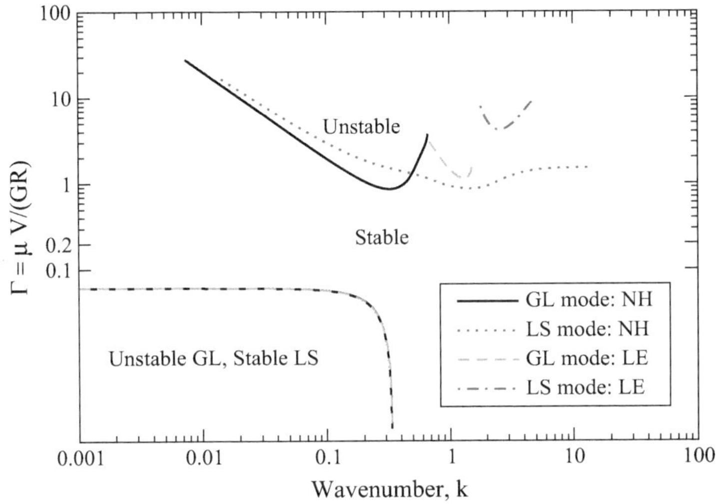

Fig. 8 Complete suppression of free-surface instability and comparison between results from linear elastic and neo-Hookean models. Neutral curves for GL and LS modes: Γ vs. k for $H = 1, Re = 1.5, \theta = 50°$, and $\Sigma_{gl} = \Sigma_{ls} = 0$

curve, there is a transition from unstable to stable region for the free-surface GL mode. This lower neutral curve is obtained by extending the low-k asymptotic result (Eq. 81) to finite wavenumbers and is referred as lower GL-mode neutral curve. Further, note that the lower GL-mode neutral curve, which gives the threshold value of Γ above which the GL- mode perturbations are stabilized, is identical for both neo-Hookean and linear elastic solid model. This implies that both linear and nonlinear constitutive relations predict free-surface instability suppression for finite wavenumbers as well. With further increase in Γ, we obtain a set of two neutral curves, for both linear elastic and neo-Hookean solid models, corresponding to the destabilization of GL and LS interfaces. Importantly, there is a range of Γ values between the lower and upper neutral curves where both GL and LS interfaces remain stable for all wavenumbers. Thus, there exists a stable gap in terms of parameter Γ (or alternatively in terms of shear modulus of solid) where falling film is stabilized by the deformable solid coating. The low-k asymptotic expression for $c^{(1)}$ contained two stabilizing contributions: the first one which is present due to wall deformability and is proportional to Γ while the second which is rigid contribution and is proportional to $\cot\theta$. For a vertical incline, this rigid stabilizing contribution vanishes. We have verified that there exists a sufficiently wide gap in terms of parameter Γ where the free- surface instability is suppressed without exciting the LS interfacial mode for a vertical incline as well. Note that, even though, the details for upper neutral curves for two solid models differ significantly (like a range of unstable wavenumbers), but both models predict a sufficiently wide stability window in terms of parameter Γ.

For linear elastic solid, the upper LS-mode neutral curve corresponds to the continuation of creeping flow viscous LS interfacial instability, while for neo-Hookean solid, it is a combination of inertial LS mode and the short-wave unstable LS mode first predicted by Gkanis and Kumar [20] for flow past a neo-Hookean deformable surface. This short-wave instability for neo-Hookean solid is a consequence of nonzero first normal stress difference in the base state for neo-Hookean constitutive relation and this instability remains absent for flow past linear elastic solid. Also, note that the upper neutral curves predicted by two solid model differ significantly from each other and the solid deformability parameter $\Gamma \sim O(1)$ for these upper neutral curves. This is outside the validity domain of linear elastic solid model, and we expect the results predicted by neo-Hookean model to be more accurate for these upper neutral curves. In this connection, note that while Fig. 8 shows a stable gap for $H = 1$ for both solid models, Fig. 9 for $H = 5$ shows that the stability window vanishes for linear elastic model. The neo-Hookean solid model still predicts a sufficient gap in Γ where stable flow configuration could be achieved. Since we expect neo-Hookean model to be more accurate, particularly in capturing upper neutral curves for which $\Gamma \sim O(1)$ (proportional to deformation gradients in solid layer), it can be concluded that the instability suppression would be valid for higher values of solid thickness as well. This difference in prediction of upper neutral curves also highlights the importance of using a nonlinear constitutive relation even for linear stability analysis.

Figures 8 and 9 illustrated the stabilization of free- surface mode for the case of Newtonian falling film down an inclined plane. We next present the neutral stability curves when the falling liquid film is represented by an Oldroyd-B

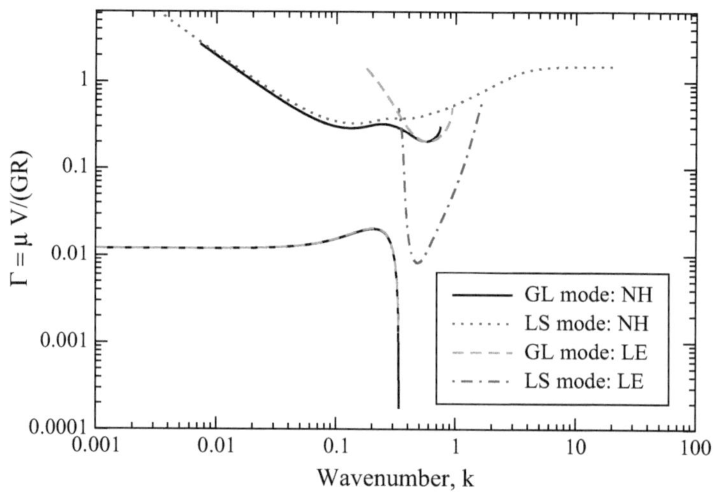

Fig. 9 Neutral stability data showing the differences in predicting the stability window for higher solid thickness H for linear and nonlinear solid models: Γ vs. k for $H = 5$, $Re = 1.5$, $\theta = 50°$, and $\Sigma_{gl} = \Sigma_{ls} = 0$

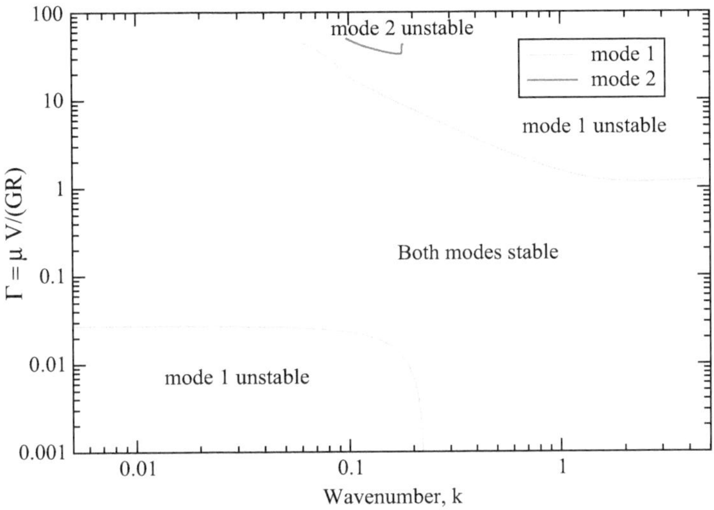

Fig. 10 Neutral curves showing the suppression of free-surface instability for UCM fluid: Γ vs. k for $H = 1, S = 0, W = 0.5, Re = 0, \theta = 50°$, and $\Sigma_{gl} = \Sigma_{ls} = 1$

constitutive relation. Recall that two additional parameters appear: Weissenberg number W which represents the elasticity of fluid or is a measure of relaxation time of the viscoelastic liquid and second the ratio of solvent to total viscosity S. When $S = 0$, the solvent is absent and the model reduces to UCM fluid which describes the behavior of polymer melts. The low-k asymptotic result (Eq. 82) shows that the elasticity or Weissenberg number has an additional destabilizing effect on falling liquid film, while the soft solid layer still has a stabilizing effect (Eq. 83) as was observed in the Newtonian film case. Figure 10 shows the neutral stability diagrams for pure polymer with no solvent ($S = 0$) and Fig. 11 shows neutral curves for polymer solution with solvent $S = 0.5$. Both these figures show a wide stability window in parameter Γ where none of the mode becomes unstable. Figure 11 shows neutral stability data for a vertical incline for which the stabilizing contribution proportional to $\cot\theta$ remains absent. In this case as well, the stabilizing contribution of the deformable solid alone is sufficient to obtain stable flow configuration. Therefore, the conclusion of free-surface instability suppression holds for viscoelastic liquid films as well. Note that in Fig. 11 several additional unstable modes are depicted as compared to Fig. 10. These additional modes correspond to the inertial LS modes which become unstable at very high values of Γ. The neutral curves which are critical for determining stability window are the lower GL-mode neutral curve and the upper GL- or creeping flow viscous LS mode.

The Weissenberg number, which is a measure of elasticity of fluid, is kept at 0.5 in both neutral stability diagrams of Figs. 10 and 11. The numerical results in terms of neutral stability curves show that increasing W has destabilizing effect on the

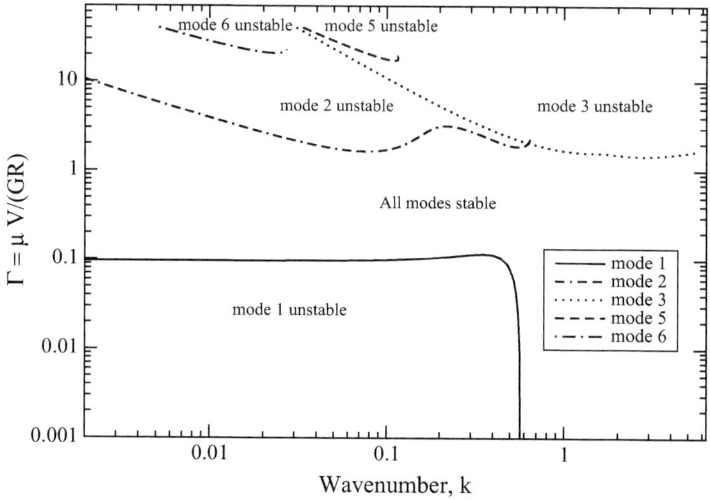

Fig. 11 Neutral curves showing the suppression of free-surface instability for the Oldroyd-B fluid: Γ vs. k for $H = 1, S = 0.5, W = 0.5, Re = 0, \theta = 90°, \Sigma_{gl} = 2$, and $\Sigma_{ls} = 1$

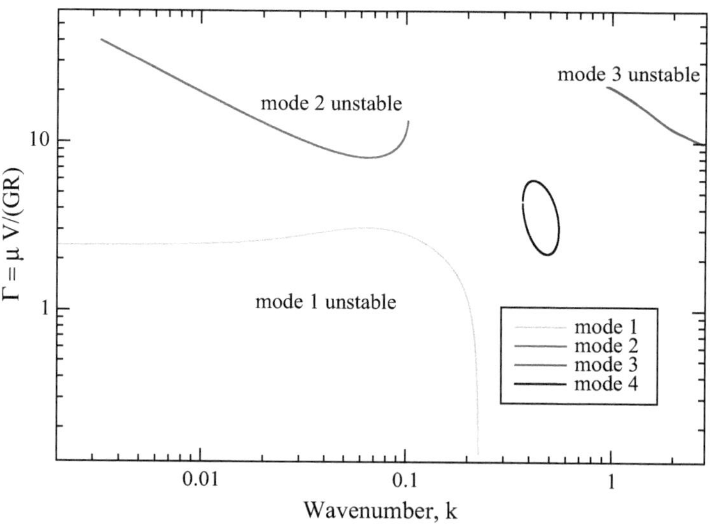

Fig. 12 Neutral stability diagram showing the closing of stability window for higher values of W. Data for the Oldroyd-B fluid: $Re = 0.1, W = 8, S = 0.5, \theta = 50°, H = 0.5, \Sigma_{ls} = 1, \Sigma_{gl} = 2$, and $\mu_{rs} = 0$

configuration under consideration. The width of stability window decreases as W is increased above 0.5 and finally vanishes above a particular value of Weissenberg number. One such case is shown in Fig. 12 which depicts neutral stability diagram for a particular set of parameters with sufficiently high Weissenberg number ($W = 8$). This figure shows that an island of instability appears on increasing W

which closes the stability gap. Thus, while instability suppression for viscoelastic film is possible for sufficiently low values of W, the free-surface instability cannot be stabilized when W increases above a critical value.

The results presented above demonstrated that the single liquid film, Newtonian or viscoelastic, could be stabilized by using a soft solid coating for a wide variety of parameters. We next consider the possibility of simultaneously suppressing the free-surface GL mode instability in the presence of an additional LL interface. The configuration of interest is the gravity-driven flow of two Newtonian liquid falling films down an inclined plane. The inclined plane is again assumed to be coated with a deformable solid layer (refer to Fig. 4) to investigate the possibility of obtaining stable flows when the flow is otherwise unstable in rigid limit.

Before discussing the effect of deformable wall on the stability of two-layer flow, it is instructive to briefly summarize earlier results for gravity-driven flow of two liquid layers down a rigid inclined plane [9]. The stability of this two-layered configuration depends on the relative placement of two liquid layers with respect to the rigid surface and the relative thickness of two liquid layers. The parameter β denotes the dimensionless thickness of the top liquid layer. The relative placement of the two liquid layers with respect to the wall is characterized by viscosity ratio μ_r: when $\mu_r > 1$ ($\mu_r < 1$) the less (more) viscous liquid is adjacent to the wall. The stability characteristics of the GL mode are qualitatively independent of μ_r and β, and for any given μ_r and β, the GL mode becomes unstable above a critical Reynolds number. On the other hand, when the less viscous liquid is adjacent to the wall ($\mu_r > 1$), the LL interface can become unstable even in the absence of inertia [9]. This is also clear from the expression of $c_{ll}^{(1)}$ for $\mu_r = 2$ given in Sect. 3.1.4 (see Eq. 91). In fact, this low-k expression for $c_{ll}^{(1)}$ shows that inertia is stabilizing, while the term proportional to $\cot\theta$ is destabilizing. Chen [9] demonstrated that the LL interface can be completely stabilized at sufficiently higher values of Re when the LL interfacial tension increases above a critical value. However, at such higher values of Re, the GL interface becomes unstable and hence simultaneous stabilization of both GL and LL interfacial modes is not possible for $\mu_r > 1$. For $\mu_r < 1$, both GL and LL mode become unstable above a critical value of Re. The long-wave results presented in Sect. 3.1.4 illustrated that the soft solid contribution is stabilizing for $\mu_r < 1$ and destabilizing for $\mu_r > 1$ for both GL and LL modes irrespective of the value of β. We first discuss the results corresponding to $\mu_r < 1$ for which the effect of deformable solid is stabilizing for both LL and GL interfaces.

Figure 13 shows the neutral stability curves for all three interfaces for $\mu_r = 0.5, Re = 0.1, \theta = 90°, \Sigma_{gl} = \Sigma_{ll} = 0.25, \Sigma_{ls} = 0.1$, and $H = 2$. Earlier results [9] and low-k results show that both GL and LL interfaces become unstable for any nonzero Re for $\mu_r = 0.5$ and $\theta = 90°$ (also see Eqs. 85 and 87). The solid layer has a stabilizing effect on both the modes for $\mu_r = 0.5$ in the long-wave limit. Therefore, in $\Gamma - k$ plane, two lower neutral curves exist corresponding to the stabilization of LL and GL interfaces as Γ is increased from zero (rigid limit) beyond the respective critical value for LL and GL modes. This figure shows that the LL mode is first stabilized followed by stabilization of GL mode on further increasing Γ above the lower GL- mode neutral curve. Similar to previous results for single

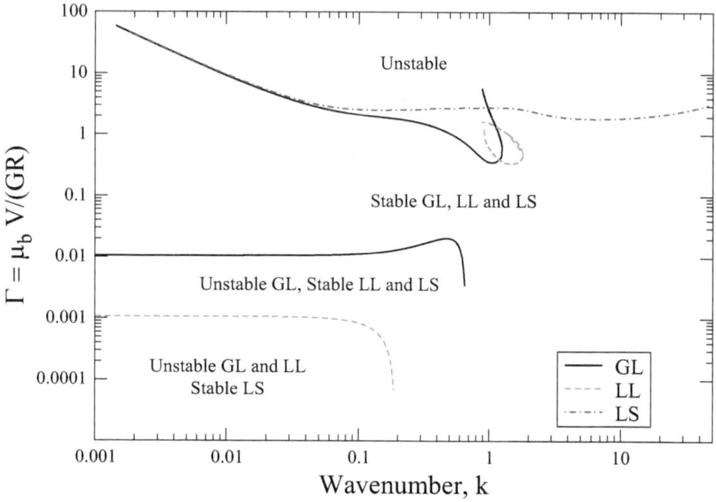

Fig. 13 Neutral stability diagram illustrating the simultaneous suppression for LL and GL interfaces for the case of two-layer film flow. Data for $H = 2, Re = 0.1, \mu_r = 0.5, \theta = 90°$, $\Sigma_{gl} = \Sigma_{ll} = 0.25$, $\Sigma_{ls} = 0.1$, and $\mu_{rs} = 0$

liquid layer falling under the action of gravity, when Γ is sufficiently increased, all three interfaces (GL, LL, and LS) become unstable due to wall deformability as depicted by the presence of three upper neutral curves showing the destabilization of LL, GL, and LS interfaces. For this three-layered configuration as well, there exists a sufficiently wide gap in terms of parameter Γ where all three interfaces remain stable. For the parameter set selected in Fig. 13, this stability window is determined by lower GL- and upper GL- or LL-mode neutral curves. Figure 14 shows the neutral stability diagram for the same parameter set, but with a lower value of solid thickness. This figure shows that the critical value of Γ above which both GL and LL modes become stable increases with decrease in solid thickness. However, decreasing solid thickness has an overall stabilizing effect as in evident by increase in width of stability window. Note that the upper GL-mode neutral curve shifts upward and the upper LL-mode neutral curve vanishes for $H = 0.5$. As a result of this, the stability gap is now determined by lower GL and LS mode neutral curves. A comparison of Figs. 13 and 14 clearly illustrates that there is a wider region in Γ (from $\Gamma \sim O(0.01)$ to $\Gamma \sim O(1)$) for $H = 0.5$ as compared for $H = 2$ (from $\Gamma \sim O(0.01)$ to $\Gamma \sim O(0.1)$) where all three interfaces remain stable. Thus, solid thickness can be varied to increase the width of the stable region. We have verified a similar presence of sufficiently wide stability window for different values of β and for various angles of inclination. It is observed that increasing the thickness of the top liquid layer, i.e., β, reduces the width of stability window, and for certain parameter regimes with sufficiently higher values of β, it is not possible to obtain stable flow configuration.

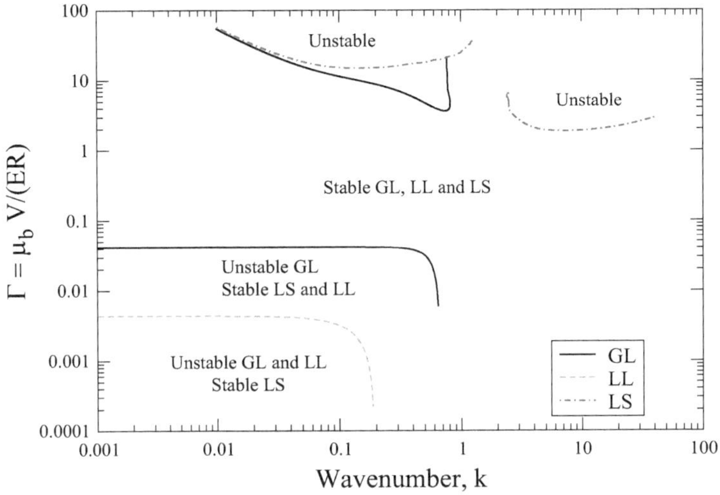

Fig. 14 Neutral stability diagram illustrating the simultaneous suppression for LL and GL interfaces for the case of two-layer film flow. Data for $H = 0.5, Re = 0.1, \mu_r = 0.5, \theta = 90°, \Sigma_{gl} = \Sigma_{ll} = 0.25, \Sigma_{ls} = 0.1$, and $\mu_{rs} = 0$

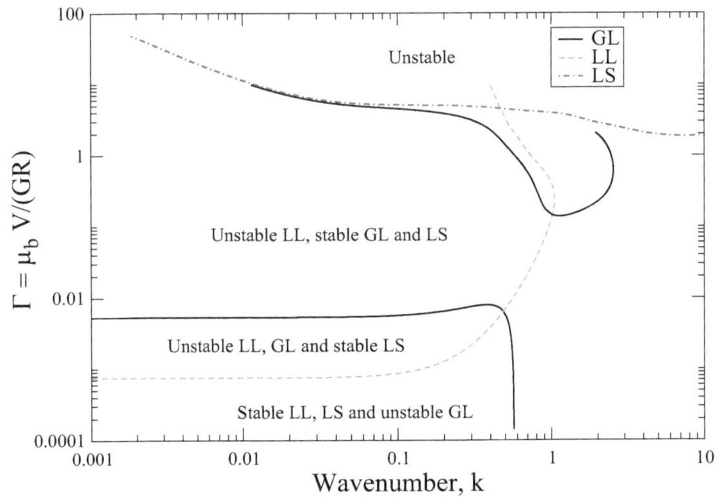

Fig. 15 Neutral stability diagram: Γ vs. k for $\mu_r = 2, \beta = 0.5, \theta = 90°, H = 2, Re = 0.1, \Sigma_{gl} = \Sigma_{ll} = \Sigma_{ls} = 0.1$, and $\eta_r = 0$

We next turn our attention to the case when the less viscous liquid is adjacent to the soft solid layer ($\mu_r > 1$). In the low-wavenumber limit, the deformable solid layer has a stabilizing effect on the GL interface and destabilizing effect on the LL interface. Figure 15 shows the neutral stability diagram for $\theta = \pi/2$, $\mu_r = 2, \beta = 0.5, Re = 0.1, H = 2, \Sigma_{gl} = \Sigma_{ll} = 0.1, \Sigma_{ls} = 0.1$, and $\mu_{rs} = 0$. In the rigid limit for $\mu_r = 2$, the long-wave results show that the LL interface remains stable, while the GL mode is unstable due to inertia. Since wall deformability is now destabilizing for LL mode, there is a transition from stable to unstable perturbations for LL interface as Γ increases above LL-mode neutral curve. With further increase in Γ beyond the lower GL neutral curve, GL-mode perturbations are suppressed due to wall elasticity. On further increasing Γ, both GL and LS modes become unstable similar to the case when $\mu_r < 1$. The LL-mode neutral curve is now a single continuous curve with wavenumbers lying above and left of the neutral curve being unstable. Figure 15 clearly shows that it is not possible to simultaneously stabilize all three interfaces for $\mu_r = 2$.

4.2 Suppression of Fluid–Fluid Interfacial Instabilities in Channel Flows

The results presented in the preceding section demonstrated the capability of deformable solid coating in suppressing interfacial instabilities for both Newtonian and viscoelastic liquid film flows down an inclined plane. We now focus our attention on plane-Couette flow of two liquid layers which could be either Newtonian or viscoelastic fluid layers. The configuration under consideration is shown in Fig. 5, and it shows that the top plate is moving with velocity V and the bottom plate, which is coated with a deformable solid layer, is stationary. When the two fluid layers are Newtonian, this configuration in the absence of the soft solid layer becomes unstable when the less viscous fluid occupies more space. In terms of the nomenclature used here, it translates as: flow is unstable when $\mu_r < 1, \beta < 0.5$ or when $\mu_r > 1, \beta > 0.5$. The long-wave results presented in Sect. 3.2.1 illustrated that the effect of deformable solid coating is always stabilizing when the contribution present due to viscosity stratification is destabilizing, i.e., when $\mu_r < 1, \beta < 0.5$ or when $\mu_r > 1, \beta > 0.5$. Figure 16 shows neutral stability data for $\mu_r = 0.5, \beta = 0.4$, and Fig. 17 shows data when the two fluids are switched with one another, i.e., $\mu_r = 2, \beta = 0.6$ for the case of plane-Couette flow of two Newtonian fluid layers flowing past a deformable solid surface. For both cases, the flow is unstable in the rigid limit ($\Gamma \to 0$). This figure clearly shows that the suppression predicted in low-k limit holds for finite wavenumbers as well and there exists a wide stability window where the fluid–fluid interfacial mode instability is completely suppressed by deformable solid coating. We have verified the existence of such a stable gap for a wide variety of parameters where the two-fluid configuration otherwise remains unstable in rigid limit.

Recall that the long-wave results presented in Sect. 3.2.1 show that the solid layer contribution could be destabilizing or stabilizing depending on the solid thickness when the viscosity stratification contribution is stabilizing. For example, Table 1 illustrates that for $\mu_r = 0.5$ and $\beta = 0.6$, the contribution proportional to Re is stabilizing, while the soft solid has stabilizing effect for $H < 7.92$ and

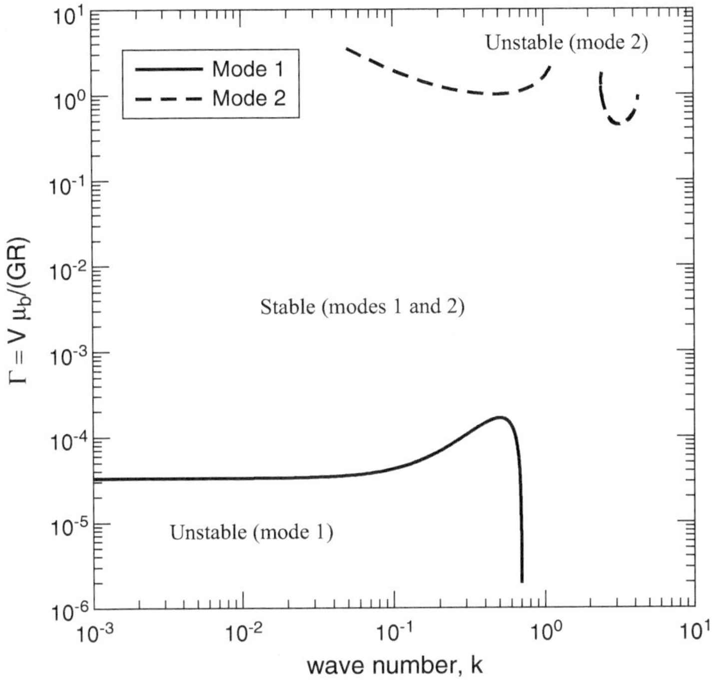

Fig. 16 Neutral stability diagrams illustrating the instability suppression for two-layer Newtonian plane-Couette flow when more viscous fluid is near the deformable solid surface: Γ vs. k for $\mu_r = 0.5, \beta = 0.4, H = 4, Re = 1,$ and $\Sigma_{II} = 0.04$

destabilizing effect for $H > 7.92$. On the other hand, when $\beta \geq 0.8$, the solid layer contribution is always destabilizing for any value of solid thickness, while the viscosity discontinuity across the interface is stabilizing. For such a case (i.e., when $\beta \geq 0.8$), the two-fluid interfacial mode remains stable in rigid limit, and as solid deformability parameter Γ increases beyond a critical value, the fluid–fluid interfacial mode becomes unstable due to wall deformability. We have verified that the destabilization continues to finite wavenumbers and the neutral curves obtained in this case will be qualitatively similar to the neutral curve presented in Fig. 15 for gravity-driven flow of two Newtonian liquid films falling down an inclined plane coated with a soft solid layer. The only difference would be that the GL-mode neutral curve will be absent because there is no gas–liquid interface present in the case of two-layer plane-Couette flow. The neutral stability diagram obtained for $\mu_r = 0.5, \beta = 0.6$, and for $H > 7.92$ will also be qualitatively similar to Fig. 15 with GL- mode neutral curve remaining absent and a presence of lower neutral curve for two-fluid interfacial mode showing destabilization of this interfacial mode by soft solid coating. For $H < 7.92$, when both soft solid and rigid contributions are stabilizing, lower neutral curves remain absent. However, in this case, LL mode and LS mode can become unstable when Γ is sufficiently increased. Thus, while lower

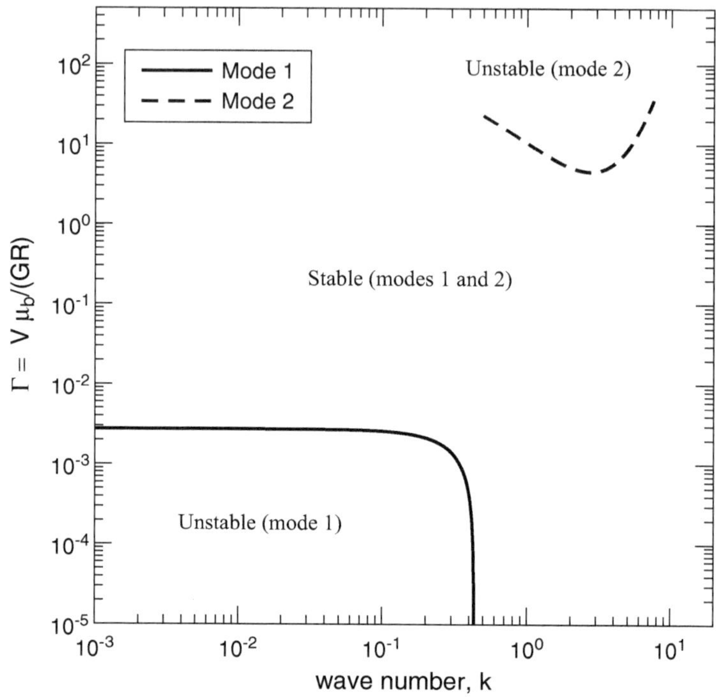

Fig. 17 Neutral stability diagrams illustrating the instability suppression for two-layer Newtonian plane-Couette flow when less viscous fluid is near the deformable solid surface: Γ vs. k for $\mu_r = 2, \beta = 0.6, H = 0.5, Re = 1$, and $\Sigma_{ll} = 0.1$

neutral curves remain absent, upper LL- and LS-mode neutral curves will be present showing the destabilization of the configuration due to excitation of fluid–fluid and fluid–solid interfacial mode.

We now discuss whether a similar instability suppression holds for the flow of two viscoelastic fluid layers as well or not? It was shown in the limit of low wavenumbers that stabilizing or destabilizing contribution of the soft solid layer is dictated solely by viscosity stratification and remains independent of discontinuity in elasticity across the interface. Thus, the soft solid contribution is again stabilizing when the (viscoelastic) fluid with lower viscosity occupies more space. However, an additional destabilizing contribution proportional to difference in Weissenberg numbers of two fluid layers is present due to the elastic nature of the fluid layers. For example, the long-wave results presented in Table 3 for $\mu_r = 0.5$ and $\beta = 0.4$ show that the term proportional to Re is destabilizing, soft solid contribution is stabilizing, and contribution present due to elastic stratification is destabilizing when $W_b > 1.13 W_a$. A comparison of different rigid contributions in the expression of $c^{(1)}$ (i.e., $c^{(1)}_{\text{visc}}$ and $c^{(1)}_{\text{elas}}$) for $W \gtrsim O(0.1)$ reveals that the destabilizing contribution proportional to the difference in Weissenberg numbers is at least an order of magnitude larger than the term proportional to Re. The low-k results show that it is possible to suppress the interfacial instability by sufficiently increasing the wall deformability for a given solid thickness. However, due to the presence of an additional destabilizing contribution, the critical value above which flow becomes stable due to wall elasticity increases as compared to the critical value of Γ required for stability of two Newtonian fluid layers. Also, it is known that both fluid–fluid and fluid–solid modes become unstable when Γ is sufficiently increased. Consequently, it is important to find out whether there exists a sufficiently wide stability window for two-layer UCM fluid flow as was observed in the case of two-layer Newtonian fluid flow. Figure 18 shows the neutral stability diagram for one such case of plane-Couette flow of two viscoelastic fluid layers. This figure clearly shows the presence of a wide gap between lower and upper neutral curves where the two-fluid mode and the fluid–solid mode remain stable. We have verified the existence of such stable gap for a variety of parameter regimes for the case of plane-Couette flow of two viscoelastic fluid layers. Thus, the results presented in this section show that it is possible to use a soft solid coating for suppressing interfacial instabilities for two-layer plane-Couette flow of both Newtonian and viscoelastic fluids.

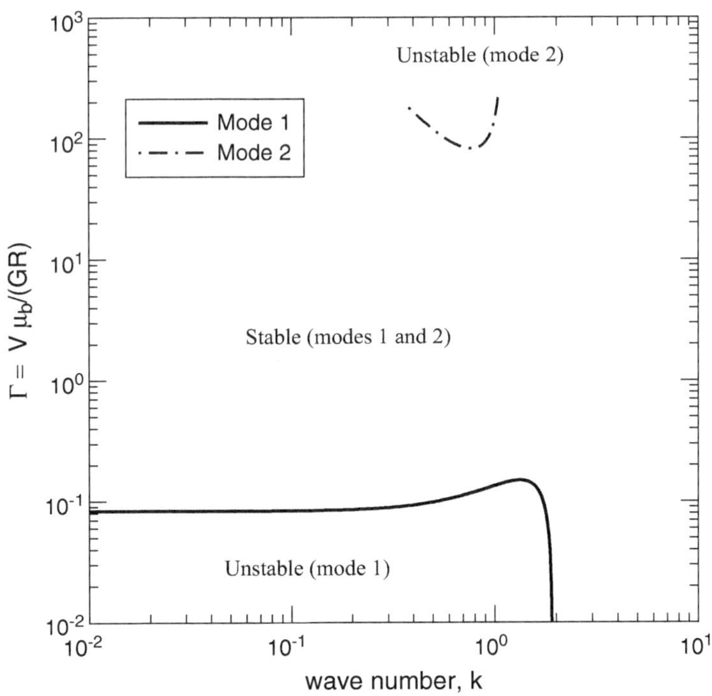

Fig. 18 Neutral stability diagram illustrating the stabilization of two-layer viscoelastic plane-Couette flow by deformable solid layer: Γ vs. k for $\mu_r = 0.5, W_a = 0.1, W_b = 0.5, \beta = 0.4, H = 0.7, Re = 1, \Sigma_{II} = 0.25$, and $\mu_{rs} = 0.6$

4.3 Suppression of Capillary Instability in the Presence of Interfacial Instabilities

Thus far, we have demonstrated the potential of the deformable solid layer in suppressing free-surface instabilities, liquid–liquid interfacial instabilities in the presence of free-surface instability, and fluid–fluid interfacial instabilities present in different configurations due to viscosity or elasticity contrast across an interface (liquid–liquid or gas–liquid). We now demonstrate the efficacy of the soft solid layer in suppressing capillary instability induced due to surface tension in the presence of free-surface and fluid–fluid interfacial instabilities, again present due to viscosity contrast across the interface. It is well known that the cylindrical fluid–fluid interfaces become unstable due to surface tension-induced capillary instability both in the presence or absence of inertia [23, 50–52, 61]. We consider two cases: first, the gravity-driven annular liquid film flow outside the wires and inside the tubes where the free surface becomes unstable in addition to instability induced due to surface tension, and second, the pressure-driven core–annular flow (CAF) of two immiscible liquids in a tube where capillary instability is present along with two-fluid interfacial mode due to a viscosity jump across the fluid–fluid interface.

The schematics showing the flow outside the wire and inside the tube where the outer and inner walls, respectively, are coated with a deformable solid layer are shown in Fig. 6. The results in low-k limit show that it is not possible to obtain stable flow configuration in rigid limit for both flow outside the wire and for cylindrical film flow inside the tube. However, the soft solid has a stabilizing effect and it is possible to suppress long-wave capillary instability in the presence of free-surface instability by manipulating the deformable wall properties (wall elasticity, i.e., Γ and soft solid thickness H). For $Re = 0$ and $\Sigma_{\rm gl} \neq 0$, only capillary instability is present, and the neutral stability data in Fig. 19 shows the presence of a stable gap illustrating essentially the suppression of capillary instability by deformable solid coating. We have verified that such suppression holds for a different values of S and H which correspond to different values of radius of curvature and hence variation in the destabilizing contribution proportional to gas–liquid interfacial tension.

Figure 20 shows that the width of stability window decreases significantly on increasing Re from 0 to $O(1)$ value. This implies that increasing destabilizing contribution due to the presence of free surface reduces the width of stable gap, and for sufficiently high $Re \sim O(1)$, the stable gap vanishes and it will not be possible to obtain stable configuration by altering wall elasticity. However, note that the film flow in rigid limit is unstable for any nonzero value of Re. We have verified that it is possible to suppress both capillary and free-surface instability for $\Sigma_{\rm gl} \sim O(1)$ and $Re \sim O(0.1)$. The above results are shown for outside the wire. The results for flow inside the tube coated with a soft solid layer and with a presence of free surface also show similar qualitative features, i.e., it is possible to obtain stable flow configuration for $Re \sim O(0.1)$, $\Sigma_{\rm gl} \sim O(1)$ for inside tube flow by using a deformable solid coating. Thus, it is possible to suppress capillary instability for free-surface film flows for $Re \sim O(0.1)$ and $\Sigma_{\rm gl} \sim O(1)$.

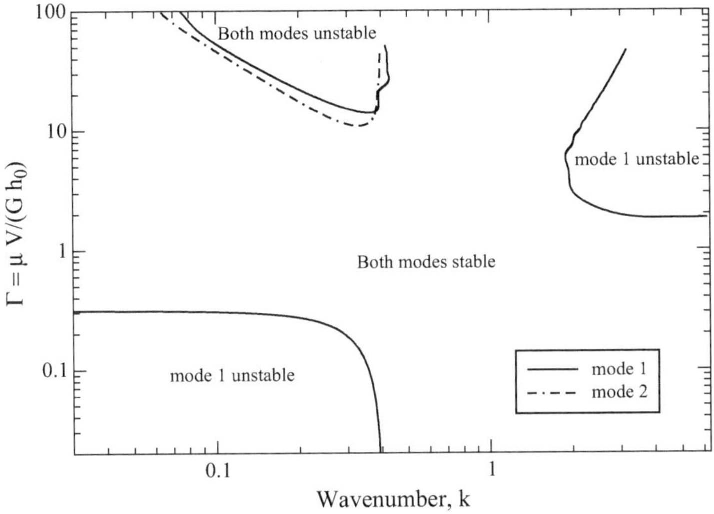

Fig. 19 Neutral stability diagram illustrating the stabilization of capillary instability: Γ vs. k for $Re = 0, S = 1, H = 0.5, \Sigma_{\mathrm{gl}} = 8.385$, and $\Sigma_{\mathrm{ls}} = 0$

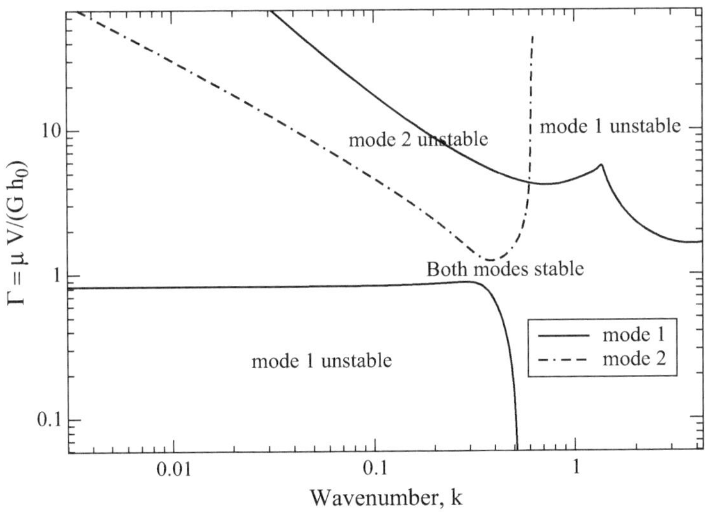

Fig. 20 Neutral stability diagram showing that the stability window closes for $Re \sim O(1)$: Γ vs. k for $Re = 1.192, S = 1, H = 0.5, \Sigma_{\mathrm{gl}} = 8.385$, and $\Sigma_{\mathrm{ls}} = 0$

We next consider the core–annular flow of two immiscible fluids in a flexible tube (see Fig. 7). Here, capillary forces render the flow unstable even in the absence of inertia and the qualitative nature of this surface tension-induced instability remains independent of the viscosity of core and annular liquids. Further, the fluid–fluid interface could also become unstable due to viscosity contrast. Preziosi et al. [50]

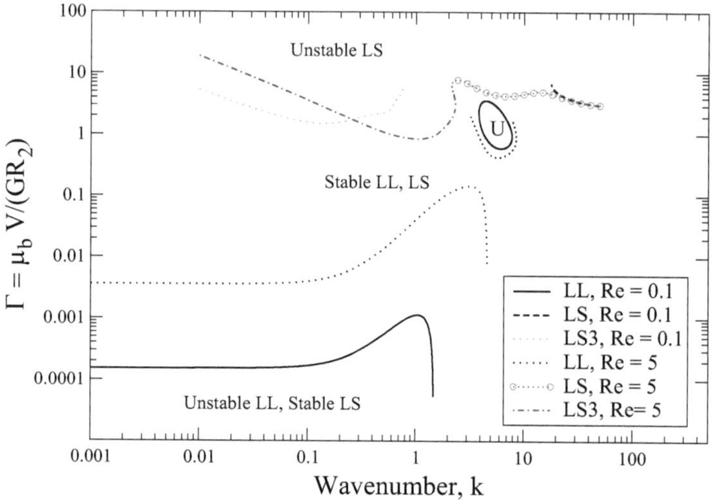

Fig. 21 Neutral stability diagram showing the stabilization of core–annular flow by the deformable solid layer: Γ vs. k for $\mu_r = 0.5, a = 0.9, H = 2, \Sigma_{ll} = 0.1, \Sigma_{ls} = 0$, and $\mu_{rs} = 0$

have shown for CAFs in the rigid tubes that the viscosity stratification is stabilizing when the less viscous liquid occupies thin annular space compared to more viscous core fluid (i.e., $\mu_r > 1$ and higher values of mean fluid–fluid interface a). It was demonstrated that for such a case, there exists a window of stability in terms of Re where stable core–annular flow can be achieved. On the other hand, when $\mu_r < 1$, it is not possible to obtain stable core–annular flow in rigid tubes. It was illustrated in the long-wave asymptotic results above that the soft solid layer has a stabilizing effect for $\mu_r < 1$ and destabilizing effect for $\mu_r > 1$. We present results for $\mu_r < 1$ in more detail when the effect of deformability of the solid layer is stabilizing on fluid–fluid interface in long-wave limit, and hence, there is a possibility of obtaining stable CAFs in a flexible tube when the flow is otherwise unstable in a rigid tube.

Figure 21 shows the neutral stability curves for two different values of Reynolds number. Let us first focus on neutral curves corresponding to $Re = 0.1$. For $\mu_r = 0.5, a = 0.9, Re = 0.1$, and $\Sigma_{ll} = 0.1$, the low-k results clearly illustrate that the flow remains unstable in rigid limit ($\Gamma \to 0$) because of the contribution proportional to Re (present due to viscosity jump) and the term proportional to Σ_{ll} (capillary instability). Similar to the results presented above for suppression of interfacial instabilities in different configurations, as Γ increases above the lower LL-mode neutral curve, there is transition from unstable to stable LL-mode perturbations. There exist a range of values of Γ for which both LL and LS interfaces remain stable and with further increase in Γ both LL and LS modes become unstable because of wall deformability. There is a significant gap between lower and upper neutral curves where stable core–annular flow configuration is achieved for $\mu_r = 0.5$ which was unstable in rigid limit. A similar stable gap is present for $Re = 5$ as well; however, the stable region is quite small as compared to the $Re = 0.1$ case. Note that the stable gap for both values of Re is determined by lower and upper LL-mode neutral curves. When Re is increased from 0.1 to 5, it is observed that the upper LL-mode curve remains largely unaffected, while the lower LL-mode neutral curve, which gives transition value of Γ above which the flow becomes stable, shifts upward by almost two orders of magnitudes for $Re = 5$ as compared to $Re = 0.1$. Therefore, there is much wider gap for lower Re and the width of stability window decreases with increase in Re. Our results show that the stable gap is absent for $Re \sim O(10)$ and stable CAF arrangement in a deformable tube could be attained only for $Re \lesssim O(1)$.

The numerical results presented above are shown for the case when annular liquid occupies much less space as compared to the core fluid ($a = 0.9$ in Fig. 21). Our numerical results for $\mu_r = 0.5$ and $a = 0.7$ also demonstrate the capability of deformable solid coating in achieving stable CAF configuration for $Re \lesssim 1$ and $\Sigma_{ll} \sim 0.1 - -1$. However, the low-$k$ results demonstrated that the qualitative nature of term proportional to Re changes from destabilizing to stabilizing with decrease in mean LL interface position, while the term proportional to Σ_{ll} still remains destabilizing. Further, it is observed that the capillary instability becomes more and more stronger as compared to term proportional to Re with decrease in a. Figure 22 depicts neutral stability curves for $a = 0.5$ and $\Sigma_{ll} = 0.1$. For $a = 0.5$, the low-k results show that the term proportional to Re becomes stabilizing. This figure shows that for $\Sigma_{ll} = 0.1$ and $a = 0.5$, there exists a sufficiently wide stable gap for both zero and nonzero Re. Thus, the stabilizing contribution of the soft solid layer is adequate for suppressing the capillary as well as two-fluid interfacial

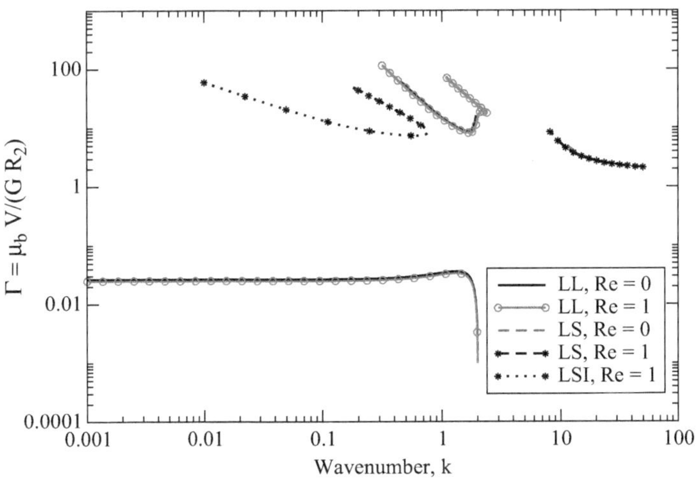

Fig. 22 Neutral stability diagram illustrating the stabilization of capillary instability for the case of CAF of two immiscible liquids: Γ vs. k for $\mu_r = 0.5, a = 0.5, H = 0.5, \Sigma_{ll} = 0.1, \Sigma_{ls} = 0$, and $\mu_{rs} = 0$

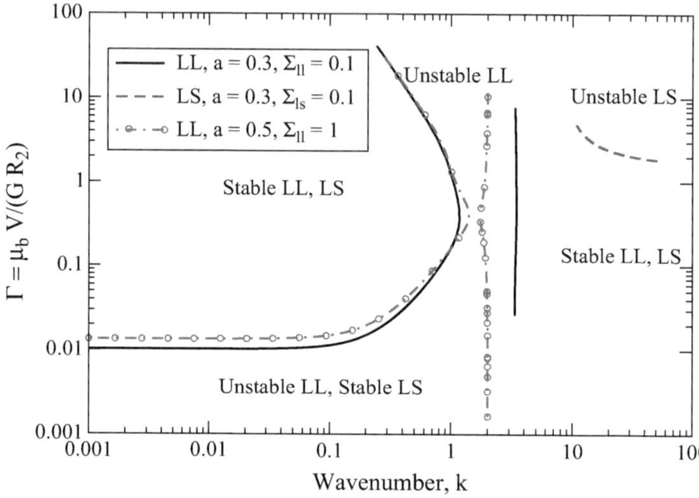

Fig. 23 Neutral stability diagrams illustrating the role of capillary instability in obtaining stable CAF configuration: Γ vs. k for $\mu_r = 0.5$, $H = 2$, $Re = 0$, and $\mu_{rs} = 0$

instabilities. Figure 23 presents neutral curves for $a = 0.3$ and $\Sigma_{ll} = 0.1$. The stabilizing contribution due to Re is suppressed by putting $Re = 0$ and the only destabilizing contribution present is due to capillary forces. This figure shows that while it is possible to suppress the LL- mode perturbations in low-wavenumber limit, it is not possible to suppress the fluctuation with $k \sim 1$ by manipulating the wall elasticity. For $a = 0.3$, the strong capillary forces dominate over the stabilizing soft solid contribution and it is not possible to obtain stable flow configuration for lower values of a for $\Sigma_{ll} \gtrsim O(0.1)$. This is in contrast to the results presented in Fig. 22 for $a = 0.5$ and $\Sigma = 0.1$, where stable CAF configuration is achieved by manipulating wall elasticity. However, Fig. 23 shows that the stable gap disappears when Σ_{ll} is increased from 0.1 to 1 for $a = 0.5$. We investigated the effect of varying Σ_{ll} for different values of a and observed the presence of stability window at $a = 0.9$ for Σ_{ll} values as high as 10. On the other hand, for $a = 0.6$, stability window disappears when Σ_{ll} is increased from 2 to 4.5. Thus, for $Re \sim O(1)$, the presence or absence of a stable region depends on the competition between the stabilizing contribution of the soft solid layer and destabilizing contribution due to capillary forces. This discussion suggests that it is possible to attain stable flow configurations for sufficiently higher values of a by using a deformable solid coating. On the other hand, the presence of strong destabilizing capillary forces prevents the stabilization of LL interface at lower values of a.

At this point, we provide some estimates of the various dimensional parameters for which the predicted suppression of instability may be realized in experiments. As an illustration, we choose the data given in Fig. 21 which presents neutral stability curves for core–annular flow of two immiscible Newtonian liquids in a flexible tube. The density of common liquids is usually $\rho \sim 10^3\ \mathrm{kg/m^3}$. The interfacial tension between different immiscible liquids and between different liquids and soft solids like PDMS is typically $\sim$0.01 N/m; thus, we set γ_{ll} and $\gamma_{ls} \sim O(0.01)$ N/m to estimate nondimensional surface tension parameter. We further set $R \sim 10^{-3}$ m, $\mu_b \sim 1$ Pa.s and centerline velocity $V \sim 0.1$ m/s. These choices of parameter implies that $Re \sim 0.1$, Σ_{ll} and $\Sigma_{ls} \sim 0.1$. For these set of parameters, Fig. 21 clearly shows that CAF remains unstable in the rigid limit $\Gamma \rightarrow 0$ and this interfacial instability is suppressed for a range of Γ values ($\sim O(0.001-0.1)$). This corresponds to a shear modulus values of $10^3 < G < 10^5$ Pa. Thus, the predicted suppression can be realized in experiments for very viscous liquid layers flowing past a soft deformable solid wall.

5 Conclusions

The effect of deformable solid layer coatings on interfacial instabilities in liquid–liquid and liquid–gas interfaces was analyzed for a variety of configurations and for both Newtonian and viscoelastic fluids. A long-wave (low-k) asymptotic analysis is carried out first which clearly elucidates the stabilizing or destabilizing role of solid deformability. We find that in the low-k analysis, dissipative effects in the solid layer do not play any role. Interestingly, the effect of the solid layer appears at the same order (i.e., $O(k)$) at which the other effects such as viscosity stratification, elasticity stratification, and interfacial tension (for curved interfaces) appear. Thus, by appropriately choosing the shear modulus and thickness of the solid layer, it is possible to completely suppress the interfacial instability when it is present without the deformable solid or even to destabilize the interface when it is stable without the deformable solid. Numerical computations were carried out using spectral method in order to verify whether the low-k predictions extend to all values of k and to investigate whether the deformability of the solid layer induces new instabilities at finite wavenumber. Using our numerical computations, we presented neutral stability diagrams wherein we show regions (in nondimensional shear modulus) where the interfacial instabilities are completely suppressed at all wavenumbers. These predictions were shown to be generic and were demonstrated for both Newtonian and viscoelastic fluids and for different geometric configurations. Thus, the present study demonstrates that soft deformable coatings could be used for passive manipulation and control of interfacial instabilities.

References

1. Achenbach DJ (1973) Wave propagation in elastic solids, 3rd edn. Wiley, North-Holland
2. Barrero A, Loscertales IG (2007) Micro and nano particles via capillary flows. Ann Rev Fluid Mech 39:89–106
3. Bassom AP, Blyth MG, Papageorgiou DT (2012) Using surfactants to stabilize two-phase pipe flows of core-annular type. J Fluid Mech 704:333–359

4. Blyth MG, Pozrikidis C (2004) Effect of surfactant on the stability of flow down an inclined plane. J Fluid Mech 521:241–250
5. Carpenter PW, Gajjar JSB (1990) A general theory for two and three dimensional wall-mode instabilities in boundary layers over isotropic and anisotropic compliant walls. Theor Comput Fluid Dyn 1:349–378
6. Carpenter PW, Garrad AD (1985) The hydrodynamic stability of flows over Kramer-type compliant surfaces. Part 1. Tollmien-Schlichting instabilities. J Fluid Mech 155:465–510
7. Carpenter PW, Garrad AD (1986) The hydrodynamic stability of flows over Kramer-type compliant surfaces. Part 2. Flow induced surface instabilities. J Fluid Mech 170:199–232
8. Chen KP (1991) Elastic instability of the interface in Couette flow of two viscoelastic liquids. J Non-Newton Fluid Mech 40:261–267
9. Chen KP (1993) Wave formation in the gravity-driven low-Reynolds number flow of two liquid films down an inclined plane. Phys Fluids A 5(12):3038–3048
10. Chokshi P, Kumaran V (2008) Weakly nonlinear stability analysis of flow past a neo-Hookean solid at arbitrary Reynolds number. Phys Rev E 77:056303
11. Davies C, Carpenter PW (1997) Instabilities in a plane channel flow between compliant walls. J Fluid Mech 352:205–243
12. Demekhin EA, Kalliadasis S, Velarde MG (2006) Suppressing falling film instabilities by Marangoni forces. Phys Fluids 18:042111 (1–16)
13. Destrade M, Saccocmandi G (2004) Finite-amplitude inhomogeneous waves in Mooney-Rivlin viscoelastic solids. Wave Motion 40:251–262
14. Fosdick RL, Yu JH (1996) Thermodynamics, stability and non-linear oscillations of viscoelastic solids – I. Differential type solids of second grade. Int J Non-Linear Mech 31:495–516
15. Gajjar JSB, Sibanda SK (1996) The hydrodynamic stability of channel flow with compliant boundaries. Theor Comput Fluid Dyn 8:105–129
16. Ganpule H, Khomami B (1999) An investigation of interfacial instabilities in the superposed channel flow of viscoelastic fluids. J Non-Newton Fluid Mech 81:27–69
17. Gaurav, Shankar V (2007) Stability of gravity-driven free-surface flow past a deformable solid layer at zero and finite Reynolds number. Phys Fluids 19:024105
18. Gaurav, Shankar V (2009) Stability of fluid flow through deformable neo-Hookean tubes. J Fluid Mech 627:291–322
19. Gaurav, Shankar V (2010) Stability of pressure-driven flow in a deformable neo-Hookean channel. J Fluid Mech 659:318–350
20. Gkanis V, Kumar S (2003) Instability of creeping Couette flow past a neo-Hookean solid. Phys Fluids 15:2864–2471
21. Gkanis V, Kumar S (2005) Stability of pressure-driven creeping flows in channels lined with a nonlinear elastic solid. J Fluid Mech 524:357–375
22. Gkanis V, Kumar S (2006) Instability of gravity-driven free-surface flow past a deformable elastic solid. Phys Fluids 18:044103
23. Goren SL (1961) The instability of an annular liquid thread. J Fluid Mech 12:309–319
24. Gupta AS, Rai L (1967) Stability of Oldroyd-B fluid flowing down an inclined plane. Proc Camb Philos Soc 33:87
25. Hains FD, Price JF (1962) Effect of flexible wall on stability of Poiseuille flow. Phys Fluids 5:365
26. Hayes MA, Saccocmandi G (2002) Finite-amplitude waves superimposed on pseudoplanar motions for Mooney-Rivlin viscoelastic solids. Non Linear Mech. 37:1139–1146
27. Hickox CE (1971) Instability due to viscosity and density stratification in axisymmetric pipe flow. Phys Fluids 14(2):251–262
28. Holzapfel GA (2000) Nonlinear solid mechanics. Wiley, Chichester
29. Huang CT, Khomami B (2000) The role of dynamic modulation in the stability of viscoelastic flow down an inclined plane. J Fluid Mech 425:213–233
30. Huang CT, Khomami B (2001) Role of dynamic modulation on stability of multilayer Newtonian and viscoelastic flows down an inclined plane. J Non-Newton Fluid Mech 97(1): 67–86

31. Huang L (1998) Reversal of Bernoulli effect and channel flutter. J Fluids Struct 12:131
32. Jain A, Shankar V (2007) Instability suppression in viscoelastic film flows down an inclined plane lined with a deformable solid layer. Phys Rev E 76:046314:1–14
33. Jiang WY, Lin SP (2005) Enhancement or suppression of instability in a two-layered liquid film flow. Phys Fluids 17:054105
34. Joseph DD, Renardy YY (1993) Fundamentals of two-fluid dynamics: part 1, Mathematical theory and applications. Springer, New York
35. Joseph DD, Renardy YY (1993) Fundamentals of two-fluid dynamics: part 2, Lubricated transport, drops and miscible liquids. Springer, New York
36. Joseph DD, Bai R, Chen KP, Renardy YY (1997) Core-annular flows. Ann Rev Fluid Mech 29(1):65–90
37. Kao TW (1968) Role of viscosity stratification in the instability of two-layer flow down an incline. J Fluid Mech 33:561–572
38. Kramer MO (1957) Boundary-layer stabilization by distributed damping. J Aeronaut Sci 24:459–460
39. Kumaran V (1995) Stability of the viscous flow of a fluid through a flexible tube. J Fluid Mech 294:259–281
40. Kumaran V (2003) Hydrodynamic stability of flow through compliant channels and tubes. In: Carpenter PW, Pedley TJ (eds) IUTAM symposium on flow past highly compliant boundaries and in collapsible tubes, University of Warwick, Kluwer Academic, Dordrecht
41. Kumaran V, Muralikrishnan R (2000) Spontaneous growth of fluctuations in the viscous flow of a fluid past a soft interface. Phys Rev Lett 84:3310–3313
42. Kumaran V, Fredrickson GH, Pincus P (1994) Flow induced instability of the interface between a fluid and a gel at low Reynolds number. J. Phys II Fr 4:893–904
43. Lai W (1967) Stability of a elasto-viscous fluid flowing down an inclined plane. Phys Fluids 10:844–846
44. Larson RG (1992) Instabilities in viscoelastic flows. Rheol Acta 31:213–263
45. Lin SP, Chen JN (1997) Elimination of three-dimensional waves in a film flow. Phys Fluids 9:3926–3928
46. Lin SP, Chen JN, Woods DR (1996) Suppression of instability in a liquid film flow. Phys Fluids 8:3247–3252
47. Macosko CW (1994) Rheology: principles, measurements, and applications. VCH, New York
48. Malvern LE (1969) Introduction to the mechanics of a continuous medium. Prentice-Hall, Englewood Cliffs
49. Lucey AD, Carpenter PW (1995) Boundary layer instability over compliant walls: comparison between theory and experiments. Phys Fluids 7:2355–2363
50. Preziosi L, Chen KP, Joseph DD (1989) Lubricated pipelining: stability of core-annular flow. J Fluid Mech 201:323–356
51. Rayleigh L (1878) On the instability of jets. Proc Lond Math Soc 10:4
52. Rayleigh L (1892) On the stability of cylinder of viscous liquid under capillary force. Philos Mag 34:145
53. Renardy Y (1988) Stability of the interface in two-layer Couette flow of upper convected Maxwell liquids. J Non-Newton Fluid Mech 28:99–115
54. Rottenberry JM, Saffman PG (1990) Effect of compliant boundaries on weakly nonlinear shear waves in channel flow. SIAM J Appl Math 50:361–394
55. Shankar V (2004) Stability of two-layer viscoelastic plane Couette flow past a deformable solid layer. J Non-Newton Fluid Mech 117:163–182
56. Shankar V, Kumaran V (1999) Stability of non-parabolic flow in a flexible tube. J Fluid Mech 395:211–236
57. Shankar V, Kumaran V (2001) Asymptotic analysis of wall modes in a flexible tube revisited. Eur Phys J B 19:607–622
58. Shankar V, Kumaran V (2002) Stability of wall modes in fluid flow past a flexible surface. Phys Fluids 14:2324–2338

59. Shankar V, Sahu AK (2006) Suppression of instability in liquid flow down an inclined plane by a deformable solid layer. Phys Rev E 73:016301 (1–12)
60. Shaqfeh ESG, Larson RG, Fredrickson GH (1989) The stability of gravity driven viscoelastic film flow at low to moderate Reynolds number. J Non-Newton Fluid Mech 21:87
61. Solorio FJ, Sen M (1987) Linear stability of cylindrical falling film. J Fluid Mech 183:365–377
62. Su Y-Y, Khomami B (1992) Purely elastic interfacial instabilities in superposed flow of polymeric fluids. Rheol Acta 31:413–420
63. Tomotika S (1935) On the instability of cylindrical thread of a viscous liquid surrounded by another viscous fluid. Proc R Soc A 150:322–337
64. Wei H-H (2005) Effect of surfactant on the long-wave instability of a shear-imposed liquid flow down an inclined plane. Phys Fluids 17:012103
65. Wei H-H (2005) Stability of a viscoelastic falling film with surfactant subjected to interfacial shear. Phys Rev E 71:066306
66. Wei H-H, Rumshitzki DS (2005) The effects of insoluble surfactants on the linear stability of a core and annular flow. J Fluid Mech 541:115–142
67. Wenstein SJ, Chen KP (1999) large growth rate instabilities in three layer flow down an incline in the limit of zero Reynolds number. Phys Fluids 11:29–53
68. Wenstein SJ, Ruschak KJ (2004) Coating flows. Ann Rev Fluid Mech 36:29–53
69. Yih C-S (1963) Stability of liquid flow down an inclined plane. Phys Fluids 6:321–334
70. Yih C-S (1967) Instability due to viscosity stratification. J Fluid Mech 27:337–350

Part III
Soft Matter Dynamics

Effect of Asymmetry on Adhesion and Locomotion of Soft-Bodied Objects

Subrata Mondal and Animangsu Ghatak

Abstract Limbless locomotion of soft-bodied objects has been achieved by variety of mechanisms: volume phase transition of hydrogels, periodic alterations of electric and photochemical field, rectification of noise on a substrate having asymmetric friction, and so on. In this report, we have presented a new method of inducing locomotion of soft elastomeric objects by generating geometric asymmetry in it. The asymmetry is generated by incomplete swelling of the material with an organic solvent. In particular, we have prepared cross-linked elastomeric cylinders which adhere weakly to a substrate. We have swelled these cylinders partially by dispensing a small quantity of solvent leading to their bending along length; the solvent however evaporates from portion of it exposed to the atmosphere. The bending causes a forward rolling motion which is sustained via continued supply of solvent from one side and its evaporation from the other. The velocity of rolling depends on several geometric and material properties: diameter and elastic modulus of the cylinder, surface tension, density and vapor pressure of the liquid, and its ability to swell the cross-linked network. The above mechanism enables the cylinder to locomote not only on a horizontal substrate but also up an inclined plane.

Closely associated with locomotion is adhesion on a substrate, specifically the ability to adhere to it strongly from one direction but to separate easily from the other, which we have achieved by preparing adhesives embedded with closely spaced fluid-filled microchannels. The microchannels are arranged in pairs with a liquid filling on only one of them which generates spatial anisotropy in respect to surface topography and shear modulus of the adhesive. As a result, when an adherent is lifted off the adhesive, it requires different lift-off load to be exerted for separating it from two different directions.

Keywords Asymmetry • Rolling motion • Directional adhesion

S. Mondal • A. Ghatak (✉)
Department of Chemical Engineering and DST Unit on Soft Nanofabrication, Indian Institute of Technology Kanpur, Kanpur, 208016 UP, India
e-mail: aghatak@iitk.ac.in

Y.M. Joshi, S. Khandekar (eds.), *Nanoscale and Microscale Phenomena*,
Springer Tracts in Mechanical Engineering, DOI 10.1007/978-81-322-2289-7_10

1 Introduction

Direction specificity is prevalent in many natural systems, e.g., many plants and insects make use of direction-specific hydrophobicity or hydrophilicity to channelize water condensed on their outer skin in a preferred direction, soft-bodied animals use directional friction for their locomotion on any substrate, many bacteria have developed methods for rectification of thermal noise to power their cilia used for swimming in water, and so on [1–3]. The directional effect is generated by a variety of symmetry-breaking mechanisms. For example, slender-bodied animals such as earthworms and snails make use of elongation-contraction waves that propagate along their body axis for their locomotion. Both the locomotion speed and direction can be controlled by these animals by asymmetric friction with the substrate at different parts of their body [4]. Similarly inside, unicellular animals like amoeba locomote via direction-specific conversion of specific proteins in cytosol from more gelatinous to liquid-like state [5].

Generation of directed motion in micro-mechanical systems is however a difficult task to accomplish with currently available micro-fabrication techniques, because conversion of energy into motion becomes increasingly difficult with reduction in the dimension of the systems. As an example, electromagnetic [6, 7] or pneumatic signals [8] which are often used to create motion in micro-systems are restricted because of their essential need of wires, onboard energy supply arrangements, etc. Therefore, efforts have been made to develop processes and systems which help direct conversion of chemical energy to mechanical energy without involving any transitional steps and motion-bearing parts. More than 100 years ago, J. H. van't Hoff showed theoretically that chemical energy can be converted in mechanical work using osmotic cells equipped with pistons. van't Hoff's theoretical osmotic engine is however very different from living organisms which produces the motility through the appropriate accumulation of protein molecules [9]. More than 50 years ago, Katchalsky et al. [10, 11] showed that when collagen fibers are cyclically immersed in salt solution and water, their morphology alters reversibly from cyclic helices to random coils resulting in change in dimension. Katchalsky et al. referred to this as a "mechanochemical system," defined as a thermodynamic system which can convert chemical energy into mechanical work. Since then, many approaches to create chemomechanical motion are currently being investigated, several of them being specifically based on soft material. These materials are capable of deforming their shape by external stimuli such as electric field [9, 12], temperature [13, 14], photo [15, 16], pH [17], and solvent [18].

1.1 Various Mechanisms of Creating Asymmetry and Locomotion

There have been innumerable attempts to generate asymmetry in soft objects leading to net forward motion. For example, when light is illuminated asymmetrically on porous polymer microstructures of NVOC-Gly TRIM/HEMA, the resultant photochemical alterations in surface chemical processes lead to substantial asymmetric change in volume, which causes motion of the structure [19]. Light also induces shape variation of gel which can result in wormlike motion of poly (NIPA-Ru complex) gel as shown in Fig. 1a [15]. Chemomechanical actuators have also been designed using ionizable polymeric gel placed within an electrolyte solution. When an electrical potential is applied across two electrodes, surfactant molecules bind selectively to one side of the gel, resulting in its contraction along its axis; reversing the electric field leads to contraction of the opposite side of the gel [9]. Thus, alternating electric field causes alternate expansion and contraction which is rectified leading to forward motion, which is shown in Fig. 1b.

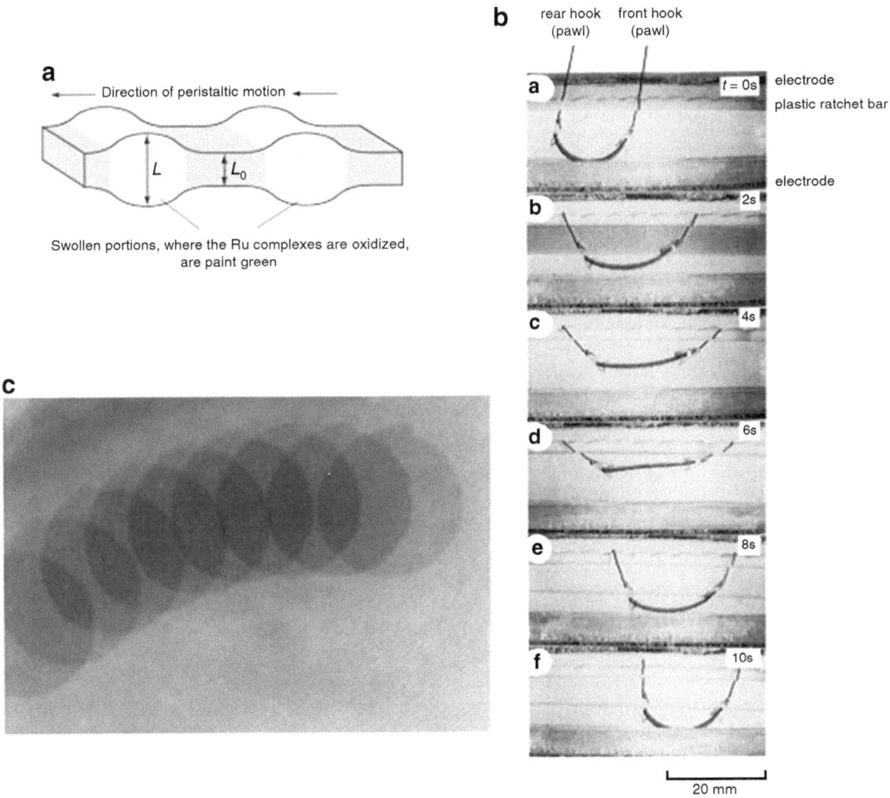

Fig. 1 (**a**) Model of peristaltic motion of the surface of poly (NIPA-Ru complex) gel: represents spatiotemporal color patterns of oscillating behavior for rectangular bulk poly (NIPA-Ru complex) gel can be observed because of the difference in color between the reduced state and the oxidized state of Ru complex [15]. (**b**) A series of images that represents the motion of polymer gel is induced by external electric field [9]. (**c**) Motion of PSA gels: translational motion of a disk-shaped gel is induced by spreading of organic solvent on water [18] (Reprinted from Refs. [9, 15, 18] with permission)

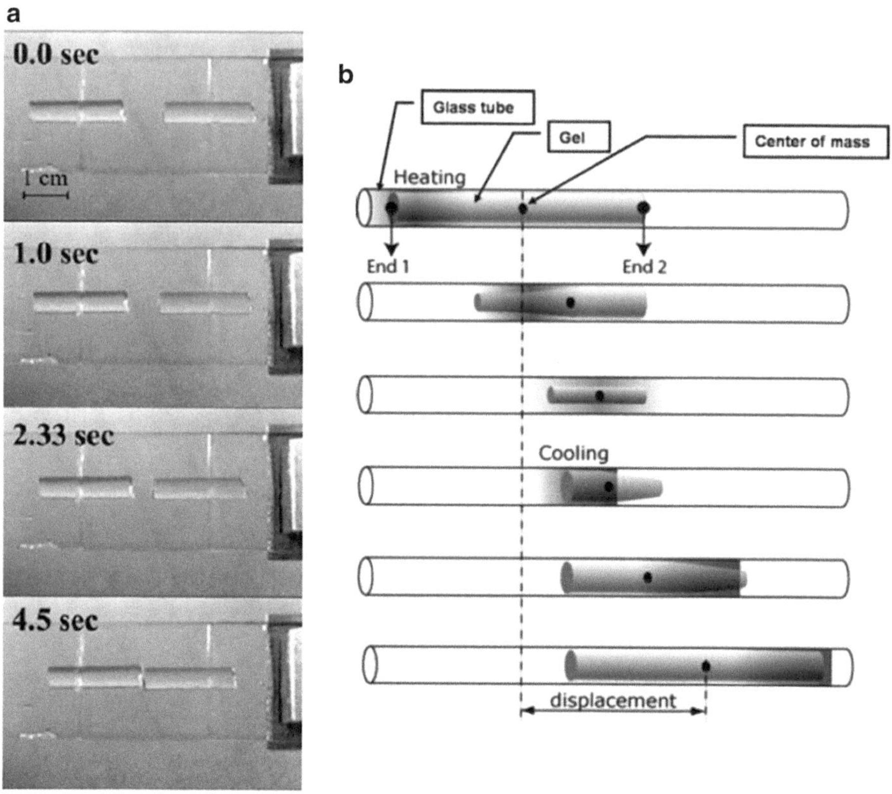

Fig. 2 A series of images represents (**a**) motion of polyacrylamide hydrogel cylinder on a thin elastomeric film of polydimethysiloxane, which is subject to vibration. Each cylinder is placed on a single slanted incision with opposite orientation so that the motion of the gel gets rectified because of preferential lubrication in one direction [20]. (**b**) Directional motion of gel is induced by temperature [21] (Reprinted from Refs. [20, 21] with permission)

Similar rectified motion of hydrogel cylinder on a soft substrate is also generated by vibrating the substrate and by using asymmetric interfacial interaction between the gel and the substrate, which is shown in Fig. 2a [20]. Vibration has been used also to induce bidirectional locomotion of the cylindrical body on top of a lubricated platform using ratchet-shaped surfaces. The cylindrical body consists of two PDMS rods with their ratchet-shaped legs having different densities and facing opposite directions [22]. In thermosensitive gels, direction-specific motion ensues in response to infinitesimal change in temperature of the environment triggering spatially varying propagation of volume phase transition along their length, which is shown in Fig. 2b [21]. Similar phase transition has been achieved by application of an electric field across polyelectrolyte gel [23]. Osmotic pressure has also been used to generate locomotion. For example, when a polymeric gel swollen by an organic solvent is placed on the surface of water, it instantaneously forms a thin skin across

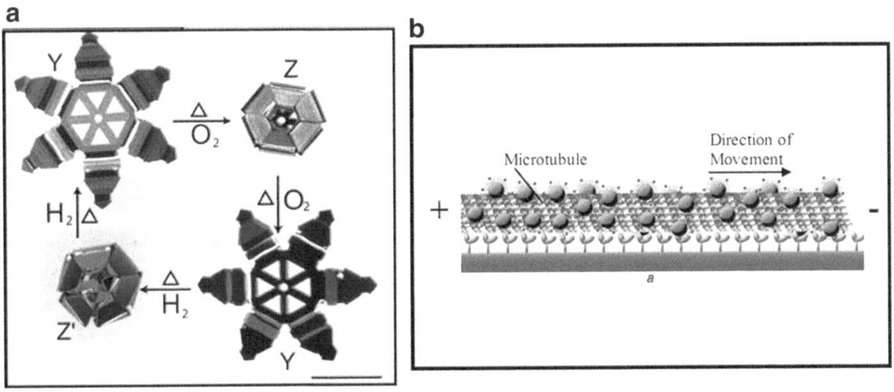

Fig. 3 (**a**) Images represent reversible actuation of microstructure with chromium/copper hinges based on stress reversal obtained in oxidative or reductive environment [24]. (**b**) The scheme of a gliding assay in which a microtubule with attached magnetic nanoparticles is transported by a kinesin motor functionalized glass surface [25] (Reprinted from Refs. [24, 25] with permission)

which an osmotic pressure develops. Large difference in pressure between inside and outside of the gel ejects the organic solvent out of the gel which imparts reactive thrust on it making it to move forward (Fig. 1c) [18].

Asymmetry can also be generated using chemical reaction, which induces reversible curling of metallic bilayers when exposed to oxidative or reductive environment. These metallic bilayers can be easily deposited and patterned on microstructures such as bidirectional microgrippers and container. As a result, it is used as hinge to reversibly open and close microstructures which is shown in Fig. 3a [24]. In addition, there are devices also which are more akin to biological systems, for example, it has been shown that direction-specific motion can be induced with the help of adenosine triphosphate (ATP)-powered biomolecules like motor proteins. Microtubules, i.e., protein filaments, undergo directed motion when the kinesin molecules, after acquiring energy from the hydrolysis of ATP molecules, move along these microtubules in discrete steps of 8 nm [26, 27]. Functionalization of these microtubules with the help of magnetic nanoparticles and the application of external magnetic fields can also cause directed motion which is shown in Fig. 3b [25]. These mechanisms for directional motion and transport in variety of microfluidics, microelectromechanical system (MEMS), and drug delivery systems can bring about new advancements leading to variety of smart applications.

While the above modes of locomotion all occur on a horizontal surface, locomotion on a vertical wall requires also strong adhesion on a vertical wall, rapid release, direction-specific adhesion, and reusability. Rapid locomotion on a smooth vertical wall requires directionality along with strong adhesion and reusability. Directionality in adhesion implies that the adhesive strength depends on the direction in which the detachment load is applied. In a particular direction, adhesion strength may be large enough to strongly resist separation, whereas in some other direction, it may be quite low resulting in easy separation or pull off. Development of engineering material with such reversible adhesion capabilities is motivated by many potential applications such as robotic manipulation, transport of delicate cargo, and fabrication of biomedical devices with specific features [28–31]. For example, climbing robots are required to operate in complex and harsh environmental conditions, such as vertical wall, ceiling, outer surface of space shuttle, and volcanoes. One of the most challenging tasks in climbing robot design is to develop a proper adhesion mechanism to ensure that the robot sticks to the wall surface reliably without sacrificing mobility. Previously vacuum suction technique [32], magnetic devices [33], and aerodynamic principle [34] were often used to adhere to a variety of smooth, flat, vertical surfaces, but they are not applicable for nonsmooth surface. In essence, the man-made wall climbers have fallen far short of their natural counterpart, e.g., the geckos, crickets, ants, and beetles which have strong climbing ability irrespective of the roughness of the contact surface.

The basic principle of adhesion of biological system has been the subject of many recent studies. It is now known that animals do not use one single adhesion mechanism but a gamut of physical principles using which they can stand or walk on smooth a vertical surface, even upside down, in dry condition or under water as the environment demands. Dry adhesion, involving micro- to nanoscale filamentous structures, is used by many reptiles, and adhesion force mainly engenders from van der Waals force (Fig. 4a) [35]. van der Waals force is the sum of attractive force between two atoms or molecules or surfaces other than those due to covalent bond, hydrogen bond, or electrostatic interaction between ions, neutral molecules, or charged molecules. Similarly, wet adhesion, involving thin liquid film between two surfaces, is also used by many animals like arthropods, vertebrates, and Mollusca, and the adhesion force essentially engenders from surface tension or capillary force

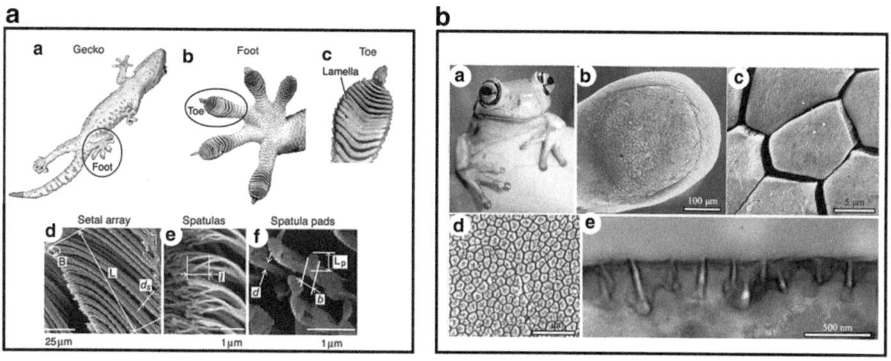

Fig. 4 (**a**) Images represent the hierarchical structures of gecko foot: (*a–f*) represent structures in the order of decreasing size [35] (Reprinted from Ref. [35] with permission). (**b**) Images represent morphology of tree frog toe pads. (*a*) White tree frog. SEM images of (*b*) toe pad, (*c*) epidermis with hexagonal epithelial cells, (*d*) high-magnification image of surface of a single hexagonal cell showing peg-like projections, and (*e*) transmission electron microscope image of cross section through cell surface (Reproduced by permission of the original source of publication and Taylor & Francis Ltd [36])

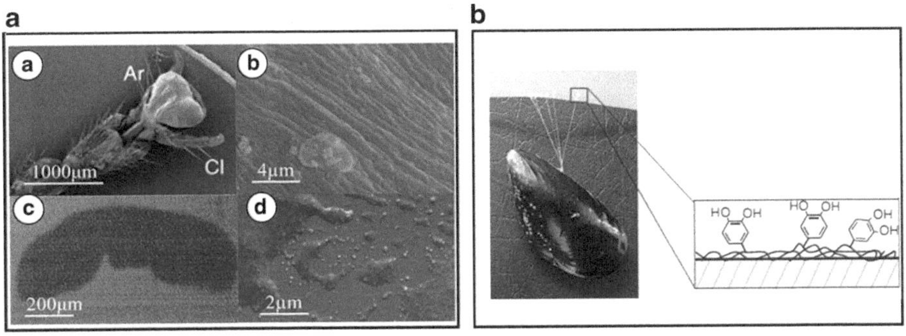

Fig. 5 (**a**) Images represent smooth pad and secretion of insects. (*a*) Smooth pad of stick insect. Arolium (*Ar*), claws (*Cl*). (*b*) SEM image of the microstructure of the pad. (*c*) SEM image of a footprint of stick insect [28]. Reproduced by permission of the original source of publication [67, 28] (*d*) Secretion. (**b**) Images represent flexible muscular foot and structure of adhesive protein of mussel [39] (Reprinted from Ref. [39] with permission)

generated by fluid-filled joint or fluidic interface between the pad and substrate [28, 37, 38]. For example, the tree frog can stand or walk on smooth vertical surface, even upside down under the surface, due to their wet adhesion pads. The toe pads of the tree frog are covered with regular hexagonal microstructures of epidermal cells separated by a deep channel into which mucus glands open (Fig.4b); the secretion from these glands augments the adhesive force at the interface which allows the animal to stand or walk on smooth vertical surface [37]. Arthropods like insects use tarsal attachment pads that adhere via small volume of fluid secreted in the contact zone that aid in wet adhesion (Fig. 5a) [28, 38, 40]. Molluscs like snails and mussels adhere to a surface using its flexible muscular foot and adhesive mucus. The mucus is composed of water, mucopolysaccharides, and glycol proteins (Fig. 5b). It forms a strong yet reversible bond with the surface [39]. Recently, biomimetic adhesive pad with good adhesion and direction dependence has been developed based on mechanical micro-fabrication, etching, or carbon nanotube growth technologies [29, 40–42].

In what follows we will summarize some of the recent works done on asymmetry-driven locomotion and direction-specific adhesion of soft-bodied objects made of cross-linked elastomer.

1.2 Method

Cylindrical rods of PDMS were prepared by following the procedure described in Ref. [43]. Briefly, PDMS precursor and the curing agent mixed at 10:1 w/w were cross-linked inside the steel shells at 80 °C for 1 h. After curing, the PDMS cylinders were pulled gently out and were cut into pieces of length $l = 10$ mm using surgical blade.

PDMS films embedded with pairs of channels were prepared by assisted method described in Ref. [44]. Briefly, the channels of circular cross section were prepared using cylindrical shells as templates, which were located in a pool of PDMS monomer mixed with curing agent (10:1 by weight). Here, also prepolymer liquid was cross-linked between two glass plates by two-step heating. The templates were withdrawn from the film after cross-linking at 60 °C for 1-h. Finally, the film was cross-linked at 80 °C for 1 h.

2 Rolling Motion of an Elastomeric Cylinder

While the mechanisms described above all lead to rectilinear motion, in this section we show autonomous "rolling" of a soft elastomeric cylinder powered by a solvent [43]. Figure 6a shows the experiment in which a cylinder made of cross-linked elastomer ($\mu \sim 1$ MPa) remains weakly adhered to a rigid substrate while a small quantity of a solvent is released at one side of it. The quantity of solvent is small enough that it does not entirely submerge the cylinder, but only partially, the portion where it remains in contact with the solvent. Solvent evaporates from the portion of cylinder exposed to the atmosphere where it shrinks. This asymmetric swelling and shrinking leads to bending of the cylinder with its convex side in contact with

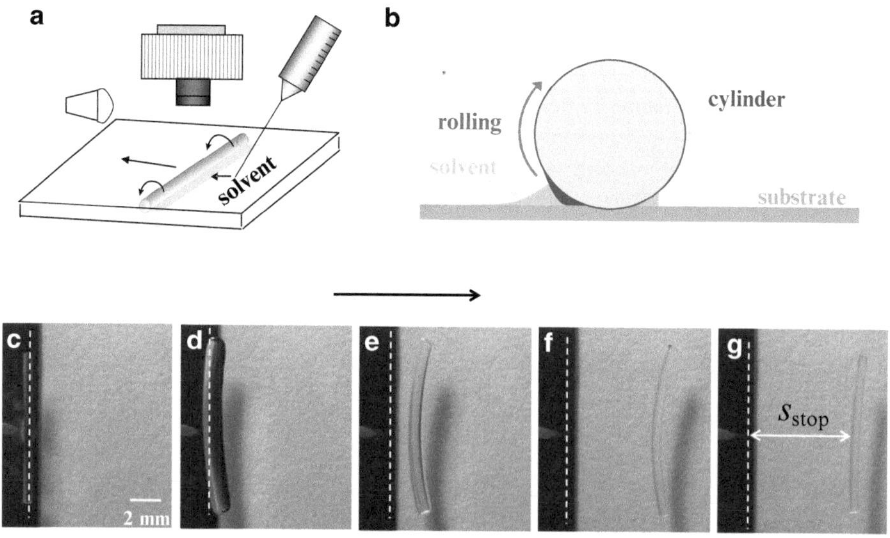

Fig. 6 (**a**) Represent schematic of the experimental setup. Here a micro cylinder made of PDMS kept on a glass plate and solvent was applied using a micropipette. (**b**) The magnified side view of the cylinder. (**c**–**g**) Video micrographs depict a typical sequence of rolling of a cylinder of diameter $d = 520$ µm and length $l = 10$ mm following release of 2 µl of chloroform (Reproduced by permission of the Royal Society of Chemistry [43])

the solvent, with the extent of bending depending on the extent of swelling of the solvent-rich side of the cylinder with respect to the solvent-indigent side.

While bending results in forward motion of the center of mass of the cylinder, the rest of the cylinder tends to catch up with it and the cylinder starts to roll. Therefore, evaporation of the solvent is necessary for maintaining the asymmetric swelling and consequently for the rolling motion to occur. Experiments show that the cylinder does not roll, if the solvent does not evaporate from its surface. Solvents like chloroform, n-hexane, *n*-heptane, toluene, and triethylamine can all drive the cylinder to roll. Boiling point (T), vapor pressure (p), surface tension (γ), density (ρ), and solubility parameters (δ) are the physical properties of the solvent that determine the rolling velocity.

In Table 1 we present a list of solvents and the value of their different physical properties. Boiling point and vapor pressure are important as they define the dynamic balance of quantity of solvent that swells the network and that evaporates from the surface of it; higher rate of evaporation of the solvent results in larger asymmetry in swelling. Surface tension is important as it determines wetting of the cylinder surface and rate of wetting by the solvent and thereby the extent of swelling. Density of solvent defines the quantity that goes to swell the elastomer and evaporates out of it. The solubility parameter is a measure of the intermolecular interactions between the solvent and the cross-linked network. The solvent swells the solid most when its solubility parameter δ matches with that of the solid. For example, PDMS swells most in solvents with δ similar to that of PDMS: $\delta_{PDMS} = 7.3$ [45]. The experiment can be carried out at ambient or elevated temperature resulting in varying velocity of rolling.

Another important condition for rolling to occur is the interfacial adhesion between the cylinder and the plate. A cylinder with smooth outer surface adheres

Table 1 Table shows different solvents used in the experiment and their physical properties

Solvent	Boiling point (°C)	Vapor pressure at 27 °C (kPa)	Surface tension (mN/m)	Solubility parameter ($cal^{1/2}\ cm^{-3/2}$)	Swelling ratio	Density (gm/cm^3)
Acetone	56.53	33.33	25.20	9.9	1.06	0.785
Chloroform	60	26.14	27.5	9.2	1.39	1.489
n-hexane	69	20.25	18.4	7.3	1.35	0.655
Ethanol	78.4	8.75	22.39	12.7	1.04	0.789
Cyclohexane	80.7	13.17	24.9	8.2	1.33	0.779
TEA	90	9.13	23.4	7.5	1.58	0.730
n-heptane	98	6.07	20.1	7.4	1.34	0.680
Water	100	3.51	72.86	23.4	1.00	1.000
Toluene	111	3.8	28.4	8.9	1.31	0.862

Reproduced by permission of the Royal Society of Chemistry [43]

Here vapor pressure is calculated using Antoine equation. Antoine constant data of liquid are taken from Ref. [68]. Solubility parameter and swelling ratio data are taken from Ref. [45] of the manuscript. Surface tension data of liquid are obtained from Ref. [46]. The density data are taken from Ref. [47]

strongly to the substrate forming a large, continuous area of contact as shown in Fig. 7b. In order to smoothen the rough surface of the cylinder, it is first coated with a thin layer of PDMS mixed with the cross-linking agent and then cross-linked as done previously. When such a cylinder is gently placed on the substrate, it forms a contact of constant width which does not debond at this interface nor does the rolling motion sets in. A rough surface of the cylinder forms only partial contact which easily releases from the substrate. Therefore, rough surface is necessary for the cylinder to roll on the substrate. On the other extreme, if the cylinder does not adhere at all on the substrate, it rotates with slippage at one place without rolling forward.

Figure 8a shows a schematic of the experiment in which the cylinder is placed within the confined gap l between two parallel plates. A PDMS cylinder of diameter $d = 870$ μm is driven to roll using n-heptane as the solvent. When the gap is large, the rolling velocity of the cylinder coincides with that when the cylinder placed in open atmosphere, i.e., 1.4 mm/s. The rolling velocity decreases as the gap is diminished, finally the velocity decreasing to zero at a threshold gap between plates (Fig. 8b).

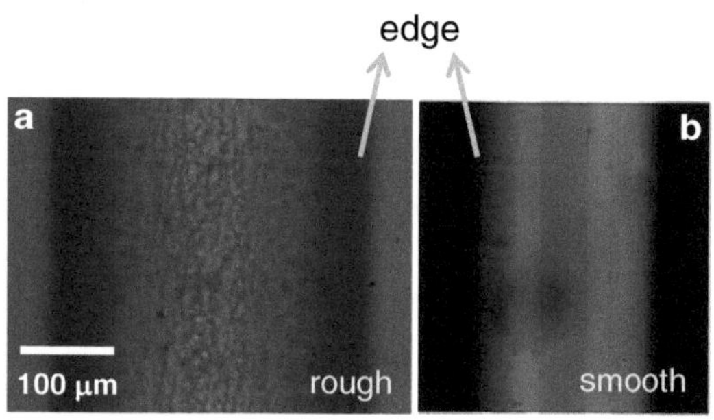

Fig. 7 Optical micrographs represent contact area of cylinder. (**a**) Cylinder with rough outer surface does not form continuous area of contact. (**b**) Cylinder with smooth surface forms a rectangular area of contact. (Reproduced by permission of the Royal Society of Chemistry [43])

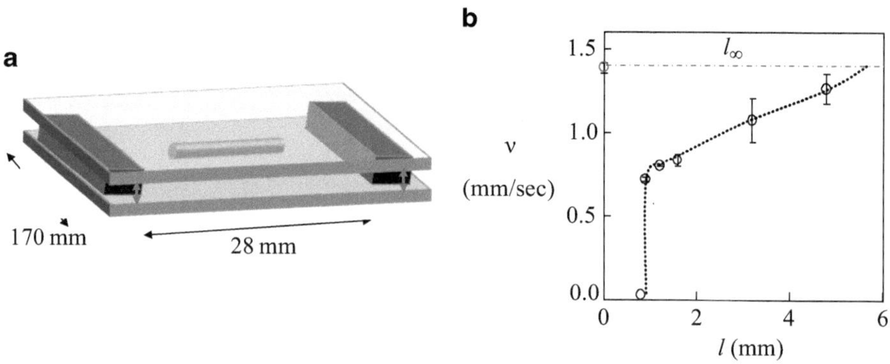

Fig. 8 (**a**) Schematic diagram of the experiment in which a cylinder was induced to roll in the confined spaced between two parallel glass plates. The confined space is used to control solvent evaporation rate from cylinder surface. (**b**) The rolling velocity v of cylinder of diameter 870 μm is plotted against distance l between two glass plates for solvent n-heptane

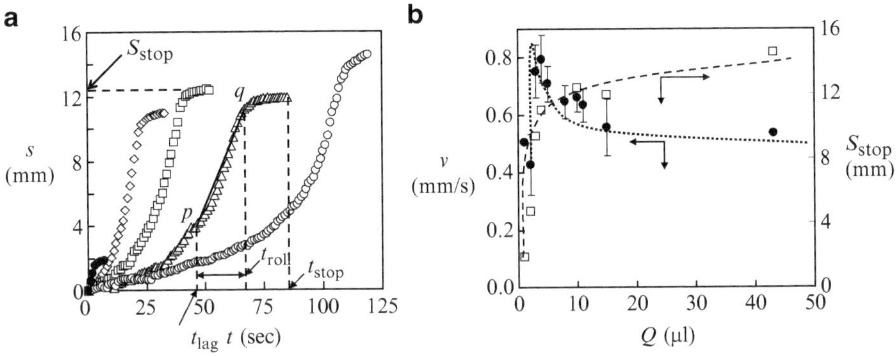

Fig. 9 (**a**) A PDMS cylinder of diameter $d = 550$ µm is made to roll by addition of chloroform as the solvent. The distance travelled is plotted against time for addition of different quantities of solvent. The slope of the data defines the rolling velocity which increases with progress in rolling, eventually attaining a constant value. The symbols ◇, □, Δ, and ○ represent drop sizes 3, 4, 10, and 15 μl, respectively. (**b**) Typical data of constant rolling velocity v and the final distance travelled S_{stop} are plotted against the quantity Q of the solvent (Reproduced by permission of the Royal Society of Chemistry [43])

Quantity of solvent is another parameter which controls the degree of asymmetry in swelling. Figure 9a shows the distance to which a cylinder of typical diameter rolls with respect to time for different quantity of solvent used. A cylinder of diameter 550 µm is made to roll for 3, 4, 10, and 15 µl chloroform. For large quantity of solvent used in experiment, for example, asymmetry in swelling of the cylinder diminishes; therefore, the cylinder does not begin to roll. The rolling motion sets in only when some of the liquid gets evaporated. The rolling velocity increases till a constant velocity is reached. Since the solvent remains affine to the cross-linked network, the pool of solvent remains adhered to the cylinder as it rolls and continues to supply the solvent required for maintaining the asymmetric swelling. The cylinder continues to roll till all the solvent is consumed. This observation is important as it brings out a significant point in regard to the rolling mechanism. The inertia of the liquid does not drive the rolling motion; on the contrary, the liquid pool exerts backward drag on the motion of the cylinder.

The data presented in Fig. 10 shows how rolling velocity varies with cylinder diameter for different solvents demonstrating that multiple properties become important. For example, chloroform and toluene having similar solubility and surface tension but different boiling point or vapor pressure (Table 1) result in different rolling velocities. Although both the solvents swell the cylinder to almost similar extent, higher vapor pressure of chloroform than toluene leads to higher rate of evaporation and faster deswelling. As a result, the asymmetry in swelling becomes more pronounced for chloroform than toluene, resulting in larger rolling velocity v of the cylinder. Difference in surface tension and solubility too affects v. For example, n-hexane has lower surface tension and higher solubility parameter than chloroform so that it spreads faster onto the surface of the cylinder. As a result,

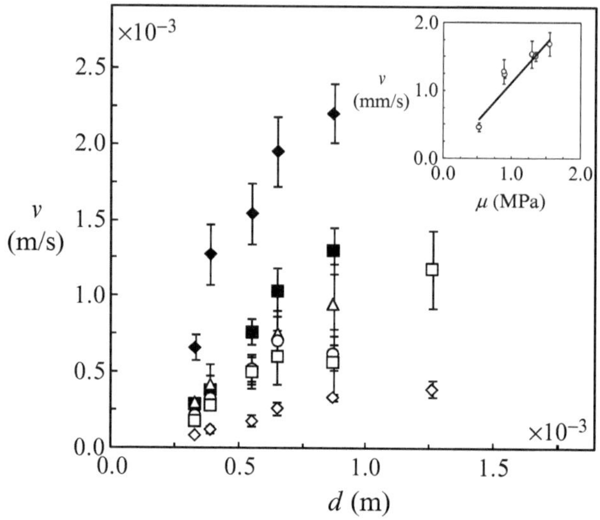

Fig. 10 The rolling velocity is plotted against diameter of cylinder for 4 μl of various solvents. The symbols ♦, ■, △, ○, ◇, and □ represent the following solvents: n-hexane, chloroform, TEA, cyclohexane, toluene, and n-heptane, respectively. The inset depicts the velocity of cylinders (d = 550 μm, l = 10 mm) plotted with respect to their shear modulus. The cylinders are driven by 4 μl of hexane. The error bars represent standard deviation of data obtained from several experiments (Reproduced by permission of the Royal Society of Chemistry [43])

despite having similar boiling temperature, n-hexane spreads over larger surface area and evaporates at a rate faster than that of chloroform.

Here too asymmetric effect becomes more pronounced for n-hexane, resulting in higher rolling velocity. It is then natural to expect that solvents with similar γ, p, and δ should bring about similar rolling velocity of cylinders, which is found with n-heptane and triethylamine, both result in similar rolling velocity of the cylinder. Effect of elastic modulus of cylinder on rolling velocity was also studied. Elastic modulus of the cylinder was varied by using different weight fractions of the cross-linker: 5–20 % by weight of the prepolymer liquid. As a result, shear modulus μ of the material varies from 1.5 to 5.0 MPa. Inset in Fig. 10 shows that with hexane, rolling velocity v of the cylinder increases linearly with μ.

2.1 Scaling of the Rolling Velocity

An expression of rolling velocity can be derived by equating the interfacial energy released due to wetting of the solid surface by the liquid to the strain energy associated with swelling-shrinking of the cylinder. Assuming that the cylinder swells uniformly along its length, the elastic strain energy of swelling of the cylinder over one complete cycle of rotation is written as $\sim\frac{\pi d^2 L}{4}\mu(d'/d-1)^2$, where d' is the effective diameter of an equivalent uniformly swelled cylinder and μ is its

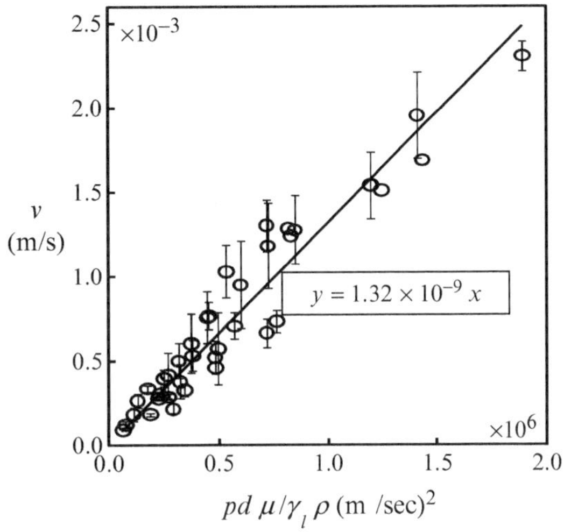

Fig. 11 Plot of linear velocity v of cylinders against $pd\mu/\gamma\rho$ shows that data from all solvents can be fitted to a master curve represented by the *solid line* (Reproduced by permission of the Royal Society of Chemistry [43])

shear modulus. The surface energy released because of wetting of the cylinder over time τ through which the cylinder undergoes one complete rotation is expressed as $(v\tau)L\gamma_l$, where γ_l is the surface tension of the liquid and v is the rolling velocity. Balancing these two expressions yields an expression for the velocity of rolling, $v \sim \frac{\pi d^2 \mu}{4\tau\gamma_l}(d'/d - 1)^2$. An expression for τ can be estimated by equating the quantity of solvent lost from the swollen cylinder to that evaporated over one complete cycle of rotation from its surface. The former is estimated by multiplying the change in volume of the cylinder with the density of solvent $\sim\rho\frac{\pi d^2 L}{4}\left((d'/d)^2 - 1\right)$. The latter is obtained by considering the rate of mass transfer per unit time and per unit differential pressure given as the difference in vapor pressure of the solvent $p(T)$ to its partial pressure p_0 at the surroundings: $(p(T) - p_0)$, where T is the surrounding temperature. Over time τ, this quantity is estimated as $\kappa\pi dL\,(p - p_0)\,\tau$, in which κ is the mass transfer coefficient. Noting that partial pressure p_0 is negligibly small and that swelling increases the cylinder diameter only infinitesimally, the expression for τ is obtained as $\tau \sim \frac{\rho d}{4\kappa p}$. Substituting this expression for τ into that of v, we obtain the following simple scaling relation: $v = \kappa\frac{dp\mu}{\gamma\rho}$. The plot in Fig. 11 shows that the data of linear velocity v for different solvents and cylinders indeed follow the above scaling: $v = 1.32 \times 10^{-9}\frac{pd\mu}{\gamma\rho}$ m/s.

2.2 Rolling up an Inclined Plane

Since the rolling motion of the cylinder is essentially driven by asymmetric swelling, it is natural to think that the degree of asymmetry can be increased by placing it on a plane inclined at an angle with the horizontal surface so that solvent accumulates

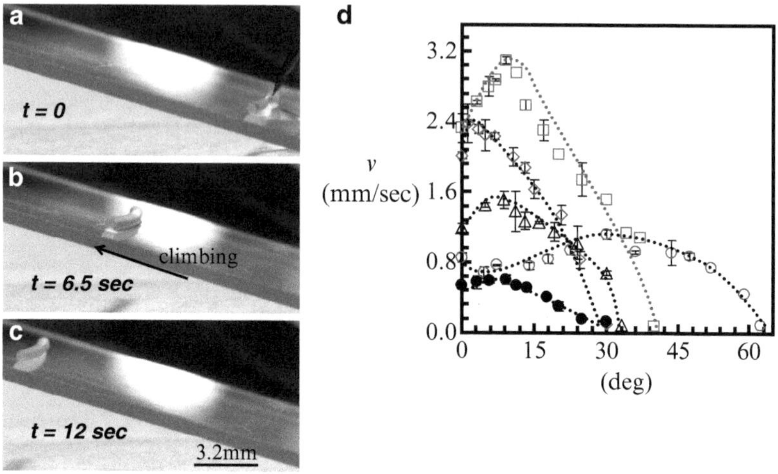

Fig. 12 Video micrographs (**a–c**) show a chloroform-driven PDMS cylinder ($d = 870$ μm) rolling up a plane, inclined at an angle 16° to the horizon. (**d**) The velocity v of cylinder of $d = 870$ μm is plotted against angle of inclination θ of a substrate. Symbols □, ◊, Δ, ∘, and • represent the following solvents: hexane, chloroform, TEA, *n*-heptane, and toluene, respectively (Reproduced by permission of the Royal Society of Chemistry [43])

more in the rear side, resulting in larger differential swelling at the rear side of the cylinder than that the front side. Indeed the cylinder can roll up a stiff inclined plane on application of the solvent as depicted by the sequence of video micrographs A–C in Fig. 12. Here chloroform is the solvent that drives a cylinder of diameter $d = 870$ μm up an inclined plane at angle $\theta = 16°$. In Fig. 12d, v is plotted θ for five different solvents: chloroform, TEA, *n*-heptane, *n*-hexane, and TEA which show that the rolling velocity v increases with the inclination angle θ till a critical angle is reached beyond which it decreases. The cylinder and the accompanying solvent are subjected to two opposing forces: gravitational pull drags both the solid and the liquid down the inclined plate, whereas asymmetric swelling-induced torque pulls the cylinder up the plane. At a small value of θ, the liquid accumulates more in the rear side, so that the difference in meniscus height of the solvent at the two sides of the cylinder increases more compared to that on the horizontal plane. The enhanced effect of asymmetry following addition of both these effects increases the rolling velocity more than what is achieved on a horizontal plane. The rolling velocity continues to increase with θ till a threshold θ_{max} is reached beyond which the gravitational pull dominates, thereby exerting an arresting effect on the cylinder which slows down. Eventually, the cylinder stops rolling. This novel mechanism of increasing the velocity while moving against gravity distinguishes it from other known physical mechanisms that can drive limbless locomotion. Different solvents (Fig. 12d) increase the rolling velocity and the range of θ through which this mechanism becomes effective to different extents. For example, the maximum achieved rolling velocity remains larger for chloroform than for TEA and *n*-heptane, yet v decreases to zero within $0^{\circ}<\theta<30^{\circ}$ for chloroform with $\rho = 1.39$ gm/cc, but for n-heptane with $\rho = 1.34$ gm/cc, the cylinder continues to roll for θ as high as 55°. This result suggests that velocity decreases more when solvent density increases.

3 Direction-Specific Adhesion of an Adhesive Pad

We will now turn our attention to another aspect associated with locomotion, e.g., directional adhesion which is important because locomotion in many situations requires strong adhesion in forward direction but easy release from the back. Attaining both the demands using the same adhesive layer has not been easy as described earlier. Conventionally, adhesives have been prepared using viscoelastic materials in the form of thin layers which can follow the roughness of an adherent when subjected to external normal or shear loads and come in intimate contact with it. While van der Waals force of attraction causes adhesion between the adhesive and the adherent surface, viscous dissipation results in amplification of this adhesion strength. Viscoelasticity also renders the adhesive susceptible to easy fouling by particulate matters which prevents repeated application of the adhesive. However, locomotion, artificial or natural, requires repeated attachment and detachment of the adhesive and in a specific direction. Diminishing the viscous component of the material, i.e., by turning the adhesive more elastic, overcomes this problem of reusability but decreases the adhesion strength quite significantly. Application of elastic adhesive surfaces patterned with hairs, pillars, and islands has proved to be a possible alternative. Several studies on model-textured adhesives have shown that surface patterns suitably designed can enhance adhesion strength remarkably [48–50]. This is because propagation of a crack at the interface of the adherents is arrested when it meets a surface discontinuity and has to be reinitiated again. The maneuverability of these micro-nanoscopic patterns however depends strongly on the hierarchical structure of the adhesive, the aspect ratio, and thickness, stiffness, and structure of the patterns, all of which appear to be perfectly optimized for foot-hairs at the feet of gecko which allow them not only to adhere to any surface but also to achieve both strong adhesion, easy release, resistance against contamination, and many other functions. Several attempts have been made to mimic these biological adhesives beginning from adhesive surfaces decorated with pillars [30, 51, 52], ones with circular cross section [53], but also ones fitted with directional flaps [54], hierarchical pillars [55, 56], to adhesive layers patterned with discontinuities [57, 58], finally the ones grafted with vertically aligned multiwall carbon nanotubes. Dhinojwala et al. showed a dry adhesive that was based on construction on polymer surfaces with strong nanometer-level adhesion [42]. J. Tuma et al. showed adhesive microstructure that was based on polymer fibers which amplify adhesion of flat contact [59]. These adhesives have the potential to be self-cleaning owing to the reason that they employ stiff, hydrophobic materials. Careful surface preparation and high preloads are a must to obtain the required levels of adhesion in most of these cases. Cutkosky et al. showed a different approach using structured arrays of moderately soft elastomeric material, termed directional polymer stalks, which are designed to exhibit adhesion only in that particular direction where load is applied [31]. Lee et al. showed that arrays of angled microfibers, with structure similar to that of gecko structure, which were fabricated from a stiff thermoplastic polymer (polypropylene) with elastomeric properties similar to those of β-keratin of natural setae, exhibit directional adhesion [60]. In addition to chemical modification and texturing of the surface of an adhesive layer, it has been shown also that subsurface microstructures too can enhance several characteristics of an adhesive. For example, air and viscous domains buried within the bulk of the adhesive can have strong dissipative effect on the work of adhesion while offering also a clean reversible separation [61–63]. For a soft solid, filling in subsurface channels with a suitable liquid can add additional features via alteration of the interfacial tension. Importantly change in surface tension can be strong enough that it can cause elastic deformation of the wall of the channel leading to change in topography of the layer [64]. Calculations show that adhesive layers, decorated with small amplitude perturbation in surface topography and oscillatory shear modulus, are expected to enhance adhesion to a flexible adherent via crack pinning effect. Besides alteration in solid-liquid interfacial tension for liquid-filled channels, change in solid-solid interfacial tension in the way of self-adhesion of subsurface microstructures too can influence adhesion [65]. It has been shown that for hierarchically structured adhesives embedded with multilayer channels, adhesion enhances significantly via hysteresis associated with self-adhesion of walls of microchannels and can cause also direction-dependent adhesion. In fact small modifications of the arrangement of subsurface microchannels lead to different adhesive strengths when peeled along two different/opposing directions [44]. In what follows we will describe some of these results to some detail.

Here we describe another mechanism for creating directional adhesion in which the surface of the adhesive remains smooth, but directional adhesion is generated by embedding fine microchannels in pairs with desired intra-pair and inter-pair lateral spacing [62–64]. Figure 13a shows the schematic of this adhesive: a thin PSMS film, $h = 0.55 - 1.05$ mm, embedded with cylindrical channels, $d = 0.5 - 1.0$ mm such that the channels remain at different vertical heights (skin thickness, t μm). The channels incorporate heterogeneity in two different ways: the effective shear modulus of the adhesive gets altered from being completely uniform for a featureless adhesive to a value which remains minimum where the thickness of the channel wall remains smallest. The spatial variation in modulus can be altered as desired by embedding a suitable network of channels. Furthermore, when these channels are filled with a liquid, the effective deformability of the film and the rate of deformation both alter depending on the viscosity of the liquid and geometric characteristics of the channel. The liquid inside the channels has another effect; as it wets the inner surface of the channel, interfacial energy gets released which causes the thin skin at the vicinity of the channel to bulge out. As a result, the topography of the adhesive no longer remains uniform but undulating with periodicity depending on the distance between the embedded channels. Excess

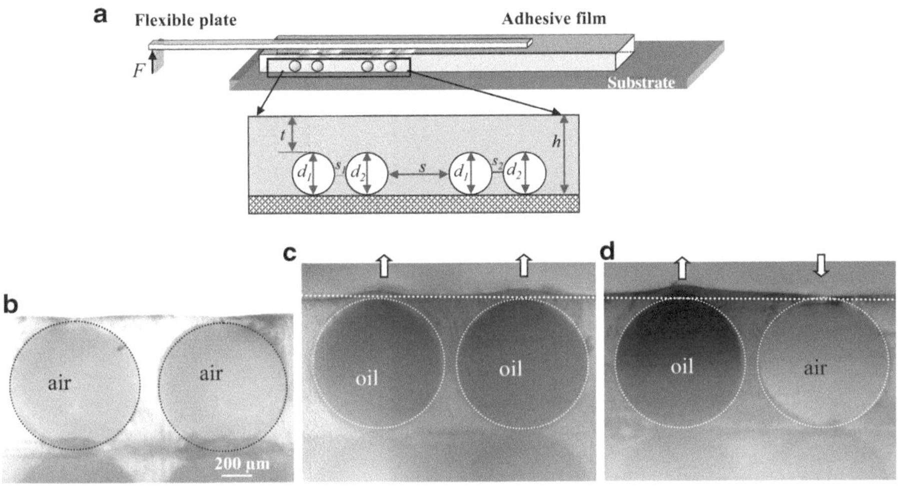

Fig. 13 (**a**) Schematic of the experiment shows a model adhesive in which PDMS layer bonded to a rigid substrate and embedded with pairs of microchannels of same or different diameter. A flexible adherent is lifted off at a constant rate ~5 μm/s, which is in contact with the adhesive surface [62]. A load cell attached to the substrate and interfaced with a computer is used to measure the lifting load. (**b–d**) Optical micrographs show the side view of a pair of channels ($d = 550$ μm), embedded inside the adhesive with skin thickness $t = 20$ μm and inter channel spacing $s = 70$ μm. The images (**b–d**) represent both channels filled with air, both filled with oil, and either one of the channels filled with oil, respectively. The bulge of the thin skin above the channel is indicated by *arrow* (Reproduced by permission of the Royal Society of Chemistry [44])

energy of $\Delta\gamma = (\gamma_{sl} - \gamma_s - \gamma_l)$ per unit wetted area is released, which causes a bulging height of $\delta \approx d\sqrt{2\pi\Delta\gamma/\mu\,(h - d/2)}$. In addition, the channel cross section alters from circular to oval shaped [64, 66]. Figure 13c, d shows optical micrographs of a typical bulging profile of the adhesive surface at the vicinity of a single channel, which changes to a double hump when a pair of channels is used in close vicinity. In order to capture the side view, a cross-linkable liquid was used which also spontaneously filled in the channel, e.g., Sylgard 170 which was then cured at 65 °C to permanently fix the deformed shape of the channel and the adhesive. Nevertheless, the bulged out profile is found to remain symmetric for all these cases. It is expected that variation in modulus too remains symmetric for all these cases with periodicity equal to that of the channel spacing. The coupled effect of variation in effective shear modulus and bulging profile of the adhesive surface introduces combined material and geometric heterogeneity which has pronounced effect on adhesion and several other properties. Heterogeneity remains symmetric when uniformly spaced channels are embedded inside the adhesive and are all filled with the same fluid, however turns asymmetric when spatial distance between channels remains nonuniform, and in addition, they are filled with different fluids, the degree of asymmetry being spatially tunable as desired (Fig. 14).

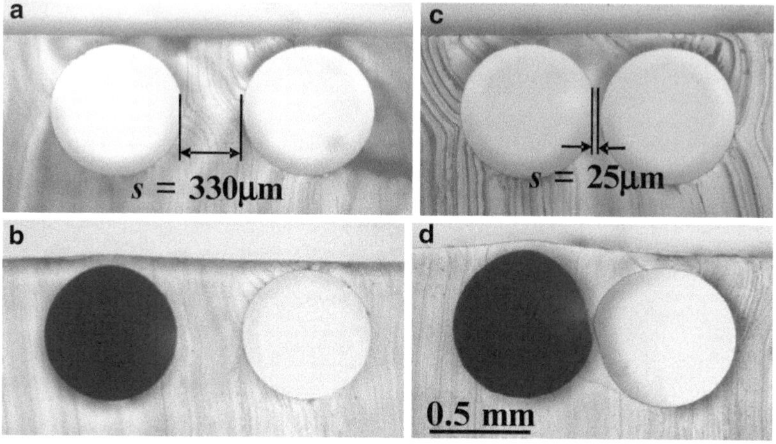

Fig. 14 Optical micrographs show the side view of channels of diameter 700 μm which are placed in pairs inside a layer of adhesive with skin thickness of 50 μm. Micrographs (**a, b**) represent one sample in which the channels remain far apart. Micrographs (**c, d**) represent the case when channels remain closely spaced. The figures show that for widely spaced channels, channels deform independently. However, when the channels are closely spaced, filling one of the channels deforms both the channels from being circular and also results in the asymmetric bulging of the adhesive surface (Reproduced by permission of the Royal Society of Chemistry [44])

Figure 15b shows such adhesive profile in which one of the channels in a pair is filled with the wetting liquid. The channel surface bulges out at the vicinity of the liquid-filled channel but bulges in for the one filled with oil. The cross section of the respective channels too alters, but differently for the two channels leading to highly asymmetric topography. Since the effective modulus of the adhesive surface depends strongly on geometry, it too remains asymmetric, in effect resulting in asymmetric heterogeneity both in topography and shear modulus.

We have shown that this asymmetric heterogeneity can be used to yield directional adhesion, i.e., different adhesion strength when an adherent separated from the adhesive from different directions. The experiment is shown in Fig. 13a in which a flexible contactor is first placed in contact with the adhesive layer such that it remains oriented normal to the channel axis. A lifting load is then exerted on the hanging end of the plate in a displacement-controlled experiment so that the contact line between the adhesive and the adherent propagates in a direction normal to the channel. It has been shown earlier that an adhesive with heterogeneity of the above kind does not allow the contact line to propagate continuously but with intermittent arrests at the vicinity of the channels. With continued lifting of the plate, the lifting torque, calculated by multiplying the lift-off load P with the distance a between the point of application of load $M \sim Pa$, increases while the contact line remains arrested, till a critical torque M_{max} is reached. For an adhesive with symmetric heterogeneity, M_{max} remains almost constant for all the channels. The situation however changes when adhesive with asymmetric heterogeneity is used.

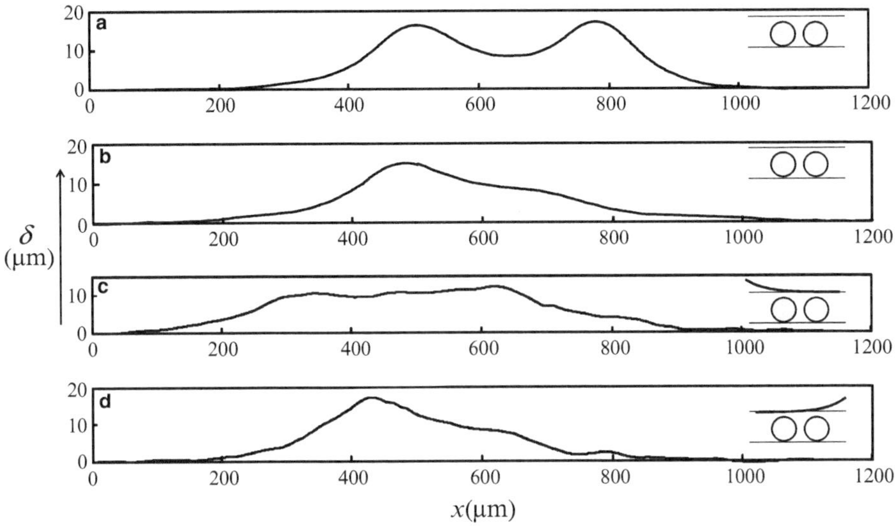

Fig. 15 (**a–d**) Represent surface profile of the adhesive embedded with liquid-filled channels. The figures (**a, b**) correspond to when both the channels in a pair and only one of the channels are filled with oil. These profiles were obtained by analyzing the side view of channels as presented in Fig. 14c, d (Reproduced by permission of the Royal Society of Chemistry [44])

In one case, the direction of peeling is such that the contact line between the two adherents encounters the oil fill channel first followed by the one which remains empty. When the flexible plate is lifted off from the opposite direction, the contact line encounters sequentially first the empty channel followed by the one filled with oil.

The data in Fig. 16a correspond to two pairs of channels with an inter-pair distance of $s = 7$ mm and intra-pair spacing of $s_1 = 56 \pm 3$ μm and 77 ± 3 μm, respectively. For all cases the flexible plate is lifted off from complete contact with the adhesive; as a result, a cusp-shaped crack is first required to be initiated. The first peak in the torque vs. displacement plot corresponds to this initiation from the edge of the film. Case 1 representing the experiment with both channels of a pair filled with air shows two additional peaks corresponding to crack arrest and initiation at the vicinity of each of these two pairs of channels. When both the channels are filled with oil, close spacing between the channels results in a single combined peak, instead of individual peaks for each channel in a pair. The height of this peak far exceeds the individual peaks, suggesting that crack arresting effect of this pair of channels remains more pronounced than when they remain far apart. Nevertheless, the maximum torque M_{max} is obtained same when the flexible adherent is lifted off the adhesive from either side of the adhesive showing no directionality. Situation changes when only one of the channels in a pair is filled with oil. When the plate is lifted such that the contact line encounters the filled channel first then the empty channel, corresponding to pairs of channels, the M vs. Δ plot shows

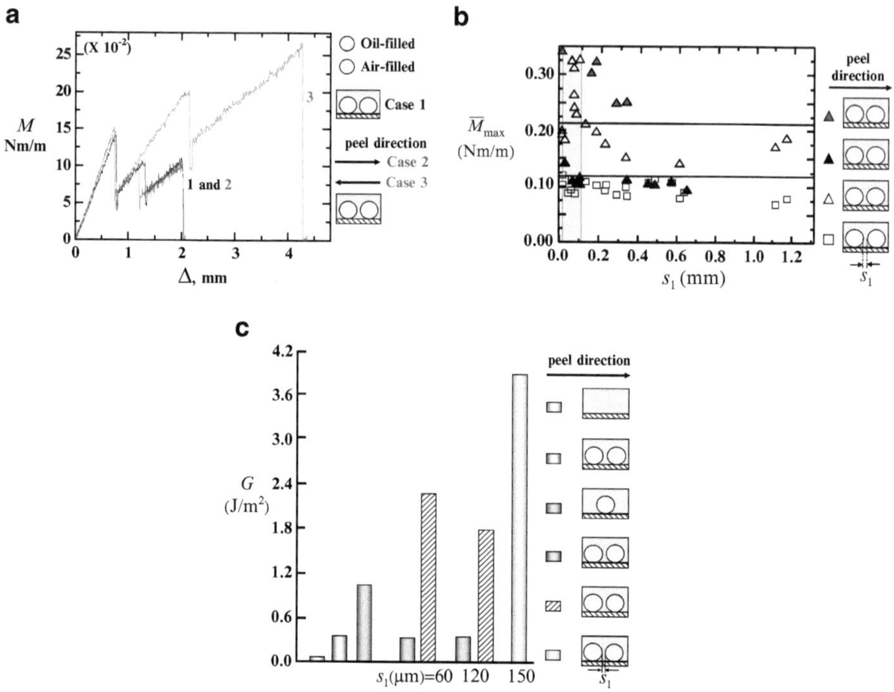

Fig. 16 (**a**) An adherent of flexural rigidity 0.02 Nm is lifted off an adhesive of thickness 750 μm embedded with channels of equal diameter ($d = 710$ μm). The torque M is plotted as a function of lifting height Δ of plate. While case 1 is obtained for both channels in the pair kept empty, 2 and 3 are obtained with one in pair filled with oil. (**b**) Represents plot of the maximum lifting torque M_{max} against the spacing s_1 between the channels of equal diameters in a pair. (**c**) The bar chart depicts the adhesion strength G for different cases with respect to the filling status of the channels and direction of lifting of the adherent (Reproduced by permission of the Royal Society of Chemistry [44])

two different peaks, the height of which does not show any increase over what is observed in case 1, i.e., corresponding to the single channels. A significantly larger M_{max} is however required to be exerted for initiating a crack from the vicinity of the channel pair when the plate is lifted from the reverse direction such that the contact line encounters the empty and filled channels in sequence. For adhesive embedded with series of large number of such channel pairs, the separation of the flexible plate off it requires significantly more effort in case 3 than in case 1. This directional effect on adhesion can be rationalized by considering that the pressure at the vicinity of the empty channel is zero while that closed to the oil-filled channel remains compressive. The zero pressure region behaves like a virtual discontinuity preventing transfer of elastic energy behind the crack tip with consequent trapping of the crack, whereas the positive pressure region arrests the crack by diminishing the stress concentration at the crack tip. When a surface crack propagates from zero to positive pressure regime, both these effects add up (Fig. 15d). For the reverse case of crack propagating from positive to zero pressure regime, their combined effect diminishes as the channel pair behaves like a single channel but with reduced compressive pressure (Fig. 15c). As a result, the stress concentration gets frustrated but to a lesser extent, so that now smaller M_{max} is required to initiate the crack. Figure 16b depicts typical numbers: for spacing $s_1 = 60\ \mu m$, M_{max} is found to be 0.25 ± 0.015 Nm/m and 0.11 Nm/m, respectively. Question arises if this directionality in adhesion remains prominent for all spacing between channels. The data presented in Fig. 16b show that maximum effect of directionality is achieved at an intermediate spacing, for very small spacing the individual effect of channels in a pair is no longer felt, whereas for very large spacing, the channels in a pair behave as single units. In either case, the asymmetry in geometric and material heterogeneity diminishes, eliminating the directional effect of adhesion.

Beside M_{max}, these adhesives are characterized also by estimating the fracture strength of the interface: $G = \int F\, d\Delta / A_{contact}$, the numerator representing the area under the lift-off load vs. displacement plot and the denominator representing the area of contact between the two adherents. Typical data presented in bar chart of Fig. 16c corresponds to the following: $d = 710\ \mu m$, $t = 40\ \mu m$, and $s_1 = 60\ \mu m$. For cases 3 and 2, G is obtained as $2,200$ and 300 mJ/m^2, respectively, i.e., about an order of magnitude difference in adhesion strength in two directions. Similar directional effect on adhesion is observed for different other geometric parameters of the adhesive. When both the channels within a pair are filled with oil, adhesion strength increases to $3,800$ mJ/m^2 which is more than 60-fold increase over what is observed on a smooth adhesive layer $\sim$60 mJ/m^2 without any heterogeneity. Thus, heterogeneity in geometry and material properties can enhance the adhesion strength of the adhesive and also impart directionality to it.

4 Summary

To summarize, we have described in this chapter few novel strategies of generating geometric and material asymmetry in soft-bodied objects which drives their locomotion and adhesion on a substrate towards a desired direction. In the first section, we have described an autonomous rolling motion of an elastomeric cylinder on a glass slide powered by a solvent which swells the cross-linked network of the elastomer. By using a small quantity of solvent, we have shown that the cylinder can be swelled asymmetrically leading to its bending. As the bent cylinder tends to straightens out, it debonds from the substrate at the rear side but adheres onto it at the front, thereby ensuing a rolling motion. The rolling motion continues as fresh portion of the cylinder continues to come in contact with the solvent while swollen portion rolls out of the solvent which evaporates out to the atmosphere from the cylinder surface. The solvent remains stuck to the cylinder surface and is pulled along with it while continuously powering the cylinder to roll. What makes this mechanism distinct is that effect of gravity can be utilized to increase the degree of asymmetry;

as a result, in contrast to all known physical mechanisms of locomotion, the velocity increases while rolling against gravity up an inclined plane. The rolling motion does not require any expensive equipment or complex synthetic routes involving special chemicals nor does it involve any rotating parts or delicate electrical connections as is commonly observed with several micro-mechanical systems. Since the cylinder is made of environmentally benign elastomer of sufficient mechanical strength, it can be used over and over again without getting destroyed unlike several soft gels which are susceptible to fracture during handling and even moderate mechanical impact. The rolling can occur in even harsh environmental conditions, e.g., against dead load: the cylinder can drag a body weight eight to ten times its body weight. The rolling motion occurs also over large range of temperature (80 °C) which allows it to roll on a hot surface not possible with several other mechanisms of locomotion and materials. Such rolling motion may specially be useful in pushing around soft sessile solids, or precious solids which any way is difficult to handle in conventional tweezing and gripping mechanisms. Besides, the mechanism of rolling is expected to inspire design of more efficient machines in macroscopic scale. In the second part of this chapter, we have extended the idea of asymmetry in designing adhesives which exhibit different adhesion strengths in different directions. The motivation of this work stems from the requirement of developing adhesives which can simultaneously produce strong adhesion and quick release during locomotion up an inclined or vertical wall along with reusability. Here, we have described adhesive layers which are embedded with channels in pairs and filled suitably with a wetting liquid. We have shown that geometric and material asymmetry results when only one of the channels in the pair is filled with a wetting liquid which diminishes the interfacial energy from a finite value. Systematic adhesion experiments on such adhesive layers not only bring out the anisotropic nature of the adhesive, but also adhesion strength varied reversibly over large range. All these are achieved without requiring the rheology of the adhesive material to be altered. This principle may better predict and explain the behavior of geckos and guide the design and control of climbing robots. Besides adhesion, the strategy described here also creates the possibility of generating a new, unique class of solid-liquid composite soft materials whose physical properties like modulus, compressibility, lubrication, refractive index, dielectric constant, and several others can be controlled spatially and reversibly by selecting a network of embedded channels filled with an appropriate liquid and by regulating its filling status.

Acknowledgment AG acknowledges also the Department of Science and Technology, Government of India for financial assistance via grant no. SB/S3/CE/036/2013.

References

1. Parker AR, Lawrence CR (2009) Water capture by desert beetle. Nature 414:33
2. Sokolov A, Apodaca MM, Grzybowski BA, Aranson LS (2010) Swimming bacteria power microscopic gears. Proc Natl Acad Sci U S A 107:969

3. Guo ZV, Mahadevan L (2008) Limbless undulatory propulsion on land. Proc Natl Acad Sci U S A 105:3179
4. Tanaka Y, Ito K, Nakagaki T, Kobayashi R (2012) Mechanics of peristaltic locomotion and role of anchoring. J R Soc Interface 9:222
5. Mast SO (1931) Locomotion in amoeba proteus. Protoplasma 14:21
6. Koch M, Harris N, Evans A, White N, Brunnschweiler A (1998) A novel micromachined pump based on thick-film piezoelectric actuation. Sens Actuators A 70:98
7. Cho I, Song T, Baek S, Yoon E (2005) A low voltage and low power RF MEMS series and shunt switches actuated by combination of electromagnetic and electrostatic forces. IEEE Trans Microw Theory Tech 53:2450
8. Lu YW, Kim CJ (2006) Microhand for biological application. Appl Phys Lett 89:164101
9. Osada Y, Okuzaki H, Hori H (1992) A polymer gel with electrically driven motility. Nature 355:242–243
10. Kuhn W, Hargitay B, Katchalsky A, Eisenberg H (1950) Reversible dilation and contraction by changing the state of ionization of high – polymer acid networks. Nature 165:514
11. Steinberg IZ, Oplatka A, Katchalsky A (1966) Mechanochemical engines. Nature 210:568
12. Ueoka Y, Gong J, Osaada Y (1997) Chemomechanical polymer gel with fish like motion. J Intell Mater Syst Struct 8:465–471
13. Arora H, Malik R, Yeghiazarian L, Cohen C, Wiesner U (2009) Earthworm inspired locomotive motion from fast swelling hybrid hydrogel. J Polym Sci 47:502712
14. Okuzaki H, Kuwabara T (2005) Temperature dependence of humid-responsive bending of conducting polymer films. Synth Met 153:101
15. Shinohara S, Seki T, Sakai T, Yoshida R, Takeoka Y (2008) Photoregulated wormlike motion of a gel. Angew Chem Int ED Engl 47:9039
16. Maeda S, Hara Y, Yoshida R, Hashimoto S (2008) Self walking gel. Angew Chem Int Ed 47:6690
17. Shishmanova I, Bastiaansen C, Schenning A, Broer D (2012) Two dimensional pH-responsive printable smectic hydrogel. Chem Commun 48:4555
18. Gong J, Matsumoto PS, Uchida M, Isogai N, Osada Y (1996) Motion of polymer gel by spreading organic fluid on water. J Phys Chem 100:11092
19. Northen TR, Woodbury NW (2005) Light directed movement of polymer microstructures. Langmuir 21:4949
20. Mahadevan L, Daniel S, Chaudhury MK (2004) Biomimetic ratcheting motion of a soft, slender, sessile gel. Proc Natl Acad Sci U S A 101:23
21. Yeghiazarian L, Mahajan S, Montemagno C, Cohen C, Wiesner U (2005) Directed motion and cargo transport through propagation of polymer-gel volume phase transitions. Adv Mater 17:1869
22. Ding Z, Ziaie B (2009) Vibration-induced frequency-controllable bidirectional locomotion for assembly and microrobotic application. IEEE Trans Robot 25:1192
23. Tanaka T, Nishio I, Sun ST, Nishio SU (1982) Collapse of gels in an electric field. Science 218:467
24. Randhawa JS, Keung MD, Tyagi P, Gracias DH (2010) Reversible actuation of microstructure by surface-chemical modification of thin-film bilayers. Adv Mater 22:407
25. Hutchins B, Platt MM, Hancock WO, Williams ME (2007) Directing transport of CoFe2o4-functionalized microtubules with magnetic fields. Small 3:126
26. Vale RD, Reese TS, Sheetz MP (1985) Identification of a novel force generating protein, Kinesin, involved in microtubule-based motility. Cell 42:39
27. Schnitzer MJ, Block SM (1997) Kinesin hydrolyses one ATP per 8-nm step. Nature 388:386–390
28. He B, Wang Z, Li M, Wang K, Shen R, Hu S (2012) Wet adhesion inspired bionic climbing robot, IEEE/ASME Trans Mechatron 19:312–320
29. Unver O, Uneri A, Aydemir A, Sitti M (2006) Geckobot: a geckoinspired climbing robot using elastomer adhesives. IEEE Int Conf Robot Autom. 2329–2335. Rome, Italy, May 2006

30. Greiner C, Campo AD, Arzt E (2007) Adhesion of bioinspired micropatterned: effects of pillar radius, aspect ratio, preload. Langmuir 23:3495
31. Santos D, Kim S, Spenko M, Parness A, Cutkosky M (2007) Directional adhesive structures for controlled climbing on smooth vertical surfaces, IEEE Int Conf Robot Autom. 1162. Rome, Italy, April 2007
32. Rosa G, Messina M, Muscato G, Sinatra R (2002) A lowcost lightweight climbing robot for the inspection of vertical surfaces. Mech Mater 12:71
33. Xu Z, Ma P (2002) A wall-climbing robot for labelling scale of oil tank's volume. Robotica 20:203
34. Illingworth L, Reinfeld D (2003) Vortex attractor for planar and non-planar surfaces. US Patents #6619922, Sept 2003
35. Tian Y, Pesika N, Zeng H, Rosenberg K, Zhao B, Mcguiggn P, Autumn K, Israelachvili J (2006) Adhesion and friction in gecko toe attachment and detachment. Proc Natl Acad Sci U S A 103:19320
36. Favi P, Yi S, Lenaghan SC, Xia L, Zhang M (2014) Inspiration from the natural world: from bio-adhesives to bio-inspired adhesives. J Adhes Sci Technol 28:290–313
37. Federle W, Barnes WJP, Baumgartner W, Drechsler P, Smith JM (2006) Wet but not slippery: boundary friction in tree frog adhesive toe pads. J R Soc Interface 3:689
38. Dirks JH, Clemente CJ, Federle W (2010) Insect tricks: two-physic foot pad secretion prevents slipping. J R Soc Interface 7:587
39. Lee H, Lee BP, Messersmith PB (2007) A reversible wet/dry adhesive inspired by mussels and geckos. Nature 448:338
40. Vogel MJ, Steen PH (2009) Capillarity-based switchable adhesion. Proc Natl Acad Sci U S A 107:3377
41. Zhang H, Guo D, Dai Z (2010) Progress on gecko-inspired micro/nano adhesion arrays. Chin Sci Bull 42:1843
42. Yurdumakan B, Raravikar R, Ajayanb P, Dhinojwala A (2005) Synthetic gecko foot-hairs from multiwalled carbon nanotubes. Chem Commun 13:3799–3801
43. Hore D, Majumder A, Mondal S, Roy A, Ghatak A (2012) How to make a cylinder roll uphill. Soft Matter 8:5038
44. Majumder A, Mondal S, Tiwari AK, Ghatak A, Sharma A (2012) Direction specific adhesion induced by subsurface liquid filled microchannels. Soft Matter 8:7228
45. Lee JN, Park C, Whiitesides GM (2003) Solvent compatibility of (polydimethyl siloxane)-based microfluidic devices. Anal Chem 75:6544
46. Adamson W, Gast AP (1997) Physical chemistry of surface, 6th edn. Wiley-International publication, New York
47. Green D, Perry P (1984) Perry's chemical engineers handbook, 6th edn. McGraw-Hill Professional, New York
48. Ghatak A, Mahadevan L, Chung JY, Chaudhury MK, Shenoy V (2004) Peeling from a biomimetically patterned thin elastic film. Proc R Soc Lond Ser A 460:2725
49. Chung JY, Chaudhry MK (2005) Roles of discontinuities in bio-inspired adhesive pads. J R Soc Interface 2:55
50. Thomas T, Crosby AJ (2006) Controlling adhesion with surface hole patterns. J Adhes 82:311
51. Geim AK, Dubonos SV, Grigorive IV, Novoselov KS, Zhukov AA, Shapoval SY (2003) Microfabricated adhesive mimicking gecko foot hair. Letters 2:461–463
52. Glassmaker NJ, Jagota A, Hui CY (2005) Adhesion enhancement in a biomimetic fibrillar interface. Acta Biomater 1:367
53. Crosby AJ, Hageman M, Duncan A (2005) Controlling polymer adhesion with pancakes. Langmuir 21:11738
54. Reddy S, Arzt E, del Campo A (2007) Bioinspired surface with switchable adhesion. Adv Mater 19:3833
55. Greiner C, Arzt E, del Campo A (2009) Hierarchical gecko-like adhesives. Adv Mater 21:479
56. Boesel LF, Greiner C, Campo AD, Arzt E (2010) Gecko-inspired surface: a path to strong and reversible dry adhesives. Adv Mater 22:2125

57. Arul EP, Ghatak A (2011) Adhesive with patterned sub-surface microstructure. Mater Sci 46:832
58. Arul EP, Ghatak A (2009) Bioinspired design of hierarchically structure adhesive. Langmuir 25:611
59. Gorb S, Varenberg M, Peressadko A, Tuma J (2006) Biomimetic mushroom-shaped fibrillar adhesive microstructure. J R Soc Interface 4:271
60. Lee J, Fearing RS, Komvopoulos K (2008) Directional adhesion of gecko-inspired angled microfiber arrays. Appl Phys Lett 93:191910
61. Arul EP, Ghatak A (2012) Control of adhesion via internally pressurized sub surface micro channels. Langmuir 28:4339
62. Verma MKS, Majumder A, Ghatak A (2006) Embedded template-assisted fabrication of complex microchannels in PDMS and design of a microfluidic adhesive. Langmuir 22:10291
63. Majumder A, Sharma A, Ghatak A (2009) A bioinspired wet/dry microfludic adhesive for aqueous environments. Langmuir 26:521
64. Majumder A, Ghatak A, Sharma A (2007) Microfluidic adhesion induced by subsurface microstructures. Science 318:258
65. Ghatak A (2010) Peeling of an adhesive layer with spatially varying modulus. Phys Rev E 81:021603
66. Majumder A, Tiwari AK, Korada K, Ghatak A (2010) Microchannel induced surface bulging of a soft elastomeric layer. J Adhe Sci Technol 24:2681
67. Bullock J, Drechsler P, Federle W (2008) Comparison of smooth and hairy attachment pads in insects: friction, adhesion and mechanisms for direction-dependence. J Exp Biol 211:3333
68. Smith J, VanNess H, Abbott M (2005) Introduction to chemical engineering thermodynamics, 7th edn. Tata McGraw Hill, India

Dynamics of Soft Glassy Materials under Tensile and Squeeze Flow Fields

Asima Shaukat, Ashutosh Sharma, and Yogesh M. Joshi

Abstract In this chapter, we study the deformation and physical aging behavior of model soft glassy materials. In the first part of this work, we systematically investigate the effect of the initial time scale of deformation and material viscoelasticity on the energy and strain response of the material under a tensile flow field. We discuss the results in the light of increased importance of either viscous dissipation at high deformation rates or brittleness induced in the material at a longer aging period. We further assess the validity of the process time–aging time–stress deformation field under tensile creep flow field. We next study the behavior of various types of soft glassy materials under the simultaneous application of tensile and rotational stress fields. We determine the yielding criterion for these materials when more than one stress field is present. Finally, we study the squeeze flow dynamics of a soft glassy material sandwiched between a rough and a smooth plate under application of a deformation field of varying strength. The effect of aging and rejuvenation is also analyzed under squeeze flow field. The results are compared with a time-independent Herschel–Bulkley model for yield stress fluids.

Keywords Soft glassy materials • Yield stress • Tensile flow • Squeeze flow

1 Introduction

Soft solid-like or pasty materials are thermodynamically out of equilibrium and are referred to as "soft glassy materials". The primary constituents of these materials are structurally arrested by the neighboring particles. Since the translational mobility of these particles is restricted, they are unable to explore the total available phase space over the practical time scales [12, 37]. In order to attain a lower-energy

A. Shaukat • Y.M. Joshi (✉)
Department of Chemical Engineering, Indian Institute of Technology Kanpur, Kanpur, 208016 UP, India
e-mail: joshi@iitk.ac.in

A. Sharma
Department of Chemical Engineering and DST Unit on Soft Nanofabrication, Indian Institute of Technology Kanpur, 208016 UP, India

Y.M. Joshi, S. Khandekar (eds.), *Nanoscale and Microscale Phenomena*,
Springer Tracts in Mechanical Engineering, DOI 10.1007/978-81-322-2289-7_11

state, the arrested particles continuously rearrange themselves, reorganizing the structure [13, 14, 38, 116, 127, 137, 163], leading to change in the physical properties of the system with time [12, 36, 37, 41, 42, 95, 99, 100, 116, 127, 133, 139, 157, 163]. However, owing to kinetic constraint, since thermodynamically driven lowering of free energy cannot lead to complete structural relaxation over practical time scales, these materials adopt a disordered, metastable configuration. This phenomenon of an incessant evolution of the system with respect to time is known as physical aging [69, 157, 163] and is affected by temperature [6, 118, 157] and applied stress/deformation field [38, 51, 52, 73, 130, 133]. Some examples of these materials are colloidal suspensions, emulsions, gels, biopolymers, pastes, foams, slurries, etc. These materials find wide usage in everyday life as well as in several industrial applications. The thixotropic behavior associated with these materials imparts strong history dependence and often poses challenges in their commercial processing as well as in application. Under an application of a strong deformation field, a structural breakdown takes place leading to the plastic flow in the material [28, 38, 73, 74]. Thus, these materials, in a broad sense, can be classified as rheologically simple thixotropic yield stress fluids which flow when a critical stress is overcome. The yield stress of these materials depends upon the deformation history as well as the aging time subsequent to mechanical quenching [38, 41, 54, 74, 150].

A fairly common problem involving soft glassy materials is that of elongational flow or tensile flow of these materials confined between two surfaces. Some examples of this problem are extensional flow of creams [30] and cosmetic pastes [96], application of coatings [77], processing of foodstuffs like jams and jellies [19], elongational flow of oil drilling fluids while passing through drill bit [93], stretching flow in lubricants [107], deformation of biological films like corneal tear films [142], biomechanics of hip joints [25], calendaring of highly filled polymer melts, mixing and sheeting of pastry dough, etc. Further, the problem of elongational flow of soft solids has attracted a lot of interest in the field of adhesives [70, 128]. Often, a technique called as the probe-tack test [22, 31, 46, 47, 55, 63, 87, 151] is employed to assess the tensile deformation in viscoelastic fluids. In this test, a rigid probe is brought into contact with a fluid film on a substrate and then detached from it by moving the probe away from the film at a constant rate. The output of this test is a force–displacement curve which is analyzed to obtain information about the work of separation, peak force, and details of defects formed during plate separation like finger propagation and cavitation [46–48, 66, 115]. In literature, this class of test has been widely employed to investigate the Saffman–Taylor [1, 5, 85, 88, 89, 120, 126, 134, 152, 160, 161, 165] and contact instabilities [64, 65, 67, 68, 114, 131, 141] in viscous and viscoelastic fluids under tensile flow. Though a lot of attention has been paid to the characterization of the debonding mechanisms in several types of fluids [18, 35, 39, 53, 60, 90, 91, 94, 124–126, 135, 136, 164, 167] and the eventual adhesive [46, 47, 53, 115, 151, 164] or cohesive [46, 47, 53, 91, 115, 126, 151, 153, 164] breakage in the material film due to tensile deformation, a relatively fewer number of studies have focused on carrying out a detailed interpretation of the force–displacement curves [159] and determining the total work of separation as a function of various experimental and geometrical variables [18, 53, 62]. Further, as explained above, most of the commonly encountered soft materials display thixotropic behavior [15, 109], that is, their properties change significantly with passage of time as well as under the application of deformation field. Therefore, apart from the flow dynamics arising from the presence of the yield stress in soft glassy materials, it is also important to assess the influence of time dependency or aging in these materials.

In several applications, a combination of deformation fields rather than a single stress field acts on soft solids. For instance, in polymer extrusion, drilling of mud, fiber spinning, spiral vortex flow under axial sliding, etc. [16, 56, 123, 158], a combination of tensile and rotational deformation fields acts simultaneously. As the response of soft glassy materials is a strong function of the nature of the applied stress/strain field, it is expected to be different in this case as compared to unidirectional deformation fields. Despite a common prevalence of situations in which more than one stress/strain fields are present, only a few studies so far have been carried out in this direction. Further, since the soft glassy materials possess a three-dimensional microstructure, the yield stress of these materials is expected to be strongly influenced not only by the magnitude but the type and direction of the stress/strain field as well. Therefore, the criterion that could successfully describe the unjamming of these materials under the scenario of more than one deformation field should possess information regarding the three-dimensional resistance to flow and would essentially be more complex than the case of simple one-dimensional flow. Ovarlez and coworkers [119] experimentally studied the yielding behavior under a combination of rotational and squeeze flow fields by carrying out shear rate-controlled experiments. They found that if the unjamming in the material takes place in one direction due to an applied stress, it simultaneously yields in all other directions. Further, they observed that the overall viscous flow behavior post yielding is dominated by the primary flow in the material. Moreover, they experimentally determined the yield surface for the case of simultaneous rotational and squeeze flow and found that it could be closely described by the Von Mises criterion. Brader and coworkers [29, 59] recently proposed a single-mode MCT (mode-coupling theory) model with a tensorial structure. In their work, the dynamic yield surface was presented as a function of various combinations of components of applied stress tensors. Interestingly, their estimated yield surface also closely matched the Von Mises criterion.

Another well-studied problem owing to numerous applications is the squeeze flow of soft glassy materials wherein the reverse of tensile flow takes place. This flow field is often encountered in material processing [56] and biology [78]. In previous studies, a squeeze flow between two parallel plates has either been produced by applying a constant compressive force [102] or by moving the top plate at a constant rate [33, 76]. Owing to the industrial and academic importance of this field, an extensive amount of work consisting of theory, simulations, and experiments has been carried out for various types of inelastic and viscoelastic fluids. In particular, detailed attention has been paid in determining the size and shape of the unyielded regions within the material undergoing squeeze deformation [45, 50, 92], the effect of slip at material–plate interface [57, 104, 166], and the measurement of various rheological properties [45, 101, 129]. Though this flow is similar to the tensile flow with the exception of direction, however, defect formation like fingering, cavitation, etc., is absent in this type of flow. This flow field is often considered as a class of rheometry and is used to determine various material properties like the yield stress. In particular, it is employed for those materials which are too stiff to be characterized using conventional rheological techniques. However, the presence of a wall slip [82, 83, 149], unyielded pockets of material [50, 84, 92], and phase separation due to nonmatching radial motion of constituent phases [105] make the interpretations more complex than conventional methods, thereby limiting the usage of this technique.

In the first part of this chapter, we present the results for the deformation behavior of an aqueous Laponite suspension, a model soft glass, under tensile flow field rendered by progressive separation of the two plates holding the material film between them. In this work, the deformation field intensity was varied by changing the magnitude of velocity of separation of the plates/applied force, and the gap between the plates and its effect on the response of the material were investigated. Further, the effect of the aging time duration on the tensile deformation behavior of this material was studied. In the second part of this work, compliance was measured as a function of time under the application of a tensile creep flow field for various samples of a Laponite suspension aged for different time periods. We report the effect of the variation in the aging time as well as the magnitude of the tensile stress on the behavior exhibited by the Laponite suspension. Further, the validity of the process time–aging time–stress correspondence was assessed under this flow field. In the third part of this chapter, we discuss the deformation behavior of soft glassy materials simultaneously acted upon by two stress fields, namely, tensile and rotational stress fields. A yielding phase diagram was also constructed for various categories of soft glassy materials under a combined application of these two stress fields. Further, the validity of the Von Mises criterion was tested for the case of combined tensile and rotational stress fields. The final part of this chapter is concerned with the squeeze flow behavior of a Laponite suspension. In this case also, like the first problem, the deformation field intensity and the aging time of the sample were varied. We compare and contrast the results with the theoretical prediction of a Herschel–Bulkley fluid model for squeeze flow field.

2 Materials and Experimental Procedure

2.1 Materials

A material which has largely been employed in this work is a 3.5 wt% aqueous Laponite suspension. Laponite (Laponite RD®) was procured from Southern Clay Products, Inc., and the suspension was prepared by dispersing the requisite amount of Laponite in water. Laponite is a smectite–hectorite clay which is prepared synthetically [162]. Laponite powder is composed of stacks or aggregates of circular disk-like layers which are around 1 nm thick and 25 nm in diameter [81]. When these particles are dispersed in water, the negative charge on the surface of Laponite layer leads to repulsion between the adjacent layers [111]; however, an attractive interaction between the layer edge and the surface of other particles, due to a dissimilarity in charge, might also be present [110]. The dispersion of Laponite particles in water undergoes ergodicity breaking with passage of time to convert into a soft solid if the concentration of Laponite is above 1 vol.% ($\sim$2.6 wt%) [72]. Beyond ergodicity breaking, a Laponite suspension demonstrates physical aging with time [14, 137]. This material possesses all the generic characteristic features of a model soft glassy material [2, 3, 7–9, 12, 20, 26, 74, 117, 140] like the essential nonequilibrium state of existence with a continuously evolving microstructure which results in an aging time and deformation field-dependent relaxation time and elastic modulus [6, 73], weak frequency dependence of modulus [139], and incomplete stress relaxation upon application of step strain [14]. The details of the method followed for the preparation of the Laponite suspension are mentioned in Ref. 133. A few samples were prepared by following this method and were left undisturbed for a period ranging from 1 to 3 months in a sealed polypropylene bottle.

Other soft glassy materials employed in the studies discussed in this chapter are a commercial hair gel, a foam, and a paint. The hair gel Brylcreem Ultra Hard gel® is a water-based, yield stress fluid. Unlike a clay suspension where interactions between anisotropic particles lead to jamming, in this case, the stretching out of constituent polyelectrolyte polymer chains causes resistance to flow. We also employ a commercial foam called Gillette® regular shaving foam. This foam has an aqueous-based formulation composed of closely packed gas bubbles dispersed in a surfactant solution [86]. When the deformation field is stronger than the yield stress, the constituent bubbles are deformed which causes microstructural rearrangements within the material eventually leading to flow [58]. The paint used in this work was purchased from Berger Paints® (British Paints). Emulsion paints commonly constitute acrylic polymer droplets dispersed in an aqueous continuous phase. These materials are also classified as soft glassy materials as they show all the characteristic features of the same like physical aging and thixotropy. Further, like yield stress fluids, they too undergo a visco-plastic flow above a threshold stress [10].

2.2 Rheological, Tensile, and Squeeze Tests

In all the problems discussed in this chapter, a parallel plate geometry (50 mm diameter) of an Anton Paar MCR 501 rheometer was employed. The sample was shear melted before the start of each experiment by first passing it through a syringe needle several times followed by application of an oscillatory rotational shear stress much larger in amplitude than the yield stress of the material after loading it on the lower plate of the rheometer. This step was necessary in order to ensure the same initial state of the material before the beginning of every experiment. Subsequent to shear melting, the sample was left unperturbed between the parallel plates for a period of time equal to the requisite aging time of the material.

The tensile tests which were performed in this study were divided into two categories: constant velocity mode experiments in which the top plate moves at a constant rate while the bottom plate stays stationary and constant force experiments in which a constant normal force was applied to the top plate. Both the top and the bottom plates were smooth (roughness: 0.8 μm) for all the tensile experiments. In the part of this work which relates to the squeeze flow of soft glassy materials, the oscillatory shear and squeeze flow experiments (both constant force and velocity mode) were carried out using a rough sandblasted top plate (roughness: 5.74 μm), unless mentioned otherwise where a smooth top plate (roughness: 0.8 μm) was used. The bottom plate was a smooth plate (roughness: 0.8 μm) in all the experiments. In order to avoid aging during the course of both tensile and squeeze tests, the experimental time was kept considerably smaller (around 10 %) than the waiting time (aging time) of the sample. In all the experiments, the temperature was maintained at 25 °C. The exposed surface of the sample which was not in contact with the plates was coated with a small amount of low-viscosity silicone oil to prevent evaporation as well as contamination with CO_2 [112]. The rheometer used in this study has a finite compliance equal to 5.9×10^{-7} m/N. The compliance can cause considerable deviation of the actual value of the gap between the plates from the set/displayed value. In these tests, the compliance was taken into account while measuring the position of the top plate; therefore, the values of gap reported in this work are the true values of the gap between the top and bottom plate.

3 Tensile Deformation Behavior of an Aging Laponite Suspension

In this problem, we carry out tensile tests to study the deformation and breakage of thin films of the Laponite suspension. As explained above, an aqueous suspension of Laponite is a soft glass whose microstructure evolves with time, leading to a continuous variation in its physical properties. We investigate the force and energy response of this material to applied tensile deformation field by varying the field intensity (initial thickness of the material film and the magnitude of the velocity/force applied to the top plate) and material viscoelasticity (aging time). We carry out experiments in either constant velocity or constant force modes. We also discuss the effects of the aging time and the strength of the deformation field intensity on the mechanism of failure of the Laponite films.

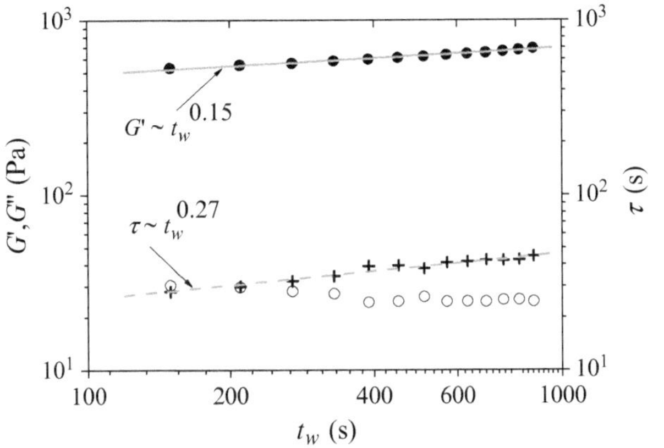

Fig. 1 The evolution of elastic modulus (G', *filled circles*), viscous modulus (G'', *open circles*), and relaxation time (τ, +) with respect to waiting time ($f = 0.1$ Hz, $\gamma_0 = 1$ %). Power law fits to elastic modulus–waiting time data $\left(G' \sim t_{\mathrm{w}}^{0.15}\right)$ and relaxation time–waiting time data $\left(\tau \sim t_{\mathrm{w}}^{0.27}\right)$ are shown with the help of *thick gray line* and *dashed gray line*, respectively (Reprinted with permission from [143]. Copyright (2009) American Chemical Society)

3.1 Physical Aging in the Laponite Suspension

Figure 1 shows the aging dynamics of a 3.5 wt% aqueous suspension of Laponite (idle time equal to 3 months) [143]. This material undergoes physical aging which causes an increase in the elastic modulus with waiting time (aging time) [74]. The figure shows an evolution of elastic G' and viscous modulus G'' with waiting time t_{w} in a small amplitude oscillatory experiment ($\gamma_0 = 1$ %, and frequency $f = 0.1$ Hz). As mentioned before, the oscillatory test was carried out using the same parallel plate geometry which was employed for carrying out the tensile tests. The figure clearly shows that G' increases with waiting time, and the dependence is approximately given by a power law: $G' \sim t_{\mathrm{w}}^{0.15}$. If the material response is approximated with a time-dependent single-mode Maxwell model [73, 74], the dominating relaxation time of the material is given by $\tau = G'/\omega G''$ [74]. We have also plotted the relaxation time in Fig. 1, which obeyed a stronger power law dependence on the waiting time ($\tau \sim t_{\mathrm{w}}^{0.27}$) compared to the elastic modulus. Although we employ a single mode of the Maxwell model to evaluate the relaxation time, it gives a good first estimate [74].

3.2 Constant Velocity Mode Experiments

The movement of the top plate away from the bottom plate at a constant velocity leads to the contraction of the fluid–air interface leading to the eventual failure in the fluid film. A typical response to applied constant velocity (V) is shown in Fig. 2. The normal force (F) shows a rapid increase in the beginning until it reaches a peak value (F_f) followed by a slow decline in its magnitude as a function of time (t). As explained before, the rheometer is not infinitely rigid; therefore, the initial increase in the force can be attributed to the elastic stretching of the rheometer rather than deformation in the material film. At the start of the experiment, the traverse starts to move at a constant velocity; however, the true gap between the plates does not increase as the top plate remains more or less stationary. This happens because it is easier to stretch the apparatus than to move the upper plate owing to a high value of the lubrication force. Once the upper plate begins to move, the lubrication force declines rapidly with further increase in the gap between the plates. This results in acceleration of the top plate to a high velocity in a short time period before decreasing again to the desired value of the velocity. Therefore, the experiment in the very beginning does not follow the constant velocity mode in a strict sense. Figure 2 shows the deviation of the velocity [rate of change of gap (d)] from the desired value at the start of the experiment. Nevertheless, excluding the short period of time in the beginning, just after starting the experiment (<1 s), the rheometer maintained the desired constant velocity of the upper plate for the rest of the experiment. The energy of separation was determined by measuring the area under the force–displacement curves. It was observed that as plates separated, the cross-sectional area of the material reduced to a fractal-like branched pattern which eventually underwent cohesive breakage in the bulk of the material. We characterize the basic features of these fractal-like branched patterns later in this section.

3.2.1 Aging Time Variation

In this set of experiments, we have varied the material viscoelasticity by employing samples which were aged for different periods of time. The initial gap between the plates d_i and the velocity of the top plate V were the same in each experiment. The

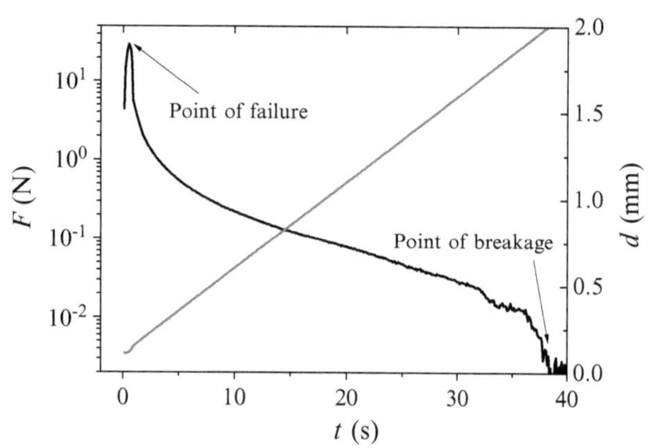

Fig. 2 The normal force F (*thick black line*) and the gap between the plates (*thick gray line*) as a function of experimental time t. For this test, the velocity $V = 50$ μm/s, aging time $t_w = 15$ min, and initial gap between the plates $d_i = 107$ μm (Reprinted with permission from [143]. Copyright (2009) American Chemical Society)

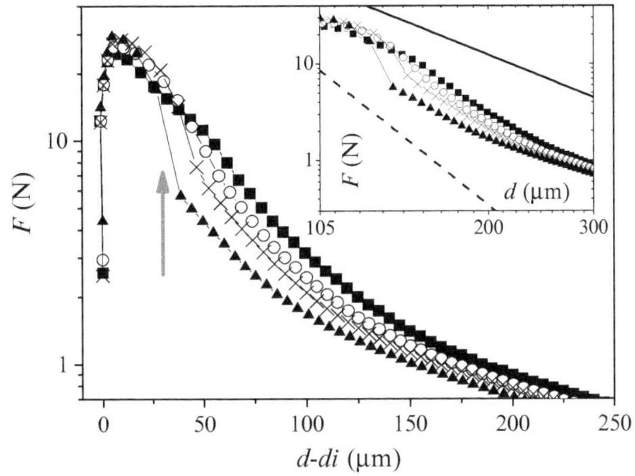

Fig. 3 Force–displacement curves for experiments carried out at various aging times ($t_w = 0$ min, *filled squares*; 5 min, *open circles*; 10 min, cross; and 15 min, *filled triangles*) for $V = 50$ μm/s and $d_i = 106 \pm 1$ μm. At higher aging times, a sharp decline in force is observed over a small displacement, which is pointed out with the help of a *gray arrow*. The *inset* shows the same plot on a double logarithmic scale. Also shown in the *inset* are *dashed* and *dotted lines* which represent the prediction for a Newtonian $[F \sim d^{-5}]$ and a yield stress fluid (Herschel–Bulkley model for small deformation rates) $[F \sim d^{-5/2}]$, respectively (Adapted with permission from [143]. Copyright (2009) American Chemical Society)

aging time was always kept significantly larger than the test duration ($t_w >> t$), so that further aging during the experiment is small enough to be neglected. Figure 3 shows variation of force–top plate displacement curves for various ages. The inset shows a double logarithmic plot of the data shown in the main figure.

The normal force F for Newtonian fluids can be determined by solving Navier–Stokes equations for a tensile deformation. The expression thus obtained is given by [22, 53]

$$F = 3\pi\eta R^4 d_i^2 V/2d^5 \tag{1}$$

where R is the initial radius of the fluid film (equal to radius of the top plate) and η is viscosity of the fluid. Soft glassy yield stress fluids are often modeled using a Herschel–Bulkley constitutive relation given by $\tau = \tau_y + m\dot{\gamma}^n$ where τ is applied stress, m is the consistency, and τ_y is the yield stress. Derks et al. [53] derived an expression for normal force for a Herschel–Bulkley model *in the limiting case of small deformation rates* given by

$$F = 2\pi m R^3 d_i^{3/2} \tau_y/3d^{5/2} \tag{2}$$

The inset in Fig. 3 also shows the variation of force with the gap between the plates for a Newtonian fluid and a Herschel–Bulkley model (for small deformation rates where the shear stress can be approximated with the yield stress) for which

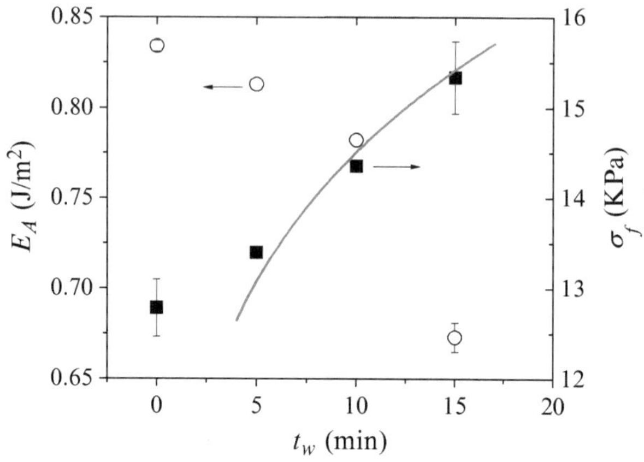

Fig. 4 Energy dissipated per unit initial area (*open circles*) and stress at failure (*filled squares*) as a function of waiting time for $V = 50$ μm/s and $d_i = 106 \pm 1$ μm. A power law fit to the engineering stress at failure data, $\sigma_f \sim t_w^{0.15}$, is shown with a *thick gray line* (Reprinted with permission from [143]. Copyright (2009) American Chemical Society)

decay in force follows power law relations, $F \sim d^{-5}$ and $F \sim d^{-5/2}$, respectively. The Laponite suspension shows a weaker dependence on the gap in comparison to a Newtonian fluid but stronger than a yield stress fluid. The dependence progressively changed from closer to the Newtonian fluid to the yield stress fluid with an increase in the aging time.

The peak stress σ_f and energy dissipated per unit initial area E_A are plotted against aging time in Fig. 4. Interestingly, the tensile stress at failure σ_f (or force at failure F_f) follows the same dependence on aging time as that of elastic modulus (as shown in Fig. 1). We do not consider the data point corresponding to the test with no waiting time because aging is expected to take place during this test, and therefore, the viscoelastic properties of suspension would change over the duration of the experiment.

For a Herschel–Bulkley model prediction [53], energy dissipated increases with increase in yield stress or elastic modulus. In the present case, the energy dissipated in the material decreased with increased aging time (or elastic modulus). This trend is essentially due to a decrease in strain at complete breakage with increase in aging time. Further, it can be seen in Fig. 3 that for the force–displacement curves at higher aging times of 10 and 15 min, there is a sharp fall in the magnitude of force just after attaining its peak value. A more severe drop is observed for the higher aging time. A gray arrow in Fig. 3 points toward this sharp drop in force. Therefore, it is evident from the figure that the response of the material becomes more brittle with increase in aging time. We believe that such behavior results from an enhancement in the elastic modulus which limits the dissipation of energy while an increasingly slower relaxation process (an enhanced relaxation time) causes progressively more limited modes of energy dissipation, thereby leading to an enhanced brittle response.

In viscoelastic materials, the failure takes place by finger/crack formation or through cavitation in the bulk of the fluid or the fluid–substrate interface. The fingering occurs due to Saffman–Taylor instability which causes the interface to distort away from its initial circular form. This instability comes into play when

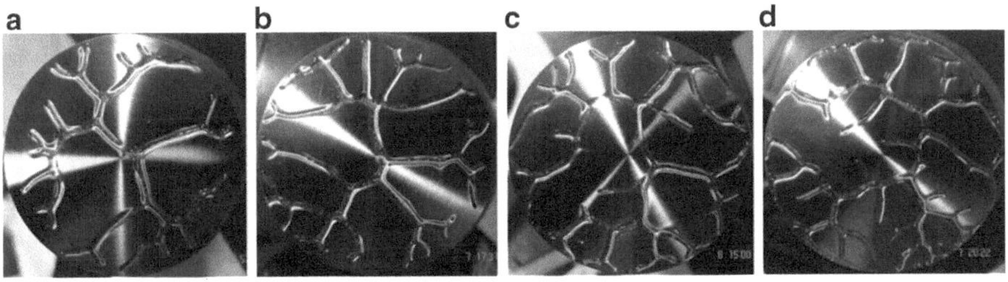

Fig. 5 The photographs of the top plate just after complete breakage. (**a**) $t_w = 0$ min (perimeter of the suspension–air interface per unit area = 0.2458 /mm, number of branches = 35), (**b**) $t_w = 5$ min (perimeter of the suspension–air interface per unit area = 0.3215 /mm, number of branches = 47), and (**c**) $t_w = 10$ min (perimeter of the suspension–air interface per unit area = 0.3492 /mm, number of branches = 54), and (**d**) $t_w = 15$ min (perimeter of the suspension–air interface per unit area = 0.4034 /mm, number of branches = 63). The initial gap $d_i = 106 \pm 1$ μm and velocity $V = 50$ μm/s for all experiments (Reprinted with permission from [143]. Copyright (2009) American Chemical Society)

a less viscous fluid (like air) displaces a more viscous fluid [85, 90, 91, 115, 153, 164]. We also observe fractal branched patterns formed due to fingering/cavitation on both the plates upon complete failure of the material film. The failure was cohesive; therefore, identical patterns which were mirror images of each other were obtained on both the plates on complete breakage. Figure 5 shows photographs of the top plate immediately after the completion of cohesive failure in the aging time-dependent experiments. The perimeter of the fluid–air interface per unit initial area of the film, as well as the number of branches in the fractal pattern, for each experiment was also determined. We find that both of these parameters which characterize the intensity of branching in the structure increase with the aging time. An increased branching was also observed with an increase in the plate separation velocity and with decrease in the initial gap. Interestingly, for the aging time experiments, an increase in the intensity of branches was accompanied with a decrease in the dissipated energy. In contrast to this, for the set of experiments corresponding to various velocity and initial gap values but fixed aging time, we observe an increase in the intensity of branches along with an increase in dissipated energy. We find that increasing the suspension elasticity and decreasing the time scale of initial deformation (d_i/V) both induce more frequently bifurcating and shorter fingers/cracks. A similar effect of higher failure speed on fingering patterns has also been studied in simple liquids [53, 75, 90, 115, 126].

3.2.2 Rate of Deformation (Velocity) and Initial Thickness of the Film

In this group of experiments, we investigate the effect of the time scale of initial deformation (d_i/V) on the force–displacement curves. In the first set of experiments, we varied the velocity of the top plate while keeping the initial gap around 100 μm and aging time equal to 15 min for each experiment. It was observed that the energy of separation E_A is a weak function of velocity and varies as $E_A \sim V^{0.2}$. In the second set, we varied the initial gap ($V = 50\ \mu m/s$ and $t_w = 15$ min) and found that the dependence of the energy dissipated on the initial gap is given as $E_A \sim d_i^{-0.36}$. The experimentally observed dependence of energy on the velocity and the initial gap is much weaker than that for a Newtonian fluid ($E_A \sim d_i^{-2},\ V^{-1}$), while it is stronger than that given by Hershel–Bulkley fluid model for small shear rates for which the energy dissipated is independent of both the initial gap and velocity. We further found that the strain at complete breakage ε_b (the instantaneous strain is evaluated by employing the relation $\varepsilon = (d(t)/d_i) - 1$) decreases with an increase in the top plate velocity or a decrease in the initial gap. We assess the role of the material viscoelasticity by plotting the energy dissipated in the separation process and the strain at breakage as a function of the Deborah number, $De = (\tau V/d_i)$, where d_i/V represents the time scale of initial deformation and τ is the relaxation time of the suspension, in Fig. 6. A high *De* (>1) denotes a predominantly elastic behavior, whereas $De < 1$ corresponds to viscous behavior. We find that both the energy dissipated and the strain at breakage increase with an increase in the Deborah number.

According to the mechanical energy balance, the dissipated energy is proportional to $\nabla_j u_i$ [24], where ∇_j is a vector differential operator and u_i is a velocity vector. An increase in the Deborah number is equivalent to an increase in $\nabla_j u_i$, and therefore, we find that viscous dissipation increases with an increase in *De*. Further, a higher strain before breakage is observed at smaller time scales of deformation because of a decreased elastic storage of the energy.

We also observe from Fig. 6 that both strain at breakage and dissipated energy do not depend solely on *De* because the data points do not collapse on a single

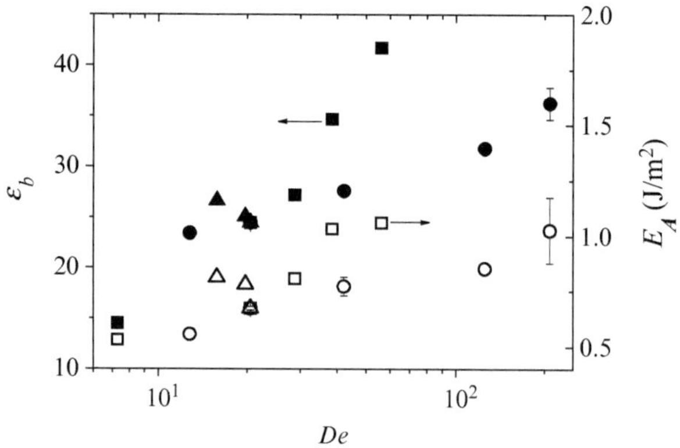

Fig. 6 A plot of strain at break (*filled symbols*) and energy dissipated (*open symbols*) as a function of the Deborah number, *De*, for various velocity (*circle*), initial thickness (*square*), and aging time (*triangle*)-dependent experiments. The *open symbols* have been plotted according to the scale shown on the right axis, while the *filled symbols* correspond to that on the left axis (Reprinted with permission from [143]. Copyright (2009) American Chemical Society)

curve when plotted against this dimensionless number. We believe that the following factors could be responsible for such behavior. Apparently, the energy dissipated does not follow a dependence of the form d_i/V but rather has a different dependence on the velocity and initial gap between the plates. The second possibility could be the continuous increase in the deformation time scale as distance between the plates increases, which causes a continuous decrease in *De* with time. Therefore, even the experiments corresponding to the same value ratio of d_i/V may lead to different strain and dissipated energy values. The third reason could be the thixotropic character of the present system. The partial rejuvenation of the material might reduce its viscosity as the deformation progresses. Due to the abovementioned reasons, the behavior might not be solely defined in terms of the initial time scale of deformation, d_i/V or *De*.

We have also shown the data points corresponding to aging time experiments in Fig. 6. As can be seen in the figure, both the strain at breakage and energy dissipated decrease with *De* or the relaxation time of the material. This trend is opposite in nature when compared to that of the velocity and initial gap-dependent tests carried out at the same age. The aging time data showed enhanced brittle failure, while the velocity/initial gap data showed pronounced viscous dissipation with increase in time scale of deformation d_i/V. The observations corresponding to the aging time experiments further support the claim that the tensile deformation of the suspension does not solely depend on the Deborah number but could also be influenced by various other reasons discussed above.

3.3 Constant Force Mode Experiments

In this group of experiments, the displacement of the top plate was measured under the application of a constant tensile force. A constant stress is maintained by the rheometer by controlling the movement of the top plate through a feedback mechanism which causes the force to fluctuate around its mean value. However, at a critical strain, the material starts to fail subsequent to which a constant stress can no longer be maintained by the rheometer. Therefore, after the initiation of failure, the value of normal force drops sharply, as shown in Fig. 7. For this reason, we have considered the data points only until the point of failure.

3.3.1 Variation of Aging Time and Deformation Field Strength

We carry out identical experiments, that is, under application of a constant force of the same magnitude ($F = 15$ N) and the same initial gap ($d_i \approx 100\ \mu$m) for samples which were aged for different time periods. Figure 8 shows a plot of time to failure t_f which refers to the time at which failure commences in the material which is accompanied by a sharp decrease in the force value. We find that t_f increases with increase in aging time following a power law dependence, $t_f \sim t_w^{0.3}$. This

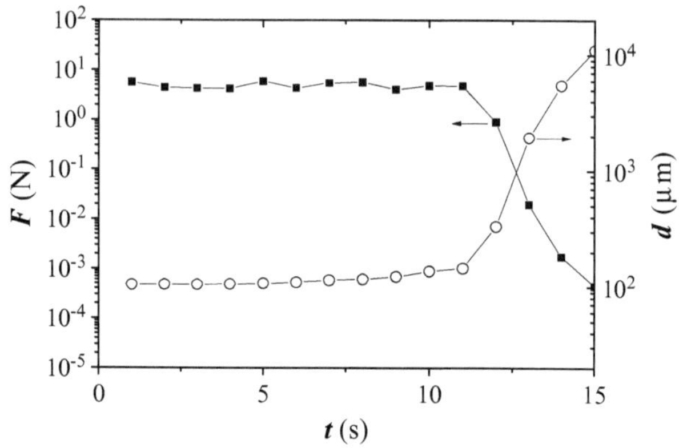

Fig. 7 Normal force F measured (*filled squares*) when the set value is equal to 5 N and variation of gap (*open circles*) between the plates as a function of experimental time t for $t_w = 15$ min, $d_i = 100$ μm (Reprinted with permission from [143]. Copyright (2009) American Chemical Society)

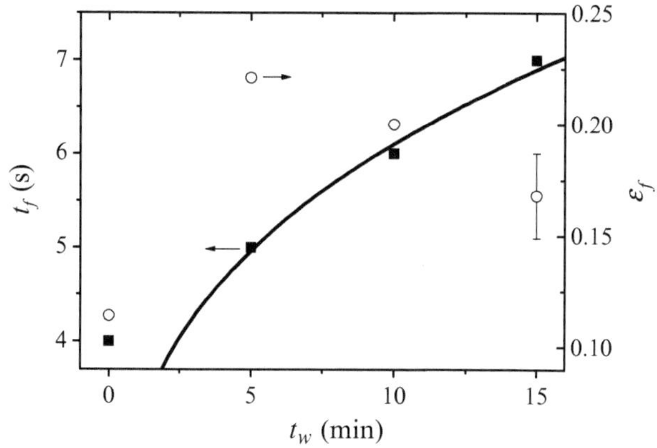

Fig. 8 The variation of time to failure (*filled squares*) and strain at failure (*open circles*) with aging time for $F = 15$ N and $d_i = 104 \pm 1$ μm. A power law fit to time to failure dependence on aging time, given by $t_f \sim t_w^{0.3}$, is shown with the help of a *solid gray line* (Reprinted with permission from [143]. Copyright (2009) American Chemical Society)

dependence is almost the same as that of relaxation time on aging time (as shown in Fig. 1). This suggests that the time to failure varies linearly with relaxation time ($t_f \sim \tau$). The figure also shows that the strain at failure ε_f decreases with the aging time (the point associated with zero age, for which case $t << t_w$, is not taken into consideration). For a constant force, strain at failure should be directly proportional to the energy dissipated in the process. Therefore, we find that strain at failure, like the energy dissipated, decreases with the aging time. This observation based on the constant velocity mode experiments also suggests that increase in elastic modulus induces brittleness in the system.

We next study the effect of the magnitude of the applied force and the initial gap between the plates on the tensile deformation of the Laponite suspension. We carried out force-dependent experiments by applying a constant tensile force of varying magnitude while keeping the initial gap ($d_i \approx 100$ μm) and aging time ($t_w = 15$ min) identical for each experiment. Similarly, for initial gap-dependent experiments, we varied the initial gap between the plates, but the aging time ($t_w = 15$ min) and the applied force ($F = 15$ N) were the same in every experiment. We observe that both the time to failure and strain at failure decrease with increase in applied normal force and the initial gap between the plates. The time of separation (which is approximately equal to time to failure) for a Newtonian fluid under application of a constant normal force is given by [21, 97, 153]: $t_s = 3\pi\eta R^4/2Fd_i^2$. In similarity to the constant velocity experiments, we find that response of the Laponite suspension is weaker in comparison to a Newtonian fluid.

Thus, in this study, we find that a soft glass exhibits interesting tensile flow dynamics which is greatly affected both by the extent of physical aging in the system and the deformation field intensity. Further, both the velocity and force mode experiments indicate that the tensile behavior of this material is far more complicated than what can be explained by simple fluid models. Therefore, we believe that a further detailed study would be required to identify the important dimensionless groups in this problem and to develop a suitable theoretical model which could predict the experimental behavior of this material.

4 Time–Aging Time–Stress Superposition Under Tensile Deformation Field

The characteristic relaxation time of a soft glass like a Laponite suspension follows a power law dependence on the aging time given by the following equation:

$$\tau \sim \tau_m^{1-\mu} t_w^{\mu} \tag{3}$$

where τ is the dominant relaxation time of the material, τ_m is a microscopic time scale which represents the time scale of attempt made by a particle to escape the energy well, t_w is the aging time, and $\mu = d \ln \tau / d \ln t_w$ represents the rate of aging and is a function of the applied stress [61, 157]. The parameter μ tends to unity under the limit of weak stress fields, while it tends to zero when the applied stress exceeds the yield stress of the material causing complete destruction of the microstructure. Under this limit, the system is in an unjammed state, and therefore no aging takes place in this situation [38].

A dependence of the relaxation time of the material on the aging time forms the basis of time–aging time correspondence which can be used to predict long-time behavior of a soft glass by carrying out short-time tests [69, 157]. Sollich et al. [61, 154, 155] developed a theoretical model known as soft glassy rheology (SGR) model to predict the rheological behavior of these materials. Further, Fielding et al. [61] developed the mathematical equivalent of the experimentally observed "time–aging time" superposition procedure which was originally developed by Struik for amorphous polymeric systems. Moreover, as explained above, the strength of the applied stress/strain field greatly affects the rate of aging. This suggests a presence of time–stress/strain correspondence which has indeed been validated in several studies [4, 71, 80, 156]. However, McKenna [98] claimed that application of stress does not lead to mechanical rejuvenation (or reversal of aging) in the material; rather, it leads the system to polyamorphism or a new deformation-induced phase. This report puts a question mark on the validity of time–stress superposition for these materials.

As the relaxation time is affected by both the aging time and the stress/strain, the time–aging time superposition was extended to include stress in the work carried out by Joshi and Reddy [73, 130] where they proposed the possibility of time–aging time–stress superposition. Owing to a strong influence of temperature on the relaxation behavior, a time–aging time–temperature superposition has also been shown to exist for this class of materials [6].

4.1 *Linear Oscillatory and Tensile Creep Tests*

A linear small amplitude oscillatory test on the sample used in this problem (3.5 wt% Laponite suspension with idle time equal to 3 months) revealed that the elastic modulus varied as $G' \sim t_{\mathrm{w}}^{0.22}$. The average relaxation time was calculated by employing a time-dependent single-mode Maxwell model ($\tau = G'/\omega G''$). The relaxation time thus determined follows a power law dependence on the aging time given by $\tau \sim t_{\mathrm{w}}^{0.67}$.

Under the application of a constant normal force, the top plate moves in the upward direction causing an enhancement in the tensile strain. As explained in the previous section, after the commencement of the failure in the film, the rheometer is not able to maintain a constant force. Therefore, we have considered data points only in the range in which the force remains constant. However, the tensile stress decreases even when the force is constant due to shrinking of the cross-sectional area as the material elongates in the normal direction. Thus, these tensile creep experiments can be termed as constant force or constant engineering stress experiments.

4.1.1 Tensile Compliance vs. Time Curves for Various Aging Times

Figure 9 shows the tensile compliance ($J\,(t + t_{\mathrm{w}}) = \varepsilon\,(t + t_{\mathrm{w}})\,/\sigma_{\mathrm{N}}$ where ε is the tensile strain and σ_{N} is the constant engineering stress) as a function of creep time (t) for four different aging times under the application of a constant normal force equal to 10 N ($\sigma_{\mathrm{N}} = 5.1$ kPa) [145]. We find that an increase in aging time reduces the compliance induced in the material. As evident from the figure, the compliance curve shifts toward larger creep times with increase in the aging time. We believe that this happens due to an enhancement in the elastic modulus and the relaxation time of the material with increase in aging time. It has been suggested in literature that the increase in both of these variables with aging time can be compensated by normalizing the compliance and the creep time by the elastic modulus and the aging

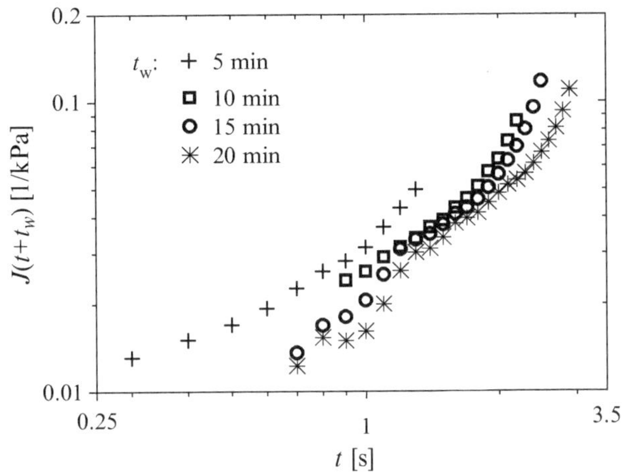

Fig. 9 Tensile compliance as a function of time for different aging times at a constant force of 10 N ($\sigma_N = 5.1$ kPa, $d_i = 100 \pm 3$ μm) (Reprinted from [145], with permission from Springer)

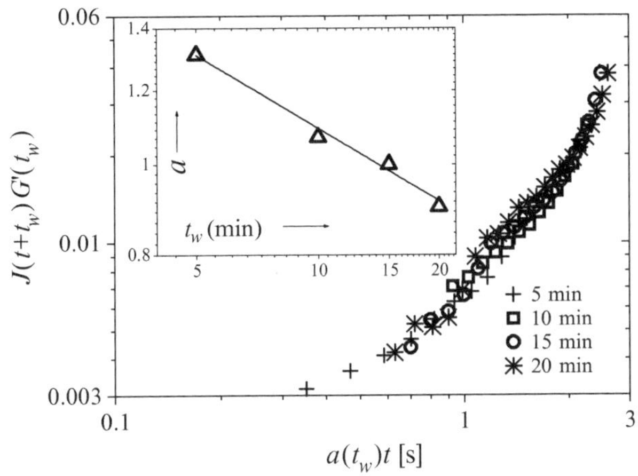

Fig. 10 Various aging time curves shown in Fig. 9 superposed onto 15-min data after carrying out horizontal and vertical shifting to obtain creep time–aging time superposition. *Inset* shows the dependence of horizontal shift factor on aging time given by $a \sim t_w^{-0.26}$ (Reprinted from [145], with permission from Springer)

time-dependent factor of the relaxation time, respectively. This leads to a collapse of all the data points corresponding to different aging time experiments on a single curve. The superposition of data, thus obtained, is known as the creep (process) time–aging time superposition [157]. Following a similar procedure, we shifted all the compliance curves horizontally by multiplying with a shift factor a and vertically by multiplication with $G'(t_w)$ to obtain a superposition as shown in Fig. 10. The value of the vertical shift factor $G'(t_w)$ was determined from the linear oscillatory shear experiment data. The inset of Fig. 10 shows a plot of the horizontal shift factor a as a function of the aging time t_w. The shift factor a represents that part of the relaxation time which depends on the aging time. From Eq. 3, the dependence of $1/a$ on t_w should lead to the dependence of the relaxation time τ on t_w. Figure 10 suggests that for a constant force of 10 N, the dependence of relaxation time on the aging time is given by $\tau \sim t_w^{0.26}$.

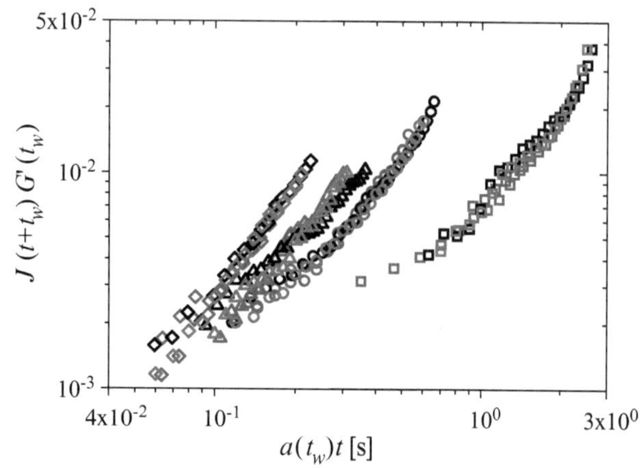

Fig. 11 Time–aging time superpositions from *left* to *right* correspond to the constant normal forces of 20 N ($\sigma_N = 10.191$ kPa, *open diamonds*), 17 N ($\sigma_N = 8.662$ kPa, *open triangles*), 15 N ($\sigma_N = 7.643$ kPa, *open circles*), and 10 N ($\sigma_N = 5.095$ kPa, *open squares*), respectively (initial gap $d_i = 100 \pm 3$ μm) (Reprinted from [145], with permission from Springer)

4.1.2 Time–Aging Time Superposition for Various Normal Stresses

We repeated a similar exercise to obtain process time–aging time superpositions at three more values of normal force: 15, 17, and 20 N. The master curves are shown in Fig. 11. Each of these superposed curves corresponding to a creep stress suggests a unique value of the shift rate μ which indicates a different dependence of relaxation time on the aging time at each stress. Figure 12 shows that μ decreases with an increase in stress which implies that the rate of aging is smaller at larger stresses. This decrease in μ can be attributed to an enhanced rejuvenation of the material under stronger deformation fields which could weaken the dependence of relaxation time on the aging time. The value of μ almost reduces to zero for the largest stress which indicates a near-complete rejuvenation of the material. Under this situation, the relaxation time becomes independent of the aging time, or in other words, the process of aging stops in the material. These observations are consistent with the qualitative trend of μ reported in literature for a shear flow field, in which case too a decrease in μ is observed with an enhancement in the strength of the stress field [38, 73].

4.1.3 Master Curve (Time–Aging Time–Normal Stress Superposition)

As the relaxation time depends both on the aging time and applied stress (through μ), it is expected that the superposed curves shown in Fig. 11 should collapse on a single master curve when the creep time is normalized by the relaxation time. We indeed find that a master curve can be obtained by shifting the superposed curves horizontally by multiplication with shift factor $b(\sigma_N)$ as shown in Fig. 13. Therefore, this shift factor represents the stress-dependent part of the relaxation time which can be determined as follows:

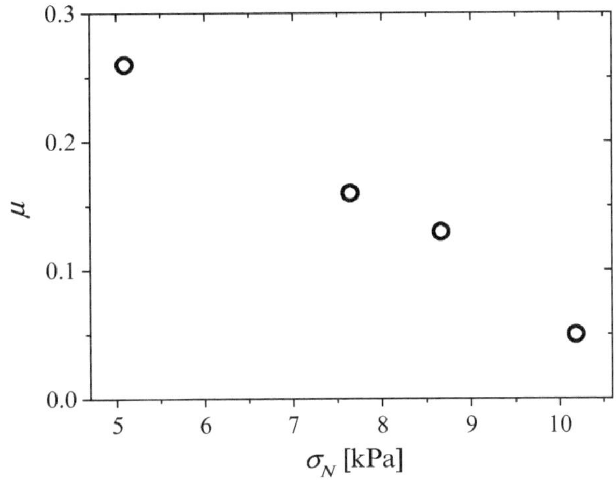

Fig. 12 The variation of power law index $\mu = d \ln \tau / d \ln t_w$ with normal stress (Reprinted from [145], with permission from Springer)

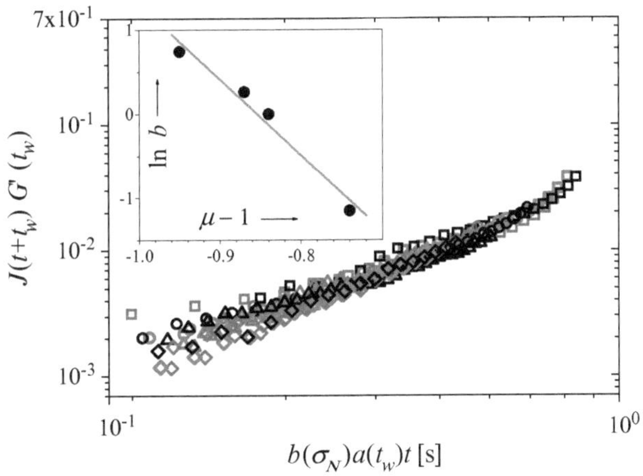

Fig. 13 The superposed curves shown in Fig. 11 are shifted on the horizontal axis to obtain a master superposition. A plot of the horizontal shift factor b with respect to $\mu - 1$ is shown in the *inset* (Reprinted from [145], with permission from Springer)

$$\frac{t}{\tau} \sim \frac{t}{\tau_m^{1-\mu} t_w^{\mu}} = b\left(\sigma_N\right) a\left(t_w\right) t \tag{4}$$

Since $a\left(t_w\right) \sim t_w^{-\mu}$, $b(\sigma_N)$ should depend on μ according to $b\left(\sigma_N\right) \sim \tau_m^{\mu-1}$ [or $\ln b \sim (\mu - 1) \ln \tau_m$]. The inset in Fig. 13 indeed shows that $\ln b$ varies linearly with $\mu - 1$ which verifies this dependence. Therefore, the master curve shown in Fig. 13 demonstrates the validity of *creep time–aging time–stress superposition* under tensile deformation field. The slope of $\ln b$ vs. $\mu - 1$ curve gives an estimate of the microscopic time scale $\tau_m = 1.23 \times 10^{-4}$ s. In the limit of complete rejuvenation, the dependence of relaxation time on aging time disappears, and it varies as $\tau \sim \tau_m$.

The process time–aging time–stress superposition has several implications. An important one is that the aging that occurs under a strong deformation field over a long duration of time would be equal to what would occur under a weaker deformation field over a shorter duration. This suggests that long-time rheological behavior of these materials can be predicted by carrying out short-time tests. This concept had been introduced for the behavior of soft glassy materials under shear flow [73]. Despite several fundamental differences between shear and tensile flow fields, in this work, we have shown that this superposition is valid for tensile deformation fields as well. Although time–stress superposition has been reported to be valid for amorphous polymers in literature [4, 71, 80, 156], we have validated, for the first time in our knowledge, time–aging time–stress superposition under tensile stress field.

5 Shear-Mediated Elongational Flow and Yielding in Soft Glassy Materials

Most of the previous studies concerned with deformation behavior of soft glassy materials are confined to the case of a stress/strain field which acts in a single direction, for instance, pure rotational shear flow [38, 73], elongational flow [4, 71, 80, 143, 145, 156], squeeze flow field [56, 102, 132], etc. However, one can find numerous examples in industrial applications and everyday life in which more than one stress field act simultaneously on these materials. In this problem, we study the deformation and yielding behavior of various soft glassy materials under a combined application of tensile and rotational shear stresses. We have chosen four soft glasses in this study: a commercial hair gel, a Laponite suspension (3.5 wt% with idle time of 3 months), a shaving foam, and an emulsion paint. All of these materials belong to a distinct category of soft glassy materials as they have different underlying micro-dynamics which leads to unjamming in the system. The quantity of hair gel in a commercially available container is around 100 ml which is not sufficient to perform the whole set of experiments. Due to this reason, we have used four samples of hair gel (from different batches) which we have labeled as hair gels 1, 2, 3, and 4 for performing different sets of experiments in this study. The rheological behavior of all these samples was identical qualitatively, but the material properties varied to some extent from batch to batch.

In the first part of this work, we study the deformation behavior of these materials under simultaneous application of two stress fields in directions perpendicular to each other, namely, a tensile force field (which produces a radial shear stress field) and a rotational stress field. In the second part, we study the yielding dynamics of soft glassy materials. Based on the experimentally determined values of yield stress for different combinations of applied stresses, we construct a universal yielding phase diagram for these materials.

5.1 Deformation Under a Combination of Tensile and Rotational Stress Fields

Under a combined application of constant tensile force F and torque T, the top plate starts to move upward while also undergoing rotation. This causes an enhancement in both the tensile ε and the rotational strain $\gamma_{\theta z}$ with time. As discussed earlier, the rheometer maintains a constant force only until the initiation of failure in the film. Thereafter, the normal force starts to decrease sharply. Hence, we consider the data points only until the commencement of failure in the film.

As the top plate moves up under the application of a normal force, a pressure gradient is created in the radial direction. This induces a radial shear stress τ_{rz} which causes the material to move radially inward, thereby reducing the cross-sectional area of the film. As the instantaneous radial dimension of the film r_{in} is much larger than the vertical one ($d << r_{in}$), the radial velocity v_r should also be much greater than the vertical velocity v_z ($v_r >> v_z$). In the limit of small gap, the lubrication approximation holds valid, and therefore, the only nonzero components of the stress tensors are the radial shear stress τ_{rz} and the rotational shear stress $\tau_{\theta z}$. For very small Reynolds number ($Re << 1$), the radial stress value can be determined from the normal force by employing the following relation:

$$\tau_{rz} = \frac{3Fd}{2\pi {r_{in}}^3} \tag{5}$$

and the rotational stress is related to torque as follows:

$$\tau_{\theta z} = \frac{3T}{2\pi {r_{in}}^3} \tag{6}$$

The radial strain rate is given by [119]

$$\dot{\gamma}_{rz} = \frac{V r_{in}}{[d(t)]^2} \tag{7}$$

Replacing the velocity V by $d[d(t)]/dt$ in the above equation and integrating, we obtain the radial strain:

$$\gamma_{rz} = \frac{2Rd_i^{1/2}}{3}\left[\frac{1}{d_i^{3/2}} - \frac{1}{[d(t)]^{3/2}}\right] \tag{8}$$

The true rotational strain is given by

$$\gamma_{\theta z} = \int_0^t \frac{\Omega(t) r_{in}(t)}{d(t)} dt \tag{9}$$

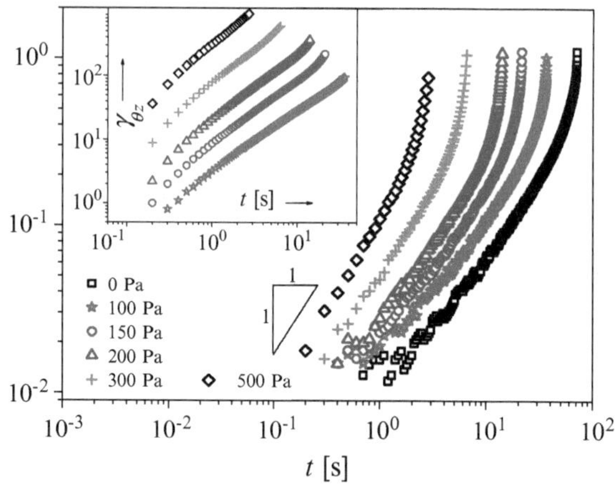

Fig. 14 Tensile strain variation with respect to time for a constant normal force $F = 10$ N and various rotational shear stresses for hair gel 1. The corresponding rotational strain $\gamma_{\theta z}$ as a function of time, for each rotational shear stress $\tau_{\theta z}$, is shown in the *inset* ([144] - Reproduced by permission of The Royal Society of Chemistry)

where Ω is angular velocity and $\Omega r_{\text{in}}/d$ is the true rotational strain rate ($\dot{\gamma}_{\theta z}$) experienced by the material [23].

In Fig. 14, we plot the tensile strain as a function of time under application of a constant normal force ($F = 10$ N) but different rotational shear stresses $\tau_{\theta z}$ for hair gel 1 sample [144]. As mentioned above, the radius of the material film reduces with increase in the tensile strain. Therefore, the true values of the radial and rotational shear stresses change with time even though the values of normal force and torque remain constant. As can be seen in the figure, the temporal evolution of the tensile strain has a linear slope which increases sharply close to failure. Remarkably, we observe that an increment in rotational shear stress not only leads to an increase in the rotational shear strain as expected, but it also causes an enhancement in the tensile strain. We believe that this behavior is extraordinary, as application of a stress field results in an enhancement of strain in a direction perpendicular to itself.

We also investigated the response when the rotational shear stress is held constant and the normal force is varied. Similar to the observation in Fig. 14, in this case too, an enhancement in the rotational strain was observed with increase in the tensile force. However, the increment observed in the rotational strain is small relative to the increase in the value of the normal force. This is probably because the value of the rotational stress $\tau_{\theta z}$ is much greater in magnitude as compared to τ_{rz} induced due to normal force. Also, the rotational stress applied in this problem is either comparable or greater than the yield stress of the hair gel. This causes unjamming of the material in all directions. Therefore, an increase in rotational stress leads to an enhanced tensile strain and vice versa. Further, it is evident from Figs. 14 and 15 that all the tensile and rotational strain time curves have a similar curvature. This suggests that an increment in the magnitude of the resultant stress field merely shifts the curves toward smaller values of time. Apparently, the same path is followed in each case during deformation; however, depending on the strength of stress fields,

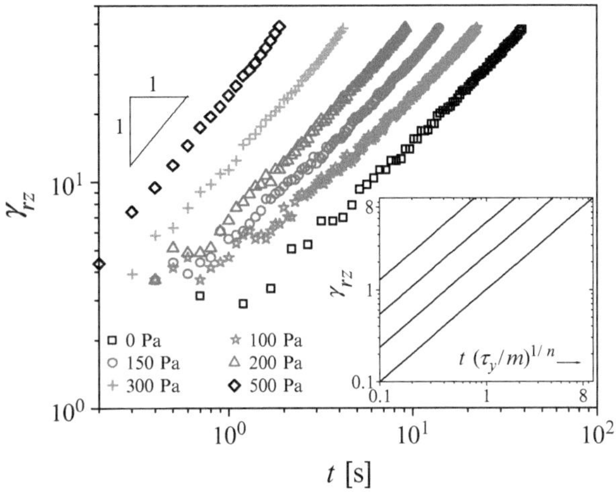

Fig. 15 Radial shear strain γ_{rz} as a function of time for the data shown in Fig. 14 for constant normal force of 10 N ($\tau_{\mathrm{rz}} = 30.6$ Pa) but different $\tau_{\theta\mathrm{z}}$ values. The *inset* shows the theoretical prediction of γ_{rz} (Eq. 12) with respect to normalized time for $n = 0.5$, $\tau_{\mathrm{rz}}/\tau_{\mathrm{y}} = 2$, and varying $\tau_{\theta\mathrm{z}}/\tau_{\mathrm{y}} = 0,\ 2,\ 4,\ 8$ (from *right* to *left*) ([144] - Reproduced by permission of The Royal Society of Chemistry)

the time scale involved in the process might be different. We further carried out a similar exercise for the Laponite suspension and the foam. We find that both of these soft glassy materials also demonstrate a deformation behavior qualitatively similar to that of the hair gel under a combination of stress fields. As mentioned above, all of these materials possess very different microstructures; therefore, we believe that these findings are consistent with the behavior of soft glassy materials in general.

A general constitutive relation for a yield stress fluid is given below [23, 50]:

$$\begin{aligned} \underset{\approx}{\tau} &= G\underset{\approx}{\gamma}, && \text{for } \sqrt{\underset{\approx}{\tau} : \underset{\approx}{\tau}/2} < \tau_{\mathrm{y}} \\ \underset{\approx}{\tau} &= \left(\frac{\tau_y}{\dot{\gamma}} + \mu\right)\underset{\approx}{\dot{\gamma}}, && \text{for } \sqrt{\underset{\approx}{\tau} : \underset{\approx}{\tau}/2} \geq \tau_{\mathrm{y}} \end{aligned} \tag{10}$$

where $\underset{\approx}{\tau}$ is the deviatoric stress tensor, G is elastic modulus, $\underset{\approx}{\gamma}$ is strain tensor, $\underset{\approx}{\dot{\gamma}}$ is rate of strain tensor, τ_{y} is the yield stress, $\dot{\gamma}$ is the second invariant of rate of strain tensor $\left(\sqrt{\underset{\approx}{\dot{\gamma}} : \underset{\approx}{\dot{\gamma}}/2}\right)$, and μ is the consistency which is a function of the shear rate $\dot{\gamma}$. For the case of Herschel–Bulkley fluid model, $\mu = m\dot{\gamma}^{n-1}$ where m and n are model parameters [40]. Therefore, in this case, Eq. 5 can be written as $\underset{\approx}{\tau} = \tau_{\mathrm{y}} + m\dot{\gamma}^{n-1}\underset{\approx}{\dot{\gamma}}$ for $\sqrt{\underset{\approx}{\tau} : \underset{\approx}{\tau}/2} \geq \tau_{\mathrm{y}}$. The rotational and radial shear strains for a Herschel–Bulkley model can be expressed as follows:

$$\gamma_{\theta\mathrm{z}} = \frac{\left(\tilde{\tau}_{\theta\mathrm{z}}\sqrt{1 + \left(\tilde{\tau}_{\mathrm{rz}}^2/\tilde{\tau}_{\theta\mathrm{z}}^2\right)} - 1\right)^{1/n}}{\sqrt{1 + \left(\tilde{\tau}_{\mathrm{rz}}^2/\tilde{\tau}_{\theta\mathrm{z}}^2\right)}} \left\{t\left(\frac{\tau_{\mathrm{y}}}{m}\right)^{1/n}\right\} \quad \text{and} \tag{11}$$

$$\gamma_{\mathrm{rz}} = \frac{\left(\tilde{\tau}_{\mathrm{rz}}\sqrt{1+\left(\tilde{\tau}_{\theta\mathrm{z}}^2/\tilde{\tau}_{\mathrm{rz}}^2\right)}-1\right)^{1/n}}{\sqrt{1+\left(\tilde{\tau}_{\theta\mathrm{z}}^2/\tilde{\tau}_{\mathrm{rz}}^2\right)}}\left\{t\left(\frac{\tau_{\mathrm{y}}}{m}\right)^{1/n}\right\}, \tag{12}$$

where $\tilde{\tau}_{\theta z} = \tau_{\theta z}/\tau_y$ and $\tilde{\tau}_{rz} = \tau_{rz}/\tau_y$.

We determined the radial shear strain in the limit of small gaps $(d(t)/R_0 \leq 0.05)$ from the tensile strain values shown in Fig. 14 by using Eq. 5. In Fig. 15, we plot the radial shear strain as a function of time. As expected, the temporal evolution of the radial strain shows the same trend as that of the tensile strain. In the inset of Fig. 15, we plot the radial shear strain calculated from Eq. 12, with respect to normalized time, for a constant value of $\tilde{\tau}_{rz}$ but different values of $\tilde{\tau}_{\theta z}$. It can be seen that the radial strain–time curves shift toward lower times with an increase in $\tau_{\theta z}$ even though τ_{rz} is constant; therefore, the Herschel–Bulkley fluid model describes the experimental behavior qualitatively. Similarly, Eq. 11 predicts an enhancement in the rotational strain with increase in the normal force (or τ_{rz}) which also matches with the experimental observations.

In Fig. 16, we plot the time to failure t_f for the experiments carried out for hair gel 1, foam, and the Laponite suspension as a function of the second invariant of the applied stress tensor. Interestingly, we find that the time to failure does not only depend on the value of the normal force as expected, but it also decreases with an increase in the value of the rotational stress. We believe that the reduction in the time to failure occurs, even when the normal force is unchanged, due to a greater unjamming of the material caused by a larger rotational stress. We observe from Fig. 16 that for approximately the same value of the invariant of the stress tensor, the time to failure varies significantly with a change in normal force. It has been reported in literature that t_f varies inversely with applied normal stress [153]. Remarkably, we find that when more than one stress field is applied, time to failure follows an inverse dependence on invariant of the stress tensor.

5.2 *Yielding Phase Diagram*

Under application of a stress/strain field, partial or complete unjamming of the material takes place depending on the yield stress of the material. These materials deform elastically when the stress field is weaker than the yield stress of the material. If the yield stress is exceeded, a structural breakdown in the material takes place resulting in plastic flow [11, 17, 27, 32, 36, 38, 44, 73, 79, 121, 122, 133]. Hence, the yield stress plays an important role in determining the response of the material to the applied deformation field; however, due to the shear localization and thixotropic behavior of the material, an accurate determination of the yield stress value is considered to be a difficult exercise [15, 108, 109]. Further, the yield stress is also greatly influenced by the deformation history and the aging time of the material [38, 41, 54].

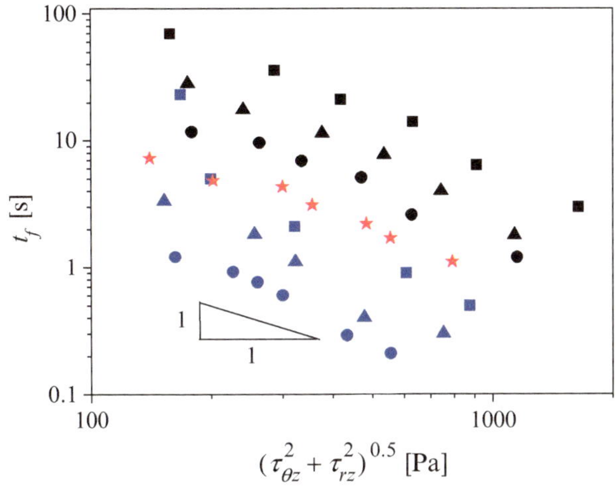

Fig. 16 The variation of time to failure with the invariant of true stress tensor corresponding to various normal force values, 5 N (τ_{rz} = 15.3 Pa, *stars*), 10 N (τ_{rz} = 30.6 Pa, *squares*), 15 N (τ_{rz} = 45.9 Pa, *triangles*), and 20 N (τ_{rz} = 61.2 Pa, *circles*), and various shear stress values ($\tau_{\theta z}$ = 0 − 500 Pa). *Black*, *red*, and *blue symbols* represent hair gel 1, shaving foam, and the Laponite suspension, respectively ([144] - Reproduced by permission of The Royal Society of Chemistry)

In order to determine the yielding curve for soft glassy materials, we determine various combinations of radial and rotational shear stresses which lead to the yielding event in the material. For determining these combinations, we carry out tests in which a constant normal force was applied to the material, but the rotational stress is varied. The value of rotational stress at which the yielding event takes place (which is evident from a sharp decrease in the value of the viscosity) is considered to be the critical rotational stress $\tau_{\theta zy}$ for that force or radial stress. However, as the soft glassy materials employed in this study are thixotropic in nature, the critical rotational stress may depend on the rate at which the rotational stress was varied. To investigate this point, we carry out a separate set of experiments on another hair gel sample, hair gel 2, in which we vary the rotational stress at different rates under application of a constant normal force and determine the critical rotational stress in each case. We find that the critical rotational stress is almost independent of the rate of variation of the rotational stress in the range explored in the experiments.

In Fig. 17, we plot the viscosity as a function of rotational shear stress for various values of the normal force for hair gel sample 3. The applied $\tau_{\theta z}$ ramp in these experiments was within the explored range wherein the critical rotational stress was found to be independent of the applied ramp. We repeat this procedure for four more soft glassy materials: hair gel 4, the Laponite suspension, foam, and emulsion paint. In Fig. 18, we plot the yielding phase diagram for all the soft glassy materials under study. This plot constitutes the normalized combinations of the true values (based on the instant value of the film area) of the radial and critical rotational shear stresses at which the yielding event takes place. Interestingly, we find that all the yielding

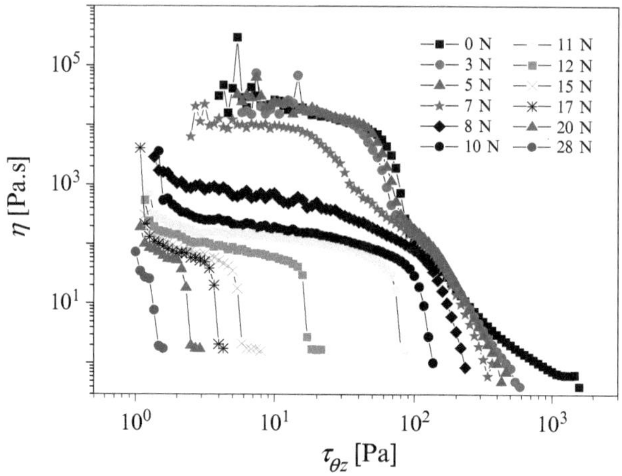

Fig. 17 The variation of viscosity η with respect to rotational shear stress under the application of various normal forces for hair gel 3. The rotational stress at which the yielding in the material takes place is considered to be the critical rotational stress $\tau_{\theta zy}$ for that normal force ([144] - Reproduced by permission of The Royal Society of Chemistry)

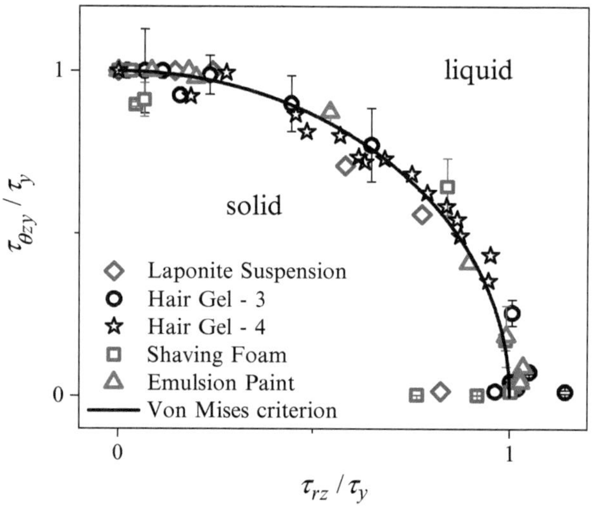

Fig. 18 The yielding phase diagram for various soft glassy materials (the Laponite suspension, $\tau_y = 80$ Pa; hair gel 3, $\tau_y = 135$ Pa; hair gel 4, $\tau_y = 18.3$ Pa; foam, $\tau_y = 68$ Pa; paint, $\tau_y = 35$ Pa) where τ_y is the yield stress in simple shear. The *symbols* represent various combinations of the normalized rotational and radial shear stress values at which the yielding takes place. The behavior can be predicted closely with the Von Mises criterion which is shown as a *solid line* ([144] - Reproduced by permission of The Royal Society of Chemistry)

data points fall on the curve representing the Von Mises criterion. According to this criterion, the yielding event takes place in a material when the invariant of the stress tensor exceeds the yield stress of the material [119]. However, some scatter in the data can be observed in the limit of tensile stress-dominated flows. We believe that this results due to the formation and propagation of fingers and/or cavitation which is usually present in pure tensile flows.

As mentioned before in the Introduction section, in a previous study, the Von Mises criterion for yielding was observed to work well for the experimentally determined data for the case of simultaneous rotational and squeeze flow [119]. On the other hand, we find that this criterion also works well in the case of combined rotational and tensile flow fields. This result is interesting because a tensile flow between two separating parallel plates is not merely a reversal of direction of squeeze flow but is much more complicated in comparison to a squeeze flow. As explained earlier, this type of flow can become increasingly complex with an increased strength of deformation field because of increased possibility of finger formation or cavitation due to the presence of a large pressure gradient in the radial direction which can significantly alter the behavior.

6 Squeeze Flow Behavior of Thixotropic Soft Glassy Material

A Herschel–Bulkley or Bingham plastic model has been employed in numerous studies in literature to predict the behavior of soft glassy materials possessing yield stress [23, 33, 40, 102, 103, 119, 129]. However, under some situations, these models, owing to their simple nature, fall short of satisfactorily predicting the behavior of soft glassy materials. An important drawback associated with these models is the absence of a time-dependent term. Such a term becomes mandatory when the fluid properties vary with time. Most of the yield stress fluids possess a thixotropic character as well because of the fact that the same underlying phenomenon of the rearrangement/destruction of the constituent microstructure gives rise to both [109]. Several models have been developed to predict the thixotropic behavior. These models usually employ a structural parameter which evolves with time. This parameter is a function of the degree of flocculation and jamming or the fraction of particles trapped in energy wells [15, 34, 43, 113].

In this work, we study the behavior of a soft glass (3.5 wt% Laponite suspension with idle time equal to 1 month) when squeezed between two parallel plates. The plates employed in this study had a different surface roughness: the top plate was comparatively rough, while the bottom plate was smooth. In literature, plates having dissimilar characteristics have been employed for shear flow studies [138]. Many commercially important soft glassy materials undergo squeeze flow between two surfaces, for example, highly filled polymer melt, wheat dough, etc. In applications where preferential sticking of the material on one of the surfaces is desired, solid substrates with a dissimilar surface roughness are employed. Therefore, the present study finds relevance in this context.

6.1 *Elastic Modulus and the Yield Stress as a Function of Aging Time*

As mentioned in detail earlier, the Laponite suspension demonstrates physical aging, and therefore, various properties of this material like the elastic modulus and the yield stress are expected to evolve with time. In order to determine the

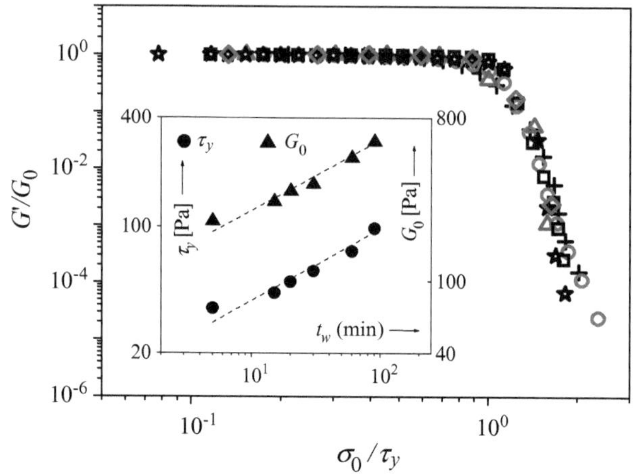

Fig. 19 Normalized elastic modulus as a function of normalized shear stress amplitude for the Laponite suspension in oscillatory tests at frequency $f = 1$ Hz for various aging times (*square*, 5 min; *circle*, 15 min; *triangle*, 20 min; *diamond*, 30 min; *star*, 60 min; and *plus*, 90 min). The variation of yield stress (τ_y) and linear regime elastic modulus (G_0) with aging time (t_w) is shown in the *inset*. Both τ_y and G_0 exhibit the same dependence on t_w given by $\tau_y, G_0 \propto t_w^{0.4}$ (Reprinted from [146], with permission from Elsevier)

dependence of the yield stress on the aging time, we carry out a set of shear stress ramp oscillatory experiments for samples aged for different durations of time and determine the yield stress in each case. In Fig. 19, we plot the elastic modulus (G') normalized with the elastic modulus associated with linear response regime (G_0), as a function of shear stress amplitude (σ_0) normalized with yield stress (τ_y), for various aging times [146]. Interestingly, all the curves corresponding to different aging times show superposition indicating a similarity in the yielding transition for all the aging times. Moreover, at the instance of yielding, the elastic modulus decreases to a very low value over a very narrow range of stress amplitudes, thus enabling the determination of the yield stress. As the material is thixotropic in nature, an error in the value of the yield stress might get introduced due to the same; however, the rate of change in shear stress was high which allowed the completion of the whole experiment in a few seconds. Also, as this time period was much smaller than the aging time of material, the structural evolution that takes place in duration of the experiment is negligibly small. The inset of the figure shows that τ_y and G_0 follow the same dependence on the aging time ($\tau_y,\ G_0 \propto t_w^{0.4}$).

In order to find the dependence of stress on shear rate and to determine material parameters for the Laponite suspension under study, we carry out a steady shear rate ramp test. We find that the Herschel–Bulkley fluid model satisfactorily describes the behavior shown by this material. Upon fitting the data points with this model, we obtained the value of the power law index n and the consistency k (we have used this notation for consistency in place of m to differentiate it from the plate frictional coefficient which we have employed in the squeeze flow analysis later in this problem) equal to 0.28 and 8 Pa.s^n, respectively, for a sample aged for 15 min. However, as explained above, due to thixotropy, the behavior is expected to be greatly dependent on the flow field and the history. Therefore, it is possible that one might obtain a different set of Herschel–Bulkley model parameters for different rates of variation of shear rate.

Unlike the tensile experiments discussed in the previous sections, in all of the squeeze flow experiments (both constant force and velocity modes) performed in this study, the cross-sectional area of the film remains constant throughout the experiment. As the squeeze flow progresses, the sample volume has to continuously decrease due to the decreasing gap between the plates. The excess sample is pushed out from between the plates which gets accumulated on the periphery of the plates. This may lead to a small pressure buildup [56, 106]; nevertheless, a constant sample area is still preferred over the constant volume because it gives an advantage of a known sample area at all times.

6.2 Constant Force Experiments

In the constant force mode, a constant compressive force was applied to the top plate, and the variation of gap between the plates was observed as a function of time. Figure 20 shows the gap vs. time for two cases: a rough top plate with a smooth bottom plate and both smooth top and bottom plates, respectively. It can be

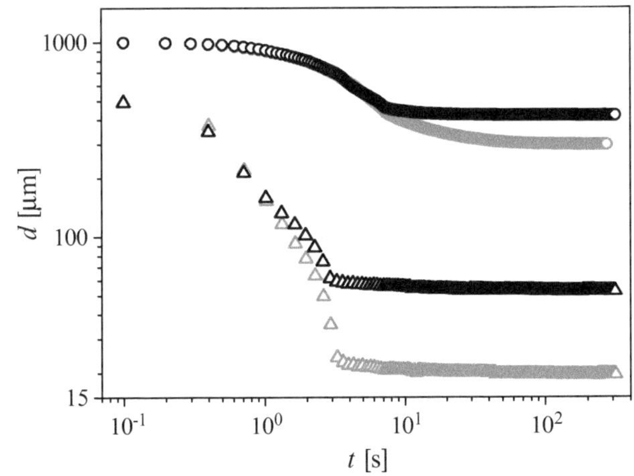

Fig. 20 The variation of gap (d) between the plates with respect to time (t) for rough (*black symbols*) and smooth top plate (*gray symbols*) for two normal compressive forces: $F = 2.5$ N (*circles*) and $F = 10$ N (*triangles*). For all the experiments shown in the figure, the bottom plate is smooth and the aging time is $t_w = 15$ min. The limiting gap d_L subsequent to which no further deformation takes place is always higher for the rough plate (Reprinted from [146], with permission from Elsevier)

seen in the figure that in both the cases, the gap decreases with time in the beginning, but it eventually reaches a plateau. In literature, it has been shown that in contrast to viscous fluids where the plate separation continuously decreases with time, for yield stress fluids, the gap between the plates ceases to reduce any further after a certain time. This plateau value of the gap is dependent on the yield stress of the material and therefore can be used to determine the value of the same [45, 129]. Further, it can be observed in the figure that the plateau value of the gap for smooth top plate is smaller than that for the rough plate. This suggests a considerable influence of the boundary condition (slip between material and plate surface) on the squeeze flow behavior of the material.

An important dimensionless group often mentioned in squeeze flow literature for yield stress fluids is the plasticity number $S = \left(Vk^{1/n}R\right) / \left(\tau_y^{1/n}d^2\right)$ which represents the contribution of viscous stress to the yield stress. In the limit of small plasticity number ($S << 1$) and under no-slip boundary condition, the limiting gap d_L (or plateau value of the gap) is given by [45]

$$d_L = \frac{2\pi R^3 \tau_y}{3F} \tag{13}$$

The value of yield stress determined using Eq. 13 by substituting the value of d_L for $F = 2.5$ N was around 0.74 times of the value predicted from the oscillatory experiments shown in Fig. 19. As the value of the limiting gap is smaller for the case of smooth top plate in comparison to the rough plate at this force, the value of the yield stress determined in this case is even smaller. We believe that this discrepancy arises due to the presence of a slip between the material and the plate surface whose effect intensifies for the case of smooth top plate. Sherwood and Durban [148] developed an expression for a Herschel–Bulkley fluid under squeeze flow when a partial slip is present at the fluid–plate interface and the friction coefficient m (≤ 1) or the roughness of both the plates is the same. For $d << R$, $m^2 << 1$, and $m >> d/R$, the limiting gap is given by [101]

$$d_L = \frac{2\pi R^3 m \tau_y}{3F} \tag{14}$$

For the case of smooth top and bottom plates, Eq. 14 gives $m = 0.53$ for $F = 2.5$ N.

6.2.1 The Expression for Limiting Gap for the Case of Dissimilar Plates

We derived an expression for the case of squeeze flow of Herschel–Bulkley fluid confined between plates with dissimilar surface roughnesses by carrying out an exercise on the same lines as that of Sherwood and Durban [147, 148] with the exception of different boundary conditions at the two confining surfaces. In the limit of $d << R$, $m_1^2, m_2^2 << 1$, and $m_1, m_2 >> d/R$, where m_1 and m_2 are the friction coefficients of the top and bottom plates, respectively, the limiting gap is given by

$$d_{\mathrm{L}} = \frac{\pi R^3 (m_1 + m_2)\,\tau_{\mathrm{y}}}{3F} \tag{15}$$

On substitution of the value of friction coefficient for the smooth plate ($m_2 = 0.53$), we determine the friction coefficient for the rough plate, $m_1 = 0.94$ for $F = 2.5$ N. Interestingly, for $F = 10$ N, Eqs. 14 and 15 give $m_1 = 0.4$ and $m_2 = 0.14$. As the friction coefficients of the plates should be independent of the applied force, we believe that a much smaller value of the limiting gap is observed due to the thixotropic behavior of the fluid which leads to partial shear melting (rejuvenation) at larger forces, causing a reduction in the value of the elastic modulus and the yield stress. As a Herschel–Bulkley model is time independent in nature, it fails to completely describe the observed behavior of a soft glassy fluid under a squeeze flow field. Further, we believe that the value of the limiting gap is smaller for the case of a smooth top plate as compared to a rough plate possibly due to an enhanced shear melting in the former case. A squeeze flow for no slip at the material–plate interface should mainly comprise a radial shear flow in the limit of $d << R$ (lubrication approximation) [40, 49]. However, when slip is present between the plate and sample surfaces, an elongational flow takes place in addition to the shear flow [104]. As elongational flow field is a much stronger flow field in comparison to shear flow field, a greater extent of shear melting is expected in the presence of slip, which results in a smaller value of the limiting gap.

6.2.2 Dependence of Limiting Gap on Initial Gap, Aging Time, and Force

For these set of experiments, we have employed a rough top plate. Figure 21 shows the variation of the gap between the plates with respect to time under the application of the same constant force $F = 2.5$ N but for different initial values of the gap. The inset of the figure shows the force as a function of time. Though a constant value of force equal to 2.5 N was set in the rheometer program, we find that the value of the force fluctuates around the set value. This is because the rheometer maintains a constant value of the force by a feedback mechanism, and therefore, there is a slight delay before the value is corrected following a deviation from the desired value. Interestingly, the main figure shows that the limiting gap is nearly the same irrespective of the value of the initial gap between the plates. However, the rate of change of the gap with time until the plateau value is attained is different for each initial gap. We believe that a difference in the approach of the gap to the limiting value is observed due to the fluctuations in the value of the applied force which can possibly effect the deformation of the sample. The Herschel–Bulkley model (Eq. 15) predicts the limiting gap to be independent of the initial gap, and therefore, it is in agreement with the experimental observations.

We also observe that the limiting gap increases with increase in the aging time with the dependence: $d_{\mathrm{L}} \sim t_{\mathrm{w}}^{0.42}$ (at $F = 2.5$ N and initial gap $d_{\mathrm{i}} = 1{,}000$ μm). We believe that this behavior originates from an increased yield stress with an increase in aging time. Indeed, it can be seen in Fig. 19 that the yield stress follows

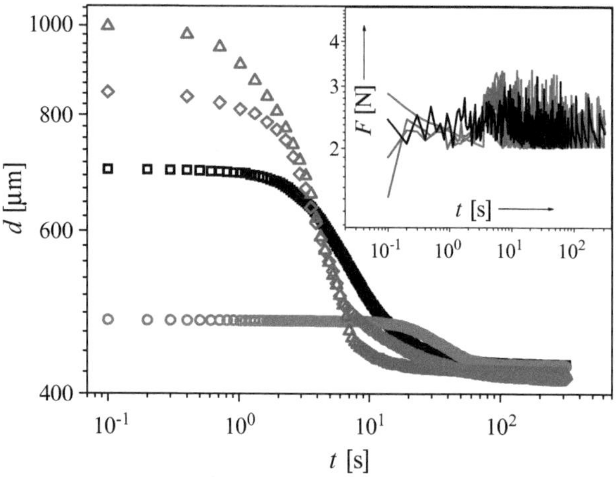

Fig. 21 The variation of gap between the rough top and smooth bottom plates (*d*) with time (*t*) for different initial gaps (*circle*, 500 μm; *square*, 700 μm; *diamond*, 850 μm; *triangle*, 1,000 μm) under the application of a constant force $F = 2.5$ N. The aging time is $t_w = 15$ min for all the experiments shown in the figure. The *inset* shows the fluctuations in the magnitude of the normal force *F* around the set value of 2.5 N as a function of time. In the *inset*, the *colors* represent the same initial gap as in the main figure (Reprinted from [146], with permission from Elsevier)

approximately the same dependence on the aging time ($\tau_y \sim t_w^{0.4}$); therefore, it can be inferred from these experimental observations that $d_L \sim \tau_y$. This dependence of the limiting gap on the yield stress is the same as that predicted by the Herschel–Bulkley model (Eq. 15). Therefore, for a relatively weak deformation field ($F = 2.5$ N), when the thixotropic effects are not pronounced, the Herschel–Bulkley model can correctly predict the squeeze flow behavior exhibited by this material. However, when we vary the magnitude of the applied force ($t_w = 15$ min and $d_i = 500$ μm for each test), we find that the limiting gap varies as $d_L \sim F^{-1.5}$. This dependence is stronger than what is predicted by a Herschel–Bulkley model, according to which d_L should vary inversely with the force *F* (Eq. 15). A possible reason for a smaller value of the limiting gap observed at larger forces, than what is theoretically predicted, could be an enhanced thixotropic effect which may cause shear melting in the sample to a greater extent in comparison to the case of smaller forces. Further, an enhanced shear melting would cause a reduction in the yield stress value which may lead to a much stronger reduction in the limiting gap with increase in force than what is theoretically predicted by this model.

6.3 *Constant Velocity Experiments*

In this group of experiments, the top plate moves at a constant velocity *V* toward the bottom plate leading to a squeeze flow in the sample. In Fig. 22, we plot the normal force as a function of gap between the plates for a rough and a smooth top

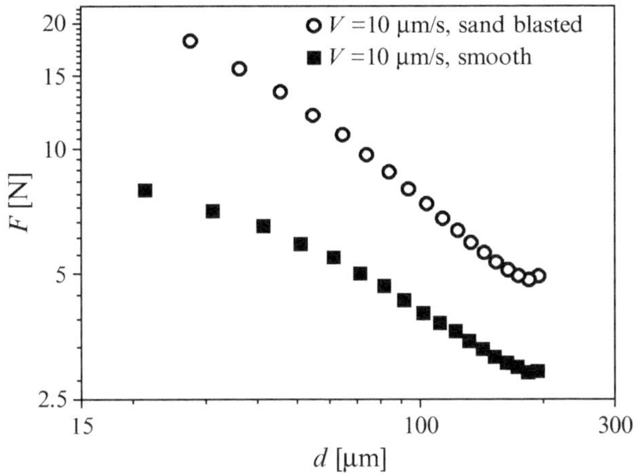

Fig. 22 The normal force (F) as a function of gap (d) between the plates for different top plate roughnesses for an initial gap of 200 μm. For both the tests shown in the figure, the velocity (V) is 10 μm/s and the aging time $t_w = 15$ min. The value of slope is different for the force–gap curves corresponding to smooth and rough top plates (Reprinted from [146], with permission from Elsevier)

plate for the same constant velocity $V = 10$ μm/s and aging time $t_w = 15$ min. As pointed out earlier, the bottom plate in both these cases is smooth. We observe that as the experiment progresses, a decrease in the gap between the plates leads to an enhancement in the normal force. However, a difference can be noticed in the manner of the variation of the force with the gap for rough and smooth top plates. The value of the normal force is significantly larger for the rough plate at all the values of the gap. Further, the slope of the force–gap curve is smaller in the case of the smooth top plate. These observations point toward a greater extent of shear melting prevalent in the case of smooth plate due to a pronounced thixotropic effect. This is in agreement with the observations made for the constant force experiments for rough and smooth top plates. As discussed earlier, we think that an increase in slip at the material–plate interface introduces an elongational flow in addition to the shear flow which leads to an enhanced shear melting or rejuvenation in the sample. This causes a weaker dependence of force on the gap for the case of smooth top plate.

6.3.1 Effect of Initial Gap, Aging Time, and Force

Figure 23 shows the variation of normal force with the gap for various initial gaps between the plates but the same top plate velocity $V = 10$ μm/s and aging time $t_w = 15$ min. Interestingly, we find that all the curves corresponding to various initial gaps fall on top of each other. Therefore, the normal force follows a unique dependence on the gap (given that the velocity and the aging time are not changed). This suggests that under squeeze deformation, the system has an identical state for a certain plate separation irrespective of the initial gap. As mentioned before, we derived an expression for the normal force for squeeze flow of a Herschel–Bulkley fluid in the constant velocity mode between parallel plates of dissimilar roughnesses (in the limit of $d << R$, $m_1^2, m_2^2 << 1$, and $m_1, m_2 >> d/R$) which is given as follows:

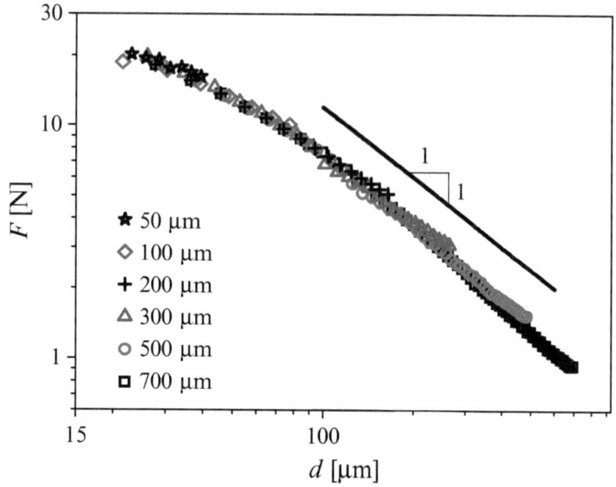

Fig. 23 The variation of the normal force (F) with distance between the rough top and smooth bottom plates (d) for different initial gaps but the same squeeze velocity $V = 10$ µm/s and waiting time $t_w = 15$ min (Reprinted from [146], with permission from Elsevier)

$$F = \frac{\pi \tau_y (m_1 + m_2) R^3}{3d} + \frac{\pi k R^3}{3^{1-n/2} d}\left(\frac{V}{d}\right)^n \left[\frac{m_1}{\left(1 - m_1^2\right)^{n/2}} + \frac{m_2}{\left(1 - m_2^2\right)^{n/2}}\right] \quad (16)$$

This equation reduces to Eq. 13 by substituting $V = 0$ in this expression. According to Eq. 16, the normal force does not depend on the initial gap between the plates; therefore, our experimental observation is in agreement with the prediction for a Herschel–Bulkley fluid. Further, Fig. 23 shows that for large gaps between the plates (or small values of normal force), the normal force varies as $F \sim d^{-1}$. In the limit of low plasticity number, the yield stress has a greater contribution toward the normal stress in comparison to viscous stress. In this scenario, the first term of the equation would be dominant which also predicts an inverse dependence of force on the gap. It can be observed in the figure that the dependence of the normal force on gap progressively weakens with decrease in gap as evident from the decreasing slope of the F vs. d curve. At smaller gaps between the plates, the normal force has a high value, and therefore, in this situation, a greater extent of shear melting is expected. Hence, we believe that a weaker dependence of force is observed due to partial yielding of the sample which causes a decrease in the yield stress [50]. The observations made in the case of velocity-controlled experiments support the argument made based on the constant force experiments that the squeeze flow behavior of the soft glassy fluid under study closely follows the theoretical prediction of a Herschel–Bulkley model under weak deformation fields; however, it considerably deviates from the same due to thixotropic effects under strong deformation fields.

Figure 24 shows the normal force as a function of gap between the plates for various aging times but for the same top plate velocity $V = 10$ µm/s and initial gap $d_i = 200$ µm. We observe that the value of the normal force is higher for a sample aged for longer duration at all the values of the gap. This is expected as

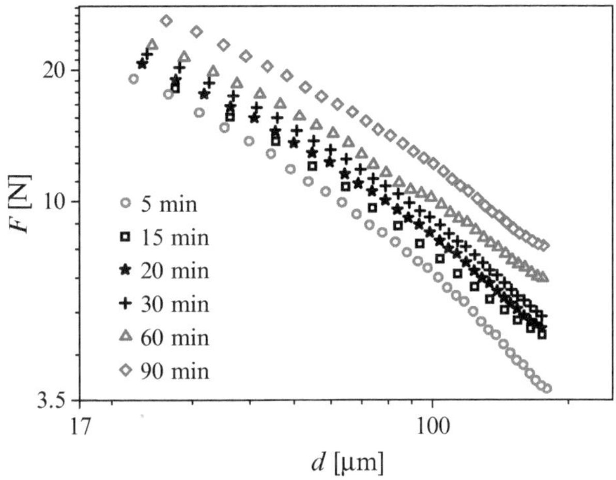

Fig. 24 The normal force (F) as a function of distance between the rough top and smooth bottom plates (d) for different aging times (t_w). The squeeze velocity V is 10 μm/s, and initial gap d_i is about 200 μm (Reprinted from [146], with permission from Elsevier)

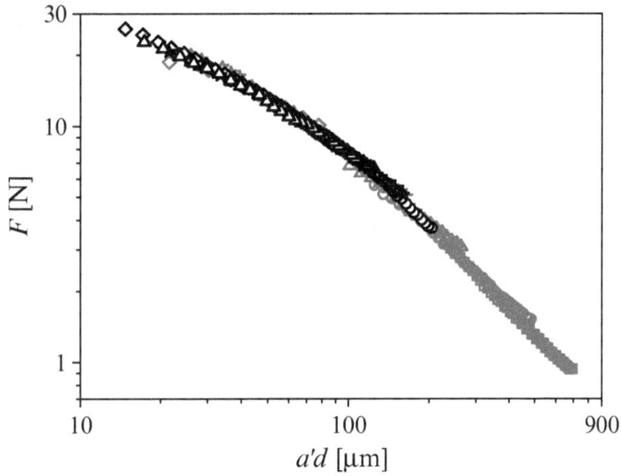

Fig. 25 A master curve comprising of superposed initial gap-dependent data shown in Fig. 23 and age-dependent data shown in Fig. 24. In this figure, all the age-dependent data curves have been shifted onto 15-min data curve (Reprinted from [146], with permission from Elsevier)

the yield stress increases with age which should cause an increase in the normal force. However, the slope of the $F - d$ curve decreases with an increase in age. This is probably because a sample aged for a longer duration is at a more advanced stage in the process of time-dependent microstructure buildup and therefore has a greater potential to undergo deformation-induced rejuvenation or breakage of the constituent microstructure. Interestingly, the dependence of force on gap at higher ages and larger gaps matches closely with that in the case of smaller ages and smaller gaps. In such a scenario, there is a possibility of obtaining a superposition of all the aging time $F - d$ curves. We indeed find that all the curves collapse to form a single curve when horizontal shifting was carried out by multiplication with a shift factor a. This shift factor follows a dependence on the aging time given by $a \sim t_w^{-0.26}$. The superposed curves for the aging time-dependent data are shown in Fig. 25 (black symbols).

The observation of superposed $F - d$ curves for different initial gaps in Fig. 23 and for different aging times strongly suggests self-similarity in the squeeze flow behavior. Therefore, an overall superposition of all the initial gap-independent and aging time-dependent curves is expected. In order to check this point, we plot all the $F - d$ curves belonging to these two sets of experiments in Fig. 25. We find that indeed a comprehensive superposition is obtained, thus suggesting a similar path followed and the same material states attained owing to structural changes due to deformation under squeeze flow.

In another set of experiments, a higher value of normal force was observed at all plate separations for larger values of plate velocities (the initial gap and the aging time were 200 μm and 15 min, respectively, in each experiment). The first term of Eq. 16 is the yield stress-dependent term which is independent of the plate velocity. The second term of the equation is the viscous stress-dependent term which has velocity dependence. Therefore, in agreement with experimental observation, this model predicts an increase in the normal force with an increment in the velocity. The dependence on velocity would be especially prominent at larger velocities when the deformation-induced melting decreases the value of the first term in Eq. 16 owing to a reduced yield stress. Thus, the relative importance of the second term in the equation increases which suggests an enhanced viscous contribution to the normal force in this case. Hence, based on the observations made in this study, we conclude that under the instance of squeeze flow of a soft glassy material, the Herschel–Bulkley model for yield stress fluids works satisfactorily well for weak deformation fields; however, it fails under strong deformation fields owing to a strong thixotropic behavior demonstrated by the material in this condition. A modified model which includes a structural parameter, whose evolution with time accounts for aging/rejuvenation in the material, could perhaps qualitatively predict the behavior in the limit of strong deformation fields.

7 Conclusions

In this chapter, we present the results for the deformation and aging behavior of soft glassy materials under tensile deformation field. In the first problem, we study the response of a model soft glass, a Laponite suspension, to tensile deformation. We find that the energy dissipated in the deformation and breakage of the suspension films increases with an increase in deformation field strength, that is, an increase in the plate separation velocity or with decrease of the initial gap between the plates. An increase in the aging time leads to a larger peak force; however, the energy dissipated decreases with increase in age of the sample. We believe that the system demonstrates an increasingly brittle behavior with increase in age due to an increased elastic modulus and relaxation time, which restrict the modes for dissipation of energy, thereby reducing the strain at which breakage occurs. This claim is further supported by a steeper drop in the force subsequent to the peak, which also indicates a greater amount of brittleness present in the material at higher ages.

Interestingly, the dependence of aging time-dependent data and deformation field (velocity and initial gap)-dependent data demonstrates discrepancy when plotted as a function of the Deborah number. Further, all the data points corresponding to various aging time, velocity, and initial gap experiments do not superpose, which implies that the tensile deformation behavior is not solely governed by the Deborah number, but apparently, other factors also play an important role. The constant force experiments show that the strain at failure decreases for older samples which also suggests an enhancement in brittle behavior with increase in age.

In the second problem, the aging time-dependent tensile creep flow experiments for the Laponite suspension demonstrate a shift in the data due to the evolution of microstructure with time, which causes an increase in elastic modulus and relaxation time with aging time. An increase in the applied normal stress causes a destruction of the microstructure that leads to a decreased rate of aging. As the rheological behavior exhibited by the material is largely governed by the dominant relaxation time, an increase in the same, either due to longer aging time period or a decrease in the magnitude of the applied stress field, is expected to affect the behavior in the same fashion. We validated the abovementioned argument by showing a process time–aging time–stress superposition under tensile deformation field.

In the third part of this work, we find that when a soft glassy material is subjected to tensile and rotational flow fields, an increment in the magnitude of either one of the stresses causes an increase in strain in a direction perpendicular to that of the stress in addition to an enhancement in the same direction as that of the stress. We find that this behavior is universally true for all the soft glassy materials employed in the study. A Herschel–Bulkley model for yield stress fluids qualitatively predicts the experimentally observed deformation induced in the material for combined tensile and rotational fields. We further determined the yielding curve which separates the regions in which the material exists in the solid and liquid phases. This curve is closely predicted by the Von Mises criterion, according to which the yielding event takes place in a soft glassy material when the invariant of the stress tensor becomes greater than the yield stress of the material.

In the final part of this chapter, we find that the Herschel–Bulkley model for squeeze flow between plates with dissimilar surface roughnesses correctly predicts the behavior for both constant velocity and force experiments when the applied deformation field is weak. When the magnitude of the deformation field is large, the sample exhibits thixotropic behavior due to which a time-dependent reduction in elastic modulus and yield stress takes place. As the Herschel–Bulkley model is time independent in nature, this leads to a significant deviation of the behavior from the theoretical prediction. We also observe that enhanced thixotropic effects are present when a smooth top plate is employed instead of a rough one due to increase in slip at the fluid–plate interface. Further, we find that in constant velocity experiments, a sample aged for a longer duration shows a weakened dependence of normal compressive force on the gap between the plates as it demonstrates a greater ability to undergo deformation-induced rejuvenation. The curvature of the force vs. gap curve is similar for the case of a greater age at smaller gap and lower age at larger gap, which enable us to successfully superpose all the aging time-dependent curves on a single curve. Moreover, due to a unique variation of force with change in gap, we obtain a comprehensive superposition for all aging time and initial gap data, which indicates the same path followed under squeeze deformation flow irrespective of the initial experimental, geometrical, and material parameters.

We believe that an analysis of the results presented in this chapter for tensile, squeeze, and tensile combined with rotational deformation fields provides novel information regarding the deformation dynamics of this class of materials. We also discuss in detail the effect of thixotropy (both physical aging and rejuvenation) and, under certain set of conditions, were able to successfully define the experimental observations with theoretical predictions. Therefore, we hope that this work would further our understanding of rheological dynamics exhibited by soft glassy materials in regard to the geometrical parameters of the material film, confining surfaces, material properties, and time elapsed since preparation.

Acknowledgment Financial support from the Department of Science and Technology through IRHPA scheme is greatly acknowledged for carrying out the work presented in this chapter. YMJ also acknowledges partial support from the Department of Atomic Energy, BRNS young scientist award scheme.

References

1. Abdelhaye YOM, Chaouche M, Van Damme H (2008) The tackiness of smectite muds. 1. The dilute regime. Appl Clay Sci 42(1–2):163–167
2. Abou B, Bonn D, Meunier J (2001) Aging dynamics in a colloidal glass. Phys Rev E 64(2 I):215101–215106
3. Abou B, Gallet F (2004) Probing a nonequilibrium Einstein relation in an aging colloidal glass. Phys Rev Lett 93(16):160603
4. Akinay AE, Brostow W (2001) Long-term service performance of polymeric materials from short-term tests: prediction of the stress shift factor from a minimum of data. Polymer 42(10):4527–4532
5. Amar MB, Bonn D (2005) Fingering instabilities in adhesive failure. Phys D Nonlin Phenom 209(1–4):1–16
6. Awasthi V, Joshi YM (2009) Effect of temperature on aging and time–temperature superposition in nonergodic laponite suspensions. Soft Matter 5:4991–4996
7. Baghdadi HA, Bhatia SR, Jensen EEC, Easwar N (2005) Evidence of re-entrant behavior in laponite-PEO systems. Mater Res Soc Symp Proc 899:188–192
8. Baghdadi HA, Jensen EC, Easwar N, Bhatia SR (2008) Evidence for re-entrant behavior in laponite-PEO systems. Rheol Acta 47(2):121–127
9. Baghdadi HA, Parrella J, Bhatia SR (2008) Long-term aging effects on the rheology of neat laponite and laponite – PEO dispersions. Rheol Acta 47(3):349–357
10. Baldewa B, Joshi YM (2011) Thixotropy and physical aging in acrylic emulsion paint. Polym Eng Sci 51(10):2084–2091
11. Baldewa B, Joshi YM (2012) Delayed yielding in creep, time-stress superposition and effective time theory for a soft glass. Soft Matter 8(3):789–796
12. Bandyopadhyay R, Liang D, Harden JL, Leheny RL (2006) Slow dynamics, aging, and glassy rheology in soft and living matter. Solid State Commun 139(11–12):589–598

13. Bandyopadhyay R, Liang D, Yardimci H, Sessoms DA, Borthwick MA, Mochrie SGJ, Harden JL, Leheny RL (2004) Evolution of particle-scale dynamics in an aging clay suspension. Phys Rev Lett 93(22):228302
14. Bandyopadhyay R, Mohan H, Joshi YM (2010) Stress relaxation in aging soft colloidal glasses. Soft Matter 6:1462–1466
15. Barnes HA (1997) Thixotropy – a review. J N-Newt Fluid Mech 70(1–2):1–33
16. Barnes HA (2000) A handbook of elementary rheology. Institute of Non-Newtonian Fluid Mechanics, Aberystwyth
17. Barnes HA, Hutton JF, Walters K (1989) An introduction to rheology. Elsevier, Amsterdam
18. Barral Q, Ovarlez G, Chateau X, Boujlel J, Rabideau B, Coussot P (2010) Adhesion of yield stress fluids. Soft Matter 6(6):1343–1351. doi:10.1039/b922162j
19. Beckett ST (ed) (1996) Physico-chemical aspects of food processing, 2nd edn. Blackie Academic and Professional, London
20. Bhatia S, Barker J, Mourchid A (2003) Scattering of disklike particle suspensions: evidence for repulsive interactions and large length scale structure from static light scattering and ultra-small-angle neutron scattering. Langmuir 19(3):532–535
21. Bhushan B (2003) Adhesion and stiction: mechanisms, measurement techniques, and methods for reduction. J Vac Sci Tech B 21(6):2262–2296
22. Bikerman JJ (1947) The fundamentals of tackiness and adhesion. J Colloid Sci 2(1):163–175. doi:10.1016/0095-8522(47)90017-2
23. Bird RB, Armstrong RC, Hassager O (1987) Dynamics of polymeric liquids, vol I, Fluid mechanics. Wiley-Interscience, New York
24. Bird RB, Stewart WE, Lightfoot EN (2002) Transport phenomena, 2nd edn. Wiley, New York
25. Blamey J, Rajan S, Unsworth A, Dawber R (1991) Soft layered prostheses for arthritic hip joints: a study of materials degradation. J Biomed Eng 13(3):180–184
26. Bonn D, Coussot P, Huynh HT, Bertrand F, Debregeas G (2002) Rheology of soft glassy materials. Europhys Lett 59(5):786–792
27. Bonn D, Denn MM (2009) Yield stress fluids slowly yield to analysis. Science 324(5933):1401–1402
28. Bonn D, Tanasc S, Abou B, Tanaka H, Meunier J (2002) Laponite: aging and shear rejuvenation of a colloidal glass. Phys Rev Lett 89(1):157011–157014
29. Brader JM, Voigtmann T, Fuchs M, Larson RG, Cates ME (2009) Glass rheology: from mode-coupling theory to a dynamical yield criterion. Proc Natl Acad Sci U S A 106(36):15186–15191
30. Brummer R (2006) Rheology essentials of cosmetic and food emulsions. Springer, Berlin
31. Carelli C, Déplace F, Boissonnet L, Creton C (2007) Effect of a gradient in viscoelastic properties on the debonding mechanisms of soft adhesives. J Adhes 83(5):491–505. doi:10.1080/00218460701377701
32. Cates ME, Evans MR (eds) (2000) Soft and fragile matter. The Institute of Physics Publishing, London
33. Chan TW, Baird DG (2002) An evaluation of a squeeze flow rheometer for the rheological characterization of a filled polymer with a yield stress. Rheol Acta 41(3):245–256
34. Cheng DC-H (2003) Characterisation of thixotropy revisited. Rheol Acta 42:372–382
35. Chung JY, Kim KH, Chaudhury MK, Sarkar J, Sharma A (2006) Confinement-induced instability and adhesive failure between dissimilar thin elastic films. Eur Phys J E 20(1):47–53. doi:10.1140/epje/i2005-10080-0
36. Cipelletti L, Ramos L (2002) Slow dynamics in glasses, gels and foams. Curr Opin Colloid Interface Sci 7(3–4):228–234
37. Cipelletti L, Ramos L (2005) Slow dynamics in glassy soft matter. J Phys Cond Matter 17(6):R253–R285
38. Cloitre M, Borrega R, Leibler L (2000) Rheological aging and rejuvenation in microgel pastes. Phys Rev Lett 85(22):4819–4822
39. Coussot P (1999) Saffman-Taylor instability in yield-stress fluids. J Fluid Mech 380:363–376. doi:10.1017/s002211209800370x

40. Coussot P (2005) Rheometry of pastes, suspensions and granular materials-application in industry and environment. Wiley, Hoboken
41. Coussot P (2006) Rheological aspects of the solid-liquid transition in jammed systems. Lect Notes Phys 688:69–90
42. Coussot P (2007) Rheophysics of pastes: a review of microscopic modelling approaches. Soft Matter 3(5):528–540
43. Coussot P, Nguyen QD, Huynh HT, Bonn D (2002) Viscosity bifurcation in thixotropic, yielding fluids. J Rheol 46(3):573–589
44. Coussot P, Tabuteau H, Chateau X, Tocquer L, Ovarlez G (2006) Aging and solid or liquid behavior in pastes. J Rheol 50(6):975–994
45. Covey GH, Stanmore BR (1981) Use of the parallel-plate plastometer for the characterisation of viscous fluids with a yield stress. J N-Newt Fluid Mech 8(3–4):249–260
46. Creton C, Hooker J, Shull KR (2001) Bulk and interfacial contributions to the debonding mechanisms of soft adhesives: extension to large strains. Langmuir 17(16):4948–4954
47. Creton C, Lakrout H (2000) Micromechanics of flat-probe adhesion tests of soft viscoelastic polymer films. J Polym Sci B 38(7):965–979
48. Crosby AJ, Shull KR, Lakrout H, Creton C (2000) Deformation and failure modes of adhesively bonded elastic layers. J Appl Phys 88(5):2956–2966
49. Deen WM (1998) Analysis of transport phenomena. Oxford University Press, New York
50. Denn MM (1998) Are plug-flow regions possible in fluids exhibiting a yield stress? In: Adams MJ, Mashelkar RA, Pearson JRA, Rennie AR (eds) Dynamics of complex fluids. Imperial College Press, London, pp 372–378
51. Derec C, Ajdari A, Ducouret G, Lequeux F (2000) Rheological characterization of aging in a concentrated colloidal suspension. C R Acad Sci Ser IV Phys Astrophys 1(8):1115–1119
52. Derec C, Ducouret G, Ajdari A, Lequeux F (2003) Aging and nonlinear rheology in suspensions of polyethylene oxide-protected silica particles. Phys Rev E 67(6):061403
53. Derks D, Lindner A, Creton C, Bonn D (2003) Cohesive failure of thin layers of soft model adhesives under tension. J Appl Phys 93(3):1557–1566
54. Di Leonardo R, Ianni F, Ruocco G (2005) Aging under shear: structural relaxation of a non-Newtonian fluid. Phys Rev E 71(1):011505
55. Doi M, Yamaguchi T (2007) Analytical solution for the deformation of pressure sensitive adhesives confined between two rigid plates. J N-Newt Fluid Mech 145(1):52–56. doi:10.1016/j.jnnfm.2006.12.006
56. Engmann J, Servais C, Burbidge AS (2005) Squeeze flow theory and applications to rheometry: a review. J N-Newt Fluid Mech 132(1–3):1–27
57. Estellé P, Lanos C (2007) Squeeze flow of Bingham fluids under slip with friction boundary condition. Rheol Acta 46(3):397–404
58. Rouyer F, Cohen-Addad S, Höhler R, Sollich P, Fielding SM (2008) The large amplitude oscillatory strain response of aqueous foam: strain localization and full stress Fourier spectrum. Eur Phys J E 27(3):309–321
59. Farage TFF, Brader JM (2012) Three-dimensional flow of colloidal glasses. J Rheol 56(2):259–278
60. Fast P, Kondic L, Shelley MJ, Palffy-Muhoray P (2001) Pattern formation in non-Newtonian Hele-Shaw flow. Phys Fluids 13(5):1191–1212. doi:10.1063/1.1359417
61. Fielding SM, Sollich P, Cates ME (2000) Aging and rheology in soft materials. J Rheol 44(2):323–369
62. Francis BA, Horn RG (2001) Apparatus-specific analysis of fluid adhesion measurements. J Appl Phys 89(7):4167–4174
63. Gay C, Leibler L (1999) Theory of tackiness. Phys Rev Lett 82(5):936–939
64. Ghatak A, Chaudhury MK (2003) Adhesion-induced instability patterns in thin confined elastic film. Langmuir 19(7):2621–2631. doi:10.1021/la026932t
65. Ghatak A, Chaudhury MK, Shenoy V, Sharma A (2000) Meniscus instability in a thin elastic film. Phys Rev Lett 85(20):4329–4332. doi:10.1103/PhysRevLett.85.4329

66. Glassmaker NJ, Hui CY, Yamaguchi T, Creton C (2008) Detachment of stretched viscoelastic fibrils. Eur Phys J E 25(3):253–266
67. Gonuguntla M, Sharma A, Mukherjee R, Subramanian SA (2006) Control of self-organized contact instability and patterning in soft elastic films. Langmuir 22(16):7066–7071. doi:10.1021/la0600696
68. Gonuguntla M, Sharma A, Sarkar J, Subramanian SA, Ghosh M, Shenoy V (2006b) Contact instability in adhesion and debonding of thin elastic films. Phys Rev Lett 97(1). doi:10.1103/PhysRevLett.97.018303
69. Hodge IM (1995) Physical aging in polymer glasses. Science 267(5206):1945–1947
70. Horgnies M, Darque-Ceretti E, Felder E (2007) Relationship between the fracture energy and the mechanical behaviour of pressure-sensitive adhesives. Int J Adhes Adhesives 27(8):661–668
71. Jazouli S, Luo W, Bremand F, Vu-Khanh T (2005) Application of time-stress equivalence to nonlinear creep of polycarbonate. Polym Test 24(4):463–467
72. Joshi YM (2007) Model for cage formation in colloidal suspension of laponite. J Chem Phys 127(8):081102
73. Joshi YM, Reddy GRK (2008) Aging in a colloidal glass in creep flow: time-stress superposition. Phys Rev E 77(2):021501–021504
74. Joshi YM, Reddy GRK, Kulkarni AL, Kumar N, Chhabra RP (2008) Rheological behavior of aqueous suspensions of laponite: new insights into the ageing phenomena. Proc R Soc A 464:469–489
75. Kabiraj SK, Tarafdar S (2003) Finger velocities in the lifting Hele-Shaw cell. Phys A 328(3–4):305–314
76. Karapetsas G, Tsamopoulos J (2006) Transient squeeze flow of viscoplastic materials. J N-Newt Fluid Mech 133(1):35–56
77. Kheirandish S, Gubaydullin I, Willenbacher N (2009) Shear and elongational flow behavior of acrylic thickener solutions. Part II: effect of gel content. Rheol Acta 48(4):397–407
78. Kieweg SL, Katz DF (2006) Squeezing flows of vaginal gel formulations relevant to microbicide drug delivery. J Biomech Eng 128(4):540–553
79. Kobelev V, Schweizer KS (2005) Strain softening, yielding, and shear thinning in glassy colloidal suspensions. Phys Rev E Stat Nonlinear Soft Matter Phys 71(2):021401/021401–021401/021416
80. Kolarik J, Pegoretti A (2006) Non-linear tensile creep of polypropylene: time-strain superposition and creep prediction. Polymer 47(1):346–356
81. Kroon M, Vos WL, Wegdam GH (1998) Structure and formation of a gel of colloidal disks. Phys Rev E 57:1962–1970
82. Laun HM, Rady M, Hassager O (1999) Analytical solutions for squeeze flow with partial wall slip. J N-Newt Fluid Mech 81(1–2):1–15
83. Lawal A, Kalyon DM (1998) Squeezing flow of viscoplastic fluids subject to wall slip. Polym Eng Sci 38(11):1793–1804
84. Lawrence CJ, Corfield GM (1998) Non-viscometric flow of viscoplastic materials: squeeze flow. In: Adams MJ, Mashelkar RA, Pearson JRA, Rennie AR (eds) Dynamics of complex fluids. Imperial College Press, London, pp 379–393
85. Lemaire E, Levitz P, Daccord G, Van Damme H (1991) From viscous fingering to viscoelastic fracturing in colloidal fluids. Phys Rev Lett 67(15):2009–2012
86. Lespiat R, Hohler R, Biance A-L, Cohen-Addad S (2010) Experimental study of foam jets. Phys Fluids 22(3):033302–033308
87. Lin YY, Hui CY, Conway HD (2000) A detailed elastic analysis of the flat punch (tack) test for pressure-sensitive adhesives. J Polym Sci B 38(21):2769–2784. doi:10.1002/1099-0488(20001101)38:21{\mathsurround=\opskip$<$}2769::aid-polb60{\mathsurround=\opskip$>$}3.0.co;2-j
88. Lindner A, Bonn D, Meunier J (2000) Viscous fingering in a shear-thinning fluid. Phys Fluids 12(2):256–261. doi:10.1063/1.870303
89. Lindner A, Bonn D, Poire EC, Ben Amar M, Meunier J (2002) Viscous fingering in non-Newtonian fluids. J Fluid Mech 469:237–256. doi:10.1017/s0022112002001714

90. Lindner A, Coussot P, Bonn D (2000) Viscous fingering in a yield stress fluid. Phys Rev Lett 85(2):314–317, Woodbury, NY, United States
91. Lindner A, Derks D, Shelley MJ (2005) Stretch flow of thin layers of Newtonian liquids: fingering patterns and lifting forces. Phys Fluids 17(7):072107. doi:10.1063/1.1939927
92. Lipscomb GG, Denn MM (1984) Flow of bingham fluids in complex geometries. J N-Newt Fluid Mech 14(C):337–346
93. Lozano J, Miska SZ, Yu M, Takach NE (2006) The effect of elongational flow through the drill bit on the rheology of polymeric drilling fluids. Paper presented at the proceedings of the IADC/SPE drilling conference, Miami, 21–23 Feb
94. Maleki-Jirsaraei N, Lindner A, Rouhani S, Bonn D (2005) Saffman-Taylor instability in yield stress fluids. J Phys Condensed Matter 17(14):S1209–S1218. doi:10.1088/0953-8984/17/14/010
95. Mamane A, Fretigny C, Lequeux F, Talini L (2009) Surface fluctuations of an aging colloidal suspension: evidence for intermittent quakes. Europhy Lett 88(5):58002
96. Mascia S, Wilson DI (2008) Biaxial extensional rheology of granular suspensions: the HBP (Herschel-Bulkley for Pastes) model. J Rheol 52(4):981–998
97. McFarlane JS, Tabor D (1950) Adhesion of solids and the effect of surface films. Proc R Soc A 202(1069):224–243
98. McKenna GB (2003) Mechanical rejuvenation in polymer glasses: fact or fallacy? J Phys Condens Matter 15:S737–S763
99. McKenna GB, Leterrier Y, Schultheisz CR (1995) Evolution of material properties during physical aging. Polym Eng Sci 35(5):403–410
100. McKenna GB, Narita T, Lequeux F (2009) Soft colloidal matter: a phenomenological comparison of the aging and mechanical responses with those of molecular glasses. J Rheol 53(3):489–516
101. Meeten GH (2000) Yield stress of structured fluids measured by squeeze flow. Rheol Acta 39(4):399–408
102. Meeten GH (2002) Constant-force squeeze flow of soft solids. Rheol Acta 41(6):557–566
103. Meeten GH (2004) Effects of plate roughness in squeeze-flow rheometry. J N-Newt Fluid Mech 124(1–3 Spec Iss):51–60
104. Meeten GH (2004) Squeeze flow of soft solids between rough surfaces. Rheol Acta 43(1):6–16
105. Meeten GH (2007) Radial filtration during constant-force squeeze flow of soft solids. Rheol Acta 46(6):803–813
106. Meeten GH (2010) Comparison of squeeze flow and vane rheometry for yield stress and viscous fluids. Rheol Acta 49(1):45–52
107. Metzner AB (1971) Extensional primary field approximations for viscoelastic media. Rheol Acta 10(3):434–445
108. Mewis J, Wagner NJ (2009) Thixotropy. Adv Colloid Interface Sci 147–148:214–227. doi:10.1016/j.cis.2008.09.005
109. Moller PCF, Mewis J, Bonn D (2006) Yield stress and thixotropy: on the difficulty of measuring yield stresses in practice. Soft Matter 2:274–283
110. Mongondry P, Tassin JF, Nicolai T (2005) Revised state diagram of Laponite dispersions. J Colloid Interface Sci 283(2):397–405
111. Mourchid A, Lecolier E, Van Damme H, Levitz P (1998) On viscoelastic, birefringent, and swelling properties of laponite clay suspensions: revisited phase diagram. Langmuir 14(17):4718–4723
112. Mourchid A, Levitz P (1998) Long-term gelation of laponite aqueous dispersions. Phys Rev E 57:R4887–R4890
113. Mujumdar A, Beris AN, Metzner AB (2002) Transient phenomena in thixotropic systems. J N-Newt Fluid Mech 102(2):157–178
114. Mukherjee R, Pangule R, Sharma A, Tomar G (2007) Contact instability of elastic bilayers: miniaturization of instability patterns. Adv Funct Mater 17(14):2356–2364. doi:10.1002/adfm.200600896

115. Nase J, Lindner A, Creton C (2008) Pattern formation during deformation of a confined viscoelastic layer: from a viscous liquid to a soft elastic solid. Phys Rev Lett 101(7):074503. doi:10.1103/Physrevlett.101.074503
116. Negi AS, Osuji CO (2009) Dynamics of internal stresses and scaling of strain recovery in an aging colloidal gel. Phys Rev E 80(1):010404
117. Negi AS, Osuji CO (2010) Time-resolved viscoelastic properties during structural arrest and aging of a colloidal glass. Phys Rev E 82(3):031404
118. O'Connell PA, McKenna GB (1997) Large deformation response of polycarbonate: time-temperature, time-aging time, and time-strain superposition. Polym Eng Sci 37(9):1485–1495
119. Ovarlez G, Barral Q, Coussot P (2010) Three-dimensional jamming and flows of soft glassy materials. Nat Mater 9(2):115–119
120. Paterson L (1981) Radial fingering in a Hele Shaw cell. J Fluid Mech 113:513–529. doi:10.1017/S0022112081003613
121. Petekidis G, Vlassopoulos D, Pusey PN (2003) Yielding and flow of colloidal glasses. Faraday Discuss 123:287–302
122. Petekidis G, Vlassopoulos D, Pusey PN (2004) Yielding and flow of sheared colloidal glasses. J Phys Condens Matter 16(38):S3955–S3963
123. Petrie CJS (1979) Elongational flows. Pitman, London
124. Poire EC, Ben Amar M (1998) Finger behavior of a shear thinning fluid in a Hele-Shaw cell. Phys Rev Lett 81(10):2048–2051
125. Poivet S, Nallet F, Gay C, Fabre P (2003) Cavitation-induced force transition in confined viscous liquids under traction. Europhys Lett 62(2):244–250. doi:10.1209/epl/i2003-00352-3
126. Poivet S, Nallet F, Gay C, Teisseire J, Fabre P (2004) Force response of a viscous liquid in a probe-tack geometry: fingering versus cavitation. Eur Phys J E 15(2):97–116
127. Purnomo EH, van den Ende D, Vanapalli SA, Mugele F (2008) Glass transition and aging in dense suspensions of thermosensitive microgel particles. Phys Rev Lett 101(23):238301
128. Quintanar-Guerrero D, Villalobos-García R, Alvarez-Colín E, Cornejo-Bravo JM (2001) In vitro evaluation of the bioadhesive properties of hydrophobic polybasic gels containing N, N-dimethylaminoethyl methacrylate-co-methyl methacrylate. Biomaterials 22(9):957–961
129. Rabideau BD, Lanos C, Coussot P (2009) An investigation of squeeze flow as a viable technique for determining the yield stress. Rheol Acta 48(5):517–526
130. Reddy GRK, Joshi YM (2008) Aging under stress and mechanical fragility of soft solids of laponite. J Appl Phys 104:094901
131. Reiter G, Sharma A (2001) Auto-optimization of dewetting rates by rim instabilities in slipping polymer films. Phys Rev Lett 87(16):166103. doi:10.1103/PhysRevLett.87.166103
132. Rodts S, Boujlel J, Rabideau B, Ovarlez G, Roussel N, Moucheront P, Lanos C, Bertrand F, Coussot P (2010) Solid-liquid transition and rejuvenation similarities in complex flows of thixotropic materials studied by NMR and MRI. Phys Rev E 81:021402
133. Rogers SA, Callaghan PT, Petekidis G, Vlassopoulos D (2010) Time-dependent rheology of colloidal star glasses. J Rheol 54(1):133–158
134. Saffman PG, Taylor G (1958) The penetration of a fluid into a porous medium or Hele-Shaw cell containing a more viscous liquid. Proc R Soc Lond A Math Phys Sci 245(1242):312–329. doi:10.1098/rspa.1958.0085
135. Sarkar J, Sharma A, Shenoy V (2005) Adhesion and debonding of soft elastic films: crack patterns, metastable pathways, and forces. Langmuir 21(4):1457–1469. doi:10.1021/la048061o
136. Sarkar J, Shenoy V, Sharma A (2004) Patterns, forces, and metastable pathways in debonding of elastic films. Phys Rev Lett 93(1):018302. doi:10.1103/PhysRevLett.93.018302
137. Schosseler F, Kaloun S, Skouri M, Munch JP (2006) Diagram of the aging dynamics in laponite suspensions at low ionic strength. Phys Rev E 73:021401
138. Seth JR, Bonnecaze RT, Cloitre M (2008) Influence of short-range forces on wall-slip in microgel pastes. J Rheol 52:1241

139. Shahin A, Joshi YM (2010) Irreversible aging dynamics and generic phase behavior of aqueous suspensions of laponite. Langmuir 26(6):4219–4225
140. Shahin A, Joshi YM (2011) Prediction of long and short time rheological behavior in soft glassy materials. Phys Rev Lett 106(3):038302
141. Sharma A, Gonuguntla M, Mukherjee R, Subramanian SA, Pangule RC (2007) Self-organized meso-patterning of soft solids by controlled adhesion: elastic contact lithography. J Nanosci Nanotechnol 7(6):1744–1752. doi:10.1166/jnn.2007.709
142. Sharma A, Ruckenstein E (1985) Mechanism of tear film rupture and formation of dry spots on cornea. J Colloid Interface Sci 106(1):12–27
143. Shaukat A, Joshi YM, Sharma A (2009) Tensile deformation and failure of thin films of aging laponite suspension. Ind Eng Chem Res 48(17):8211–8218
144. Shaukat A, Kaushal M, Sharma A, Joshi YM (2012) Shear mediated elongational flow and yielding in soft glassy materials. Soft Matter 8(39):10107–10114
145. Shaukat A, Sharma A, Joshi YM (2010) Time-aging time-stress superposition in soft glass under tensile deformation field. Rheol Acta 49(11):1093–1101
146. Shaukat A, Sharma A, Joshi YM (2012) Squeeze flow behavior of (soft glassy) thixotropic material. J N-Newt Fluid Mech 167–168:9–17. doi:10.1016/j.jnnfm.2011.09.006
147. Sherwood JD, Durban D (1996) Squeeze flow of a power-law viscoplastic solid. J N-Newt Fluid Mech 62(1):35–54
148. Sherwood JD, Durban D (1998) Squeeze-flow of a Herschel-Bulkley fluid. J N-Newt Fluid Mech 77(1–2):115–121
149. Shirodkar P, Bravo A, Middleman S (1982) Lubrication flows in viscoelastic liquids – 2. Effect of slip on squeezing flow between approaching parallel rigid planes. Chem Eng Commun 14(3–6):151–175
150. Shukla A, Joshi YM (2008) Ageing under shear: effect of stress and temperature field. AIP Conf Proc 1027:1018–1020
151. Shull KR, Creton C (2004) Deformation behavior of thin, compliant layers under tensile loading conditions. J Polym Sci B 42(22):4023–4043
152. Shull KR, Flanigan CM, Crosby AJ (2000) Fingering instabilities of confined elastic layers in tension. Phys Rev Lett 84(14):3057–3060
153. Sinha S, Dutta T, Tarafdar S (2008) Adhesion and fingering in the lifting Hele-Shaw cell: role of the substrate. Eur Phys J E 25(3):267–275
154. Sollich P (1998) Rheological constitutive equation for a model of soft glassy materials. Phys Rev E 58(1):738–759
155. Sollich P, Lequeux F, Hebraud P, Cates ME (1997) Rheology of soft glassy materials. Phys Rev Lett 78(10):2020–2023
156. Starkova O, Yang J, Zhang Z (2007) Application of time-stress superposition to nonlinear creep of polyamide 66 filled with nanoparticles of various sizes. Composites Sci Tech 67(13):2691–2698
157. Struik LCE (1978) Physical aging in amorphous polymers and other materials. Elsevier, Houston
158. Tirumkudulu M, Russel WB, Huang TJ (2003) Measuring the "tack" of waterborne adhesives. J Rheol 47(6):1399–1415
159. Tirumkudulu M, Russel WB, Huang TJ (2003) On the measurement of "tack" for adhesives. Phys Fluids 15(6):1588–1605
160. Tordjeman P (2007) Saffman-Taylor instability of shear thinning fluids. Phys Fluids 19(11):118102–118104
161. Van Damme H, Lemaire E, Ould Y, Abdelhaye M, Mourchid A, Levitz P (1994) Pattern formation in particulate complex fluids: a guided tour. In: Non-linearity and breakdown in soft condensed matter, vol 437, Lecture notes in physics. Springer, Berlin/Heidelberg, pp 134–150. doi:10.1007/3-540-58652-0_34
162. Van Olphen H (1977) An introduction to clay colloid chemistry. Wiley, New York
163. Wales DJ (2003) Energy landscapes. Cambridge University Press, Cambridge

164. Webber RE, Shull KR, Roos A, Creton C (2003) Effects of geometric confinement on the adhesive debonding of soft elastic solids. Phys Rev E Stat Nonlin Soft Matter Phys 68(2 Pt 1):021805/021801–021805/021811
165. Wilson SDR (1990) The Taylor-Saffman problem for a non-Newtonian liquid. J Fluid Mech 220:413–425. doi:10.1017/s0022112090003329
166. Yang F (1998) Exact solution for compressive flow of viscoplastic fluids under perfect slip wall boundary conditions. Rheol Acta 37(1):68–72
167. Yang FQ, Li JCM (2001) Adhesion of a rigid punch to an incompressible elastic film. Langmuir 17(21):6524–6529. doi:10.1021/la010409h

Part IV
Microscale Reactors: Issues and Applications

Microstructured Reactors for Hydrogen Production from Ethanol

Nageswara Rao Peela and Deepak Kunzru

Abstract The continuously increasing demand for clean and renewable energy warrants the development of renewable, nonpolluting energy resources. Hydrogen is emerging as a natural choice as a more secure and cleaner energy carrier. Fuel cells can be used to produce clean energy from hydrogen, particularly for portable applications. Hydrogen can be produced from a variety of fossil fuel sources, but to decrease the dependence on fossil fuels, hydrogen has to be produced from a renewable source. Hydrogen production from steam reforming of ethanol (a renewable fuel) has emerged as a promising alternative in recent years. For conducting this reaction on board a vehicle, a compact reactor system is required. A microchannel reactor is more efficient and attractive for this purpose, because of the high surface to volume ratio, resulting in high heat and mass transfer rates.

The reactions involved in producing CO-free hydrogen from ethanol include steam reforming of ethanol, water–gas shift reaction, and preferential oxidation of carbon monoxide. This chapter discusses the steps involved in the development of a microfuel processor for producing hydrogen from ethanol that include fabrication of the microchannels on the metal substrate, coating of the catalyst and support on the microchannels, assembly of the microchannel reactor, optimization of the catalysts for the three reactions, and, finally, heat integration of the different processes to maximize the efficiency of the fuel processor.

Keywords Microreactors • Steam reforming • Hydrogen • Ethanol • Catalysts

N.R. Peela
Department of Chemical Engineering, Indian Institute of Technology Kanpur, Kanpur, 208016 UP, India

Department of Chemical Engineering, Indian Institute of Technology Guwahati, Guwahati, India

D. Kunzru (✉)
Department of Chemical Engineering, Indian Institute of Technology Kanpur, Kanpur, 208016 UP, India
e-mail: dkunzru@iitk.ac.in

Y.M. Joshi, S. Khandekar (eds.), *Nanoscale and Microscale Phenomena*,
Springer Tracts in Mechanical Engineering, DOI 10.1007/978-81-322-2289-7_12

1 Introduction

The demand for clean and renewable energy is continuously increasing because of rapid depletion of conventional fuels and stringent environmental constraints. Therefore, renewable, nonpolluting energy resources need to be developed. In the historical background of the decarbonization pattern of primary fuel usage, hydrogen is emerging as a natural choice as a more secure and cleaner energy carrier [6]. Fuel cells can be used to produce clean energy from hydrogen, particularly for portable applications. An added incentive for using fuel cells is that, with the ongoing research, the cost per unit of power is continuously decreasing.

The production of hydrogen from methane and naphtha by steam reforming is a well-established commercial process. Hydrogen can be produced from a variety of other sources including fossil fuels (coal, natural gas, LPG, gasoline, diesel, methane, propane, wind, solar, methanol, ethanol, biofuel, etc.), but to decrease the dependence on fossil fuels, fully or partially, current situation strongly demands that hydrogen be produced from a renewable source. Although other methods, such as electrolysis of water, photochemical splitting of water, fermentation of carbohydrate-rich substrates, etc., have been proposed, the preferred method is still the steam reforming or oxidative steam reforming of hydrocarbons. In this context, ethanol/bioethanol is a promising biomass-derived liquid fuel for hydrogen production [5, 22].

Advantages of ethanol as a fuel include a relatively high hydrogen content (on molar basis), non-toxicity, ease of storage, and handling safety. Ethanol can be readily produced from renewable feedstocks such as sugarcane, municipal solid waste, and agro-waste. Another advantage of using ethanol is that the CO_2 produced during steam reforming reaction is equal to the CO_2 required for biomass growth and thus provides a closed cycle for CO_2 consumption. Therefore, no net pollutants are released into the environment. Primary demerit includes high reforming temperatures ($\sim$600 °C).

1.1 Why Use Microreactors?

Another very important aspect to be addressed for portable energy needs is the reactor design. Use of hydrogen for transportation applications can be accomplished in two ways: one is to store hydrogen and another is to produce hydrogen on board a vehicle. Storing hydrogen on board a vehicle would reduce the payload capacity and is also not safe. Therefore, producing hydrogen on board a vehicle is the preferred option. If conventional reactors are used for producing hydrogen on board a vehicle, then the payload capacity will be reduced because of the requirement of larger space to accommodate these devices. In this aspect, microreactors are more suitable. During the last two decades, micro-devices and systems (popularly known as microreactors), having their characteristic dimensions typically in the order of a few micrometers, have attracted the interest of the chemical process industry because of certain inherent advantages [15, 43, 44].

The distinctive feature of microreactors is their high surface to volume ratio compared to conventional chemical reactors, due to the small size of the channels through which the reactants flow. This results in high heat and mass transfer rates. Due to the small channel dimensions in microflow reactors, the flow is usually laminar, and, thus, the heat/mass transfer coefficients are inversely proportional to the channel hydraulic diameter. Moreover, since the catalyst is deposited as a thin layer, the pore diffusional resistances are usually negligible, and the catalyst can be utilized more effectively. Due to the compact size and low holdup, the system response is faster and results in better process control. Although there is an increase in the corresponding momentum flux transfer, i.e., pressure drop, the overall pressure drop is not appreciable.

The higher heat and mass transfer characteristics of microreactors are especially advantageous for conducting highly endothermic/exothermic reactions. Thus, local hot spot formations are avoided because of high heat removal capacity. This leads to low operating cost and higher system efficiency. Finally, because of smaller reactants and products' inventories, a high level of safety is achieved. The technical feasibility of using microreactors for fabrication of fuel processors has been demonstrated by several workers [25].

1.2 Reforming of Ethanol

Three processes can be used for producing hydrogen from ethanol: (1) steam reforming of ethanol (SRE), (2) partial oxidation (POX), and (3) oxidative steam reforming of ethanol (OSRE). Steam reforming is the reaction of ethanol with steam to produce hydrogen and carbon monoxide by the following reaction:

$$C_2H_5OH\ (g) + H_2O\ (g) \rightarrow 2CO\ (g) + 4H_2\ (g) \quad \Delta H^o_{298} = +256\ kJ/mol \quad (1)$$

This is a highly endothermic reaction and requires high temperatures to proceed. Acetaldehyde is also produced by dehydrogenation of ethanol:

$$C_2H_5OH\ (g) \rightarrow CH_3CHO + H_2 \quad \Delta H^o_{298} = +71\ kJ/mol \quad (2)$$

However, at high temperatures and high residence times, acetaldehyde decomposes to methane and CO. The CO can react with steam to give CO_2, and additional hydrogen is further produced by the water–gas shift reaction as follows:

$$CO + H_2O \leftrightarrow CO_2 + H_2 \quad \Delta H^o_{298} = -41.1\ kJ/mol \quad (3)$$

Thus, the reformate contains H_2, CO, CO_2, CH_4, and CH_3CHO, together with any unreacted ethanol and water.

In partial oxidation (POX), the fuel is reacted with insufficient oxygen to give CO and H_2:

$$C_2H_5OH\ (g) + 0.5O_2 \rightarrow 2CO\ (g) + 3H_2\ (g) \qquad \Delta H^{o}_{298} = +13\ \text{kJ/mol} \tag{4}$$

Together with the above reaction, total oxidation to give CO_2 also occurs simultaneously. POX is highly exothermic and hence does not require any external heat supply. The major drawback of POX is that hydrogen yield in this process is lower in comparison to SRE. Moreover, because of the high heat generation, hot spot formation can occur resulting in catalyst damage.

Oxidative steam reforming (OSRE) is essentially a combination of SRE and POX:

$$C_2H_5OH\ (g) + (1-2n)\,H_2O\ (g) + nO_2\ (g) \rightarrow 2CO\ (g) + (4-2n)\,H_2\ (g) \tag{5}$$

For ethanol, the OSRE process can be made autothermal at an oxygen/ethanol molar ratio of approximately 0.35. Moreover, CO yield is lower in comparison to SRE because of the CO oxidation reaction. Hydrogen yields are in between the other two processes. As the oxygen/ethanol molar ratio increases, the hydrogen yield decreases. Therefore, an optimum value of oxygen/ethanol ratio must be used to maximize H_2 yield and to minimize CO yield. Due to the presence of steam, the hot spot phenomenon is less pronounced in comparison to POX.

1.3 Integrated Microfuel Processor

The production of H_2 from ethanol for use in a polymer electrolyte membrane fuel (PEMFC) includes the reaction of ethanol with steam, followed by steps to reduce the CO content of the product stream to less than 50 ppm because the fuel cell catalysts are poisoned by even small amounts of CO. A schematic of a microfuel processor is shown in Fig. 1. The premixed stream of water and ethanol is vaporized in a preheater and then fed to the ethanol reforming reactor. Here, ethanol reacts with steam to produce hydrogen as well as other products such as CO_2, CO, and CH_4. After the reformer, the concentration of CO in the effluent is ~10 mol%, and this is reduced by further processing. As discussed by Moharana et al. [40], three routes can be used for reducing the CO content in the product stream: (a) membrane separation, where the effluent is passed through a membrane after reforming, which allows only H_2 molecules to pass through it. In this way, 99.99 % pure hydrogen can be produced [48]. The other two, (b) and (c), are chemical routes in which CO is reduced to ppm levels by WGS followed by either PROX or selective methanation of CO.

The steps involved in the development of a microfuel processor for producing hydrogen from ethanol include fabrication of the microchannels on the metal substrate; coating of the catalyst and support on the microchannels; assembly of

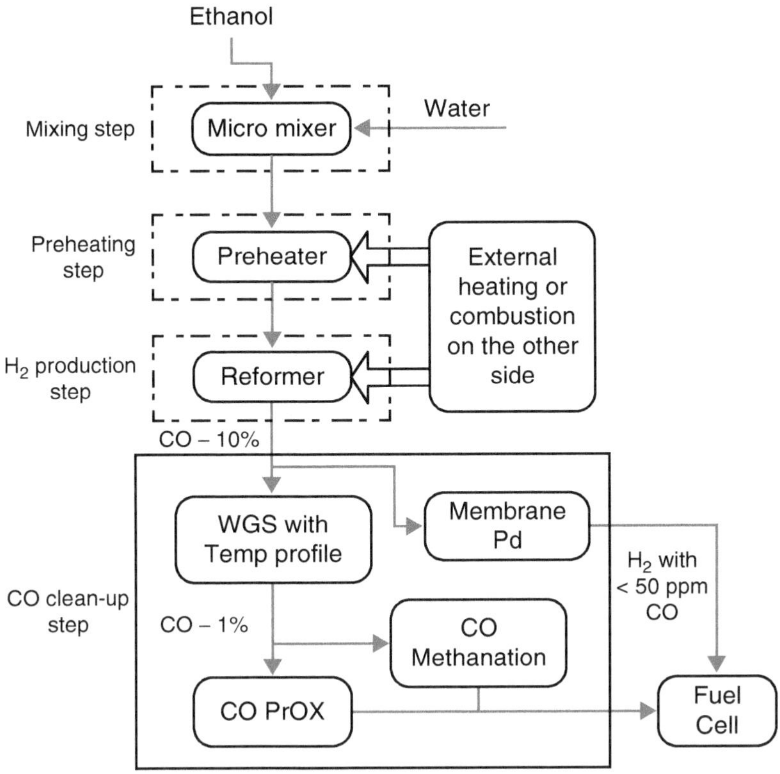

Fig. 1 Schematic of an integrated reactor system for generation of hydrogen from ethanol and water (Reproduced with permission from Moharana et al. [40], copyright © 2011 Elsevier)

the microchannel reactor; optimization of the catalysts for SRE, WGS, and PROX; and, finally, heat integration of the different processes to maximize the efficiency of the fuel processor. In subsequent sections, the different steps are discussed further.

2 Fabrication of Microchannels

In recent years, microchannel reactors have been used for conducting various catalytic and non-catalytic reactions [23, 26]. For reaction temperatures below 300 °C, microchannel devices are usually fabricated using materials such as silicon, glass, copper, etc.; however, these materials cannot withstand high temperatures. As many of the chemical processes operate at temperatures above 300 °C, it is essential to fabricate these devices with materials that can withstand higher temperatures. Several methods such as micro-molding; wet chemical etching; a combination of lithography, electroforming, and molding (LIGA process); micro-electro-discharge machining (μEDM); and mechanical techniques such as milling, punching, and embossing and laser ablation have been reported for fabrication of microchannels on metals [14].

The LIGA process, developed in Germany, is based on a combination of deep lithography, electroforming, and molding. Using the LIGA process, micro-devices with high precision and high surface quality can be fabricated. The μEDM method is very well suited for making microstructures in metals, with minimum features in the order of 10 μm. The basic equipment consists of a voltage generator, an electrode, and the part to be machined. Due to a high-energy discharge between the electrode and the workpiece, the unwanted material is removed by erosion. In this process, there is no contact between the electrode and the part being machined, and, therefore, contamination of the workpiece is minimized.

An attractive method for the fabrication of microchannels is wet chemical etching. Chemical etching is one of the oldest nontraditional machining processes. It uses a chemical solution, which is called an etchant, to remove unwanted workpiece material by controlled dissolution [9]. Besides low cost, the advantages of this method include fast processing, no burring on the edges of the walls, possibility of processing very thin materials, precise control of the channel dimensions, and a short lead time between prototyping and production [13, 35]. However, this method has the disadvantage that the etching is anisotropic so that the depth of the channel is always less than the width of the channel. Moreover, the removal of photoresist can occur during the process with concentrated etchant and also at high temperature [1].

Several publications have discussed the general aspects of wet chemical etching [19, 30, 33, 36, 68], but the information on the fabrication of microchannels on stainless steel is limited. A commonly used etchant in industry is aqueous ferric chloride solution. The disadvantage of using ferric chloride alone is that, as the etching proceeds, concentration of ferrous ion increases and these ions do not dissolve in the etching solution. Moreover, ferrous ions are deposited on the parts being etched making penetration of the etchant to the etching surface difficult. By addition of acids such as hydrochloric acid, the solubility of ferrous ion in the etchant solution can be increased [3]. In their patent, Matthews and Hanneman [36] revealed that saturated ferric chloride solution (4 parts), concentrated nitric acid (1 part), and concentrated hydrofluoric acid (1 part) will give a uniform, nonselective, and rapid etching of stainless steel. Others [30, 33, 68] have also used combinations of ferric chloride, different acids, and other agents as etchants to enhance the etch rate and uniformity of etching of stainless steels. However, for making microchannels, the width of the photoresist between the channels is very small and results in the photoresist peeling off very easily during etching. Moreover, in this situation, it is very difficult to obtain uniform etching [21]. Therefore, developing an etchant that will give reasonable microchannel depths without the photoresist peeling off is a challenging task.

Rao and Kunzru [57] fabricated microchannels in stainless steel by using solutions of various concentrations of $FeCl_3$, HCl, and HNO_3 in water as etchants. Their study showed that for obtaining smooth uniform channels on stainless steel substrates, an etchant consisting of $FeCl_3$, HCl, and HNO_3 is necessary. An increase in the concentration of HCl in the etchant increased the etch rate as well as etch factor but adversely affected the roughness. Addition of HNO_3 was necessary to obtain smooth uniform channels. The depth and etch factor were significantly affected by the composition of etchant, operating temperature, and initial width of the channel. In the range of operating conditions studied, an etchant containing 10 wt% $FeCl_3$, 10 wt% HCl, and 5 wt% HNO_3 at 40 °C and initial width of 190 μm gave the best results. At these conditions, a channel depth of 180 μm with an etch factor of 1.36 was obtained after 160 min of etching. With this etchant, the channels were smooth and uniform without any cavities on the edges of the channels.

Laser micromachining (LMM) can also be used for creating micro-features in metallic substrates. The parts to be removed are locally vaporized using a laser beam, and the vapor is sucked out with a vacuum pump. The physical principle of laser ablation uses a beam of electromagnetic waves in phase, photons focalized with optical medium and concentrated on the workpiece, thereby creating highly localized powers. When the laser beam strikes the work surface, its energy is partially reflected and partially absorbed causing fusion of the material. If the intensity exceeds a certain limit, there is instantaneous vaporization of the base metal in the zone where the beam is more concentrated. Around this zone, the metal is in the molten state. Metallic vapor at high temperature and pressure pushes the melted material laterally, and the beam proceeds to the metal further below. The hole created is defined as a keyhole. Equilibrium is created between the vapor pressure that tends to enlarge the hole and the hydrostatic pressure of the melted metal that tends to close the hole. This equilibrium determines the penetration depth of the hole in the workpiece.

Several process parameters can affect the uniformity and smoothness of the channels as well as the machining time. These include the power and frequency of the laser, the scan speed, and the number of passes. By increasing the laser power, the depth obtained per unit time increases. On increasing the laser frequency, the depth obtained per unit time decreases, but the smoothness of the channels increases. Therefore, the laser power and frequency have to be adjusted such that smooth channels are obtained in a reasonable time.

3 Deposition of Catalysts in Microchannels

The surface area of the fabricated microchannels is not sufficient for performing catalytic reactions, and it is necessary to increase the specific surface area by chemical treatment of the channels and/or by applying porous coatings on the walls. This is a crucial step. The coating has to be of uniform thickness and should adhere strongly to the metal substrate. Moreover, the active phase should be well dispersed and have long-term stability. Depending on the properties of the surface and the catalyst to be coated, several methods can be used to deposit the catalyst. These can be broadly classified as coating methods based on a liquid phase, thin film techniques such as chemical vapor deposition (CVD) and physical vapor deposition (PVD), and thermal spraying methods (powder plasma spraying, flame spray deposition, etc.).

A commonly used technique is washcoating of the channels with a suspension of the support material, followed by impregnation of the active phase or directly washcoating the ready-made catalyst [38]. For nonporous substrates, a pretreatment step is necessary in order to improve the adherence of the catalyst layer. One way is by thermal treatment of the substrate to form an oxide layer. This is often used with Fecralloy substrates in which a surface aluminum oxide layer forms when the substrate is heated to ~900 °C. In the subsequent washcoating step, the ceramic support adheres strongly to this oxide layer. Although an oxide layer also forms on metals in which Al is not present, in such cases, the adherence of the washcoat is not as good. Another pretreatment method is to oxidize the substrates with acids such as HCl or HNO_3 [71]. For aluminum-free substrate, a primer coating of a sol can also be used to form a thin layer of alumina that can enhance the adherence of the subsequent washcoat [18, 51].

In the suspension method, the adherence and uniformity of the washcoat depend on the particle size, pH of the slurry, viscosity, and slurry concentration. The washcoating slurry usually consists of powder (catalyst support or catalyst itself), binder, dispersant, acid, and solvent (usually water). The particle size of the solids has a very strong influence on the uniformity and adherence of the washcoat, and the average particle size should typically be less than 3 μm. The stability of the slurry depends on the pH, and for alumina, it should be less than 5 or more than 9.

Peela et al. [51] studied the effect of primer and slurry composition on the washcoat characteristics of γ-alumina on stainless steel substrate. Washcoating was done following a two-step procedure: primer coating followed by slurry coating. Primer deposition was done by using a mixture of boehmite sol and binder in water. The boehmite sol was prepared by adding aluminum hydroxide powder (Disperal P2, average particle size = 45 μm, Sasol®, Germany) to a 0.4 wt% HNO_3 aqueous solution. The dispersion was stirred for 30 min and then aged for 2 days. During this period, its pH and viscosity gradually increased to a constant value, respectively, and did not change thereafter. When Disperal is dispersed in 0.4 % HNO_3 aqueous solution, acid progressively breaks the powder particles. This process (peptization) is slow and requires about 48 h to reach equilibrium. The binder used was polyvinyl alcohol (PVA), and its concentration in the dispersion was varied from 0 to 4 wt%. The deposition was done by filling the microchannels with the primer and then wiping out any excess dispersion from the plate. The substrate was dried at room temperature for 3 h and then at 120 °C for 8 h, followed by calcination at 600 °C for 5 h with a ramp rate of 2 °C/min.

After primer coating, the channels were coated with a slurry containing γ-alumina, PVA, and, in some cases, an alumina sol as binder. The average particle size of the as-received γ-alumina was 50 μm, and this was reduced to less than 3 μm by wet milling in a ball mill. For milling, a 30 wt% aqueous slurry of alumina was prepared, and the pH was adjusted to around 3 by adding concentrated nitric

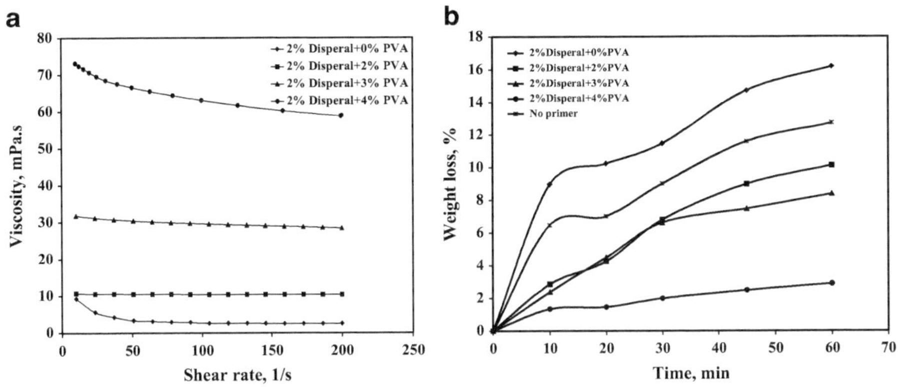

Fig. 2 (**a**) Variation of viscosity with shear rate and (**b**) weight loss with sonication time for different primer compositions (Reproduced with permission from Peela et al. [51], copyright © 2009 Elsevier)

acid, to avoid gelation. The milled slurry was then mixed with the required amount of binder and the pH adjusted to the desired value by adding HNO_3 or KOH. After adding all the required components, the slurry was stirred for 2 h. Then, the slurry was deposited onto the channels by following the same procedure used for primer coating.

To study the effect of primer composition on adherence of the washcoat, different amounts of PVA (0–4 wt%) were added to 2 % Disperal in water. In each case, the pH of the primer suspension was kept at 3.5. The effect of primer composition on the viscosity and weight loss on ultrasonication is shown in Fig. 2. The viscosity of the suspension increased drastically with increasing wt% of PVA (Fig. 2a). The increase in the weight of the substrate after the primer coating was approximately 2 mg. After the primer coating, all the samples were coated with an aqueous slurry containing 20 wt% γ-alumina and 4 wt% PVA. For comparison, this slurry was also coated on microchannels without any primer coating. Within experimental error, there was no effect of primer composition on the final washcoat loading (~24.7 mg/plate). The weight loss from the plate pre-coated with 2 wt% Disperal and no PVA as primer was slightly higher than that from the plate washcoated without any primer pre-coating (refer Fig. 2b). This was most probably due to the settling of the primer in the channels, resulting in poor adherence of the subsequent coating. In contrast, with 2 wt% Disperal + PVA, the adherence was better, and weight loss was lowest for a primer solution containing 2 wt% Disperal and 4 wt% PVA. In all cases, the weight loss increased with time of sonication till 45 min and leveled off thereafter.

Peela et al. [51] also studied the effect of slurry concentration, slurry pH, and binders on the washcoat characteristics. The washcoat loading increased almost linearly with slurry concentration. Without any binder, the weight loss was significant,

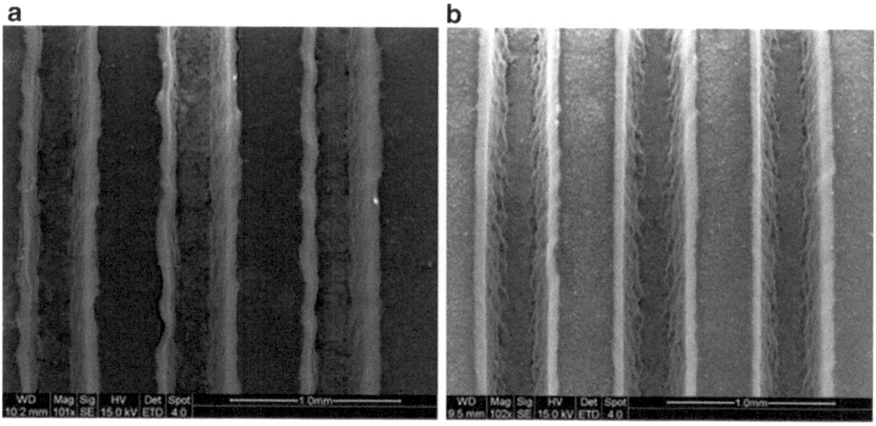

Fig. 3 SEM photographs of the washcoated microchannels for different PVA concentrations (**a**) 2 % PVA and (**b**) 4 % PVA (Reproduced with permission from Peela et al. [51], copyright © 2009 Elsevier)

and, moreover, the washcoat was not crack-free. When a binder (PVA) was added to the slurry, the slurry viscosity increased, and the cracks in the washcoat, as well as the weight loss on sonication, decreased with an increase in the concentration of PVA. The SEM images of the washcoated substrate at two binder concentrations are shown in Fig. 3.

The approximate thickness of the coating on the sidewalls of the microchannels was measured from these photographs and was found to be maximum for 4 % PVA (~65 μm) and almost negligible for coating with the slurry without any added PVA, even though the washcoat loading was similar in both the cases. When the viscosity of the slurry is low, the washcoat is not distributed uniformly over the surface of the channel, and the amount of washcoat deposited is low on the walls and more at the bottom. The surface area of the dried and calcined powder obtained from 20 % γ-alumina + 4 % PVA slurry (161 m^2/g) was relatively lower when compared to that of γ-alumina powder (183 m^2/g).

To study the effect of pH of the slurry on washcoat loading and adhesion, the pH of the slurry (20 wt% γ-alumina + 4 wt% PVA) was varied from 2 to 6.5. Slurry viscosity increased with pH. The results showed that pH affected only the adhesion and not the loading. With increasing pH, the weight loss passed through minima at a pH of 3.5. At high pH, the dispersion of the slurry was poor, whereas at low pH, there could have been dissolution of smaller alumina particles resulting in poor adhesion.

The results of this study confirmed that a primer coating and the use of binders are essential for obtaining well-adhered, uniform crack-free washcoats on stainless steel substrates.

4 Steam Reforming of Ethanol

SRE has been extensively studied in packed bed reactors and several review articles are available [18, 42, 55, 70]. In contrast to the steam reforming of methanol, the published information on steam reforming of ethanol in microreactors is limited. Although nickel-based catalysts are extensively used for reforming of hydrocarbons in packed beds, their activity is not good enough for use in microreactors. Due to the lower mass per unit volume in a microreactor, more active catalysts are required. Several investigators have studied SRE on noble metal catalysts. Out of the noble metal catalysts studied, rhodium shows the highest activity and selectivity toward H_2 due to its capability to break the C–C bond. Men et al. [39] tested nickel, rhodium, and ruthenium catalysts on different supports for ethanol steam reforming in a microchannel reactor and found $Rh/Ni/CeO_2/Al_2O_3$ to be the most active and selective catalyst. This catalyst gave 100 % conversion at 500 °C. Roh et al. [60] studied a series of Rh catalysts supported on various supports such as Al_2O_3, $MgAlO_4$, ZrO_2, and CeO_2–ZrO_2 and found that among the various catalysts tested, 2 wt%Rh/CeO_2–ZrO_2 exhibited the highest H_2 yield at 450 °C, possibly due to the high O_2 storage capacity of CeO_2–ZrO_2. For rhodium supported on alumina, Dictor and Roberts [11] reported an increase in the dispersion of Rh on addition of ceria. Romero-Sarria et al. [61] reported an increase in coke formation when Ni was added to a Rh catalyst and attributed this effect to deeper reduction, resulting in a greater amount of metallic particles of larger size. As an alternative to noble metals, Domínguez et al. [12] coated the microchannels with $Co_3[Si_2O_5]_2(OH)_2$. At 375 °C and a steam/carbon ratio of 3, complete conversion of ethanol was achieved.

Kunzru and coworkers [52, 58] investigated the steam reforming of ethanol over $Rh/CeO_2/Al_2O_3$ catalysts in a microchannel reactor. The catalyst was coated on a SS304 plate of 1 mm thickness and consisted of 25 microchannels (depth = 400 μm, width = 500 μm, width of separating fin between channels = 300 μm and length = 60 mm) along with inlet and outlet chambers. The channels were made by laser micromachining. First, the microchannels were coated with γ-Al_2O_3 by washcoating and then impregnated with ceria (precursor: cerium nitrate) followed by the active metal (Rh and/or Ni) using incipient wetness method. For preparing catalysts containing Rh and Ni, both co-impregnation and sequential impregnation were used. The catalyst-coated plate was joined to another uncoated SS plate by laser welding. The plates before and after welding are shown in Fig. 4. The reactor so fabricated was sandwiched between two guiding plates, in which four cartridge heaters, each of 400 W, were inserted to maintain the temperature of the whole block at the desired value. The temperature of the microreactor was controlled by a thermocouple, connected to a PID controller and inserted in the middle of the microreactor.

The SRE was studied in the temperature range of 400–600 °C at 1 atm. Addition of CeO_2 to a 2 wt% Rh/Al_2O_3 (2Rh/Al) catalyst increased the dispersion of the active metal. As the CeO_2 content increased, the dispersion also increased and reached about 95 % for the catalyst with 30 % CeO_2. Moreover, the TPR

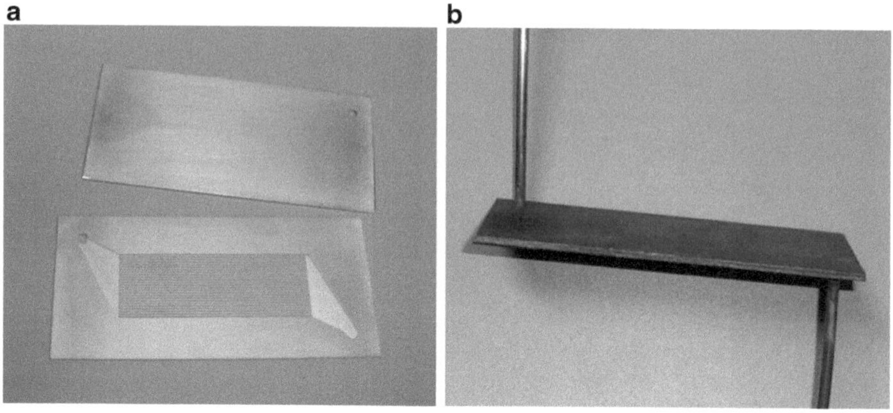

Fig. 4 Photographs of the microreactor plates: (**a**) before and (**b**) after laser spot welding (Reproduced with permission from Peela et al. [52], copyright © 2011 Elsevier)

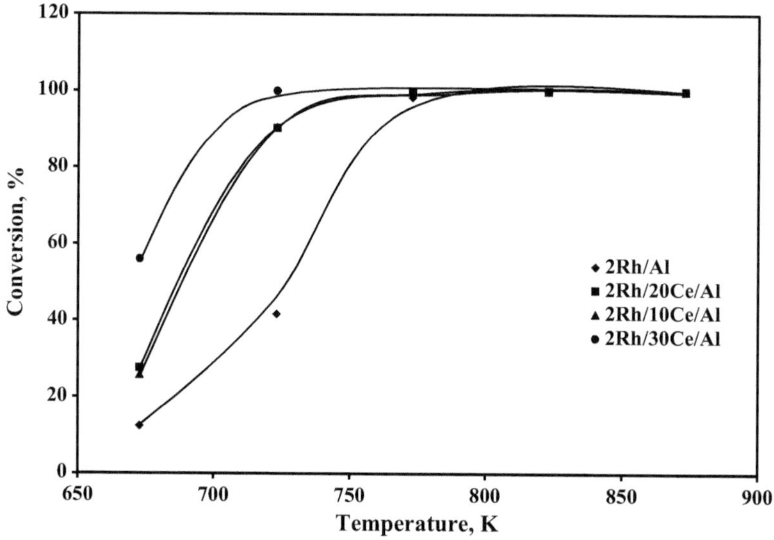

Fig. 5 Effect of ceria loading on conversion of ethanol (water/ethanol = 6 mol/mol and $W/F_{A0} = 1.6$ g h mol^{-1}) (Reproduced with permission from Peela et al. [52], copyright © 2011 Elsevier)

results showed that the strength of the interaction between the support and catalyst decreased with increasing CeO_2 content. The variation of ethanol conversion with temperature on 2Rh/Al catalysts containing different amounts of CeO_2 (0–30 wt% of alumina) is shown in Fig. 5. As can be seen from this figure, addition of CeO_2 enhanced the ethanol conversion, most probably due to the higher metal area of Rh in the presence of ceria and the easier reducibility of Ce containing Rh/Al catalysts.

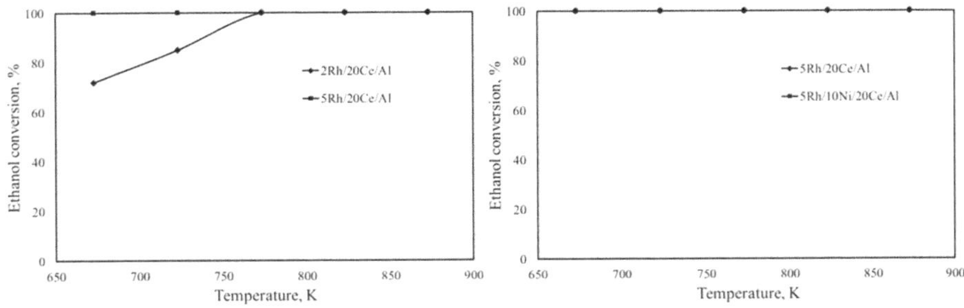

Fig. 6 Effect of Rh and Ni loading on ethanol conversion

The effect of Rh loading and nickel loading on the conversion of ethanol is shown in Fig. 6. At identical conditions, conversion increased with Rh loading. In contrast, there was no effect of Ni loading on the conversion.

It should be emphasized that the method of incorporation of nickel to the catalyst affected its performance. When Rh/Ni catalyst was prepared by co-impregnation, the conversion of ethanol was reduced considerably at low temperatures. On the other hand, when Ni was incorporated by sequential impregnation, there was no effect on the activity; however, the coke formation was significantly reduced.

The following reaction scheme was proposed to explain the product distribution obtained with $Rh/CeO_2/Al_2O_3$ catalyst at different temperatures:

$$\text{Dehydrogenation}: \ CH_3CH_2OH \rightarrow CH_3CHO + H_2 \tag{6}$$

$$\text{Decomposition of acetaldehyde}: \ CH_3CHO \rightarrow CH_4 + CO \tag{7}$$

$$\text{Steam reforming of acetaldehyde}: \ CH_3CHO + H_2O \rightarrow 2CO + 3H_2 \tag{8}$$

$$\text{Steam reforming of methane}: \ CH_4 + H_2O \leftrightarrow CO + 3H_2 \tag{9}$$

$$\text{Water gas shift reaction}: \ CO + H_2O \leftrightarrow CO_2 + H_2 \tag{10}$$

At low temperature (<773 K), the H_2 selectivity was more or less constant, whereas at high temperatures, it increased with temperature. On this catalyst, ethanol was first dehydrogenated to acetaldehyde and hydrogen by reaction (6), and at the lowest temperature (673 K), the rate of dehydrogenation was higher than that of reaction (7). With an increase in temperature, the selectivity to CO and CH_4 increased, indicating that the rate of decomposition of CH_3CHO increased. The drop in selectivity to CO and the increase in the selectivity to CO_2 in the temperature range of 723–773 K indicated that the rate of WGS reaction increased. As the temperature was further increased, the CO_2 selectivity remained constant, and CO selectivity increased while the selectivity to methane decreased, most likely

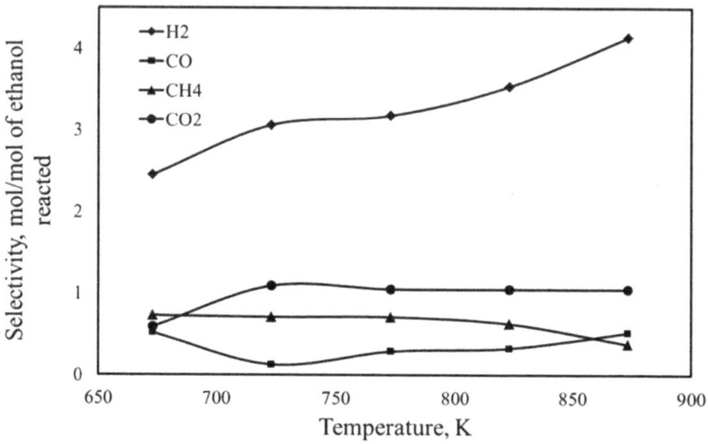

Fig. 7 Variation of selectivities of different products with temperature on catalyst 5Rh/10Ni/20Ce/Al in microchannel reactor (water/ethanol = 6 mol/mol; $W/F_{A0} = 1.6$ g h mol^{-1})

due to the onset of steam reforming of methane (reaction 9). Similar variations of product selectivities with temperature have been reported by others [17, 41]. The productivity of hydrogen on 2Rh/20Ce/Al_2O_3 at 873 K was approximately 65 l/gcat h.

The variation of selectivities to products with temperature, obtained on catalyst 5Rh–5Ni/20Ce/Al, is shown in Fig. 7. A comparison of the product selectivities obtained on catalysts with and without Ni showed that addition of Ni reduced the CO selectivity and increased CH_4 selectivity due to CO hydrogenation.

Longer-duration tests, conducted at 500 °C, showed that 2Rh/20Ce/Al catalyst was stable for ~35 h, and after that, the conversion decreased gradually with ~20 % conversion loss in 25 h. On the other hand, for the nickel-containing catalyst prepared by sequential impregnation, i.e., 5Rh/10Ni/20Ce/Al, there was no change in conversion or product distribution even after 100 h. A TGA analysis confirmed that there was no coke deposition on the catalyst.

Thus, for SRE in microchannels, Rh-based catalysts containing ceria and nickel seem to be the most active and stable.

5 Water–Gas Shift Reaction

The reactor effluent from the steam reforming reaction, in addition to hydrogen and carbon dioxide, also contains significant amounts of carbon monoxide and unreacted water. The CO and H_2O can be further converted to H_2 and CO_2 by utilizing the water–gas shift reaction (WGS). Water–gas shift reaction is a mildly exothermic reversible reaction ($\Delta H_{298} = -41.1$ kJ/mol). Compared to the other reactions in the fuel processor (steam reforming and preferential oxidation of CO), it is the slowest reaction, and hence it is essential to enhance its rate. The purpose of this reaction is to process the output of the reformer to produce additional hydrogen and also reduce the CO content of the exit gases. Since the WGS is an exothermic reaction, the equilibrium conversion increases with a decrease in temperature; however, at lower temperature, the reaction rates are low. Therefore, to fully exploit WGS for reducing CO, it is desirable to carry out the reaction at a lower temperature with high reaction rates.

Conventionally, to maximize the conversion and limit the reactor size, the reaction is carried out in two adiabatic reactors with interstage cooling. In the first reactor, the high-temperature stage (HTS) operates at 350–450 °C, the kinetics of the reaction governs the conversion, and most of the CO is converted. The conventional catalyst used for HTS reaction is Fe_2O_3/Cr_2O_3. The second stage is the low-temperature stage (LTS) that operates at 200–300 °C, and here the conversion is governed by the equilibrium effect. It is carried out by using a CuO/ZnO catalyst. Typically, at the exit of the LTS, 90 % of the inlet CO is converted. The conventional catalysts are pyrophoric, are sensitive to poisoning, and require high residence times and are therefore not suitable for use in microreactors [63]. Also, it would be preferable if the same catalyst could be used for HTS and LTS rather than using different catalysts.

Several catalysts have been tried for the combined HTS/LTS reaction, most of which are based on noble metals such as Pt, Ru, Rh, Au, or Pd supported on CeO_2, CeO_2/ZrO_2, ZrO_2, or TiO_2. Radhakrishnan et al. [56] tested various noble metals and found Pt to be more active, selective, and stable than Pd, Ru, Rh, Ir, or Au. Kolb et al. [28] tested various bimetallic catalysts such as $Pt/CeO_2/Al_2O_3$, $Pt/Rh/CeO_2/Al_2O_3$, $Pt/Pd/CeO_2/Al_2O_3$, and $Pt/Ru/Al_2O_3$ in a microreactor in the temperature range of 290–340 °C and found $Pt/CeO_2/Al_2O_3$ to be the most active and selective. The optimum platinum and ceria contents were found to range between 3–5 and 12–24 wt%, respectively.

The nature of the support can also affect the rate of reaction. Azzam et al. [4] measured the WGS activity of Pt supported on different oxides such as CeO_2, ZrO_2, CeO_2–ZrO_2, and TiO_2. Out of these, Pt/CeO_2 was reported to be the most active. In contrast to this study, Ricote et al. [59] found that Pt supported on ceria–zirconia was more active and stable than on ceria or zirconia alone. It is well-known that addition of ZrO_2 to CeO_2 increases the oxygen storage capacity of the catalyst. Sandupatla [64] investigated the water–gas shift reaction at atmospheric pressure in the temperature range of 250–400 °C on Pt-based catalysts in a microreactor. The catalyst composition used was 2 wt% Pt/10 wt% reducible oxide and 88 wt% γ-Al_2O_3. The reducible oxides tested were CeO_2, ZrO_2, TiO_2, and CeO_2–ZrO_2 ($CeO_2/ZrO_2 = 0.76$ mol/0.24 mol).The feed (in vol%) was 4 % CO, 12 % CO_2, 30 % H_2O, 40 % H_2, and balance N_2. For these runs, the weight of alumina washcoated on the microchannels was 34 mg, and the total flow to the reactor was 50.9 sccm. Further details are available elsewhere [54]. The variation of CO conversion with temperature on the different supports is shown in Fig. 8.

As can be seen from Fig. 8, the nature of the support significantly affects the conversion of CO. The highest activity was obtained with ceria–zirconia support,

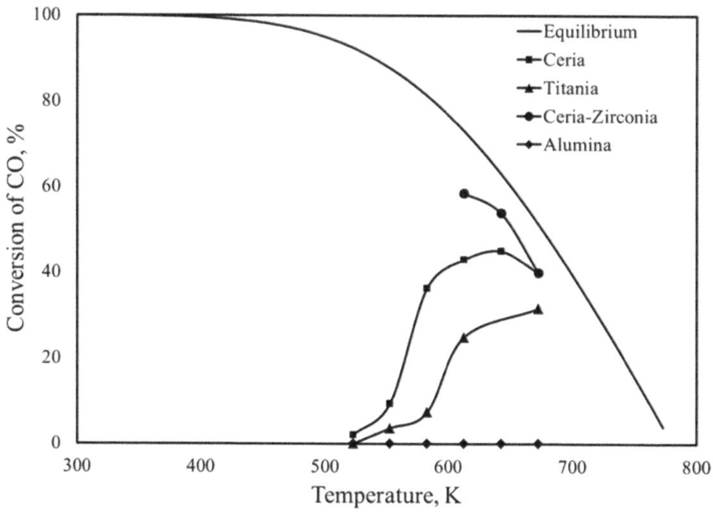

Fig. 8 Variation of conversion of CO with temperature in water–gas shift reaction for Pt supported on different oxides (Reproduced with permission from Peela et al. [54], copyright © 2014 Inderscience)

whereas alumina showed no activity. For Pt-based WGS catalysts, a hydrophilic support is essential to adsorb and activate water because platinum does not interact internally with water. It has been proposed that that CO adsorbs on Pt sites and H_2O on ceria sites [16]. The high activity of Pt/ceria–zirconia/alumina catalyst was most probably due to the better reducibility and higher oxygen storage/release capacity of this support. This facilitates the removal of carbonaceous deposits from the surface of Pt which are reported to be the cause of deactivation. This catalyst showed no deactivation in a 10 h run. Kolb et al. [29] have reported a stable activity of Pt/CeO_2 catalyst for WGS in microchannels for more than 1,000 h at 325 and 400 °C.

To fully realize the benefit of using noble metals for WGS in microreactors, it is desirable to have a decreasing temperature profile in the reactor. This can significantly improve the conversion compared to isothermal or adiabatic reactors. Numerical calculations by Zalc and Löffler [74] showed that such a profile can be obtained by countercurrent cooling. Experimental confirmation of implementing such a profile in an integrated WGS heat-exchanger reactor has been discussed by O'Connell et al. [46].

6 Preferential Oxidation of Carbon Monoxide

After steam reforming and water–gas shift reaction, the reactor effluent contains approximately 1 % CO. This CO needs to be reduced to ppm levels because the electrode of PEMFC is prone to poisoning if the CO concentration exceeds

50 ppm. The different methods available for reducing the concentration of CO to ppm levels include pressure swing adsorption, Pd-based membrane separation, catalytic CO methanation, and preferential oxidation of CO (PROX). PROX is a simple method of removing CO from H_2-rich gas streams and has been studied by several investigators. The following reactions take place during the preferential oxidation of CO:

Desired reaction:

$$CO\ (g) + \frac{1}{2}O_2\ (g) \rightarrow CO_2\ (g) \quad \Delta H_R = -283\ kJ/mol \tag{11}$$

Undesired reaction:

$$H_2\ (g) + \frac{1}{2}O_2\ (g) \rightarrow H_2O\ (g) \quad \Delta H_R = -242\ kJ/mol \tag{12}$$

Undesired reverse water–gas shift reaction can also take place in an oxygen-deficient atmosphere:

$$H_2\ (g) + CO_2\ (g) \rightarrow H_2O\ (g) + CO\ (g) \quad \Delta H_R = +41\ kJ/mol \tag{13}$$

To minimize the hydrogen oxidation, the catalyst should have a high selectivity for CO oxidation. A wide variety of catalysts have been studied for the PROX reaction such as precious metals (Rh, Ru, Pt, Pd, and Au) and transition metal oxides [7]. Typical supports are ceria, alumina, and zeolites. Oh and Sinkevitch [49] carried out the PROX reaction over noble metals supported on alumina for a feed containing 0.09 % CO, 0.08 % O_2, 0.85 % H_2, and balance N_2 and found 100 % CO conversion for all noble metals except Pd. For Rh and Ru, complete conversion was obtained below 100 °C, whereas for Pt, it occurred at 200 °C. The complete conversion at low temperature for Ru and Rh was associated with methanation which is undesirable. Ko et al. [24] used very low amount of Pt (0.05 %) for a feed stream containing 1 % CO, 1 % O_2, 80 % H_2, 2 % H_2O, and balance He and obtained only 70 % conversion along with 30 % selectivity at a temperature of 200 °C.

Ouyang et al. [50] used microkinetic modeling to show that at low CO conversions, the metal sites are mostly covered by adsorbed CO leading to high selectivity, but at high conversions (i.e., when very little CO was present in the gas phase), the metal sites are occupied by adsorbed oxygen and hydrogen resulting in loss of hydrogen by oxidation. Moreover, with an increase in temperature, the coverage of active sites by CO decreases, and more hydrogen is adsorbed which again results in lower selectivity at high temperature.

Kumar and Kunzru [31] conducted PROX on 1 % Pt/γ-Al_2O_3 in the presence of 40 % H_2 at different CO/O_2 ratios, and the results are shown in Fig. 9. CO conversion passed through a maximum with increasing temperature, and complete CO conversion was obtained for all the cases, i.e., for 1, 2, and 3 % O_2.

Both CO and H_2 compete for the active sites, and the decrease in CO conversion at high temperature is due to the onset of hydrogen oxidation which decreases

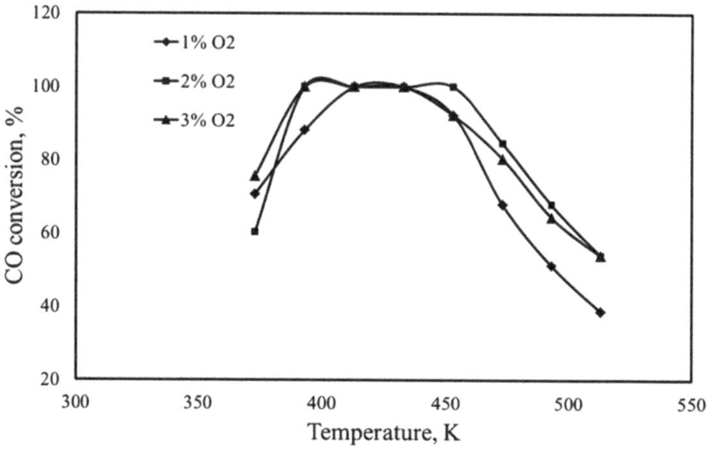

Fig. 9 Variation of conversion of CO with temperature for 1 %Pt/γ-Al_2O_3 (feed composition: 1 % CO, 1–3 % O_2, 40 % H_2, balance nitrogen)

the availability of oxygen for CO oxidation. The maximum conversion window widened with an increase in the oxygen concentration. For the same amount of O_2 concentration (i.e., for 1 % O_2), the maximum CO conversion was obtained at a lower temperature for PROX in the presence of 40 % H_2 than that for PROX in the absence of H_2. As reported by Ouyang et al. [50], the oxidation of CO is due to the reaction between adsorbed CO and adsorbed OH rather than adsorbed O species. Consequently, the rate of CO oxidation is higher in the presence of hydrogen.

The above discussion illustrates that both a suitable catalyst and the reactor design are important in PROX. Kolb (reaction 7) has discussed the different types of microreactors and operating conditions that have been tested for PROX.

7 Microstructured Fuel Processors

Fuel processors are miniaturized complex chemical plants, consisting of mixers, evaporators, reactors, separators, and heat exchangers [47]. The goal of a fuel processor is to produce fuel cell grade hydrogen. The steps shown in Fig. 1 are generally used for any fuel that requires high temperature for steam reforming and produces significant amounts of CO. However, some of these steps may be eliminated for simple fuels such as methanol. In addition, if the hydrogen is used in other types of high-temperature fuel cells such as solid oxide fuel cells (SOFC), then the CO clean-up step can be completely eliminated.

Integrated microstructured reactors can be considered as one class of multifunctional reactors. Multifunctional reactors have been developed in the past as part of process intensification. A classic example of such a reactor is reactive distillation in which both reaction and mass transfer take place simultaneously. Integrated microstructured reactors are already undergoing industrial trials. An example is the production of synthetic oil in microstructured reactors by Velocys®, a spin-out company from Battelle Memorial Institute [refer: www.velocys.com]. Microfuel processors have several advantages. For example, due to their compactness, these can be very easily integrated with fuel cell stack to generate power on board a vehicle. These can also be used for decentralized power generation for household utilization and other industrial purposes, where the needs are moderate and compactness is essential. Such devices can also be used in sub-watt power applications in electronics such as laptops, cell phones, and tablets in place of batteries.

In 2004, Holladay et al. [20] reviewed the developments in portable hydrogen production using microreactor technology. The authors presented an extensive review of various classes of fuel processors based on their power capacity. O'Connell et al. [47] reviewed the recent progress at Mainz Institute of Microtechnology (IMM) in mass production of microstructured fuel processors for application toward distributed energy generation systems. The authors, in particular, reviewed various techniques used at IMM for fabrication of microstructures, catalyst coating, and reactor sealing. As case studies, the authors have briefly discussed three fuel processors, namely, 100 W methanol, 250 W LPG, and 50 kW ethanol. Recently, Kolb [25] reviewed the progress in microstructured reactors for distributed and renewable production of fuels and electrical energy.

Till now, fuel processors have been developed with capacities from sub-watt up to kilowatt range. Different types of fuels (methane, gasoline, diesel, methanol, ethanol, ammonia, etc.) have been used as a hydrogen source. Various designs such as microstructured reactors, microchannel reactors, ceramic and metallic monoliths, etc., have been used. Several computational fluid dynamics studies have also been conducted to optimize the flow distribution and other parameters, such as channel geometry, inlet and outlet geometry, etc. Micromembrane separators have also been used to integrate the reactors and to separate the hydrogen from toxic gases such as carbon monoxide.

Due to the discovery of shale gas reserves worldwide, natural gas has become available in abundance [37], and several researchers have focused on this fuel for production of hydrogen [32, 37, 67]. Other aspects of shale gas, such as life cycle assessment and greenhouse gas footprint, have also been studied [72, 73]. Moreover, the catalysts for steam reforming of methane are well developed and are being used industrially. Infrastructure is also available reasonably well for distributing this fuel. For these reasons, methane could be a potential fuel for production of hydrogen and electricity. Different types of integrated microstructured reactors have been developed for producing hydrogen from methane [10, 45, 65, 66, 69].

A number of studies have been reported for producing hydrogen from methanol using microreactors. Methanol is extensively used in fuel processors to produce hydrogen for various applications such as laptops, military applications, etc. The major reason that methanol has been used extensively as the fuel is the simpler design and the requirement of a relatively lower temperature for reforming. Due to the relatively low CO content at the reformer exit, the WGS reactor is often not necessary. Available infrastructure for distribution of the fuel is another advantage. However, the disadvantages of methanol are that it is very toxic and is also not a renewable fuel. Asian researchers, particularly Japanese and Koreans, have developed methanol-based microfuel processors for use in mobile phones and laptops.

Till now, there are relatively few studies devoted to the development of ethanol-based fuel processors, but in recent years, the interest in using ethanol as the feed has increased [2, 34, 53]. In our group, we combined the SRE and WGS to produce hydrogen from ethanol [53]. The micro-device and various parts of it are shown in Fig. 10. In Fig. 10a, the catalyst-coated microchannel plates are shown, and a scanning electron microscope image can be seen in the inset. In the same figure, the design of the inlet and outlet chambers to maintain a uniform flow along the channels can also be observed. The integrated micro-device, consisting of evaporator, reformer, heat exchanger, and water–gas shift reactor, is shown in

Fig. 10 Micro-devices for steam reforming of ethanol: (**a**) catalyst-coated microchannel plates (*inset*: scanning electron microscope image of catalyst-coated plate); (**b**) integrated device, consisting of evaporator, reformer, heat exchanger, and water–gas shift reactor; (**c**) water-cooled heat exchanger, and (**d**) grooved stainless steel plates for thermal separation of individual parts

Fig. 10b. This figure also shows the guiding plates for supporting all the individual devices and for holding the heaters to maintain the temperature at the required level. The heat exchanger used in this study is separately shown in Fig. 10c. The heat exchanger was used to reduce the reformate temperature from 600 °C to that of the water–gas shift reactor temperature (370 °C). Water was used as the coolant. Two grooved stainless steel plates (Fig. 10d) were used to maintain an air gap between the three units to thermally separate them. The external dimensions of the integrated device were 100 × 45 × 28 mm. The microfuel processor was tested using 5%Rh/20%CeO_2/γ-Al_2O_3 at 600 °C for SRE and 2%Pt/10%CeO_2–ZrO_2/γ-Al_2O_3 at 370 °C for WGS. At these conditions, the ethanol conversion was 100 %, and no catalyst deactivation was observed for a run time of 10 h.

Aicher et al. [2] developed a 250 W capacity fuel processor with autothermal reforming of ethanol. The authors first carried out process simulation using CHEMCAD® to obtain the optimal operating parameters, expected hydrogen yield, and system efficiency. The reformer was operated at 730 °C and at an oxygen-to-steam ratio of ~0.9, with minimal methane production and no soot formation. Selective methanation was used as a CO clean-up step to maintain the CO levels below 10 ppm. The start-up and shut down procedures were also optimized to minimize the degradation of the catalysts. Authors claimed a fully automated operation of the integrated reformer fuel cell system with the use of system controls.

Bowers et al. [8] from Nuvera Fuel Cells Inc., USA, developed an on-board fuel processor for PEM fuel cells in collaboration with Renault, France. The volume of the fuel processor was 80 and 150 l with all the plant control components. The processor can be operated with ethanol, methanol, and gasoline and gave a power output of 200 kWh with hydrogen efficiencies of more than 77 %. The start-up time was below 4 min. The photograph of the fuel processor is shown in Fig. 11.

Rossetti et al. [62] demonstrated a 5 kW_e + 5kW_t bioethanol processor for PEMFC. The processor consists of a pre-reformer, reformer/combustor, high-temperature water–gas shift and low-temperature water–gas shift, and two methanation reactors. The reformate of the processor consists of CO concentration levels

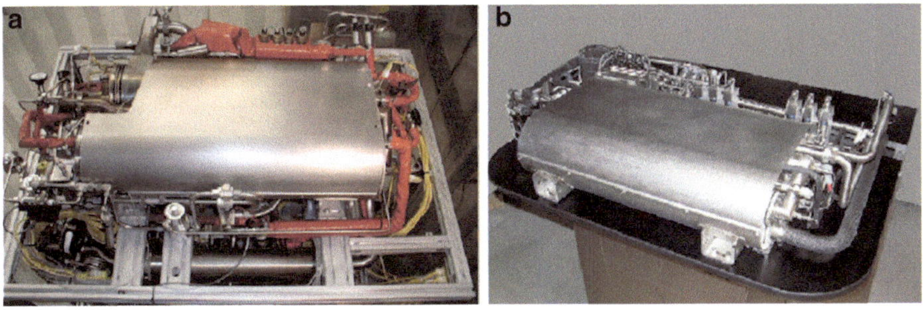

Fig. 11 Two versions of integrated ethanol fuel processor with a power generation capacity of 200 kW_{th} (Adapted, with permission, from Bowers et al [8] copyright © 2007 Elsevier)

less than 20 ppm. The system is capable of producing 5 kW electrical power and 5 kW of thermal power.

Kolb et al. [27] presented the development work on a 50 kW ethanol micro-fuel processor carried out in collaboration with Rosetti Marino, an Italian plant engineering company. The system consists of an ethanol reformer operated at 10 bar, water–gas shift reactor, and a pressure swing adsorption for separating hydrogen from the rest of the reformate. Several heat exchangers and coupled evaporators/catalytic burners were also included as the balance of plant. The work included the development of catalysts and evaluation of their stability. The entire system was simulated using Aspen Plus and Aspen Dynamics.

8 Conclusions

The use of microreactors for producing hydrogen from ethanol is potentially very attractive for on-board and distributed power applications. For the successful development of a microfuel processor, not only are active, selective, and stable catalysts required for SRE, WGS, and PROX, but the reactor design, operating parameters, and heat integration strategies also need to be optimized. Although significant advances have been made in catalyst development, particularly with the use of noble metals, more active and selective catalysts are required. Moreover, further work is required in the design, fabrication, and operation of integrated reactors.

Acknowledgment The financial support from the Department of Science and Technology, New Delhi, under the IRHPA Scheme is gratefully acknowledged.

References

1. Abate K (1997) Photochemical etching of metals. Met Finish 95(1):39–44
2. Aicher T, Full J, Schaadt A (2009) A portable fuel processor for hydrogen production from ethanol in a 250 Wel fuel cell system. Int J Hydrog Energy 34(19):8006–8015, doi:http://dx.doi.org/10.1016/j.ijhydene.2009.07.064
3. Allen DM (2004) Photochemical machining: from [']manufacturing's best kept secret' to a $6 billion per annum, rapid manufacturing process. CIRP Ann Manuf Technol 53(2):559–572
4. Azzam KG, Babich IV, Seshan K, Lefferts L (2007) A bifunctional catalyst for the single-stage water–gas shift reaction in fuel cell applications. Part 2. Roles of the support and promoter on catalyst activity and stability. J Catal 251(1):163–171, doi: http://dx.doi.org/10.1016/j.jcat.2007.07.011
5. Balat M, Balat H, Öz C (2008) Progress in bioethanol processing. Prog Energy Combust Sci 34(5):551–573, doi:http://dx.doi.org/10.1016/j.pecs.2007.11.001
6. Barreto L, Makihira A, Riahi K (2003) The hydrogen economy in the 21st century: a sustainable development scenario. Int J Hydrog Energy 28(3):267–284, doi:http://dx.doi.org/10.1016/S0360-3199(02)00074-5

7. Bion N, Epron F, Moreno M, Mariño F, Duprez D (2008) Preferential oxidation of carbon monoxide in the presence of hydrogen (PROX) over noble metals and transition metal oxides: advantages and drawbacks. Top Catal 51(1–4):76–88. doi:10.1007/s11244-008-9116-x
8. Bowers BJ, Zhao JL, Ruffo M, Khan R, Dattatraya D, Dushman N, Beziat J-C, Boudjemaa F (2007) Onboard fuel processor for PEM fuel cell vehicles. Int J Hydrog Energy 32(10–11):1437–1442, doi:http://dx.doi.org/10.1016/j.ijhydene.2006.10.045
9. ÇakIr O, Temel H, Kiyak M (2005) Chemical etching of Cu-ETP copper. J Mater Process Technol 162–163:275–279
10. Cipitì F, Pino L, Vita A, Laganà M, Recupero V (2013) Experimental investigation on a methane fuel processor for polymer electrolyte fuel cells. Int J Hydrog Energy 38(5):2387–2397, doi:http://dx.doi.org/10.1016/j.ijhydene.2012.11.143
11. Dictor R, Roberts S (1989) Influence of ceria on alumina-supported rhodium: observations of rhodium morphology made using FTIR spectroscopy. J Phys Chem 93(15):5846–5850. doi:10.1021/j100352a038
12. Domínguez M, Cristiano G, López E, Llorca J (2011) Ethanol steam reforming over cobalt talc in a plate microreactor. Chem Eng J 176–177:280–285, doi:http://dx.doi.org/10.1016/j.cej.2011.03.087
13. Drozda TJ (1949) Tool and manufacturing engineers handbook, vol 1. Society of Manufacturing Engineers, Michigan
14. Ehrfeld W, Hessel V, Lowe H (2000) Microreactors. Wiley-VCH, Weinheim
15. Freni S, Calogero G, Cavallaro S (2000) Hydrogen production from methane through catalytic partial oxidation reactions. J Power Sources 87(1–2):28–38, doi:http://dx.doi.org/10.1016/S0378-7753(99)00357-2
16. Germani G, Alphonse P, Courty M, Schuurman Y, Mirodatos C (2005) Platinum/ceria/alumina catalysts on microstructures for carbon monoxide conversion. Catal Today 110(1–2):114–120, doi:http://dx.doi.org/10.1016/j.cattod.2005.09.017
17. Goula MA, Kontou SK, Tsiakaras PE (2004) Hydrogen production by ethanol steam reforming over a commercial Pd/[gamma]-Al2O3 catalyst. Appl Catal B Environ 49(2):135–144
18. Haryanto A, Fernando S, Murali N, Adhikari S (2005) Current status of hydrogen production techniques by steam reforming of ethanol: a review. Energy Fuel 19(5):2098–2106. doi:10.1021/ef0500538
19. Hays SA (1959) Continuous chemical milling process. United States Patent No. 2890944
20. Holladay JD, Wang Y, Jones E (2004) Review of developments in portable hydrogen production using microreactor technology. Chem Rev 104(10):4767–4790. doi:10.1021/cr020721b
21. Jubb EC (1964) Comparison of copper etchants. Plating 51:311–316
22. Kim S, Dale BE (2004) Global potential bioethanol production from wasted crops and crop residues. Biomass Bioenergy 26(4):361–375, doi:http://dx.doi.org/10.1016/j.biombioe.2003.08.002
23. Kiwi-Minsker L, Renken A (2005) Microstructured reactors for catalytic reactions. Catal Today 110(1–2):2–14
24. Ko E-Y, Park ED, Seo KW, Lee HC, Lee D, Kim S (2006) A comparative study of catalysts for the preferential CO oxidation in excess hydrogen. Catal Today 116(3):377–383, doi:http://dx.doi.org/10.1016/j.cattod.2006.05.072
25. Kolb G (2013) Review: microstructured reactors for distributed and renewable production of fuels and electrical energy. Chem Eng Process Process Intensif 65:1–44, doi:http://dx.doi.org/10.1016/j.cep.2012.10.015
26. Kolb G, Hessel V (2004) Micro-structured reactors for gas phase reactions. Chem Eng J 98(1–2):1–38
27. Kolb G, Men Y, Schelhaas KP, Tiemann D, Zapf R, Wilhelm J (2010) Development work on a microstructured 50 kW ethanol fuel processor for a small-scale stationary hydrogen supply system. Ind Eng Chem Res 50(5):2554–2561. doi:10.1021/ie100602w
28. Kolb G, Pennemann H, Zapf R (2005) Water-gas shift reaction in micro-channels—Results from catalyst screening and optimisation. Catal Today 110(1–2):121–131, doi:http://dx.doi.org/10.1016/j.cattod.2005.09.012

29. Kolb G, Schürer J, Tiemann D, Wichert M, Zapf R, Hessel V, Löwe H (2007) Fuel processing in integrated micro-structured heat-exchanger reactors. J Power Sources 171(1):198–204, doi:http://dx.doi.org/10.1016/j.jpowsour.2007.01.006
30. Kreml JF (1972) Chemical milling method and bath for steel. United States Patent No. 3676094
31. Kumar J, Kunzru D (2012) Preferential oxidation of carbon monoxide on Pt/gamma-Al2O3 catalyst in a micro-packed bed reactor. Paper presented at the Chemcon-2012, Jalandhar
32. Laget O, Richard S, Serrano D, Soleri D (2012) Combining experimental and numerical investigations to explore the potential of downsized engines operating with methane/hydrogen blends. Int J Hydrog Energy 37(15):11514–11530, doi:http://dx.doi.org/10.1016/j.ijhydene.2012.03.153
33. Lindner WE, Coggins DL (1986) Steel etchant. United States Patent No. 4592854
34. Lopez E, Gepert V, Gritsch A, Nieken U, Eigenberger G (2012) Ethanol steam reforming thermally coupled with fuel combustion in a parallel plate reactor. Ind Eng Chem Res 51(11):4143–4151. doi:10.1021/ie202364y
35. Lyman T (1967) Metals handbook, vol 3. American Society of Metals, Metals Park
36. Matthews IC, Hanneman GB (1941) Etching solution. United States Patent No. 2266430
37. McGlade C, Speirs J, Sorrell S (2013) Unconventional gas – a review of regional and global resource estimates. Energy 55:571–584, doi:http://dx.doi.org/10.1016/j.energy.2013.01.048
38. Meille V (2006) Review on methods to deposit catalysts on structured surfaces. Appl Catal A Gen 315:1–17, doi:http://dx.doi.org/10.1016/j.apcata.2006.08.031
39. Men Y, Kolb G, Zapf R, Hessel V, Löwe H (2007) Ethanol steam reforming in a microchannel reactor. Process Saf Environ Prot 85(5):413–418
40. Moharana MK, Peela NR, Khandekar S, Kunzru D (2011) Distributed hydrogen production from ethanol in a microfuel processor: issues and challenges. Renew Sust Energ Rev 15(1):524–533, doi:http://dx.doi.org/10.1016/j.rser.2010.08.011
41. Montini T, De Rogatis L, Gombac V, Fornasiero P, Graziani M (2007) Rh(1 %)@CexZr1-xO2-Al2O3 nanocomposites: active and stable catalysts for ethanol steam reforming. Appl Catal B Environ 71(3–4):125–134
42. Ni M, Leung DYC, Leung MKH (2007) A review on reforming bio-ethanol for hydrogen production. Int J Hydrog Energy 32(15):3238–3247
43. Ni M, Leung MKH, Leung DYC (2008) Energy and exergy analysis of hydrogen production by a proton exchange membrane (PEM) electrolyzer plant. Energy Convers Manag 49(10):2748–2756, doi:http://dx.doi.org/10.1016/j.enconman.2008.03.018
44. Ni M, Leung MKH, Leung DYC (2008) Technological development of hydrogen production by solid oxide electrolyzer cell (SOEC). Int J Hydrog Energy 33(9):2337–2354, doi:http://dx.doi.org/10.1016/j.ijhydene.2008.02.048
45. Northrop WF, Choi SO, Thompson LT (2012) Thermally integrated fuel processor design for fuel cell applications. Int J Hydrog Energy 37(4):3447–3458, doi:http://dx.doi.org/10.1016/j.ijhydene.2011.11.034
46. O'Connell M, Kolb G, Schelhaas K-P, Schuerer J, Tiemann D, Keller S, Reinhard D, Hessel V (2010) Investigation on the combined operation of water gas shift and preferential oxidation reactor system on the kW scale. Ind Eng Chem Res 49(21):10917–10923. doi:10.1021/ie1005614
47. O'Connell M, Kolb G, Schelhaas KP, Wichert M, Tiemann D, Pennemann H, Zapf R (2012) Towards mass production of microstructured fuel processors for application in future distributed energy generation systems: a review of recent progress at IMM. Chem Eng Res Des 90(1):11–18
48. Ockwig NW, Nenoff TM (2007) Membranes for hydrogen separation. Chem Rev 107(10):4078–4110. doi:10.1021/cr0501792
49. Oh SH, Sinkevitch RM (1993) Carbon monoxide removal from hydrogen-rich fuel cell feedstreams by selective catalytic oxidation. J Catal 142(1):254–262, doi:http://dx.doi.org/10.1006/jcat.1993.1205

50. Ouyang X, Bednarova L, Besser RS, Ho P (2005) Preferential oxidation (PrOx) in a thin-film catalytic microreactor: advantages and limitations. AIChE J 51(6):1758–1772. doi:10.1002/aic.10438
51. Peela NR, Mubayi A, Kunzru D (2009) Washcoating of γ-alumina on stainless steel microchannels. Catal Today 147(Supplement):S17–S23, doi:http://dx.doi.org/10.1016/j.cattod.2009.07.026
52. Peela NR, Mubayi A, Kunzru D (2011) Steam reforming of ethanol over Rh/CeO2/Al2O3 catalysts in a microchannel reactor. Chem Eng J 167(2–3):578–587, doi:http://dx.doi.org/10.1016/j.cej.2010.09.081
53. Peela NR, Sandupatla AS, Rao PLP, Kunzru D (2012) Hydrogen production from ethanol in a microchannel reactor. In: International conference on sustainable energy and environmental protection, Dublin, 2012
54. Peela NR, Sandupatla AS, Kunzru D (2014) Development of a microfuel processor: oxidative steam reforming of ethanol and water-gas shift reaction on noble metal catalysts in a microreactor. Int J Environ Eng 6:78–90
55. Piscina PR, Homs N (2008) Use of biofuels to produce hydrogen (reformation processes). Chem Soc Rev 37:2459–2467
56. Radhakrishnan R, Willigan RR, Dardas Z, Vanderspurt TH (2006) Water gas shift activity of noble metals supported on ceria-zirconia oxides. AIChE J 52(5):1888–1894. doi:10.1002/aic.10785
57. Rao PN, Kunzru D (2007) Fabrication of microchannels on stainless steel by wet chemical etching. J Micromech Microeng 17(12):N99
58. Rao PLP, Kunzru D (2011) Compact microfuel processor for hydrogen production from ethanol. Chemcon-2011, Bangalore, 2011
59. Ricote S, Jacobs G, Milling M, Ji Y, Patterson PM, Davis BH (2006) Low temperature water-gas shift: characterization and testing of binary mixed oxides of ceria and zirconia promoted with Pt. Appl Catal A Gen 303(1):35–47
60. Roh H-S, Wang Y, King D (2008) Selective production of H2 from ethanol at low temperatures over Rh/ZrO2–CeO2 catalysts. Top Catal 49(1):32–37. doi:10.1007/s11244-008-9066-3
61. Romero-Sarria F, Vargas JC, Roger A-C, Kiennemann A (2008) Hydrogen production by steam reforming of ethanol: study of mixed oxide catalysts Ce2Zr1.5Me0.5O8: comparison of Ni/Co and effect of Rh. Catal Today 133–135:149–153
62. Rossetti I, Biffi C, Tantardini GF, Raimondi M, Vitto E, Alberti D (2012) 5kWe + 5kWt reformer-PEMFC energy generator from bioethanol first data on the fuel processor from a demonstrative project. Int J Hydrog Energy 37(10):8499–8504, doi:http://dx.doi.org/10.1016/j.ijhydene.2012.02.095
63. Ruettinger W, Ilinich O, Farrauto RJ (2003) A new generation of water gas shift catalysts for fuel cell applications. J Power Sources 118(1–2):61–65, doi:http://dx.doi.org/10.1016/S0378-7753(03)00062-4
64. Sandupatla AS (2010) Water-gas shift reaction in microstructured reactor using noble metal catalyst. M.Tech. thesis, Department of Chemical Engineering, Indian Institute of Technology, Kanpur
65. Seris ELC, Abramowitz G, Johnston AM, Haynes BS (2005) Demonstration plant for distributed production of hydrogen from steam reforming of methane. Chem Eng Res Des 83(6):619–625, doi:http://dx.doi.org/10.1205/cherd.04353
66. Seris ELC, Abramowitz G, Johnston AM, Haynes BS (2008) Scaleable, microstructured plant for steam reforming of methane. Chem Eng J 135(1):S9–S16, doi:http://dx.doi.org/10.1016/j.cej.2007.07.018
67. Shi L, Yang G, Tao K, Yoneyama Y, Tan Y, Tsubaki N (2013) An introduction of CO2 conversion by dry reforming with methane and new route of low-temperature methanol synthesis. Acc Chem Res 46(8):1838–1847. doi:10.1021/ar300217j
68. Snyder HB, Rosenberg LM (1960) Chemical milling. United States Patent No. 2940838
69. Triphob N, Wongsakulphasatch S, Kiatkittipong W, Charinpanitkul T, Praserthdam P, Assabumrungrat S (2012) Integrated methane decomposition and solid oxide fuel cell for efficient

electrical power generation and carbon capture. Chem Eng Res Des 90(12):2223–2234, doi:http://dx.doi.org/10.1016/j.cherd.2012.05.014

70. Vaidya PD, Rodrigues AE (2006) Insight into steam reforming of ethanol to produce hydrogen for fuel cells. Chem Eng J 117(1):39–49
71. Valentini M, Groppi G, Cristiani C, Levi M, Tronconi E, Forzatti P (2001) The deposition of γ-Al2O3 layers on ceramic and metallic supports for the preparation of structured catalysts. Catal Today 69(1–4):307–314, doi:http://dx.doi.org/10.1016/S0920-5861(01)00383-2
72. Wang J, Ryan D, Anthony EJ (2011) Reducing the greenhouse gas footprint of shale gas. Energ Policy 39(12):8196–8199, doi:http://dx.doi.org/10.1016/j.enpol.2011.10.013
73. Weber CL, Clavin C (2012) Life cycle carbon footprint of shale gas: review of evidence and implications. Environ Sci Technol 46(11):5688–5695. doi:10.1021/es300375n
74. Zalc JM, Löffler DG (2002) Fuel processing for PEM fuel cells: transport and kinetic issues of system design. J Power Sources 111(1):58–64, doi:http://dx.doi.org/10.1016/S0378-7753(02)00269-0

Axial Back Conduction through Channel Walls During Internal Convective Microchannel Flows

Sameer Khandekar and Manoj Kumar Moharana

Abstract Recent developments during the last decade in the field of manufacturing and development of many high-power mini-/micro-devices led to increased interest in microfluidic devices involving heat transfer. Since the pioneering work by Tuckerman and Pease (IEEE Electron Device Lett 2:126–129, 1981) on the use of microchannels for high heat flux removal, certainly a lot of developments has been witnessed through ever-increasing analytical, experimental, and highly sophisticated numerical studies by many researchers across the globe. In spite of this progress, many fundamental understandings of flow and heat transfer phenomena in mini-/microchannel systems are still obscure. One such phenomenon is the flow of heat in the solid wall of microchannel systems by means of conduction normally in a direction opposite to that of internal convective mini-/microchannel flow of fluid, called "axial wall conduction" or "axial back conduction." Axial back conduction is not a new phenomenon, rather mostly neglected unintentionally because of its convincingly smaller influence on heat transfer in conventional-size channels. As the hydraulic diameter of a channel decreases, the coupling between the substrate and bulk fluid temperatures becomes significant because of the relative size of the fluid to the solid wall. Unlike in conventional-size channels, negligence of axial back conduction along the solid walls of micro heat exchangers frequently leads to erroneous conclusions and inconsistencies in the interpretation of transport data. Thus, it is important to explicitly identify the thermofluidic parameters of interest which lead to a distortion in the boundary conditions and thus the true estimation of species transfer coefficients. In this chapter, we focus our attention on the axial back conduction in the solid substrate/channel wall as against the axial back conduction in the liquid flow domain; thus, a detailed review of the state-of-the-art on axial back conduction in both conventional as well as mini-/microchannel systems is presented.

S. Khandekar (✉)
Department of Mechanical Engineering, Indian Institute of Technology Kanpur, Kanpur, 208016 UP, India
e-mail: samkhan@iitk.ac.in

M.K. Moharana
Department of Mechanical Engineering, National Institute of Technology Rourkela, Rourkela 769008, India

Y.M. Joshi, S. Khandekar (eds.), *Nanoscale and Microscale Phenomena*,
Springer Tracts in Mechanical Engineering, DOI 10.1007/978-81-322-2289-7_13

Keywords Axial conduction • Microchannel • Microtube • Microscale • Circular channel • Rectangular channel • Axial back conduction • Single-phase • Conjugate heat transfer • Peclet number • Prandtl number • Biot number • NTU • Nusselt number • Reynolds number • Temperature gradient • Constant temperature • Constant heat flux • Laminar flow • Conduction parameter • Heat transfer coefficient • Heat exchanger • Bulk mean fluid temperature • Dimensionless wall temperature • Navier–Stokes equation • Thermally developing • Thermally fully developed • Hydrodynamically developing • Hydrodynamically fully developed • Thermal resistance • Numerical simulation • Aspect ratio • Conductivity ratio

1 Introduction

The pioneering work by Tuckerman and Pease [62] had demonstrated that internal convective flow through microchannel geometries is capable of removing very high heat flux. For a given available temperature difference, owing to their higher surface area per unit volume compared to a conventional-size channel, microchannels provide higher heat removal rate. With advances in microfabrication techniques, microchannels of different shape and size can be fabricated easily using a wide variety of materials ranging from metals to nonmetals.

The recent developments in micro-manufacturing technology and the seminal work by Tuckerman and Pease [62] have led to the widespread use of microchannel heat exchangers for a variety of applications, including, but not restricted to, electronics thermal management, micro-reaction, spacecraft thermal management, etc. A summary of developments in the last three decades with a historical perspective is available in Goodling [13] and Mehendale et al. [39]. With the growing literature in this field, quite a few topical books and monographs dedicated to microchannel fluid flow and heat transfer have also been published [5, 20, 22, 24, 66, 67, 69], including a recent update by Kandlikar [21].

It has been established that scaling effects play a vital role in microscale single-phase fluid flow and heat transfer devices [14, 47, 54]. Among these effects, axial back conduction in the solid substrate plays a critical role in the performance of a microscale heat exchanger device. In this background, we focus our attention in understanding the nuances of axial back conduction in the solid substrate and its effect on thermal transport phenomena, as applicable to single-phase internal convective flows in microchannel geometries.

2 What is Axial Back Conduction?

We consider laminar fluid flowing through a circular duct subjected to a "constant heat flux" or "constant temperature" boundary condition on the duct wall. Under non-conjugate heat transfer conditions, it is conventionally assumed that channel

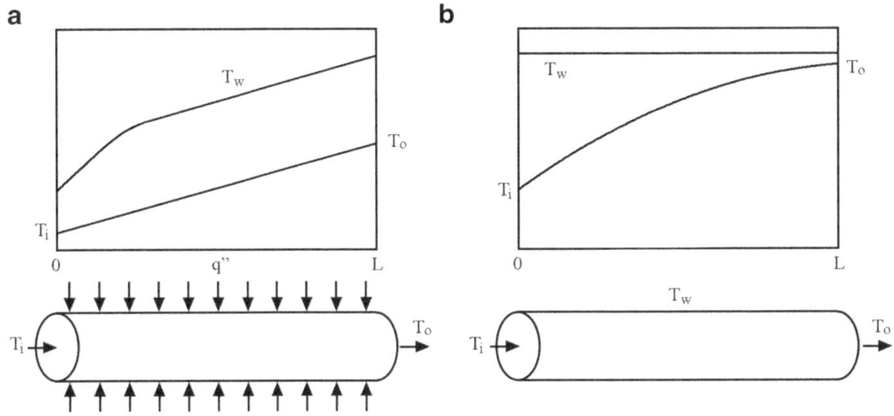

Fig. 1 Streamwise variation of bulk fluid and local temperature of the wall in a circular duct subjected to (**a**) constant wall heat flux, (**b**) constant wall temperature, boundary condition

wall does not affect the transport behavior, i.e., the channel wall has negligible thermal inertia. The axial temperature profile for the fluid and the wall, for these two respective boundary conditions, is as shown in Fig. 1.

From Fig. 1a, b, it is observed that, for both the solid as well as the fluid domain, there is a temperature gradient between any two points along the axial direction. By virtue of this temperature gradient, there is a net heat flux which also gets transported in the axial direction (from upstream towards the downstream direction), both in the fluid and the solid domain of the substrate. Under what conditions this flux associated with the axial direction starts affecting the boundary layers, so as to eventually affect the local and average heat transfer coefficient? In this work, we focus our attention on the axial back conduction in the solid substrate/channel wall as against the axial back conduction in the liquid flow domain. Alternately speaking, the Peclet numbers considered in this chapter are typically quite large so as to assume that the axial conduction effects in the fluid domain are negligible, as compared to that of the solid substrate [66]. Second, the work by Xu et al. [64], Koo and Kleinstreuer [25], and Morini [47] clearly highlights that the viscous dissipation in the liquid domain can also be neglected for the range of parameters considered in this discussion.

In all real-time engineering applications, the channel wall is not "inertia-free"; it has finite thickness and thermal diffusivity. Practically speaking, the boundary conditions can only be applied on the outer surface of the channel wall, as shown in Fig. 2, for a conventional-sized circular tube. Thus, the heat transfer is always conjugate in nature, and as will be appreciated in the later sections, the conjugate nature of the problem strongly affects the estimation of the local and average heat transfer coefficient.

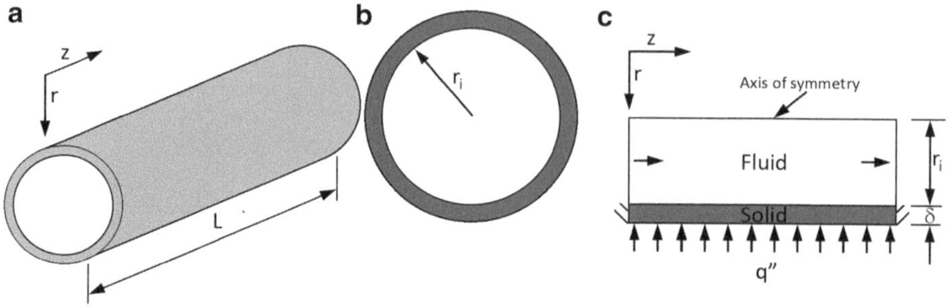

Fig. 2 Schematic diagram of a conventional circular tube. (**a**) Circular tube with conductive wall. (**b**) Cross-section view. (**c**) Transverse section view

A detailed review on early developments on axial wall conduction in conventional-size heat exchangers was presented by Peterson [49], according to which, one of the first studies in this field dates back to as early as 1926. Subsequently, a range of related issues have been addressed in the literature [2, 3, 8, 11, 12, 46, 50] for conventional-size heat exchangers.

3 Axial Back Conduction in Conventional-Size Channels

The concept of "conduction parameter" can be invoked to quantify the effect of axial conduction in the solid walls during single-phase internal convective flows. As per the existing definitions in the literature definition, it gives a relative importance of back conduction heat transfer compared to the thermal energy flow carried by the fluid. It is defined as the ratio of axial heat transfer within the solid duct or tube due to axial temperature gradient in it to the thermal energy carried by the fluid in the channel in the axial direction. Mathematically, it can be defined as

$$P = \frac{k_s A_s \Delta T_s / L}{\dot{m} c_{p,f} \Delta T_f} \tag{1}$$

The driving force for axial back conduction ΔT_s is the difference in temperature between the inlet and outlet location of the solid wall, and ΔT_f is the difference between fluid outlet and inlet temperature. Considering that $\Delta T_s \approx \Delta T_f$, Eq. 1 becomes

$$P = \frac{k_s A_s}{\dot{m} c_{p,f} L} \tag{2}$$

The parameter given by Eq. 2 was applied to the analysis of recuperators by Bahnke and Howard [2].

Petukhov [50] used the following parameter to characterize the effect of axial conduction in a conventional-size circular tube.

$$P = k_{\mathrm{sf}}\left[1 - (1 + (\delta/r_{\mathrm{i}}))^2\right] \tag{3}$$

Mori et al. [46] studied the effect of conduction in the tube wall on steady-state heat transfer to laminar flow in circular tube. For the case of a circular tube, Faghri and Sparrow [12] indicated that the wall conduction parameter measures the strength of axial conduction in its wall and defined it as

$$P = k_{\mathrm{sf}}\,(\delta/r_{\mathrm{i}}) \tag{4}$$

Chiou [8] presented a dimensionless parameter for a circular duct called "conductance number" for quantitative description of axial back conduction by rearranging Eq. 3 as follows:

$$P = \frac{k_{\mathrm{sf}} A_{\mathrm{sf}}}{Pe \cdot (L/d_{\mathrm{i}})} \tag{5}$$

From Eq. 5, it is seen that the effect of axial back conduction depends on conductivity ratio (k_{sf}), area of cross-section ratio (A_{sf}), and Prandtl number (Pr). A_{sf} is a geometric function, and for a circular tube, it depends on δ/r_{i}. Chiou [8] indicated that if the conduction parameter defined in Eq. 5 is less than 0.005, the effect of axial heat conduction along the tube wall can be neglected.

Barozzi and Pagliarini [3] studied conjugate heat transfer in fully developed laminar flow in a circular pipe. Cotton and Jackson [11] had presented dimensionless parameter for axial conduction in circular tube, given by

$$P = \frac{k_{\mathrm{sf}}}{Pe^2}\frac{\delta}{d_{\mathrm{i}}}\,(1 + (\delta/d_{\mathrm{i}})) \tag{6}$$

In the context of conventional channels, the literature suggests that sufficient information was generated for designers and open literature on the subject declined after mid-1990s (Peterson [49]). Due to the advent of microchannels and rapid maturity in associated manufacturing routes, scrutinizing this issue has again become necessary.

4 Microchannels vs. Conventional-Size Channels

To appreciate the difference between micro and macro ducts, in the context of axial conduction in the substrate, consider the case of circular ducts. In conventional-sized heat exchangers and other heat transfer application devices, the duct wall thickness is substantially smaller as compared to the inner diameter (or flow cross-sectional

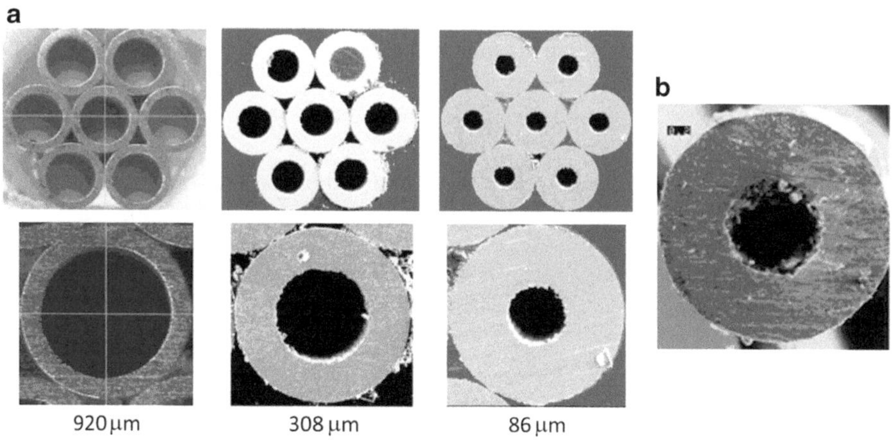

Fig. 3 Cross-sectional view of microtubes ((**a**) [65] (**b**) [35])

area). But for flow in microchannels, the solid wall thickness is normally of the same order or quite, often, larger than the channel diameter. This is exemplified in Fig. 3, which shows an example of cross-sectional views of some microtubes used for the purpose of experimental investigations or in engineering applications.

In the context of laminar duct flow in individual flat plate microchannel structures or an array of flat plate microchannels too, conjugate heat transfer leads to a strong multidimensional coupling. The small dimensions of microchannels in the transverse direction are often comparable to the thickness of the substrate. As the hydraulic diameter of microchannels decreases, this thermofluidic coupling between the substrate and the bulk fluid temperatures becomes significant. It is important to explicitly identify the thermofluidic parameters of interest which lead to a distortion in the applied boundary conditions and thus affecting the true estimation of the transfer coefficients of species.

5 Axial Back Conduction in Micro-Sized Channels

Bier et al. [4] experimentally studied heat transfer in micro heat exchangers consisting of approximately 8,000 rectangular microchannels arranged in a cube-shaped volume of 1 cm^3 in a cross flow. They found that heat exchanger made from stainless steel yields significantly higher heat transfer efficiency than copper. Bier et al. [4] attributed this to axial heat conduction in the walls of the microchannels; the axial heat conduction effect is much less evident in stainless steel compared to copper because of lower thermal conductivity of steel than copper. Stief et al. [57] numerically studied influence of the thermal conductivity of the wall on heat transfer efficiency of micro heat exchangers. They found that efficiency of heat exchanger depends on wall conductivity, and there exists optimum conductivity at which the efficiency is highest.

Peterson [49] numerically studied conduction effects in counter flow microscale heat exchangers and found that higher wall thermal conductivity leads to excessive conduction losses in the device. Peterson [49] recommended that very low thermal conductivity material should be used for the construction of micro counterflow heat exchangers in order to achieve reasonable performance in small devices.

Some of the early works on microchannel heat transfer (e.g., [9, 61]) reported that at microscale, the results are deviating significantly from conventional theory of heat transfer and fluid flow; early studies proposed this deviation to be an effect of additional, unexplored flow physics. Herwig [15] and Herwig and Hausner [16] questioned the new results in microchannel heat transfer. Practically, the bulk mean fluid temperature cannot be measured experimentally inside a microchannel along its length. Therefore, it is normally interpolated linearly between the inlet and the outlet temperature of fluid as it passes through a microchannel. As explained earlier, the axial variation of bulk fluid temperature will be linear in nature if and only if the applied "constant heat flux" is felt at the solid–fluid interface of the microchannel. This is not true when a microchannel system is under the influence of conjugate condition wherein the actual location of application of boundary condition is away from the real-solid–fluid boundary. Herwig [15] and Herwig and Hausner [16] introduced "standard macroanalysis" and "scaling effects with respect to standard macroanalysis." Based on these considerations, they postulated five different scaling effects. The prominent among them are (1) conjugate heat transfer attributed to relatively thick microchannel wall, and (2) and axial heat conduction due to very small Peclet number. Herwig [15] and Herwig and Hausner [16] also presented a critical view on the experimental work of Tso and Mahulikar [61]. Through numerical simulation, they had taken into account axial conduction and conjugate effects to explain the result of Tso and Mahulikar [61] in a conventional manner. Through numerical simulation, Herwig [15] and Herwig and Hausner [16] found that the bulk fluid temperature is not at all linearly varying between the inlet and the outlet fluid temperature values; the experimental results of Tso and Mahulikar [61] were based on linear interpolation for bulk fluid temperature. Taking these considerations, the corrected value of fully developed Nusselt number was indeed found to be in line with the conventional value of 4.36 (for constant heat flux boundary condition). Although Herwig [15] and Herwig and Hausner [16] highlighted the importance of axial conduction in microchannels, they did not propose any specific recommendations as regards selection of material or thickness of microchannel walls, vis-à-vis the flow conditions.

Kroeker et al. [27] considered three-dimensional numerical analysis of circular and rectangular parallel microchannels in a solid substrate subjected to constant heat flux on one face, while the other face was kept insulated. It was found that axial back conduction intensifies with increasing (1) interchannel distance, (2) substrate thickness, and (3) conductivity ratio of wall material and the fluid. The variation of heat flux in the circumferential direction of the circular microchannel is negligible, but it varies significantly in the axial direction. Kroeker et al. [27] compared the performance of circular and rectangular parallel microchannels on the basis of equal hydraulic diameter and equal Reynolds number and found that heat sinks with rectangular channels have less thermal resistance than their circular counterpart.

Maranzana et al. [36] considered flow between two parallel plates of which the lower plate was kept adiabatic and the upper plate (thickness δ_s and length L) was subjected to constant heat flux on its outer surface. Considering unit width and assuming one-dimensional heat transfer, Maranzana et al. [36] reintroduced the concept of conduction parameter (as defined in Eq. 2 considering $\Delta T_s \approx \Delta T_f$ and was used by Bahnke and Howard [2]) as axial conduction number (M) and rearranged Eq. 2 as follows:

$$M = \frac{\text{NTU} \cdot \alpha^2}{Bi} \tag{7}$$

where the Biot number (Bi), number of transfer units (NTU) of the system, and upper plate aspect ratio (α) are defined as

$$\text{NTU} = \frac{h \cdot L}{\dot{m} \cdot c_{p,f}}, \quad Bi = \frac{h \cdot \delta_s}{k_s}, \quad \alpha = \frac{\delta_s}{L} \tag{8}$$

Here, the heat transfer coefficient h is assumed to be uniform. Maranzana et al. [36] reported that at higher value of M (say $M = 0.05$, which occurs at lower flow rate, i.e., at $Re = 100$ in this case), the axial variation of bulk fluid temperature is not linear. The slope of this curve is higher near the inlet due to large axial conduction effects. Secondly, with increasing value of M, the deviation of axial variation of bulk fluid temperature from linearity also increases. Maranzana et al. [36] also stated that the effect of axial back conduction was negligible for all the cases considered when M was less than 0.01. Maranzana et al. [36] also studied effect of axial wall conduction on efficiency of counterflow heat exchanger and found that exchanger efficiency is a function of wall conductivity and showed that axial wall conduction results in lowering the exchanger efficiency. They also reported that there exists an optimal wall conductivity which maximizes the heat exchanger efficiency.

Parallel microchannel array heat sink requires inlet and outlet collectors for uniform flow distribution across all the microchannels. Tiselj et al. [59] studied effect of axial heat conduction in the channel wall under such conditions. They performed both experimental and numerical study of microchannel array heat sink made from silicon chip with 17 parallel triangular microchannels with inlet and outlet collectors made of steel. Tiselj et al. [59] reported that the axial heat flux has a maximum near the inlet collector and decreases in the flow direction up to zero. Therefore, heat flows in the solid walls in a direction opposite to the flow. Away from the inlet, again heat flux acts in the direction of flow. In this case, the wall temperature decreases, and the heat is transferred from the fluid to the silicon wafer.

Hetsroni et al. [17] analyzed possible sources of unexpected effects, including axial conduction, via an experimental and theoretical investigation. They reported that at $Re > 150$ axial conduction number M becomes less than 0.01; thus, axial conduction effects are negligible. But when $Re < 150$, M becomes higher than 0.01. In such case, heat transfer through the solid substrate should be considered.

As mentioned earlier, both the conduction parameter (λ) and the axial conduction number (M) are based on the assumption that the axial temperature difference between the inlet and outlet locations in the solid substrate (ΔT_s) as well as in fluid (ΔT_f) domain is equal; this assumption is, in reality, rather unrealistic. To address this shortcoming in the definition of parameter M (or λ), as defined by Maranzana et al. [36], Li et al. [32], and later Zhang et al. [68], incorporated the effect of individual temperature differences between the inlet and outlet locations in the solid substrate (ΔT_s) as well as in fluid (ΔT_f) domain. The conduction number, as revised by Li et al. [32], is given by

$$M = \frac{k_s}{\rho_f c_{p,f} \bar{u} L} \frac{A_s}{A_f} \frac{\Delta T_s}{\Delta T_f} \tag{9}$$

where ΔT_s and ΔT_f are the difference in average temperature of the cross section at inlet and outlet of solid and fluid region, respectively.

Based on the outcome of numerical simulations and investigations, Zhang et al. [68] stated that the discerning criteria ($M < 0.01$), as proposed by Maranzana et al. [36], is not always valid. It was highlighted that the specificities of the situation or the problem under investigation will affect the criteria for judging the effect of axial heat conduction in the substrate wall. The studies showed that there was a significant effect of wall axial heat conduction in the circular tube considered by them, in spite of the fact that the parameter M was less than 0.01. The study revealed that the criteria proposed by Maranzana et al. [36] is not suitable for the analysis of thermally developing fluid flow and heat transfer in microchannels.

Stutz and Poulikakos [58] numerically studied the effects of microreactor wall heat conduction on the reforming processes of an autothermal tubular methane micro-reformer. They found that conversion of methane and hydrogen yield strongly depends on the inner surface temperature of the wall, which is influenced by the heat conduction through the channel wall. They found that conversion of methane is highest for nonconductive channel wall and methane conversion is reduced by up to 16 % by using a conductive channel wall. Based on their numerical study, Stutz and Poulikakos [58] concluded that neglecting the thermal conductivity of the channel wall significantly overpredicts the amount of conversion of methane, hydrogen yield, and maximum temperature of the microreactor wall.

Liu et al. [35] experimentally studied axial wall heat conduction in stainless steel microtube and found that the axial heat conduction can reduce the convective heat transfer. They also found that the influence of axial wall heat conduction on convective heat transfer decreases with increase in applied heating power.

Microreactors assist in utilization of maximum potential of catalysis used during endothermic/exothermic reactions [43]. Axial wall conduction also affects the performance of microreactors. Therefore, selection of substrate material is an important issue in the design of microreactor systems [45]. Depending on the constraints on large thermal gradients, localized hot-spot formation, isothermal operation, etc., substrate thermal conductivity may be selected to facilitate by limiting the rate of axial heat conduction. Moreno et al. [45] presented a parametric study to highlight design challenges of thermally integrated microchannel networks for chemical and/or fuels reforming and investigated the influence of solid-phase thermal conductivity and thermal packaging on (1) thermal efficiency (2) reaction conversion and (3) steady-state multiplicity. Their results indicated that low thermal conductive substrates provide optimal reactor performance.

Mathew and Hegab [37] analytically studied the effect of axial heat conduction on the thermal performance of balanced counterflow microchannel heat exchanger. They found that effectiveness of the fluid depends on axial heat conduction parameter, in addition to other parameters such as NTU, manifold fluid temperature, and Biot number. Mathew and Hegab [38] numerically studied the effect of axial heat conduction on the hot and cold fluid effectiveness of a balanced parallel flow microchannel heat exchanger. They found that axial conduction influences the effectiveness of the heat exchanger. They also found that the effectiveness of the counter flow microchannel heat exchanger is better than that of parallel flow microchannel heat exchanger subjected to similar axial heat conduction parameter.

Nonino et al. [48] used a hybrid finite element procedure to study the effect of axial heat conduction in the solid walls of simultaneously developing flow in circular microducts. They found that higher value of k_{sf} leads to a more uniform streamwise temperature distribution at the solid–fluid interface, thus lowering the Nusselt number. They also verified the reliability of axial conduction number M and the commonly adopted criterion (proposed by Maranzana et al. [36]) and found that M is not the only criterion for judging whether the axial heat conduction can be neglected.

Kosar [26] numerically studied the effect of substrate thickness and substrate material on heat transfer in microchannel heat sinks by considering square microchannels (size 200 μm× 200 μm) on different substrate thicknesses (100–1,000 μm) and different substrate materials (polyimide, silica glass, quartz, steel, silicon, copper). For constant heat flux applied on the bottom of the substrate, Kosar [26] found that the Nusselt number is lower for low thermal conductivity materials and higher for high conductivity materials. It was concluded that for low thermal conductivity materials, a significant deviation from the constant heat flux boundary condition occurs at the solid–fluid interface. Kabar et al. [19] numerically studied simultaneously developing flow through parallel-plate microchannel heat sinks. They found that due to axial conduction, the thickness of the plate may have an influence on both thermally developing length and asymptotic Nusselt number, corresponding to fully developed region.

Moharana et al. [40], both experimentally and numerically, studied conjugate heat transfer in a minichannel array and found that for higher value of the axial conduction number (M), the experimental setup is prone to conjugate effects as a consequence of axial back conduction in the solid substrate. This causes the observed experimental local Nusselt numbers to be smaller than the actual counterparts predicted by the numerical model (without considering the axial conduction), and its axial variation is also affected.

For conjugate heat transfer in a parallel-plate microchannel, Cole and Cetin [10] have reported an analytical solution using the Green's functions. They reported that axial conduction in the channel wall is important under the conditions when (a) length over height ratio of the microchannel is very small; (b) flow Peclet number is very small; (c) thickness of the channel wall, relative to the channel height is large; and (d) conductivity of wall material is high, relative to conductivity of the fluid. Karakaya and Avci [23] simulated coupled methane oxidation/isooctane steam reforming in parallel microchannels and found the effect of axial conduction in the microreactor substrate on its axial temperature distribution and subsequent hydrogen yield. Chein et al. [7] numerically studied the effects of axial conduction in the reformer wall on performance of methanol–steam reforming. They suggested that the methanol–steam reformer with low wall thermal conductivity and thin wall thickness has small axial conduction parameter and results in higher performance.

Rahimi and Mehryar [52] numerically studied the effect of wall conductivity and wall thickness on the local Nusselt number at the entrance and end regions of a circular microduct. They found that due to axial back conduction in the duct wall, the local Nusselt number at the entrance and end region decrease, as compared to their values for circular microduct without conjugate effects.

Avci et al. [1] numerically investigated conjugate heat transfer in a microtube where flow is steady, laminar, hydrodynamically fully developed, and thermally developing in nature. They considered interactive effects of thermal conductivity ratio (solid to fluid) with varying length and wall thickness. They observed that the dimensionless wall flux (which ideally should be equal to unity along the channel) decreases gradually near the outlet. This effect becomes more pronounced for higher conductivity ratio (solid to fluid) and lower length to diameter ratio. This observation is in line with the observation of Nonino et al. [48] and Rahimi and Mehryar [52]. They also reported that a decrease in thermal resistance in the wall, which corresponds to larger value of k_{sf} and δ/r_i, leads to lower value of Nusselt number along the channel.

Lin and Kandlikar [33] have recently developed a theoretical model to analyze the effect of axial conduction on heat transfer during single-phase flow in microchannels. This model is based on the assumption that flow is thermally and hydrodynamically fully developed. They found that the axial conduction effects in the wall are severe for gas flows through any tube material. Lin and Kandlikar [33] also found that axial conduction effects in the wall are negligible for water flowing through stainless steel microtubes, while for higher conductivity materials, it is not negligible even for water.

Normally wall temperatures are measured at different axial locations of a microchannel, while only inlet and outlet bulk fluid temperature are practically measured in experimental investigation. Any intrusive technique used to measure bulk fluid temperature along the axial locations will distort the flow field, thus deviation from the ideal fluid flow condition. From numerical investigations, it is clear that axial wall conduction caused nonlinear variation in bulk fluid temperature between the inlet and the outlet fluid temperature values. However, most analyses of experimental investigations consider linear variation, thus arriving at wrong conclusions. To overcome this experimental difficulty, Huang et al. [18] used molecule-based temperature sensors in their experimental investigation to study the effect of axial heat conduction on the heat transfer analysis in microchannel flow. Rhodamine B/DI water and Ru(bpy)/dope have been respectively applied as temperature-sensitive fluid and temperature-sensitive paint for fluid and surface temperature measurements. Based on their experimental evidence, they validated the effects of axial heat conduction in microchannel flow with detailed temperature evolution.

6 Axial Wall Conduction in Square Microchannels

Bases on the detailed review of the literature on axial wall conduction in mini-/microchannels discussed in Sect. 5, it can be inferred that clear, concise, and conclusive statements on effects of axial wall conduction and its remedial suggestions are not available. While the phenomena is well acknowledged, there is a need of thorough investigation of effects of axial wall conduction, especially on the performance of microscale heat transfer devices by considering a wide parametric variation. In this backdrop, Moharana et al. [44] numerically studied the effect of axial wall conduction in a conjugate heat transfer situation involving simultaneously developing laminar flow in a square microchannel (carved on a flat substrate) with constant wall heat flux applied on the bottom of the substrate.

Figure 4 schematically shows details of the microchannel considered by Moharana et al. [44]. Constant heat flux (q') is applied at the bottom face of the substrate simulating heat generation, e.g., from an electronic chip, and all other surfaces of the substrate exposed to the surrounding are assumed to be insulated. Considering symmetry conditions and to reduce computational effort, only one half of the microchannel was considered as the computational domain as shown in Fig. 4c.

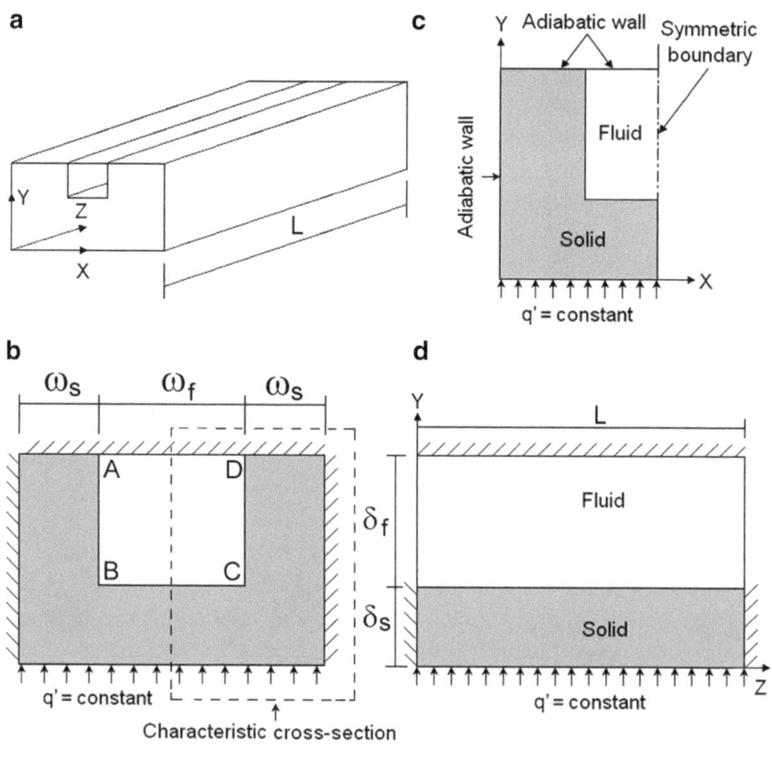

Fig. 4 Details of the microchannel. (**a**) Square microchannel carved on a solid substrate. (**b**) Substrate cross section. (**c**) Cross section of computational domain. (**d**) Transverse section view along the plane of symmetry [44]

The width (ω_f), height (δ_f), and length (L) of the microchannel in the computational model are kept constant at 0.4, 0.4, and 120 mm, respectively. As the thickness of the substrate increases, the boundary on which constant heat flux applied moves away from the actual solid–fluid interface (in this case, the three sides of the channel through which heat transfer to the fluid takes place). Therefore, the width of the substrate ($2\omega_s + \omega_f$) is kept constant (1.2 mm), and the thickness of the substrate ($\delta_s + \delta_f$) is varied in the Y direction to understand the effect of the substrate thickness on the conjugate heat transfer behavior. The values of δ_s and δ_{sf} are varied in the range of 0.4–9.6 mm and 1–24 respectively. Pure water (at an inlet temperature of 300 K, $Pr = 5.85$) is used as the working fluid, which enters the microchannel with a slug velocity profile ($\bar{u}$). The fluid flow Re is maintained in the range of 100–1,000. The substrate material thermal conductivity (k_s) is varied such that the conductivity ratio ($k_{sf} = k_s/k_f$) varies in the range of 0.34–703.0.

The species conservation equations as applicable to the fluid and the solid domain, i.e., the continuity, Navier–Stokes, and the energy equation, along with the boundary conditions are solved using commercially available software ANSYS FLUENT®. The details of numerical solution methodology and grid independence check can be found in Moharana et al. [44]. The following dimensionless variables are used for interpretation of the results:

$$\begin{gathered} \phi = \frac{q'}{\overline{q'}}, \quad \delta_{sf} = \frac{\delta_s}{\delta_f}, \quad z^* = \frac{z}{RePrD_h}, \quad Nu_z = \frac{hD_h}{k_f} \\ \Theta = \frac{T - T_i}{T_o - T_i}, \quad \Theta_f = \frac{T_f|_z - T_{fi}}{T_{fo} - T_{fi}}, \quad \Theta_w = \frac{T_w|_z - T_{fi}}{T_{fo} - T_{fi}} \end{gathered} \tag{10}$$

where the temperature of bulk fluid at location $z = 0$ (inlet) and $z = L$ (outlet) is represented by T_i and T_o, respectively; the local heat flux at any axial location z^* is represented by $\overline{q'_z}$, which is calculated by taking the peripheral average of heat flux along the three conjugate walls of the substrate (AB, BC, and CD in Fig. 4b) at the X–Y plane at that location; and $\overline{q'}$ is the ratio of the applied heat at the bottom of the substrate and available at the net area of the conjugate walls (i.e., total area of the solid–fluid interface), given by

$$\overline{q'} = q' \frac{(2\omega_s + \omega_f)}{(2\delta_f + \omega_f)} \tag{11}$$

The flow Reynolds number is estimated as $Re = \rho \cdot \bar{u} \cdot D_h/\mu$, and the peripheral average local heat transfer coefficient is denoted by h_z. The local values (h_z and $T_w|_z$) are calculated by performing peripheral averaging along the three conjugate walls (AB, BC, and CD in Fig. 4b) of the channel, at the particular location of interest. The local fluid temperature $T_f|_z$ is the average bulk fluid temperature, taken across the channel cross-sectional area, at any axial location z. The average Nusselt number over the channel length is estimated as:

$$\overline{Nu} = (1/L) \int_0^L Nu_z dz \tag{12}$$

As discussed in Sect. 5, most of the existing studies on axial wall conduction are concerned with circular microtubes. There is a crucial difference between a circular microtube and a rectangular microchannel carved in a flat substrate. In a circular microtube, the solid–fluid interface always remains parallel to the outer surface (irrespective of the tube thickness) on which constant heat flux/wall temperature boundary condition is applied. In contrast, in a square/rectangular microchannel carved in a solid substrate as shown in Fig. 4a, only one wall of the microchannel (bottom wall) is parallel to the base of the solid substrate on which heat is applied (in the present case, constant heat flux is applied); the remaining two vertical walls (solid–fluid interface) are perpendicular to the base of the substrate. This makes the heat transfer from the base to the working fluid three dimensional in nature, whereas it remains two dimensional in case of a microtube due to angular symmetry.

Constant heat flux at the solid–fluid interface leads to maximum heat transfer coefficient. When the solid–fluid interface is away from the base of the substrate on which constant heat flux is applied, the actual condition at the solid–fluid interface gets distorted depending on the situation. It is evident that the boundary condition experienced at the solid–fluid interface of the channel governs the heat transfer process from the substrate to the working fluid. The parameters that discern the effect of axial wall conduction and which leads to alteration in the boundary condition at the solid–fluid interface are axial variation of peripheral averaged wall heat flux and wall temperature and bulk fluid temperature. Therefore, the axial variation of dimensionless local heat flux ϕ (averaged over the microchannel periphery) experienced at the solid–fluid interface is presented in Fig. 5 for selected range of the parameters used by Moharana et al. [44]. It is obvious from these results that the real distribution of heat flux values experienced at the solid–fluid interface is roughly axially uniform at low conductivity ratio (k_{sf}), irrespective of the thickness ratio (δ_{sf}) and flow *Re* (see Fig. 5a, b). At higher values of conductivity ratio (k_{sf}), with increasing thickness ratio (δ_{sf}), the real heat flux experienced at the solid–fluid interface deviates from the constant value applied at the bottom surface of the substrate (see Fig. 5b). At very low thickness ratio (δ_{sf}), the actual heat flux experienced at the solid–fluid interface is identical to the actual value of flux applied at the substrate bottom (i.e., $\phi \approx 1$), except at the region very near to the inlet of the channel, where developing nature of the thermal boundary layer dominates.

Qualitatively speaking, this ensuing pattern of axial variation in the local heat flux values is attributed to the fact that low value of the thermal conductivity ratio leads to higher axial thermal resistance in the solid substrate and vice versa. Accordingly, for cases having higher value of k_{sf}, low axial thermal resistance of the substrate leads to significant back conduction; this effect naturally becomes more prominent with increasing thickness ratio δ_{sf}. It can be observed that for higher k_{sf} and δ_{sf} (in Fig. 5b), the boundary condition at the solid–fluid interface increasingly approaches the trend which is closer to an isothermal temperature boundary condition, although constant heat flux boundary condition was applied at the bottom surface of the substrate. Similar observations were also reported by Zhang et al. [68]; in their case, a constant temperature boundary condition was applied at the outer surface of a circular tube, and it was found that the dimensionless

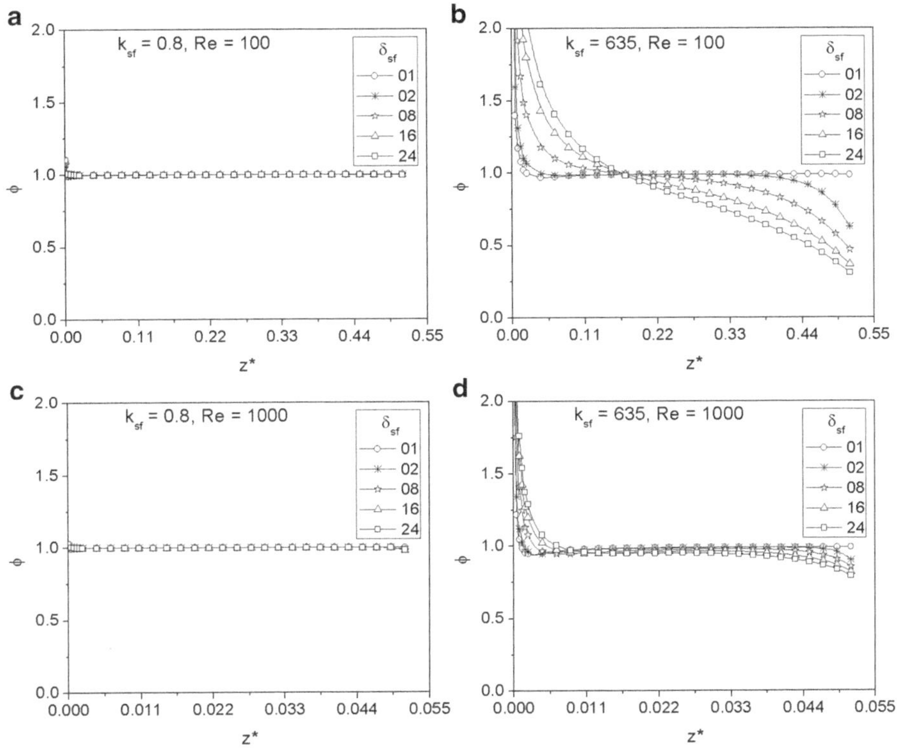

Fig. 5 Axial variation of dimensionless local surface heat flux at the solid-fluid interface [44]

heat flux at the solid–fluid interface tends to become constant when axial conduction in the tube wall dominates. With increasing flow *Re*, the value of ϕ approaches towards unity (compare Fig. 5b, d) even at higher value of k_{sf} and δ_{sf} as the fluid carries comparatively more heat, thus reducing the effect of axial conduction.

Figure 6 presents the axial (streamwise) variations of Θ_w (dimensionless wall temperature which is averaged over the periphery of the channel wall) and the bulk temperature of the fluid Θ_f. At low values of the conductivity ratio k_{sf}, irrespective of the thickness ratio (δ_{sf}), the wall and the fluid temperatures rise as per the usual conventional theory, as applicable for ducts with a zero wall thickness (as shown in Fig. 1a). Under such circumstances, after the flow has fully developed inside the duct, the temperature difference (T_w–T_f or $\Theta_w - \Theta_f$) attains a constant value (Fig. 6a). In contrast to this conventional case, as the value of the conductivity ratio k_{sf} increases, there is a clear and noticeable deviation in the "well-ordered" or "standard" behavior of the wall and the fluid temperatures, respectively. The increase in the fluid temperature is no longer linear indicating; this indicates that the condition of constant heat flux boundary is being compromised due to the prevailing conjugate nature of heat transfer occurring in the substrate (see Fig. 6b). The variation of wall temperature profile clearly indicates that heat flows from the

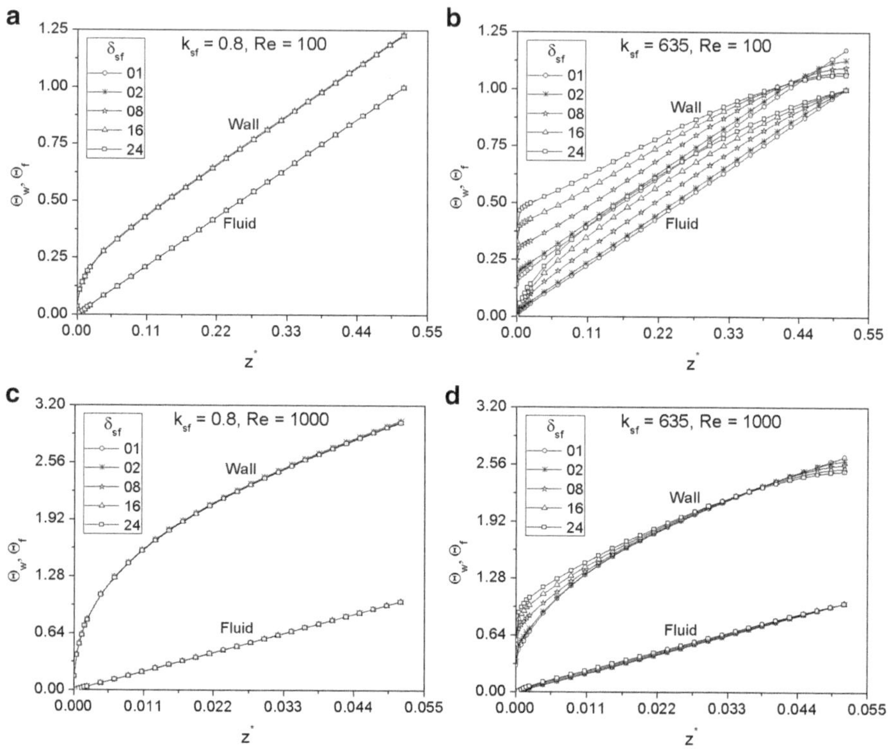

Fig. 6 Streamwise variation of the dimensionless (peripheral averaged) local wall temperature and the bulk fluid temperature, respectively, in the channel [44]

downstream location of the substrate towards the upstream direction. This thermal exchange process tends to "isothermalize" the wall temperature distribution on the solid–fluid boundary of the channel. This phenomenon is also reflected in the gradual occurrence of exponential component superimposed on the linear fluid temperature profile, indicative of a clear shift towards a pseudo-constant temperature boundary condition experienced by the fluid at the boundary interface. Such an effect is obviously more dominant as the thickness ratio δ_{sf} goes on increasing (see Fig. 6b); this further reduces the diffusional thermal resistance offered by the substrate, manifesting the conjugate nature of heat transfer.

The local Nusselt number (Nu_z) is function of local heat flux, local wall and bulk fluid temperature. Thus, the axial variation of local Nusselt number (Nu_z) is presented in Fig. 7, which corresponds to Figs. 5 and 6. In these figures, the two dotted lines indicate the two respective asymptotic values of the Nusselt number, as applicable for fully developed flow, without any conjugate heat transfer (i.e., wall thickness = 0) in a square cross-sectioned channel, which is (1) heated from three of its sides and is insulated from the fourth side; for which the fully developed $Nu = 3.556$ and, (2) heated from one side, three remaining sides being insulated;

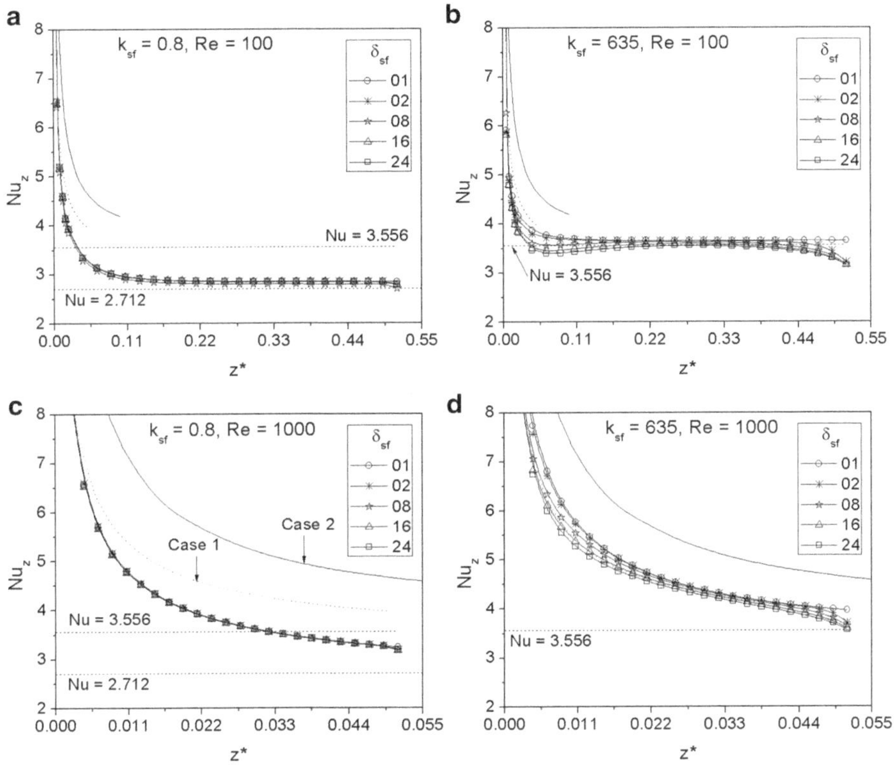

Fig. 7 Axial variation of local Nusselt number (Nu_z) [44]

for which the fully developed $Nu = 2.712$ (Shah and London [56]). For substrates having very high value of conductivity ratio (i.e., $k_{sf} = 635$) and very low thickness ratio ($\delta_{sf} = 1$), *Nu* converges to a value which is close to 3.556. As the thickness ratio (δ_{sf}) of the substrate is increased, the *Nu* for thermally developed region decreases and drifts away from 3.556. Such a behavior can be explained in the background of the fact that, as detailed earlier for these cases in Figs. 5 and 6, the applied boundary condition at the fluid–solid interface gets modified due to the prevailing conjugate effects; it tends to be experienced as a 'pseudo-isothermal' boundary condition by the flowing fluid. This is attributed to axial back conduction, as was also seen earlier in Figs. 5 and 6. For substrate materials having exceedingly low conductivity ratio ($\sim k_{sf} = 0.8$), Nu_z converges to 2.712. This suggests that under such situations heat transfer takes place prominently through the bottom channel wall, leading to the asymptotic behavior, in accordance with the data suggested in Shah and London [56] for a channel with only one side heating and the rest three sides being insulated.

For benchmarking purpose, Nu_z estimates for two more reference cases, which are not exactly as the present case under study, nevertheless are similar in some respects, are also shown in Fig. 7:

- Case 1: Correlation for thermally developing but hydrodynamically fully developed flows [29].
- Case 2: Data for simultaneously developing laminar flow in a square channel for a fluid with $Pr = 0.7$, as reported in Shah and London [56]. For the case of simultaneously developing flows, the dependency of Nusselt number on the fluid Prandtl number is much stronger (Shah and London [56]). For the present study, since $Pr = 5.85$, Nu_z is always smaller than this reference case having $Pr = 0.7$, as expected for a simultaneously developing flow.

From the above discussion, with reference to Fig. 7, it can also be concluded that the dominance of k_{sf} in affecting the alteration in the local Nusselt number due to conjugate heat transfer effects is higher than the geometric ratio of the system, δ_{sf}. Secondly, another important finding is that, while a higher value of k_{sf} increases the effect of axial conduction in the substrate, thereby lowering Nu, a lower value of k_{sf} tends to fundamentally alter the spatial distribution of heat as it gets transferred to the fluid through the three walls of the channel; this, in turn, again lowers the applicable effective local Nusselt number. Thus, it is logically expected that an optimum value of k_{sf} for maximizing the ensuing Nusselt number exists, in-between these two asymptotic mechanisms of heat transport. The occurrence of this optimum k_{sf} for maximizing the average Nusselt number can be observed in Fig. 8., which shows the variation of average Nusselt number over the channel length ($\overline{Nu}$) as a function of the conductivity ratio (k_{sf}), while Re and δ_{sf} are parametrically varied.

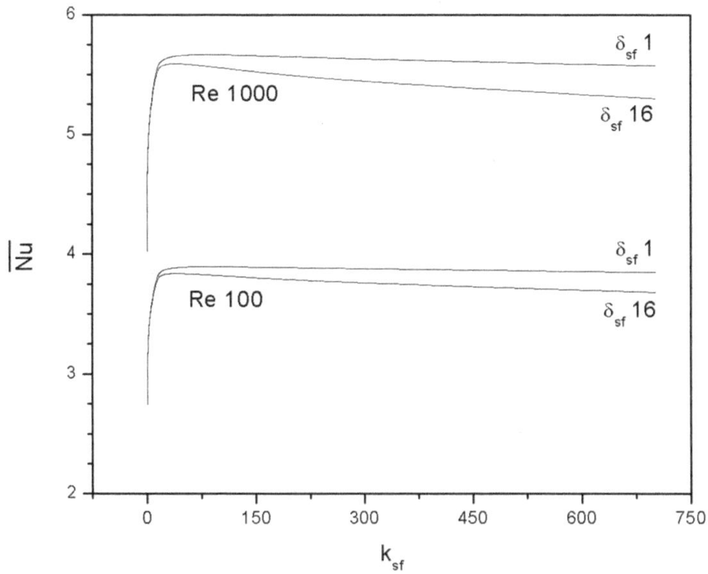

Fig. 8 Average Nusselt number, of the square microchannel, as a function of conductivity ratio (k_{sf}) at different values of thickness ratio (δ_{sf}) and flow Re [44]

At $k_{sf} = 0$, the $\overline{Nu}$ will obviously be zero, irrespective of thickness of the substrate, as the substrate becomes nonconductive (perfectly insulated). As the value of k_{sf} increases to a vanishingly small value, heat flow can now commence and $\overline{Nu}$ rapidly increases; this can be observed for the entire range of flow *Re* and δ_{sf}. $\overline{Nu}$ attains a peak maximum value and then follows a downward trend with further increase of k_{sf}; it can however be noted that the downward slope of this decrease in $\overline{Nu}$ is comparatively rather moderate. This decrease in $\overline{Nu}$ is primarily attributed to the increasing conjugate heat transfer effects in the solid substrate, leading to axial back conduction of heat in it. The numerical simulations suggest that the k_{sf} corresponding to the maximum $\overline{Nu}$ is the function of both flow *Re* and δ_{sf} – however, its dependence on the latter is not very strong. For a given flow *Re* condition, the importance of conductivity ratio k_{sf} as the dominating parameter in affecting the maximum $\overline{Nu}$, as compared to the geometric parameter δ_{sf}, is also quite obvious from these simulations. From Fig. 8, it can also be observed, in line with conventional observation for developing flows in channels and ducts, that the average Nusselt number value, for a given geometry and conductivity ratio, increases with increasing the value of flow *Re*.

To understand the effect of k_{sf} more clearly, Fig. 9 shows the peripheral variation of dimensionless heat flux passing through the solid–fluid interface for different values of k_{sf}, at $\mathrm{Re} = 100$ and $\delta_{sf} = 1$ and 16. There is a substantial change in the boundary condition at the solid–fluid interface for lower values of k_{sf}. The heat flux through the vertical interfaces of the channel (AB and CD; see inset in Fig. 9) decreases, and the heat flux through the horizontal interface (BC) increases, with decreasing k_{sf}; this shifts the situation closer to one side heating rather than three-side heating of the channel. Under such condition, resistance for heat flow through the vertical interfaces is quite high as compared to that of the horizontal interface. In contrast to this, at very high thermal conductivity, the heat flow resistance through the vertical interfaces considerably decreases and thereby manifests the heat flux distribution, as is shown on Fig. 9. In addition, increasing the thermal conductivity ratio also decreases the thermal resistance in the longitudinal direction (i.e., along the flow), leading to axial back conduction of heat; this combined effect causes $\overline{Nu}$ to again decrease with increasing k_{sf}. Qualitatively speaking, there is little difference in the behavior of $\overline{Nu}$, if the thickness ratio δ_{sf} is increased, as shown in Fig. 9b. This parameter directly affects the cross-sectional area of the substrate in the transverse and longitudinal directions and therefore the resulting thermal resistance to heat flow. Unlike the previous case, as was shown in Fig. 9a, for higher values of δ_{sf}, there is less appreciable change in the net thermal resistance between the vertical and horizontal sections of the solid–fluid interface. As a result of this, in Fig. 9b, the heat flux through the vertical interfaces, i.e., AB and CD, is not maximum for the case of highest k_{sf} – rather, it is maximum corresponding to some intermediate value of k_{sf}. This intermediate value of k_{sf} corresponds to that given in Fig. 8, at which the Nusselt number, $\overline{Nu}$, is a maximum.

Another noticeable effect seen here is the change in the nature of heat flux near the wall corners B and C, with varying value of k_{sf}. The local heat flux passing

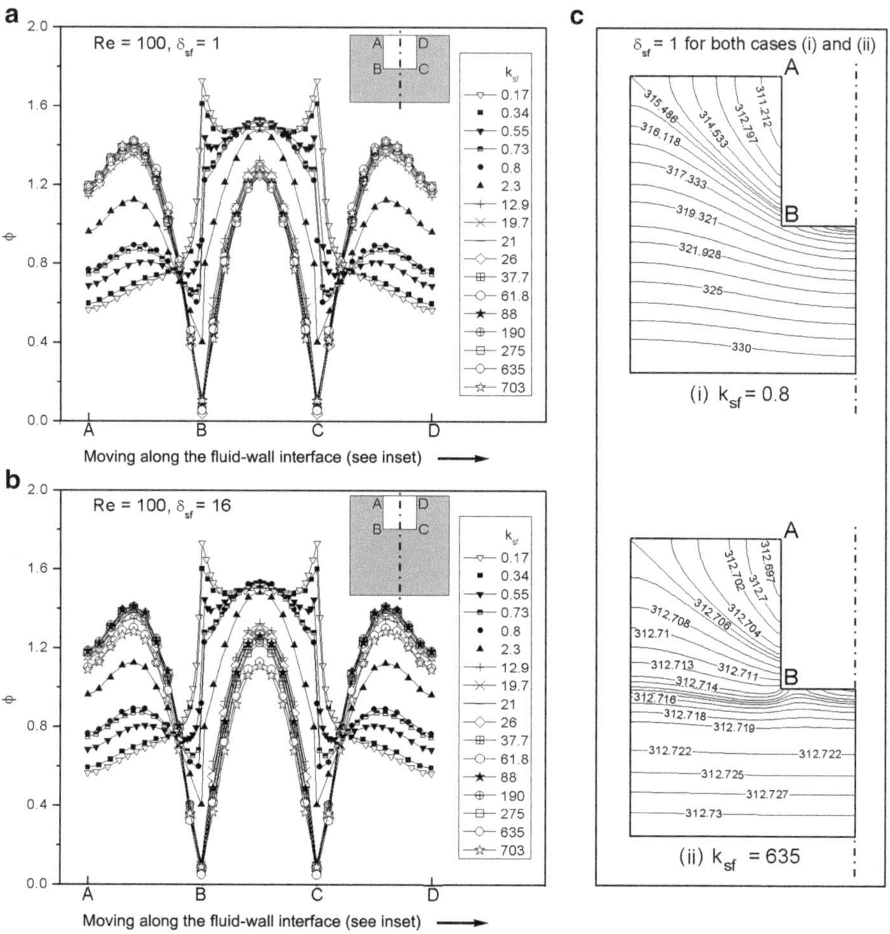

Fig. 9 Peripheral variation of dimensionless heat flux at a section in X–Y plane, midway along the length of the substrate ($z^* = 0.256$) with varying conductivity ratio, k_{sf}, when (**a**) $Re = 100$, $\delta_{sf} = 1$, and (**b**) $Re = 100$, $\delta_{sf} = 16$, and (**c**) isotherms corresponding to $Re = 100$, $\delta_{sf} = 1$, and (*i*) $k_{sf} = 0.8$ (*ii*) $k_{sf} = 635$ [44]

through these corners drastically increase with decreasing k_{sf}, asymptomatically approaching negligibly diminishing value for very high k_{sf}, and vice versa. This trend gets clarified from Fig. 9c, where the isotherms inside the computational substrate domain on the same plane, corresponding to the case of Fig. 9a, are shown, respectively, for (1) $k_{sf} = 0.8$ and (2) $k_{sf} = 635$. The shift in the local heat flux value from local minima (for the case of large k_{sf}) to local maxima (for the case of small k_{sf}) at location B (so also at C) is attributed to the interplay of local conductive resistance of the substrate to the convective resistance offered by the flowing fluid at the fluid–solid boundary. Qu and Mudawar [51] also observed this local minimum at the corners. However, their study was only limited to higher values of k_{sf}; as a reason, therefore, they did not report the entire nature of variation of local heat flux with varying k_{sf}, as done here in Fig. 9a, b. As can be seen, qualitatively similar isotherms are also obtained for the case described in Fig. 9b.

7 Effect of Channel Aspect Ratio on Axial Wall Conduction

As discussed in Sect. 6 (with reference to Fig. 8), with decreasing conductivity ratio, the thermal resistance to flow of heat to the two vertical walls of the square channel increases. From this observation, it is expected that, while all other factors remaining the same, the aspect ratio of the channel will also play a dominant role in the conjugate heat transfer process during convective flows. It is to be noted that, as the aspect ratio of the channel varies, the heating perimeter, hydraulic diameter, ratio of area of cross section of solid to fluid of the substrate, substrate thickness to channel height ratio, etc., also vary. This makes the conjugate heat transfer system more complicated. In this background, Moharana and Khandekar [42] had carried out three-dimensional numerical investigation in rectangular microchannels of different channel aspect ratio (ratio of channel width to the channel height) carved on a fixed-size substrate. The width $W = 2\omega_s + \omega_f$, thickness $H = \delta_s + \delta_f$, and length (L) of the substrate in the computational model are kept constant at 0.6, 0.4, and 60 mm, respectively (see Fig. 10a). While the length (L) of the channel is kept constant

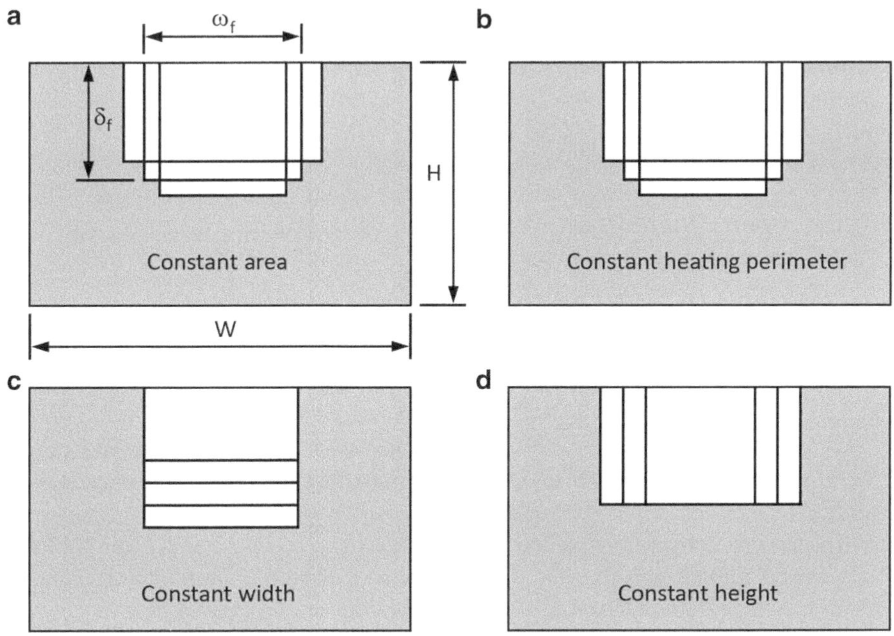

Fig. 10 Four different ways in which the aspect ratio of the microchannel flow area is varied: maintaining constant. (**a**) Area of cross section ($\omega_f \cdot \delta_f$) of microchannel. (**b**) Heating perimeter ($2\delta_f + \omega_f$). (**c**) Channel width (ω_f) and (**d**) channel height (δ_f) [42]

at 60 mm, the width (ω_f) and the height (δ_f) of the channel are varied such that the channel aspect ratio ($\varepsilon = \omega_f/\delta_f$) varies in the range of 0.45–4.0. The substrate material thermal conductivity (k_s) is varied such that k_{sf} varies in the range of 0.34–703. In all cases $Re = 100$, considering that the lower the Re, the more prone it is to axial back conduction, as indicated in Sect. 6. With increasing Re, the effect of axial back conduction decreases; thus, there is increase in magnitude of both average and local Nu. But the overall trend remains the same. The details of simulation condition used can be found in Moharana and Khandekar [42].

To understand the effect of channel aspect ratio on axial back conduction, the aspect ratio of the channels is varied in four different ways, as detailed in Fig. 10:

(a) Constant area of channel cross section ($\omega_f \cdot \delta_f$): The aspect ratio of the channels is varied such that the area of cross section of the channel is always kept constant, i.e., ($\omega_f \cdot \delta_f$) = constant. In this way, the ratio of area of cross section of solid and fluid remains constant and equals to 5.
(b) Constant heating perimeter ($2\delta_f + \omega_f$): Normally, the more the heating perimeter, the more the heat transfer. But in conjugate heat transfer situation, this may not be true. So the channel aspect ratios are varied in such a manner that the heating perimeter always remains constant.
(c) Constant channel width (ω_f): The height of the channel is varied, while the width is kept constant to understand the effect of channel height on conjugate heat transfer.
(d) Constant channel height (δ_f): The width of the channel is varied, while the height is kept constant to understand the effect of channel width.

Figure 11 highlights the variation of average Nusselt number as function of conductivity ratio, k_{sf}, as the channel aspect ratio is varied. First, the occurrence of the optimum k_{sf} for maximizing the average Nusselt number is seen in Fig. 11a–d, for every channel aspect ratio considered, which is in line with the results depicted in Fig. 8. Figure 12 shows the value of k_{sf} at which $\overline{Nu}$ is maximum at different channel aspect ratio, denoted by k_{sf}^*. It shows that as channel aspect ratio increases, the k_{sf} value at which $\overline{Nu}$ is maximum goes down. However, at higher value of the channel aspect ratio, it again slightly goes up.

In Fig. 11, it can also be seen that the value of average axial Nusselt number ($\overline{Nu}$), corresponding to any value of k_{sf}, initially decreases with increasing value of channel aspect ratio (ε). However, on very close observation, it can be seen that the value of $\overline{Nu}$ is again increasing beyond certain value of channel aspect ratio (ε). This trend is almost same in all the four cases considered in Fig. 11. This indicates the presence of a minimum for at certain value of channel aspect ratio (ε) in each of the four cases considered.

To understand the effect of channel aspect ratio more precisely, the variation of average Nusselt number, $\overline{Nu}$, as a function of the aspect ratio of the channel, at $k_{sf} = 635$, 12.19 and the k_{sf} value at which $\overline{Nu}$ is maximum in each case, respectively (as depicted in Fig. 11), is shown in Fig. 13a–d. It can be observed in Fig. 13a that the magnitude of average Nusselt number is minimum at channel aspect ratio $\varepsilon = 2.0$, which corresponds to minimum heating perimeter, i.e.,

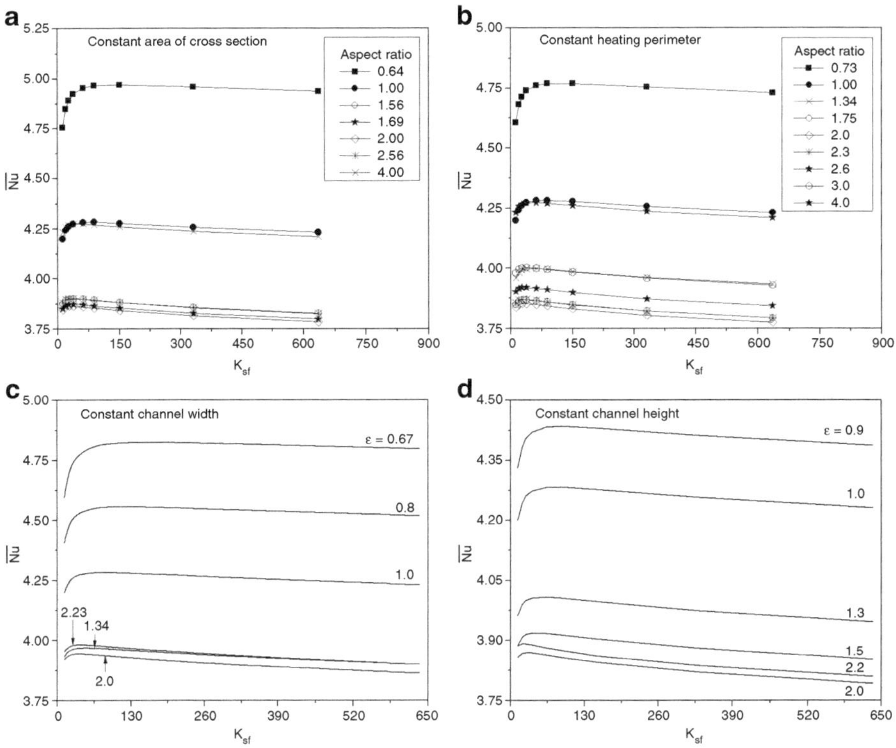

Fig. 11 Variation of average Nusselt number of the rectangular microchannel with varying channel aspect ratio as a function of conductivity ratio, k_{sf}, at (**a**) constant area of channel cross section. (**b**) Constant heating perimeter. (**c**) Constant channel width. (**d**) Constant channel height [42]

$(2\delta_{\mathrm{f}} + \omega_{\mathrm{f}}) =$ minimum. For the case of $\varepsilon < 2.0$, as the aspect ratio decreases, $\overline{Nu}$ increases. Again for $\varepsilon > 2.0$, as the aspect ratio increases, $\overline{Nu}$ increases. Looking at these results, at this point, it seems the least heating perimeter at $\varepsilon = 2.0$ causes a minima for average Nusselt Number $\overline{Nu}$.

To appreciate the role of heating perimeter on the value of $\overline{Nu}$, Fig. 13b presents the variation of average Nusselt number as a function of channel aspect ratio (ε), while the heating perimeter is constant. Here, it can again be observed that the variation of $\overline{Nu}$ as a function of channel aspect ratio ε is quite similar to the results presented in Fig. 13a. In Fig. 13a, average Nusselt number $\overline{Nu}$ varies with channel aspect ratio, and so from this result, it is assumed that this is due to change in heating perimeter. However, in Fig. 13b even when the heating perimeter is constant, variation in $\overline{Nu}$ is still observed, which indicates that the magnitude of $\overline{Nu}$ is independent of heating perimeter of the channel. Figure 13c, d presents variation of $\overline{Nu}$ with varying aspect ratio while channel width and height are maintained to be constant, respectively. In Fig. 13c, it is observed that the variation of $\overline{Nu}$ as a

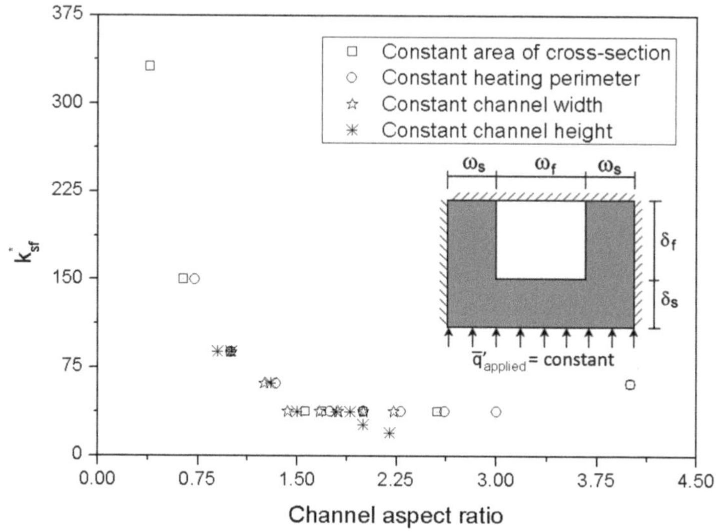

Fig. 12 Value of k_{sf} at which $\overline{Nu}$ is maximum at different channel aspect ratio [42]

function of ε is also similar as it was in Fig. 13a, b, but the minima is observed at $\varepsilon = 1.8$ instead of at $\varepsilon = 2.0$. Like in Fig. 13c, the minima in Fig. 13d is also observed at $\varepsilon < 2.0$, at $\varepsilon = 1.9$.

8 Axial Wall Conduction in Circular Microtube

There is a basic difference between a conventional tube and a microtube, which was described in Sect. 4. Secondly, the basic difference between a circular microduct and a rectangular microchannel carved in a flat substrate was also discussed in Sect. 6. Unlike channels cut on flat substrates, microtubes have angular symmetry. Therefore, a circular microtube leads to a two-dimensional conjugate heat transfer problem compared to a three-dimensional system for any noncircular microchannel. Therefore, the present authors have also investigated for occurrence of an optimum Nusselt number condition for a circular microtube (as shown in Fig. 14) similar to square or rectangular microchannel discussed in Sects. 6 and 7 respectively.

The two-dimensional computational domain, which is half of the transverse section along r–z plane, is shown in Fig. 4c. For circular microtube, δ_{sf} is interpreted as the ratio of tube wall thickness (r_o–r_i) to the inner radius (r_i). For a microtube of certain inner radius (r_i or δ_f), δ_{sf} can be increased by increasing the outer radius (r_o) of the tube, which increases the area of cross section by an order of square of the radius.

The inner radius (δ_f) and length (L) of the microtube in the computational model are kept constant at 0.2 and 60 mm, respectively. The outer radius of the microtube is varied such that the value of δ_{sf} varies in the range of 1–16. Pure water (at an

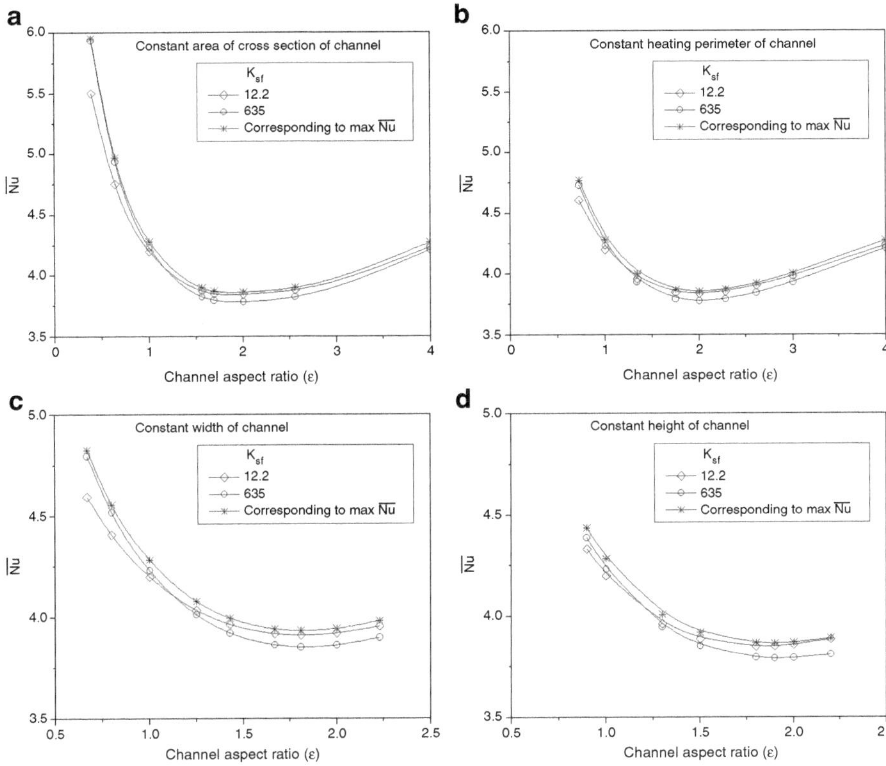

Fig. 13 Variation of average Nusselt number of the rectangular microchannel with varying channel aspect ratio at $k_{sf} = 635$, 12.19 and k_{sf} value at which $\overline{Nu}$ is maximum in each case, as shown in Fig. 12 [42]

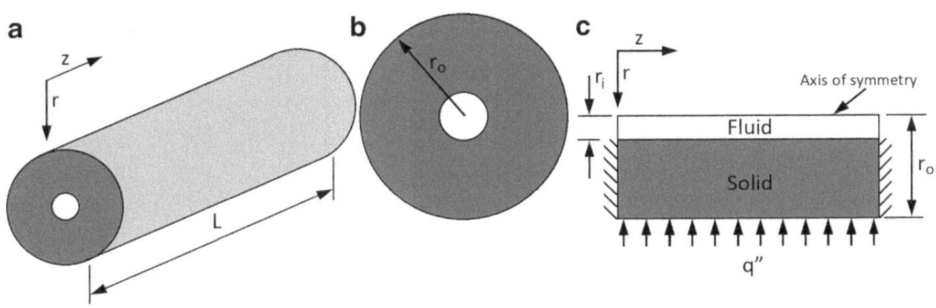

Fig. 14 Details of a microtube. (**a**) Microtube with conductive wall. (**b**) Cross-section view. (**c**) Two-dimensional computational domain (chosen from axis-symmetry conditions) [41]

inlet temperature of 300 K, $Pr = 5.85$) is used as the working fluid which enters the microtube with a slug velocity profile ($\bar{u}$). The fluid flow Re is maintained in the range of 100–1,000. The substrate material thermal conductivity (k_s) is varied such

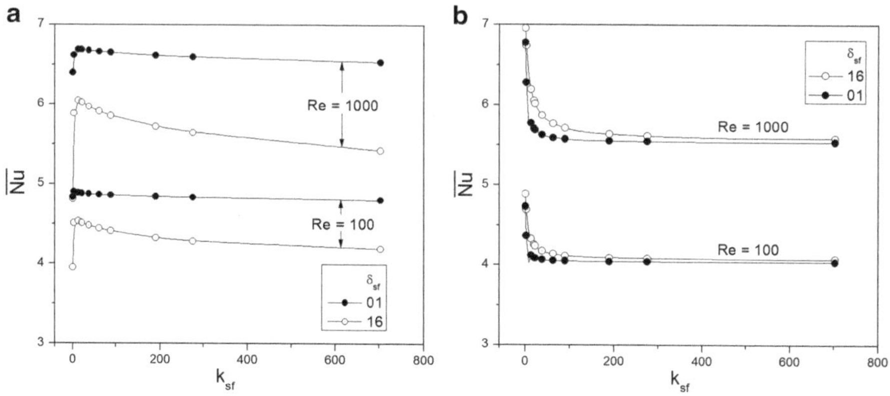

Fig. 15 Variation of $\overline{Nu}$ as a function of k_{sf} for a microtube subjected to boundary condition of (**a**) constant heat flux [44] and (**b**) constant wall temperature [41] at its outer surface

that the conductivity ratio ($k_{sf} = k_s/k_f$) varies in the range of 0.34–703. Two types of boundary condition, i.e., constant wall heat flux and constant wall temperature, are applied on the outer surface of the microtube separately.

First, the case of constant wall heat flux applied on the outer surface of the microtube is considered. The variation of average Nusselt number with k_{sf}, at different flow *Re* and δ_{sf}, subjected to constant wall heat flux is presented in Fig. 15a. For this case, the presence of optimum k_{sf} at which the average Nusselt number is maximum can be observed clearly. Secondly, all other parameters remaining constant, the average Nusselt number decreases as the value of δ_{sf} increases from 1 to 16. This plot indicates similar qualitative variation of average Nusselt number, as noted for square microchannel in a flat substrate in Fig. 8. Of course, in the case of the circular microtube, $\delta_{sf} = 16$ leads to a much larger cross-sectional area of the solid wall in the radial direction. Thus, for a circular microtube too, there lies an optimum value of k_{sf}, corresponding to which the average Nusselt number is maximum, for a given *Re* and δ_{sf}.

Next, the case of constant wall temperature maintained on the outer surface of the microtube is considered. The details of the numerical procedure and the axial variation of wall temperature at the solid–fluid interface, bulk fluid temperature, and local Nusselt number can be found in Moharana and Khandekar [41]. To understand the explicit effect of k_{sf} on heat transfer more clearly, the variation of average Nusselt number ($\overline{Nu}$) over the length of the tube, as a function of k_{sf}, while varying flow *Re* and δ_{sf}, is presented in Fig. 15b, where constant wall temperature condition is applied on the microtube outer surface. It can be seen that for fixed value of *Re* and δ_{sf}, the value of $\overline{Nu}$ increases, as the value of k_{sf} decreases. The slope of this curve changes rapidly below k_{sf} less than 100, and the value of $\overline{Nu}$ increases sharply, as the value of k_{sf} approach towards zero, suggesting that lower k_{sf} results in higher $\overline{Nu}$. With increasing flow *Re* at constant k_{sf}, the $\overline{Nu}$ value increases due to increase in flow development length. Secondly, it can also be observed that for all other parameters remaining the same, higher wall thickness leads to higher value of $\overline{Nu}$. Finally, the gap between the curves at $\delta_{sf} = 1$ and 16 increases with decreasing value of k_{sf}.

A comparison of Fig. 15a, b shows that they are exactly opposite to each other. This is because axial conduction in a microtube subjected to constant heat flux at the outer surface drifts the condition at the inner surface of the tube towards a conventional constant wall temperature boundary condition. On the other hand, axial conduction in a microtube subjected to constant wall temperature at the outer surface drifts the condition at the inner surface of the tube towards a conventional constant heat flux boundary condition.

9 Axial Wall Conduction in Partially Heated Microtubes

All the studies reported in Sects. 5, 6, 7, and 8 had considered heating over the full length of the microchannels. However, in many practical applications, partially heated conditions are widely encountered, where the heating length is not always equal to the full length of the channels in which flow is taking place. In such cases of partial heating, different situations may arise, e.g., (1) upstream heating, downstream insulated and vice versa, and (2) heating of the central portion with insulated boundary condition near the inlet and outlet sections of the channel, etc. Secondly, the length of the heated portion of the channel, as a percentage of the total channel length, can also vary. Further, the portion of the tube/channel which is heated may be subjected to constant wall heat flux or constant wall temperature-type boundary conditions, among others. Review of the available literature clearly reveals that only a very limited number of studies on thermal performance of partially heated microchannels are available.

One frequently occurring situation in many practical situations is conjugate heat transfer in partially heated microtube where half of the tube length is heated (either upstream or downstream) and the remaining half is insulated. Lelea [30] numerically studied this problem by considering constant heat flux on the outer surface along the heated portion of the microtube and found that axial wall conduction is negligible when microtube wall material conductivity is comparatively low. Their observation was based on numerical analysis on three microtube wall materials (made of steel, silicon, and copper, respectively). Later on Lelea [31] considered partially heated (constant heat flux at the bottom of the substrate) rectangular microchannel array on a flat substrate for their numerical study and found that upstream heating has a lower thermal resistance compared to central or downstream heating. Explicit effects of substrate thickness or substrate conductivity were not reported in this study.

Chaudhuri et al. [6] reported numerical study of partially heated (one third length) microchannel with nitrogen, hydrogen, and argon as the working fluids, respectively. They reported that for higher Knudsen number microscale flows, heat conduction due to molecular diffusion becomes predominant. Secondly, thermo-hydrodynamic characteristics of the partially heated microchannels greatly depend on the transport properties of the convective medium.

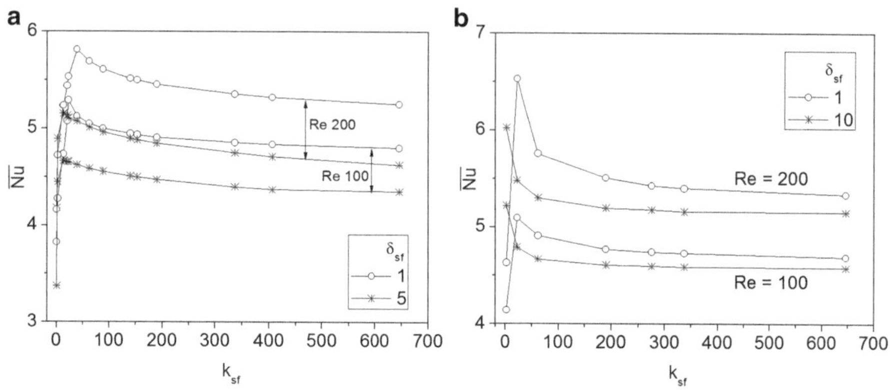

Fig. 16 Average Nusselt number varying with conductivity ratio in partially heated microtube with (**a**) constant wall heat flux [60]. (**b**) Constant wall temperature [28] along the heated portion

Raisi et al. [53] numerically studied thermal performance of partially heated (central portion heated with constant wall temperature condition) microchannel using pure water as well as water with copper nanoparticles (i.e., Cu–water nanofluid) as the coolant. The effects of Reynolds number, solid volume fraction, and slip velocity coefficient on the thermal performance of the microchannel were reported. However, the effect of microchannel wall thermal conductivity was not reported in this study.

Tiwari et al. [60] numerically studied axial wall conduction in partial heated circular microtube using water as the working fluid. A microtube of 0.4 mm inner diameter and 60 mm was considered in which the central portion of 48 mm was subjected to constant wall heat flux on the outer surface, and the remaining length of 6 mm each near the inlet and the outlet was thermally insulated. The parametric variations considered in this study were microtube wall thickness, its conductivity, and coolant flow rate. It was found that the solid wall to coolant fluid conductivity ratio (k_{sf}) and solid wall thickness to inner radius ratio (δ_{sf}) of the microtube play a dominant role in the conjugate heat transfer process. Figure 16a shows the average Nusselt number ($\overline{Nu}$) varying with k_{sf}, δ_{sf}, and flow *Re*, as predicted by Tiwari et al. [60] for the partially heated microtube. It can be observed in Fig. 16a that even for partially heated microtube, there exists an optimum k_{sf} at which $\overline{Nu}$ is maximum. Secondly, for all other parameters remaining the same, $\overline{Nu}$ is found to be lower for higher value of wall thickness (δ_{sf}). This is due to axial movement of heat along the thicker solid wall. Again, on increasing flow *Re*, it is found that $\overline{Nu}$ increased. This is due to higher thermal development length at higher flow *Re*.

Kumar and Moharana [28] also numerically studied the geometry considered by Tiwari et al. [60] by considering constant wall temperature across the heated portion of the microtube instead of constant wall heat flux. Figure 16b shows the average Nusselt number ($\overline{Nu}$) varying with k_{sf}, δ_{sf}, and flow *Re*, as predicted by Kumar and Moharana [28] for partially heated microtube. For lower wall thickness ($\delta_{sf} = 1$), the average Nusselt number ($\overline{Nu}$) is found to be increasing with decreasing value of k_{sf}; it acquires a maximum value. On further decreasing value of k_{sf}, $\overline{Nu}$ decreases drastically near very lower value of k_{sf} ($\approx$0).

At high value of k_{sf}, $\overline{Nu}$ is found to be lower for higher wall thickness ($\delta_{sf} = 10$). But with decreasing the value of k_{sf}, the rate of increase in $\overline{Nu}$ is less compared to $\delta_{sf} = 1$. Secondly, as k_{sf} approaches zero, the slope of $\overline{Nu}$ increases drastically near very low value of k_{sf} ($\approx$0), similar to fully heated condition as in Fig. 15b. Thus, the $\overline{Nu}$ curves for the thin and the thick wall microtube ($\delta_{sf} = 1$, and 10) intersect each other at lower value of k_{sf} ($\approx$0).

10 Summary and Closure

Koşar [26] numerically investigated conjugate heat transfer in rectangular microchannel carved on solid substrate and subjected to constant wall heat flux on the bottom face of the substrate; similar to the ones discussed in Sects. 6 and 7, Kosar [26] considered very low thermal conductive materials like polyamide ($k_s = 0.25$ and $k_{sf} = 0.409$), silica glass ($k_s = 1.005$ and $k_{sf} = 1.64$), and quartz ($k_s = 1.3$ and $k_{sf} = 2.13$) and found that the value of Nusselt number decreases with decreasing the thermal conductivity of the substrate. While this is certainly in accordance with the results discussed in Fig. 8, the conclusions are only partly correct; Koşar [26] did not capture the full spectrum of k_s to get its optimum value. Similarly, Li et al. [32] both experimentally and numerically studied conjugate effects on microtubes of fused silica and stainless steel but failed to capture the complete range of the effect of solid substrate on the variation of Nusselt number, as has been discussed in Sect. 8.

Depending on the type of boundary condition imposed on the outer surface and the exact condition desired at the solid–fluid interface of a microchannel system, one requires to select the material and dimension of solid substrate or thickness of the channel wall. Chein et al. [7] also numerically investigated axial heat conduction in microscale tubular methanol–steam reformer made from silica glass, steel, and copper. Their study clearly revealed that the performance of methanol–steam reformer made from high-conductive material was lower, compared to that of low-conductive materials. They therefore suggested that the material for microscale reformer fabrication should have low thermal conductivity and small thickness. While this is also certainly in line with the results discussed in Fig. 16a, their conclusions are again only partly correct; Chein et al. [7] did not capture the full spectrum of k_{sf}, as discussed in Sect. 8, to get the optimum value of Nusselt number. Secondly, as discussed in Sect. 5, Stutz and Poulikakos [58] found highest conversion of methane for nonconductive channel wall. This is in line with results discussed in Fig. 15b, where average Nusselt number is maximum corresponding to smallest k_{sf}. It is to note that microreactors require uniform wall temperature for catalytic conversion. Karakaya and Avci [23] found that higher conductive material of the microreactor wall among alumina, steel, and iron, i.e., iron walls, provides higher hydrogen yield due to better spreading of heat in the wall.

Türkakar and Okutucu-Özyurt [63] have studied geometrical optimization of rectangular-shaped heat sinks made of silicon microchannels by minimizing the total thermal resistance. Four different analytical models of optimization, valid for uniform heat load condition (without any conjugate effects) and based on the work of Liu and Garimella [34], were used and subsequently improved by considering the entrance effects and property variations. It was illustrated that for all the four models used, the thermal resistance decreases with increasing channel height (or decreasing aspect ratio), i.e., optimization geometries have much greater channel heights as compared to the channel width dimensions. While this result is certainly in line with the results discussed in Figs. 11 and 13, their conclusions are only partly correct; it must be noted that Türkakar and Okutucu-Özyurt [63] did not capture the full spectrum of channel aspect ratio so as to observe the presence of either a maxima or a minima. Their domain of findings was limited, essentially based on a single value of conductivity ratio k_{sf} and only three values of channel aspect ratios. Earlier, Ryu et al. [55] performed a three-dimensional numerical study to obtain the optimal fin channel shape that minimizes the thermal resistance and stated that among different design variables, the channel width is the most crucial quantity in deciding the performance of a microchannel heat sink.

The results in Fig. 16a are in line with that of Fig. 15a though Figs. 15a and 16a correspond to a fully and partially heated microtube, respectively. The common feature between them is that the boundary condition used along the heated portion is constant wall heat flux. There is a mismatch in the overall trend (compare Figs. 15b and 16b) when constant wall temperature is used along the heated portion of the thin microtube wall. However, for a relatively thick partially heated microtube, the overall trend agrees with that of the case having fully heated microtube.

From the results presented in Sects. 6, 7, 8, and 9 and above discussion, it is clear that axial wall conduction plays a prominent role in the heat transfer process of microchannel system as there exists an optimum k_{sf} at which the average Nusselt number over the channel length is maximum if the microchannel (be it rectangular or square or circular) is subjected to constant wall heat flux.

Finally, it is also clear that the aspect ratio of the channel influences axial back conduction, and there exists a minimum in the average Nusselt number with respect to this parameter, which is approximately near $\varepsilon = 2.0$. Secondly, channels having lower aspect ratio provide higher values of average Nusselt number compared to those having higher aspect ratios, located on the other side of the minimum value. Although decreasing the channel aspect ratio or increasing the channel height is favorable from a thermal point of view, the manufacturability of accurately dimensioned low aspect ratio channels is more involved and difficult as compared to high aspect ratio channels. In view of the fact that very high aspect ratio channels are comparatively easier to manufacture and they are also thermally favorable if $\varepsilon > 2.0$, for same thermal performance of two reciprocal valued channel aspect ratios, the channel with higher aspect ratio should be obviously preferred.

Nomenclature

A	Area of cross section, m^2
Bi	Biot number, –
c_p	Specific heat at constant pressure, J/g K
D	Diameter, m
D_h	Hydraulic diameter, m
h	Heat transfer coefficient, W/m^2 K
H	Height of substrate, m
k	Thermal conductivity, W/mK
k_{sf}^*	Value of k_{sf} at which $\overline{Nu}$ is maximum, –
L	Length of the channel/substrate, m
M	Axial conduction number, –
$\dot{m}$	Mass flow rate, kg/s
NTU	Number of transfer units, –
Nu	Nusselt number, –
$\overline{Nu}$	Average Nusselt number over the channel length, –
P	Axial conduction parameter, –
Pe	Peclet number, –
Pr	Prandtl number, –
q'	Applied heat flux, W/m^2
$\overline{q'}$	Heat flux at the solid–fluid interface, W/m^2
r	Radius, radial direction, m
Re	Reynolds number, –
T	Temperature, K
z	Length, transverse direction, m
W	Width of substrate, m
z^*	Dimensionless length, –
Symbols	
α	Ratio of δ_s and L, –
δ	Thickness or height, m
ε	Channel aspect ratio (width/height), –
μ	Dynamic viscosity, Pa.s
ω	Width, m
ϕ	Dimensionless wall heat flux as defined in Eq. 10, –
ρ	Density, kg/m^3
Θ	Dimensionless temperature as defined in Eq. 10, –
ΔT	Temperature difference, K
Subscripts	
f	Fluid
i	Inner, inlet
o	Outer, outlet
s	Solid
sf	Ratio of solid to fluid of any parameter
z	Local value

Acknowledgments This work is funded by the Department of Science and Technology, Government of India, under the IIT Kanpur sponsored project No: DST/CHE/20060304, titled "Micro-devices for Process Applications." The second author would also like to acknowledge the German Academic Exchange Program (DAAD) for supporting his research visit to University of Stuttgart, Germany, as part of his doctoral assignment at the Indian Institute of Technology Kanpur, Kanpur, India.

References

1. Avci M, Aydin O, Arici ME (2012) Conjugate heat transfer with viscous dissipation in a microtube. Int J Heat Mass Transf 55(19–20):5302–5308
2. Bahnke GD, Howard CP (1964) The effect of longitudinal heat conduction on periodic-flow heat exchanger performance. J Eng Power 86:105–120
3. Barozzi GS, Pagliarini G (1985) A method to solve conjugate heat transfer problems: the case of fully developed laminar flow in a pipe. J Heat Transf 107(1):77–83
4. Bier W, Keller W, Linder G, Seidel D, Schubert K, Martin H (1993) Gas to gas heat transfer in micro heat exchangers. Chem Eng Proc 32:33–43
5. Celata GL (2004) Heat transfer and fluid flow in microchannels, series in thermal and fluid physics and engineering. Begell House, New York
6. Chaudhuri A, Guha C, Dutta TK (2007) Numerical study of fluid flow and heat transfer in partially heated microchannels using the explicit finite volume method. Chem Eng Tech 30(4):425–430
7. Chein R, Chen YC, Chung JN (2012) Axial heat conduction and heat supply effects on methanol-steam reforming performance in micro-scale reformers. Int J Heat Mass Transf 55(11–12):3029–3042
8. Chiou JP (1980) The advancement of compact heat exchanger theory considering the effects of longitudinal heat conduction and flow non-uniformity. In: Proceedings of the symposium on compact heat exchangers-history, technological advancement and mechanical design problems. Book no. G00183, HTD vol 10, ASME, New York
9. Choi SB, Barron RF, Warrington RO (1991) Fluid flow and heat transfer in microtubes, micromechanical sensors, actuators, and systems, vol 32, ASME, DSC, pp. 123–134
10. Cole KD, Cetin B (2011) The effect of axial conduction on heat transfer in a liquid microchannel flow. Int J Heat Mass Transf 54(11–12):2542–2549
11. Cotton MA, Jackson JD (1985) The effect of heat conduction in a tube wall upon forced convection heat transfer in the thermal entry region. In: Proceedings of numerical methods in thermal problems, vol 4, Pineridge Press, Swansea, pp 504–515
12. Faghri M, Sparrow EM (1980) Simultaneous wall and fluid axial conduction in laminar pipe-flow heat transfer. J Heat Transf 102:58–63
13. Goodling JS (1993) Microchannel heat exchangers: a review. In: Proceedings of high heat flux engineering II, San Diego, 12–13 July 1997, pp 66–82
14. Guo ZY, Li ZX (2003) Size effect on single-phase channel flow and heat transfer at microscale. Int J Heat Fluid Flow 24(3):284–298
15. Herwig H (2002) Flow and heat transfer in micro systems: is everything different or just smaller? J Appl Math Mech 82:579–586
16. Herwig H, Hausner O (2003) Critical view on "new results in micro-fluid mechanics": an example. Int J Heat Mass Transf 46:935–937
17. Hetsroni G, Mosyak A, Pogrebnyak E, Yarin LP (2005) Heat transfer in micro-channels: comparison of experiments with theory and numerical results. Int J Heat Mass Transf 48(25–26):5580–5601

18. Huang CY, Wu CM, Chen YN, Liou TM (2014) The experimental investigation of axial heat conduction effect on the heat transfer analysis in microchannel flow. Int J Heat Mass Transf 70:169–173
19. Kabar Y, Rebay M, Kadja M, Padet C (2010) Numerical resolution of conjugate heat transfer problem in a parallel-plate micro-channel. Heat Transf Res 41(3):247–263
20. Kakac S, Vasiliev LL, Bayazitoglu Y, Yener Y (eds) (2005) Microscale heat transfer: fundamentals and applications. Springer, Dordrecht
21. Kandlikar SG (2012) History, advances, and challenges in liquid flow and flow boiling heat transfer in microchannels; a critical review. J Heat Transf 134(3):034001-1–034001-15
22. Kandlikar SG, Garimella S, Li D, Colin S, King MR (2006) Heat transfer and fluid flow in minichannels and microchannels. Elsevier, Oxford
23. Karakaya M, Avci AK (2011) Microchannel reactor modelling for combustion driven reforming of iso-octane. Int J Hydrogen Energ 36(11):6569–6577
24. Kim MH, Lee SY, Mehendale SS, Webb RL (2003) Microchannel heat exchanger design for evaporator and condenser applications. In: Advances in heat transfer. Elsevier, San Diego
25. Koo J, Kleinstreuer C (2004) Viscous dissipation effects in microtubes and microchannels. Int J Heat Mass Transf 47(14–16):3159–3169
26. Kosar A (2010) Effect of substrate thickness and material on heat transfer in microchannel heat sinks. Int J Therm Sci 49(4):635–642
27. Kroeker CJ, Soliman HM, Ormiston SJ (2004) Three-dimensional thermal analysis of heat sinks with circular cooling micro-channels. Int J Heat Mass Transf 47(22):4733–4744
28. Kumar M, Moharana MK (2013) Axial wall conduction in partially heated microtubes. In: Proceedings of 22nd national and 11th international ISHMT-ASME heat and mass transfer conference, IIT Kharagpur, Kharagpur, 28–31 Dec 2013
29. Lee PS, Garimella SV (2006) Thermally developing flow and heat transfer in rectangular microchannels of different aspect ratios. Int J Heat Mass Transf 49(17–18):3060–3067
30. Lelea D (2007) The conjugate heat transfer of the partially heated microchannels. Heat Mass Transf 44(1):33–41
31. Lelea D (2009) The heat transfer and fluid flow of a partially heated microchannel heat sink. Int Comm Heat Mass Transf 36(8):794–798
32. Li Z, He YL, Tang GH, Tao WQ (2007) Experimental and numerical studies of liquid flow and heat transfer in microtubes. Int J Heat Mass Transf 50(17–18):3447–3460
33. Lin TY, Kandlikar SG (2012) A theoretical model for axial heat conduction effects during single-phase flow in microchannels. J Heat Transf 134(2):020902-1–020902-6
34. Liu D, Garimella SV (2005) Analysis and optimization of the thermal performance of microchannel heat sinks. Int J Numer Methods Heat Fluid Flow 15(1):7–26
35. Liu Z, Zhao Y, Takei M (2007) Experimental study on axial wall heat conduction for conductive heat transfer in stainless steel microtube. Heat Mass Transf 43(6):587–594
36. Maranzana G, Perry I, Maillet D (2004) Mini- and micro-channels: influence of axial conduction in the walls. Int J Heat Mass Transf 47(17–18):3993–4004
37. Mathew B, Hegab H (2008) Axial heat conduction in counter flow microchannel heat exchangers. In: Proceedings of ASME heat transfer summer conference, Jacksonville, 10–14 Aug 2008
38. Mathew B, Hegab H (2009) Axial heat conduction in parallel flow microchannel heat exchangers. In: Proceedings of ASME international mechanical engineering congress and exposition, Lake Buena Vista, 13–19 Nov 2009
39. Mehendale SS, Jacobi AM, Shah RK (2000) Fluid flow and heat transfer at micro and meso scales with application to heat exchanger design. Appl Mech Rev 53(7):175–193
40. Moharana MK, Agarwal G, Khandekar S (2011) Axial conduction in single-phase simultaneously developing flow in a rectangular mini-channel array. Int J Therm Sci 50:1001–1012
41. Moharana MK, Khandekar S (2012) Numerical study of axial back conduction in microtubes. In: Proceedings of 39th national conference on fluid mechanics and fluid power, Sardar Vallabhbhai National Institute of Technology, Surat, 13–15 Dec 2012

42. Moharana MK, Khandekar S (2013) Effect of aspect ratio of rectangular microchannels on the axial back conduction in its solid substrate. Int J Microscale Nanoscale Thermal Fluid Trans Phenomena 4(3–4):211–229
43. Moharana MK, Peela NR, Khandekar S, Kunzru D (2011) Distributed hydrogen production from ethanol in a microfuel processor: issues and challenges. Renewable Sustainable Energy Rev 15(1):524–533
44. Moharana MK, Singh PK, Khandekar S (2012) Optimum Nusselt number for simultaneously developing internal flow under conjugate conditions in a square microchannel. J Heat Transf 134:071703-01–071703-10
45. Moreno A, Murphy K, Wilhite BA (2008) Parametric study of solid-phase axial heat conduction in thermally integrated microchannel networks. Ind Eng Chem Res 47:9040–9054
46. Mori S, Sakakibara M, Tanimoto A (1974) Steady state heat transfer to laminar flow in circular tube with conduction in the tube wall. Heat Transf Japanese Res 3(2):37–46
47. Morini GL (2006) Scaling effects for liquid flows in microchannels. Heat Transf Eng 27(4):64–73
48. Nonino C, Savino S, Giudice SD, Mansutti L (2009) Conjugate forced convection and heat conduction in circular microchannels. Int J Heat Fluid Flow 30(5):823–830
49. Peterson RB (1999) Numerical modeling of conduction effects in microscale counter flow heat exchangers. Microscale Thermophy Eng 3:17–30
50. Petukhov BS (1967) Heat transfer and drag of laminar flow of liquid in pipes. Energiya, Moscow
51. Qu W, Mudawar I (2002) Analysis of three-dimensional heat transfer in micro-channel heat sinks. Int J Heat Mass Transf 45(19):3973–3985
52. Rahimi M, Mehryar R (2012) Numerical study of axial heat conduction effects on the local Nusselt number at the entrance and ending regions of a circular microchannel. Int J Therm Sci 59:87–94
53. Raisi A, Ghasemi B, Aminossadati SM (2011) A numerical study on the forced convection of laminar nanofluid in a microchannel with both slip and no-slip conditions. Numerical Heat Transf: Part A 59(2):114–129
54. Rosa P, Karayiannis TG, Collins MW (2009) Single-phase heat transfer in microchannels: the importance of scaling effects. Appl Thermal Eng 29(17–18):3447–3468
55. Ryu JH, Choi DH, Kim SJ (2002) Numerical optimization of the thermal performance of a microchannel heat sink. Int J Heat Mass Transf 45(13):2823–2827
56. Shah RK, London AL (1978) Laminar flow forced convection in ducts. In: Advances in heat transfer. Academic Press, New York
57. Stief T, Langer OU, Schubert K (1999) Numerical investigation of optimal heat conductivity in micro heat exchangers. Chem Eng Technol 21:297–303
58. Stutz MJ, Poulikakos D (2005) Effects of microreactor wall heat conduction on the reforming process of methane. Chem Eng Sci 60:6983–6997
59. Tiselj I, Hetsroni G, Mavko B, Mosyak A, Pogrebnyak E, Segal Z (2004) Effect of axial conduction on the heat transfer in micro-channels. Int J Heat Mass Transf 47(12–13):2551–2565
60. Tiwari N, Moharana MK, Sarangi SK (2013) Influence of axial wall conduction in partially heated microtubes. In: Proceedings of 40th national conference on fluid mechanics and fluid power, National Institute of Technology Hamirpur, 12–14 Dec 2013
61. Tso CP, Mahulikar SP (2000) Experimental verification of the role of Brinkman number in microchannels using local parameters. Int J Heat Mass Transf 43(10):1837–1849
62. Tuckerman DB, Pease RFW (1981) High-performance heat sinking for VLSI. IEEE Electron Device Letters 2:126–129
63. Türkakar G, Okutucu-Özyurt T (2012) Dimensional optimization of microchannel heat sinks with multiple heat sources. Int J Therm Sci 62:85–92
64. Xu B, Ooi KT, Mavriplis C, Zaghloul ME (2003) Evaluation of viscous dissipation in liquid flow in microchannels. J Micromech Microeng 13(1):53–57

65. Yang CY, Chen CW, Lin TY, Kandlikar SG (2012) Heat transfer and friction characteristics of air flow in microtubes. Expt Thermal Fluid Sci 37:12–18
66. Yarin LP, Mosyak A, Hetsroni G (2009) Fluid flow, heat transfer and boiling in micro-channels. Springer, Berlin
67. Zhang LK, Goodson KE, Kenny TW (2004) Silicon microchannel heat sinks: theories and phenomena. Springer, Berlin
68. Zhang SX, He YL, Lauriat G, Tao WQ (2010) Numerical studies of simultaneously developing laminar flow and heat transfer in microtubes with thick wall and constant outside wall temperature. Int J Heat Mass Transf 53(19–20):3977–3989
69. Zohar Y (2003) Heat convection in micro ducts. Kluwer Academic Publishers, London

Index

Y.M. Joshi, S. Khandekar (eds.), *Nanoscale and Microscale Phenomena*, Springer Tracts in Mechanical Engineering, DOI 10.1007/978-81-322-2289-7

D